MESSAGES
from the
GODS

ETHNOBOTANICAL CONTRIBUTORS

Juana and Antonio Cuc, Barbara Fernandez,
Thomas Green, Winston Harris, Don Eligio Panti,
Andrew Ramcharan, Percival Hezekiah Reynolds,
Hortense Robinson, Leopoldo Romero,
Beatrice Waight, and Juana Xix

LINGUISTIC CONTRIBUTORS

Victor Cal, Liza Grandia, and Richard R. Wilk

EDITORIAL CONTRIBUTORS

Irina Adam, Jillian De Gezelle, Katherine Herrera,
Rebekka Stone, and Willa Capraro

WEB-BASED CHAPTER AUTHORS

Robert Heinzman and Conrad Reining
Gordon M. Cragg and David J. Newman

MESSAGES
from the
GODS

A GUIDE TO THE
USEFUL PLANTS OF BELIZE

*Illustrations from the LuEsther T. Mertz Library
of The New York Botanical Garden
and by Francesca Anderson*

MICHAEL J. BALICK
ROSITA ARVIGO

OXFORD
UNIVERSITY PRESS

THE NEW YORK
BOTANICAL GARDEN

OXFORD
UNIVERSITY PRESS

Oxford University Press is a department of the University of Oxford. It furthers the University's objective of excellence in research, scholarship, and education by publishing worldwide.

Oxford New York
Auckland Cape Town Dar es Salaam Hong Kong Karachi
Kuala Lumpur Madrid Melbourne Mexico City Nairobi
New Delhi Shanghai Taipei Toronto

With offices in
Argentina Austria Brazil Chile Czech Republic France Greece
Guatemala Hungary Italy Japan Poland Portugal Singapore
South Korea Switzerland Thailand Turkey Ukraine Vietnam

Oxford is a registered trade mark of Oxford University Press in the UK and certain other countries.

Published in the United States of America by
Oxford University Press
198 Madison Avenue, New York, NY 10016

Library of Congress Cataloging-in-Publication Data
Balick, Michael J., 1952–
    Messages from the gods : a guide to the useful plants of Belize / Michael J. Balick and Rosita Arvigo.
      pages cm
    Other title: Guide to the useful plants of Belize
    Includes bibliographical references and index.
    ISBN 978-0-19-996576-2 (alk. paper) — ISBN 978-0-19-996574-8 (alk. paper) 1. Medicinal plants—Belize. 2. Traditional medicine—Belize. 3. Healers—Belize. 4. Ethnobotany—Belize.
I. Arvigo, Rosita. II. Title. III. Title: Guide to the useful plants of Belize.
    QK99.B395B35 2014
    581.6'34097282—dc23   2014020343

9 8 7 6 5 4 3 2 1
Printed in China on acid-free paper

TO OUR CHILDREN,
who will tend the soil, plant the trees,
and help restore sanity to Earth.
Choose your path wisely with great awareness
of the consequences of your actions.
We are counting on you to give back more
than you ask in return.

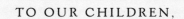

DISCLAIMER

The statements in this book are a compendium of Belizean lore and belief, gathered from interviews and the available literature, and have been compiled in an attempt to preserve and codify traditional cultural values and information. The information has been recorded as presented by the persons interviewed. In the case of plant species reported to be used traditionally for health care issues, this book does not purport to contain, nor is it intended to be, any kind of recommendation or self-treatment guide for the use of plants or traditional practices. Some of the species reported as useful are quite toxic or could cause harm when used inappropriately. Even plants that are commonly consumed as food and reported to be generally recognized as safe may have adverse effects, including drug interactions and hypersensitivity in some individuals. The authors have checked with sources believed to be reliable to confirm the accuracy and completeness of the contents of this book; however, the authors, editors, copyright holders, and the publisher disclaim all warranties, expressed and implied, to the extent permitted by law, that the contents are in every respect accurate and complete, and they are not responsible for errors, omissions, or any consequences from the application of this book's contents. As new research continues to be released, this publication represents our best efforts at the time this manual was compiled. Nothing in this book should be construed to represent an attempt to diagnose, prescribe, or administer in any manner to any physical ailments or conditions, nor should the information in this book be used in place of qualified medical advice and care.

Although the information provided in this book has been presented to promote education and scientific research, we realize that elements of this work have the potential for contributing to commercialization. Such endeavors must include a mechanism for returning benefits to the community of traditional healers who, along with their forebears, have developed their knowledge through centuries of experimentation and experience. Ethical codes of conduct and, increasingly, in many nations, legal codes as well demand nothing less. Interested parties may contact the authors and copyright holders for names and addresses of organizations in Belize that can advise in selecting appropriate means for compensating traditional communities.

CONTENTS

WEB-BASED CHAPTERS: http://www.oup.com/us/messagesfromthegods

ACKNOWLEDGMENTS

We thank the Board and staff of The New York Botanical Garden (NYBG) for their steadfast support of—and enthusiasm for—this program since it was conceived in the late 1980s. The comprehensive horticultural collections, groundbreaking public exhibitions, innovative educational programs, and unparalleled resources of NYBG's International Plant Science Center, including the Pfizer Plant Research Laboratory, Lewis B. and Dorothy Cullman Program for Molecular Systematics, William and Lynda Steere Herbarium, C. V. Starr Virtual Herbarium, LuEsther T. Mertz Library, Institute of Economic Botany, Institute of Systematic Botany, Graduate Studies Program, NYBG Press, and international field studies programs, along with the dedicated staff and graduate students, have made this institution an ideal place to continue to learn about ethnobotany—particularly how traditional cultures have used plants for healing and to promote wellness. Working with a long-term research horizon, we have been able to gather a great deal of information on the relationship between plants and people of Belize that might otherwise have been lost. Fieldwork has been the cornerstone of this research, and many people participated in plant collection activities, provided ethnobotanical information, or took part in related projects in Belize that ultimately contributed to this book, and we are grateful for the help of the following individuals:

Patty Anderson, Daniel Atha, Bill Austen, Laura Benitez, Ted Berlin, Mary Blakes, Scribe Blakes, Mark Blumenthal, Brian Boom, Joan Brittain, John L. Brown, Santos Cal, David Campbell, Elsa Carlton, Julie Chinnock, Victor Cho, Santos Coc, Eligio Cocom, Pablo Cocom, Rolando Cocom, Javier Cordero, Gordon Cragg, Antonio Cuc, Faustino Cucul, Santos Cucul, Jimmy Davis, Aaron Doolittle, Jim Duke, Laura Evans, Wayne Ewing, Maria Fadiman, José Mário Frazão, E. Gabourel, Patricia Gildea, Noah Goldstein, Rupert Gray, Thomas Green, Ruth Gutierrez, Bill Haggerty, Peggy Hansen, Winston Harris, Kim Hawkin, Christian Heckert, Jorge Jiménez, Penny King, Harlan Lahti, David Lentz, Gene Leone, Sally Armstrong Leone, Gregory Long, Matthew Long, Emery Long, Patrick Long, Lisa McDonald, E. McRae, Isabel Mai, Arcelito Manzarero, Zoe Marchal, Sharon Matola, Martin Meadows, Rustem Medora, Peter Meleady, Roberto Meléndez, James Mesh, Nathan Mesh, Manuel Mis, Antonio Morales, James Nations, Michael Nee, José Oh, Emily Ostberg, Don Eligio Panti, Galbino Pau, Juliano Pau, Isodoro Peck, Claudio Pinheiro, Antonio Pop, Ignacio Pop, Luciano Pop, Pedro Pop, Sebastian Pop, Andrew Ramcharan, Garfield Ramcharan, Percival Hezekiah Reynolds, Teresa Risso, Hortense Robinson, Leopoldo Romero, José Saki, Ernesto Sam,

Juan Sam, Manuel Sam, Silviano Camberos Sánchez, Melanie Santiago, Ernesto Saqui, Hermelindo Saqui, Lynn Schumake, Alfredo Sho, Elegio Sho, Gregorio Sho, Pablo Sho, Gregory Shropshire, Manuel Shuc, Manuel Shush, Dennis W. Stevenson, Jan W. Stevenson, José Tot, Juan Tot, Roberto Tuel, Miguel Tzib, Jay Volchok, Dan Wagner, Beatrice Waight, Jay Walker, Wilfred Warrior, Scott Wennergren, Donna White, Teo Williams, Carilyn Wilson, John Woodland, Alex Woods, Katrina Underwood, Doña Juana Xix, Eddie Xix, Manolo Xix, Class of NYBG Botany 940 (January 1990) and Class of New York University G23.1074 (May 1995).

Mee Young Choi and Daphne Christopher of the NYBG Institute of Economic Botany provided much assistance with questions on voucher specimens and database records. Jillian De Gezelle, Katherine Herrera, and Irina Adam graciously provided a thorough review of the original manuscript and made many useful suggestions. Daniel E. Atha and Michael H. Nee were very generous in identifying a great deal of our Belizean plant material and in confirming determinations from photographs. We are grateful to the entire staff of the NYBG Institute of Systematic Botany and William and Lynda Steere Herbarium, along with other colleagues at NYBG, particularly Brian M. Boom, Douglas C. Daly, Andrew J. Henderson, Jacquelyn A. Kallunki, Lawrence M. Kelly, Fabian A. Michelangeli, Robbin C. Moran, Scott A. Mori, W. Wayt Thomas, as well as Paul Maas, for kindly providing species identification for this project. Our thanks to the NYBG Operations Department, particularly Mark Cupkovic and his staff for all of their work that made our efforts possible, and to the NYBG Information Technology Department, particularly Patrick Maraj, who was always available to resolve the ever-present computer issues that are endemic to any program that manages large amounts of data. Thanks to Miguel Choco and Mike Green of the Lodge at Chaa Creek, Belize for assistance with some of the vernacular names listed in Chapter 4, and to Don Victor Cal for his guidance on Q'eqchi' plant names. In addition, we would like to thank Nigel Encalada of the National Institute of Culture and History in Belize (NICH) for facilitating the co-copyright agreement between the NICH and NYBG.

Literature on the plants used traditionally in Belize was gathered from the LuEsther T. Mertz Library of NYBG; the various libraries of the University of Oxford, UK (including the Bodleian, Rhodes House, and Forestry Department); the Royal Botanic Gardens, Kew, UK; as well as through correspondence with individuals at numerous institutions. This literature included books, scientific papers, reports, correspondence, and miscellaneous files.

The program in Belize described in this book has been supported by a number of foundations and individuals since its inception in 1987, and we offer our gratitude to the staff of the Development Office of NYBG, who worked hard over many years to help secure support for this scientific program. The research carried out in Belize and at NYBG has, ultimately, been made possible through the provision of long-term grant and contract funding as well as separate grants from the U.S. National Cancer Institute of The National Institutes of Health, The U.S. Agency for International Development, The Rockefeller Foundation, The MetLife Foundation, The Overbrook Foundation,

The Edward John Noble Foundation, The Philecology Trust, The Rex Foundation, The John D. and Catherine T. MacArthur Foundation, The Nathan Cummings Foundation, Healing Forest Conservancy, and The Gildea Foundation, and gifts from the following individuals: Tom and Anne Hubbard, Kathy Gallagher, Mary R. Kemmerer, Bruce McCowan, Mary R. Morgan, and Susannah Schroll. From 1989 to 2002, Michael Balick was a MetLife Fellow at NYBG. In addition, we would like to thank Tom and Anne Hubbard for their steadfast, decades-long support of the NYBG scientific enterprise in general and interest in this project in particular from the beginning. Their dedication and commitment has been a source of inspiration for us and many others working in related fields of study.

Special thanks go to our friends Francesca Anderson, Adam Cummings, Mickey Hart, Sharon Matola, Diana Landreth, David Sostman, and Katie Valk for believing in us from the very beginning and to Gregory Shropshire for his enthusiastic participation in nearly all of the fieldwork involved in this project, as well as the creation of much of the infrastructure needed to implement it. Cooperation of The Forest Department; Ministry of Forestry, Fisheries and Sustainable Development; Ministry of Natural Resources and Agriculture; The Belize College of Agriculture, and the NICH is gratefully acknowledged, as is the collaboration of the Belize Zoo and Tropical Education Center.

And last but certainly not least, the authors are grateful to the dedicated team at Oxford University Press for their commitment to this volume: Phyllis Cohen, Julie Fergus, Erik Hane, Natalie Johnson, Richard Johnson, Michelle Kelly, Bonni Leon-Berman, Wesley Morrison, Linda Roppolo, and Marc Schneider.

PHOTOGRAPHY AND ILLUSTRATION CREDITS

Photographs have been provided by the following and are credited with the photographers' initials in the figure captions: Pedro Acevedo, Smithsonian Institution [PA]; Irina Adam [IA]; Brett Adams, Belize Botanic Gardens [BA]; Mac Alford [MA]; Greg Allikas [GA]; Rosita Arvigo [RA]; Daniel Atha [DA]; Michael J. Balick [MB]; Scooter Cheatham [SC]; Jim Conrad [JC]; G. Arthur Cooper [GAC]; Tom Croat [TC]; Gerritt Davidse, Missouri Botanical Garden [GD]; Jan De Laet (on www.plantsystematics.org) [JDL]; Robin B. Foster [RF]; Vicki Funk, Smithsonian Institution [VF]; W. John Hayden [WJH]; Sune Holt [SH]; Richard A. Howard [RH]; Clyde Imada [CI]; Lawrence Kelly [LK]; Gertrud Konings [GK]; Richard W. Lighty [RL]; Fabian A. Michelangeli [FM]; Robbin C. Moran [RM]; Michael H. Nee [MN]; Kevin Nixon [KN]; Tim Plowman [TP]; Jackeline Salazar [JS]; Gregory Shropshire [GS]; Forest and Kim Starr [ST]; Dennis W. Stevenson [DWS]; W. Wayt Thomas [WT]; John Vandermeer [JV]; Corinne Vriesendorp [CV]; Warren L. Wagner, Smithsonian Institution [WLW]; and Hugh Wilson [HW]. Photographs provided courtesy of the Trustees and Director of Royal Botanic Gardens, Kew, are credited as [RBGKew]. Photographs provided courtesy of Belize Botanic Gardens are credited as [BBG].

We are grateful to Rusty Russell for assistance with images from the Smithsonian Institution collection [RH and GAC]. We thank Heather duPlooy for her assistance with photographs courtesy of Belize Botanic Gardens. Special thanks to Robin B. Foster for taking the time to search through his database and for providing us with several much needed plant photographs.

Original drawings were kindly prepared for this volume by the noted botanical illustrator Francesca Anderson, who is credited by her initials in the figure captions [FA]. Copies of additional botanical illustrations were obtained from the magnificent and very comprehensive collections of the NYBG LuEsther T. Mertz Library, indicated by [NYBG]. We are grateful to the Library staff who helped us with so many aspects of this project over nearly two decades.

MESSAGES
from the
GODS

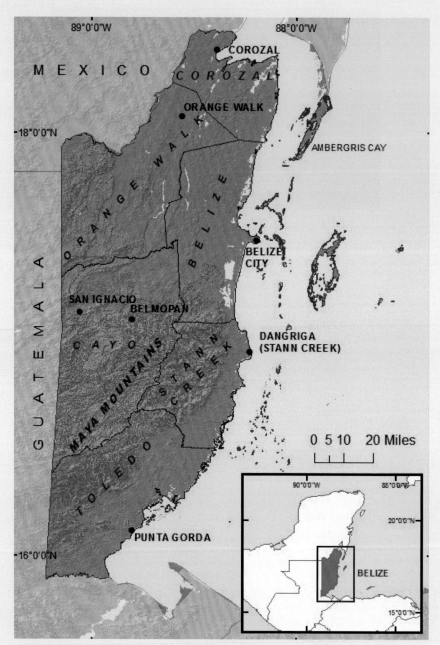

Map created by Dr. Wayne Law, Geographic Information Systems Laboratory [NYBG].

1

INTRODUCTION & BACKGROUND

AS A FIELD OF INQUIRY, ethnobotany examines the relationship between plants, people, and culture. It employs tools from many disciplines and, in contemporary times, often involves teams of individuals, bringing multiple perspectives to bear on research questions being posed. Initially, this field focused on producing lists of plants employed as food or medicine by indigenous people. In contemporary times, ethnobotanists utilize modern research techniques and tools to examine multifaceted aspects of the plant/people relationship, with a particular focus on resolving important issues of the day—topics such as resource utilization, health care practices, land management systems, economics, pharmacology, nutrition, and many other subjects.

Our research trajectory has been shaped by our own perspectives as an ethnobotanist (Michael J. Balick) and a health care professional (Rosita Arvigo), combined with the guidance, collaboration, knowledge, and input of a small group of traditional healers and bushmasters (people who are experts in the use of plants from forests and other wilderness areas). We recognize that if the research team for the Belize Ethnobotany Project had been composed differently, the results and their interpretation would be different. Ethnobotanical study, with its methodology as a science steeped in observation, is most often shaped by the perspectives of the observer. We have always viewed our core group of specialists—ethnobotanist, naprapathic physician, and traditional healers/bushmasters—as a tripartite and equal partnership. In this project, the traditional healers and bushmasters are considered to be teachers, colleagues, and friends rather than the traditional "informants." This is more consistent with the nature of transdisciplinary research, work that is integrative at its core and goes beyond the boundaries of traditional scientific disciplines. The knowledge and beliefs of people in Belize—both recent and ancient—comprise a significant element of our studies, and this realm of inquiry and data cannot always be classified according to the Western scientific construct. The partnerships developed through this work have also required additional obligations and responsibilities—for example, in the area of intellectual property—as will be discussed elsewhere in this book.

The people who chose to work with us were firm in their belief that traditional knowledge held by their communities and themselves is in danger of being lost. Each felt that by working on this project, they could contribute to the longevity of their specific traditional knowledge and, ultimately, its use by the younger generation through the rebirth of attention and interest that a nationwide project would generate. Indeed, this has been the case; there is now much more interest in Belize regarding the importance and value of traditional practices, especially those involving traditional healing or "bush medicine," as compared to the recent past. Project activities included ethnobotanical and ethnomedical fieldwork; national meetings, symposia, and workshops devoted to ethnomedical practices; video and television shows; college and high school course curricula developed by the project; a bush medicine and conservation camp for young people; creation of a forest reserve devoted to the protection of medicinal plants; production of content for ecotourism industries focused on traditional uses of plants; a line of local medicinal tinctures and other products based on traditional knowledge and local natural resources; and several books on useful plants for both children and adults. As a result, the project generated a great deal of local attention. In large measure, the goal of rekindling interest in traditional practices has been met, both through our project as well as the work of many others who were

(top) One of the first national traditional healers meetings, held in Belize City in 1991 [MB].

(middle) Participants in a traditional healing workshop examining a display of plant products used in Belizean traditional medicine [MB].

(bottom) Participants in a bush medicine camp held at Ix Chel Farm in 2005 [MB].

also concerned about the loss of traditional knowledge.

That is not to say that the project has
been carried out without controversy, as
is increasingly the case of ethnobotanical
studies in a post-modern, post-colonial,
and post-CBD (United Nations Con-
vention on Biological Diversity) world,
where inherent suspicion—at times justi-
fied—of the intentions and operations of
outsiders as well as local people is a part of
the research environment. But, as a local
woodcarver once responded to us when
asked from whom he learned his rather
extraordinary skills—"my critics have
been my best teachers"—so, too, have we

The authors discussing the use of an herb for
treating snakebite with Andrew Ramcharan of
Calcutta, Orange Walk District [GS].

learned much from our critics. The opinions of many observers from all walks of life
have helped shape our studies and other activities in Belize since 1987, for which we
express our sincere gratitude. In this book, we discuss the development of the project
and its outcomes. There is a great deal to learn from these experiences; our intention
is to provide these lessons to those who come after us in designing and implementing
their own studies.

Since 1987, we have conducted interviews with dozens of traditional healers. Hundreds of local people participated in surveys involving plants and their uses, and we made
field collection trips to over 220 different sites in Belize. Since 1987, more than 8,000
plant collections have been made in Belize by our staff and both local and international
colleagues as part of the various floristic, ecological, and ethnobotanical projects related
to this work. The largest number of collections was made in the Cayo District, with
fieldwork undertaken in all of the districts. In the Belize National Herbarium, the specimens gathered by the ethnobotany project alone that have been curated to date comprise
nearly 18 percent of the total holdings. Following a decade of intensive fieldwork,
information on nearly 900 plant taxa is presented in this book. However, we suspect
that this information comprises only a fraction—perhaps not even five percent—of the
universe of knowledge that exists in the minds and teachings of thousands of individuals
in Belize who have been raised on nature's gifts. As that generation ages, the next must
bear the responsibility of learning how to use and protect their nation's fragile biological diversity. Today, we are seeing great progress in this area, as more individuals, local
organizations, and the Government of Belize take interest in the topics of traditional
knowledge and biodiversity conservation.

This book is organized into four chapters, with two additional chapters found on
the Oxford University Press website (http://www.oup.com/us/messagesfromthegods).
Chapter 1 is an introduction to Belize and the Belize Ethnobotany Project. We discuss
project objectives, methodology, and the nature of traditional healing in Belize. Chap-

ter 2 focuses on the art of traditional healing in Belize, the illnesses treated, how medicines are prepared and administered, the risk of losing plant species utilized in primary health care, as well as how food plants and even local poisonous plants are used to heal. The chapter concludes by resolving a mystery—the identity of a plant once widely used to treat malarial and other fevers but now forgotten. In Chapter 3, we highlight the different traditional healers who made this project possible, providing short biographies and excerpts from lengthy interviews that were made during the late 1980s and early 1990s. This section, entitled "In Their Own Words," gives voice to the ideas, philosophies, and motivations of eight of the knowledgeable elders who worked with us. The final chapter in this book, Chapter 4, is an encyclopedic treatment of the useful plants of Belize—a guide to their scientific names, local names, family descriptions, and uses in ancient and modern times. These uses include the broad variety of purposes for which plants were, and are today, employed by local people. This section contains illustrations for most of the plant genera discussed, with the intent of aiding in their identification. As pointed out elsewhere, this book reports on a very small fraction of the plants used traditionally in Belize; it is intended as a resource for those wishing to learn this knowledge and lore and as a foundation for ethnobotanical, floristic, and conservation studies and teaching that we hope will continue to be carried out by Belizeans in the future.

Two additional chapters based on work arising from this project are, as previously mentioned, published electronically. The first, Chapter 5, addresses the uses, economics, and history of the harvest of timber and non-timber forest products from Belize. The second, Chapter 6, discusses the natural products research sponsored by the US National Cancer Institute that took place in Belize from 1987 to 1996 as part of this project. Background information is provided on how new compounds are discovered from plants and other natural products and on the process by which these are developed into therapeutic agents. Appendices to this particular chapter include documents outlining benefit sharing provisions and research agreements that were, and continue to be, part of this research.

THE SETTING

Belize is a small country, with 22,963 km² of land area, bordered by Mexico in the north and Guatemala in the south and west, with 280 km of coastline on the Caribbean Sea in the east. It is situated in the subtropical region at 15°53′ to 18°30′ N latitude and 87°15′ to 89°15′ W longitude, which includes the area in the Caribbean Sea. Offshore, some 450 small islands (cayes—pronounced "keys") add perhaps 690 km² to the national landmass. There is a pronounced dry season in the central and northern region from January until April or May and in the south-central region from February to April, although this pattern appears to be shifting in contemporary times. A period of less rain often occurs in August (Hartshorn et al., 1984). The coastal plain in the north receives only about a third of the rainfall of the southern Toledo District. For example, Walker (1973) reported annual average rainfall of 1,347 mm in Libertad, Corozal (northern Belize); 1,323 mm at Benque Viejo del Carmen, Cayo (central Belize); and increasing to

4,526 mm in Barranco, Toledo (southern Belize). The change in vegetation types can be observed as one travels from north to south, and the increase in rainfall is an important contributing factor (Balick, Nee, and Atha, 2000).

Hartshorn et al. (1984) delineated six ecological life zones in Belize, based on the Holdridge life zones classification (Holdridge, 1967). **Subtropical Moist Forest** is found in the northern lowlands of Belize, extending south to the lower, western section of the Mountain Pine Ridge, into the Macal River Valley, and on the Vaca Plateau. Rainfall in this forest type ranges between 1,300 and 2,000 mm per year. **Subtropical Lower Montane Moist Forest** is found in the Mountain Pine Ridge, commencing at an elevation of approximately 650 to 700 m. It is also found in the elevated parts of the Vaca Plateau and the higher western slopes of the Maya Mountains. **Subtropical Lower Montane Wet Forest** is found in the windward sections of the Mountain Pine Ridge and Maya Mountains, which have significantly more rainfall than the Subtropical Lower Montane Moist Forest. This forest is found in the higher parts of the Mountain Pine Ridge, such as Cooma Cairn, Baldy Beacon, and Baldy Sibun. **Subtropical Wet Forest** is found below approximately 600 m on the windward side of the Maya Mountains, with much higher rainfall than with the above vegetation types. This forest is found in the Upper Stann Creek Valley, Cockscomb Basin, and the bulk of the Toledo District, extending to the coast between Monkey River Town and Punta Gorda. **Tropical Moist Forest**, a zone between Subtropical Moist and Subtropical Wet, is found south of the Western Highway around the northern and northeastern foothills of the Maya Mountains and extends down the coast near to Monkey River Town. The last type of forest, **Tropical Wet Forest**, is the area of highest moisture in Belize, found in the south in the Toledo District. Here, rainfall can reach 4,000 to 4,600 mm annually.

Examples of plants characteristic of these ecological life zones are found in *Checklist of the Vascular Plants of Belize, with Common Names and Uses* (Balick, Nee, and Atha, 2000), and the reader is directed to that volume for further details on floristic composition. Based on their collections and study of herbarium material, the native and naturalized vascular flora of Belize comprise 3,408 species in 1,219 genera and 209 families. The 10 largest families (with numbers of included species) are Fabaceae *sensu latu* (295), Orchidaceae (279), Poaceae (248), Asteraceae (153), Cyperaceae (146), Rubiaceae (142), Euphorbiaceae (104), Melastomataceae (96), Myrtaceae (58), and Aspleniaceae (58). Within this flora, 41 species in 24 families, or 1.2 percent of the flora, are endemic to Belize. This low percentage is not surprising because of the habitats and flora shared with neighboring countries. One monotypic genus in the palm family, *Schippia*, is found quite commonly throughout its range in the Cayo and Stann Creek Districts.

Belize is a diverse nation, with Maya, Creole, Garifuna, East Indian, Mestizo, Mennonite, English, Chinese, Taiwanese, and Lebanese peoples. There are three Maya ethnic groups in Belize, each with their own language: the Q'eqchi' in the southern district of Toledo; the Yucatec, mostly living in the north; and the Mopan, mostly living in the western Cayo, Toledo, and Stann Creek Districts. The Creole people are descendents of enslaved Africans brought to work in the logwood, mahogany, *chicle*, and sugar cane industries and the British colonists who controlled those industries. The Garifuna people are an indigenous group descended from Amerindians, with Arawak, Carib, and African ancestry, who were exiled from the Caribbean to Central America. The East Indian people are descendants of indentured laborers brought from India around the turn of the 20th century. Chinese people were also brought as indentured laborers around that time, though most Chinese and Taiwanese in Belize arrived after Belize's independence. The Lebanese people arrived in the first quarter of the 20th century and were involved in timber and related industries. The Mennonites originated in Europe and migrated to Central America by way of North America around the middle of the 20th century. The Mestizo population has its roots in many of the surrounding countries of Central America, including Mexico, Guatemala, El Salvador, Honduras, and Nicaragua. The diverse interaction between and within so many different cultural and ethnic groups, out of a total population of around 333,000, makes the study of Belizean ethnobotany quite interesting. Each of the different groups has its own names and uses for the local flora, some of which are shared by groups and others which are particular to one individual group, community, or even a few members of an individual family—the latter, for example, in the case of healing recipes for snakebite.

There has been a great deal of interest in the plants of the Central American region and their uses. Studies by Heyder (1930), Lamb (1946), Standley and Record (1936), and Morris (1883) focused on the economic utility of the flora, particularly the forest trees. Williams (1981) provided an inventory of the useful plants of Central America, while Lundell (1938), Steggerda (1943), and Roys (1931) were among those conducting early studies of plant use by the Maya. A great deal of contemporary research has been undertaken on ethnomedical practices in Belize and its environs—for example, Ankli, Sticher, and Heinrich (1999a, 1999b); Berlin and Berlin (1996); Arnason et al. (1980); Balick et al. (2002); Comerford (1996); Flores and Ricalde (1996); Mallory (1991); and Stepp and Moerman (2001)—and examples of studies of Maya ethnobotany in this region include Alcorn (1984), Amiguet et al. (2005), Mutchnick and McCarthy (1997), and Peters (1983). There is indeed a rich literature on the useful plants of Belize, scattered throughout time and numerous publications. In this book, we report on uses of plants that we have studied and collected during the field project described below. In Chapter 4, we have included some, but not all, of the uses available in the literature; a complete compilation of published traditional information on Belizean plant utilization is far beyond the scope of this book and, perhaps, more suitable for a database project. The primary objective of this volume is to report our observations and the data on plants collected during the field phase of the project.

ORIGINS OF THE BELIZE ETHNOBOTANY PROJECT

What was to become the Belize Ethnobotany Project began with a simple letter, written by Rosita Arvigo of Ix Chel Farm in Belize to Michael J. Balick of The New York Botanical Garden on April 7, 1987:

Dear Dr. Balick,

I have just received in the mail a copy of an article that appeared in the *Atlanta Journal* on March 22, 1987, in which is explained the forthcoming efforts of the scientific community to scour the tropical forests for possible cancer cures. I am writing to you because your institution is charged with Central America where I live.

I have been working very closely this past year and a half with an old Mayan bush doctor who is 87 years old and has practiced his ancient system of medicine for 50 years. Don Eligio Panti is the man to see down here and I am his assistant. I would like to invite you and your group to our farm while you are passing through Belize on your rounds. We have a guest house and could comfortably put up four persons, more if they are willing to use their own tents. Don Eligio knows of at least three plants used for cancer. He has successfully treated skin cancer and now has a seven-year-old patient with a brain tumor and no money for an operation. He speaks only Mayan and Spanish, but I am fluent in Spanish and would translate. Also, there is an 80-year-old dory maker by the name of Thomas Green who is our neighbor and has proven on many occasions to be a fine bush doctor in his own right. He, too, would be worth talking to.

We have taken on the project of recording and chronicling Don Eligio's medicinal plants and healing systems without the benefit of a grant or anything of that like from anyone—no one asked us to do this, but my husband and I could clearly see what was about to be lost to the world because no one else seems to care about it—anywhere. We do. You can read the enclosed article and get an idea of some of the things we are involved in. Dr. Timothy Plowman and Dr. Ann Bradburn of Chicago Field Museum and Tulane University, respectively, have been providing specimen vouchers and moral support.

Hoping to hear from you in the near future, I remain,
Sincerely yours,
Rosita Arvigo, D.N.
Director, Ix Chel Farm

Michael Balick contacted Dr. Timothy Plowman, a friend and mentor from his days at the Botanical Museum of Harvard University, who advised that he had met Rosita Arvigo and her husband, Gregory Shropshire, both naprapathic physicians from Chicago. He was convinced that they were sincere and had provided them with a plant press and instructions on how to collect herbarium vouchers. Balick then wrote a letter to Arvigo, requesting information on logistics and the local environment, and proposed a trip for the end of July, on the return leg of a flight from Honduras. He met Arvigo and Shropshire as planned, and they had an initial visit of a few days with Don Eligio Panti. During the visit, Balick accompanied Panti and Arvigo on their rounds in the forest, helped in the preparation of plant remedies, and observed visits with patients who

An initial meeting with Don Eligio Panti (right) at his home in San Antonio, Cayo District, discussing his work as a healer [GS].

had come to see the Maya *h'men* (healer priest) of San Antonio, Cayo District. Late one evening, the conversation went something like this:

The local people here, the priests, the teachers say I [Don Eligio Panti] am no good, that my spells and prayers to the Maya Gods are of the devil, and that my work is evil. But I am a good man, I believe in God and the Maya spirits; they are all around us, here, now. But we ancients have been forgotten by the young people. And when the people, the priests, and the teachers get sick, they come to me, for prayers, for medicine, and for healing. So I must be doing something right ... You have traveled very far to my simple house, and told me of the sickness of your people, and that it cannot be cured by the medicine of your doctors. They are modern doctors, but still cannot cure all of the illness of the people ... I will work with you and your doctors if you wish, helping you to collect and study the plants of the ancient Mayas. I will help you with all of my knowledge of the past 50 years of work. And if you find that something of the Mayas is useful in your medicine, then it will be as a foundation stone in a bridge, building a path from the ancient Mayas to your modern world that has forgotten us, and in this way once again bring respect to the Maya spirits and ancient ways that will perhaps make our young people pay attention to their elders and their traditions, before it is too late and we are dead! And when I have completed this task, I will join my wife, whom I miss so dearly, in heaven.

The next morning, Arvigo and Balick walked the earthen path back from San Antonio to Ix Chel Farm, a five-mile trail that she traversed, rain or shine, several times a week for her work as Don Eligio's assistant. They discussed the possibility of collaboration with Panti in collecting plants and information on their uses and how this could be funded through the National Cancer Institute (NCI) contract recently awarded to The New York Botanical Garden. Upon arrival at Ix Chel, they continued the discussions and, standing upon an ancient Maya ruin site, agreed to work together on what was to become the Belize Ethnobotany Project—a decade-long field survey of the relationship between the plants and people of this nation. They would record Panti's knowledge for posterity, as he had requested, collecting herbarium specimens and photographing the process of preparing plants for medicine and treating patients. In addition, a sample of each plant they collected would be dried, shipped to Frederick, Maryland, home of the Developmental Therapeutics Program of the NCI, extracted in various solvents, and screened against many dozens of living cell lines of human cancers, such as lung, colon, ovarian, and others. In later years, two HIV/AIDS assays would be added to the battery of screens. Importantly, the results of the screening process for each plant, as well as a portion of any eventual royalties that might accrue from an important medical discovery, would be due the collaborating institutions in Belize, as well as the local government, through an agreement developed by Dr. Gordon Cragg and his colleagues at the NCI. Details of the NCI-sponsored laboratory research, intellectual property agreements, and initial results are presented in Chapter 6 (http://www.oup.com/us/messagesfromthegods).

The traditional paradigm for scientific research is the postulation of one or more hypotheses and, with the application of labor and tools, testing these hypotheses and arriving to the point where the hypotheses are either accepted or rejected. In some branches of science, however, and ethnobotany is one of these, the linear path from hypothesis to predicted outcome often develops branches and bifurcations, with additional research questions raised. The investigator is then charged with rethinking the conclusions, readjusting the hypotheses, setting off in new directions, and at the same time, ensuring that the study maintains a definable shape and end point. Such has been the history of this project during the period of our most active field studies in Belize (1987–1997), with numerous subprojects being formed and implemented as opportunities were presented and as other researchers expressed interest in becoming involved in the project. In addition, as local people heard of our work with Don Eligio Panti, dozens expressed the desire to collaborate as well. The names of the local people who contributed the most to this work are listed as the source of knowledge for this study on the title page. Scientific colleagues who prepared chapters that were edited for inclusion in this book are also listed on the title page. Others who agreed to contribute information through interviews or participation in the fieldwork are thanked in the acknowledgements at the beginning of this book.

PROJECT OBJECTIVES

From its outset, the Belize Ethnobotany Project has been an evolving work in progress. Initially, the goals included general inventories of the plant diversity of Belize, particularly in populated regions, to gather information on the uses of plants as beverages, construction materials, dyes, fibers, food, forage, fuel, gums, latex, medicines, oils, ornamentals, poisons, handicrafts, resins, components of local rituals, spices, tannins, and for other purposes. Herbarium and bulk sample collections were made, and the herbarium collections were mounted and distributed to Belizean and international research institutions for identification and study. These collections, along with previously collected herbarium specimens housed at The New York Botanical Garden and a variety of other institutions, form the basis for the previously mentioned *Checklist of the Vascular Plants of Belize, with Common Names and Uses* (Balick, Nee, and Atha, 2000). More than 120 people (the majority being from Belize) participated in the fieldwork of the current study, receiving training in botanical methodology and collection

Filmmaker Françoise Pierrot (top), working with Bertha Waight in the production of *Diary of a Belizean Girl: Learning Herbal Wisdom from Our Elders* [MB].

techniques and developing skills in ethno-botanical interviewing. Additional training was provided to Belizeans at Ix Chel Tropical Research Foundation and at The New York Botanical Garden.

To properly curate and preserve plant collections in Belize, supplies and steel herbarium storage cases were provided to the Belize College of Agriculture and the National Herbarium of Belize as part of the effort to develop local infrastructure and provide facilities for future biological studies. Another objective was a pharmacological survey of the vegetation via the collection of bulk samples, primarily for the Developmental Therapeutics Program of the NCI, and later in collaboration with various other academic laboratories. As mentioned, Chapter 6 is a report on the work with the NCI, prepared by Gordon Cragg and David Newman.

One project output—a primary health care manual to teach current and future generations about the use of plants as components of the health care delivery system—resulted from an initial meeting with the community of traditional healers. The result was *Rainforest Remedies: One Hundred Healing Herbs of Belize*, by Rosita Arvigo and Michael J. Balick, first published in 1993 and revised and expanded in 1998. Early in the history of the project, it became clear that if the work was to have lasting impact, a group of young Belizeans should be involved. We developed a college curriculum on medicinal plants for use at the Belize College of Agriculture along with materials for use in middle and high schools, public exhibits, and school field trips. Educational outreach included the production of a video, *Diary of a Belizean Girl: Learning Herbal Wisdom from Our Elders*, to be shown at schools throughout the country and on national television. We also have continued to hold an annual "bush medicine camp" for children. The campers, primarily from urban parts of Belize, live in a field setting and work with traditional healers and bushmasters, learning about environmental and cultural topics.

As interest, particularly in traditional healing, grew, two local organizations devoted to developing and empowering the community of traditional healers (the Traditional Healers Foundation [THF] and the Belize Association of Traditional Healers [BATH]) were set up. Many national and community-level meetings also were held to discuss traditional healing, plants, and conservation.

Economic valuation studies of local resource use, carried out with colleagues from the Yale School of Forestry and Environmental Studies and Grinnell College, identified the need for more sustainable harvest strategies of non-timber forest products. In addition, there was the need for a place where this could be carried out—establishment of the Neotropic ecozone's first ethnobiomedical forest reserve in the Yalbac Region of Belize was the result. Unfortunately, in later years, much of this important site was logged and otherwise degraded.

TRADITIONAL HEALING IN BELIZE

Today, traditional healing in Belize is primarily a mixture of ancient indigenous traditions and Christian ritual. Following the conquest of Latin America, Mesoamerican peoples succeeded in maintaining their cultural identity through an amalgamation of practices that combined their traditional ways with the teachings introduced by the Spanish. Just how the indigenous people in Belize escaped the "heavy colonial hand" of the Spanish *conquistadores* is clearly explained by Graham, Pendergast, and Jones (1989). Because Spanish Colonial rule following the conquest of the Yucatan in 1540 was primarily concerned with the resource-rich region of central Mexico, a large wave of Maya refugees settled in the remote forests along the southern borders of the Yucatan, including Belize. The Spanish did not have total control over these areas, and the communities there were able to resist Spanish economic and religious mastery to some degree. In discussing the ways by which the Maya resisted, Graham, Pendergast, and Jones (1989) noted,

[T]hese mechanisms included not only outright rejection of Spanish Rule but also different forms of cultural adaptation and cooperation. Part of this adaptation was Maya acceptance of the Church . . . much if not all essential Catholic ritual became part of Maya life. On the other hand, the appearance of pre-Columbian-style animal and anthropomorphic effigies in refuse deposits and in offerings in buildings indicates that the new religion had not entirely replaced the old, and references to fears of Maya apostasy pepper Spanish reports. (pp. 1255–7)

In their combined investigation of the Spanish chronicles as well as archeological record, the authors concluded:

Though never able to abandon past customs totally in order to accept Christianity and its practices on wholly European terms, the Maya of the early Colonial period managed to draw from Spanish hegemony a faith and way of life in which two worlds could survive side by side. Pre-Columbian Maya technology and economic networks likewise coexisted with the way of life introduced by the Spaniards, probably as hidden in some respects as was the union of pagan and Christian beliefs. (p. 1259)

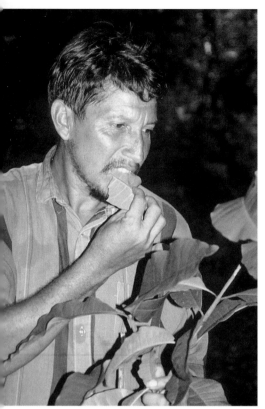

Abel Carrias eating leaves of allspice to cure an upset stomach [MB].

When observing traditional healers such as Don Eligio Panti as they say their prayers and practice their rituals, that mix of ancient Maya traditions and Catholic beliefs immediately becomes apparent. Don Eligio's prayers and incantations for sick patients always consisted of references to the Maya Gods as well as Jesus Christ. Once, with another family, we participated in a ceremony marking the day of the Holy Cross (May 3rd) in the village of Succotz, which began in the evening as a somber and sacred event, led by women dressed in black who were chanting Catholic prayers and hymns and lighting candles. Around midnight, the scene changed radically, casting off the solemnness of the occasion with festive, loud music as well as drink and dance, with men dancing wildly, one of whom entered the room carrying the just-severed and somewhat bloody head of a pig on his shoulders and eventually passing it from guest to guest, perhaps as a recreation of an earlier Maya ritual. This ceremony was reported to take place at the village level by Thompson (1930), but our participation involved a single-family household.

Through work with traditional peoples, we have observed this mixture of ancient traditions and modern practices in many places, particularly sites on the edge of an encroaching outside society. So there is no reason to believe that in such situations, traditional practices, such as agriculture and healing, would be preserved in their pure form to the present day. In fact, traditional practices such as medicine are in a constant state of evolution via ongoing clinical experimentation and observation as well as conversations with other practitioners, resulting in the addition of new modalities and "improved" ways of using known medicaments. For example, in 1988, when we noticed some of the first patients with HIV/AIDS returning to Belize to be cared for by their families, the only treatment the traditional healers would offer was burning copal incense at the foot of each of the four bedposts. The healers admitted that there was little they could do to relieve the symptoms and suffering of their patients. A decade later, we observed healers using herbal teas and other treatments made from local plants that were said to give the patients more energy and help treat their illness. While we have no evidence for the efficacy of these plant remedies in the treatment of HIV/AIDS, the healers' interest in experimentation was evident, and it was not hard to imagine this same curiosity and observational skill as being a driving force in the evolution of traditional medicine in this region. In another instance, we were walking along a forest trail and observed a man eating a small handful of the leaves of allspice (*Pimenta dioica* (L.) Merrill), a common forest plant in the Cayo District. When asked what he was using this for, he responded that it was to treat his stomach ache. This is not the common way to treat gastric problems using allspice. Normally, leaves or berries are made into a soothing tea and consumed (Arvigo and Balick, 1998). "I learned this treatment from my father," he explained. "He was a bushmaster who had worked in the *chicle* camps, and this was an important medicine." Farmers and healers in particular seem very amenable to experimentation and the adaptation of new things to their own worlds.

The earliest known book on the subject of healing with medicinal plants in Central America is known as the *Badianus Manuscript*, a fascinating text prepared in 1552 and believed to be the only such manuscript extant that does not reflect the influence of European medical practice. It was found on a shelf of the Vatican Library in 1929, and it is the work of two young Aztec Indians, Martinus de la Cruz, a healer, and Juan Badianus, the translator of the text from Aztec into Latin (Arvigo and Balick, 1998).

Following the decline of the Maya civilization, around the 14th century (the end of the Post-Classic Phase), there were still Maya people in the land now known as Belize. They continued to live in organized communities, cultivate their fields, trade with other groups, and utilize the native plants growing in the forests and fields that surrounded their dwellings. Their system of healing was (and still is) a medico-religious system in which the concepts of disease and spiritual forces are intertwined. Many gods and goddesses inhabited their spiritual world—a veritable pantheon of benevolent and malevolent spirits. Included among these was a goddess of medicine and healing—Ix Chel, which translates as Goddess or Lady Rainbow—who was central to the ancients and their medical system. She was depicted as a maiden, mother, and a crone goddess. The maiden looked after women during pregnancy and childbirth and brought new weaving patterns through dream visions. Holding a rabbit, a symbol of fertility, the mother sat on a crescent moon and watched over mothers and their children. Their counterpart, the crone, was a goddess of the moon and medicine. The maiden and the crone images of Ix Chel showed her wearing a snake sitting comfortably on her head. It is indeed interesting that so many cultures around the world and throughout history—Maya, Aztecs, Druids, Africans, Greeks, and Romans—have settled on the snake as a symbol of medicine. In contemporary times, as we shall discuss, there persist a number of illnesses unique to this region caused by malevolent forces and treated with a combination of physical agents (e.g., plant extracts) in combination with spiritual elements (e.g., prayers, incantations, and protective amulets).

METHODOLOGY

Brent Berlin (1992) discussed two approaches to ethnobiological inquiry—cognitive and economic—in his book *Ethnobiological Classification: Principles of Categorization of Plants and Animals in Traditional Societies*. Cognitive ethnobiology (and thus ethnobotany) is the study of how people view (and classify) nature, while economic ethnobiology is the broader study of how people use nature and its components such as plants. For this project, we focused on economic ethnobotany, which includes surveys of plant distribution, utilization, and abundance; traditional healing practices involving plants; the harvest and utilization of timber and non-timber forest and wilderness products; conservation practices; resource management; economic valuation; pharmacology; and education and community development. As noted, our focus—indeed, our admitted bias—has been on plants utilized in local health care practices, and many more facets of the plant/people interaction in Belize are, of course, extremely worthy of study.

Alexiades (1996) outlined basic techniques of ethnobotanical inquiry in *Selected Guidelines for Ethnobotanical Research: A Field Manual*, a comprehensive yet understandable and very useful guide to the nature and practice of ethnobotany. Many of the techniques he listed were utilized at one stage or another in our ethnobotanical research in Belize and are discussed below, with examples of how these were employed in our study.

One of the principal techniques of the ethnobotanist is **direct observation**, or **participant observation**, in which the researcher observes and participates in the activities of the people being studied. We were able to undertake this type of inquiry in several different ways. We accompanied the traditional healers and bushmasters during their work in the forest, such as the collection of medicinal plants for that day's group of patients. These included hundreds of walks through the forest and other wilderness areas of Belize with dozens of elders involved in the collection and application of plants for use in primary health care, food, fiber and construction materials, and rituals and ceremonies. In this research, we were very interested in methods of preparing the plants collected or grown for a specific use, and we documented those processes to the extent possible. For example, working with people who harvested and blended raw materials such as bark and roots into ethnomedical mixtures allowed us to observe and record data on preparation and dosage that are often ignored in ethnobotanical surveys. Some of

Learning to collect water from the plant known locally as "water vine" (*Vitis tiliifolia* Humb. & Bonpl. ex Roem. & Schult.) [MB].

this information is presented in *Rainforest Remedies: One Hundred Healing Herbs of Belize* (Arvigo and Balick, 1993, 1998), to which the reader is referred. Because of Arvigo's role as apprentice to Don Eligio Panti (Arvigo, 1994), she was able to participate directly in the treatments of his patients, as a trainee, and record the details of his and other healers' modalities for addressing many conditions presented during the later years of his practice. Both Arvigo and Balick also worked closely with numerous other healers and bushmasters, participating in their collection and utilization of plants for various purposes.

Next, Alexiades (1996) discussed the technique of **simulation**, which is "used to get participants to reenact activities that are no longer performed or to perform them out of context" (Clammer, 1984). One example of this was the fabrication of various types of rudimentary stretchers made from forest plants that were once used to carry the sick or injured over long distances from the isolated *chicle* camps in the forests to the clinics or other medical facilities that might be available in a town or village. During a walk through a forest in the Yalbac region, Leopoldo Romero, a healer and bushmaster, reminisced about how he learned many uses for plants during his time as a *chiclero*, a *chicle* harvester

Leopoldo Romero and friend making a stretcher out of palm leaves in the Yalbac Region of the Cayo District [MB].

in northern Belize and in the Yucatan. When asked about the care the sick received in the camps, he began to discuss the ways in which people were treated and, when this failed, such as in the case of traumatic injury or life-threatening illness, how they were carried out of the forest. With a machete, he quickly chopped a few saplings into six-foot poles, strung some palm leaves between them, and constructed a stretcher capable of carrying an adult. He admitted that he had not made such a carrier in many decades, which is why this action and observation would be categorized as a simulation. An example of the performance of an act out of context took place when Hortense Robinson demonstrated the use of palm leaves in making a plaster for a "patient" with a sore back. She demonstrated how the leaves are massaged into the skin and constructed a plaster that would press the leaves into the patient's back. Once again, this practice, although only a demonstration, was photographed and documented through written notes for the project files.

Also known as a "bagging interview" (Alcorn, 1984) or "ethnobotanical inventory" (Boom, 1987), the **field interview**, another technique, involves researchers walking through an area where plants are found, either marked (inventoried, as in a transect) or unmarked, and discussing the use of species recognized by either party. This technique has the risk of an inexperienced interviewer leading the interviewee, such as through the simple question of "What is that plant for?" or "Does that plant have a use?" Nevertheless, it has value in allowing an interview to proceed at the appropriate pace (often as determined by the guide or local expert) and in comparing the knowledge of different people, who can be taken on the same path. This method of inquiry was useful when the ethnobotanical team was in a particular location for a period of only a few weeks and wished to collect plants and data based on the knowledge of a limited number of local people. In particular, we used this approach in southern Belize, working with the Q'eqchi' people. Due to the remote location and rough terrain of the sites, we were

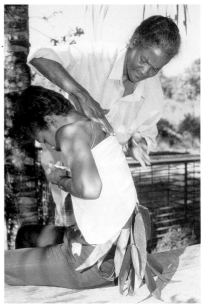

(top left) Hortense Robinson using a bottle to crush the leaf of the palm *Chamaedorea tepejilote* Liebm., used to make a healing plaster [MB].

(top right) Hortense Robinson demonstrating how to apply palm leaf plaster to a patient with back pain [MB].

(bottom) Traditional healer Victor Cho instructing ethnobotanist Julie Chinnock on the use of plants along a forest path in the Toledo District [MB].

dropped by helicopter (courtesy of the Royal Air Force, UK) for a trip to the Dolores Estate area of Toledo District. Our days there were quite limited, and the prospects for return at a later date were not good. This fieldwork was devoted to field and bagging interviews, working with several local people, including one elderly healer and two bushmasters from the community. During daylight hours, we walked along the trails that traverse this region, collecting plants and taking photographs and copious notes on what were reported to be their ethnobotanical uses and local names. Again, the risk in this type of interview is obtaining, or being given, wrong information, recording it carefully, and allowing it to dilute the verity of the overall data collection. The value of what is referred to as a "credibility rating" (Balick, 1999) is obvious in this circumstance. However, with time and patience, the experienced researcher can form impressions as to the value of the knowledge being offered in a relatively short time, particularly if he or she is familiar with the local dialect and flora.

The **plant interview** is a type of field interview, except that plants are collected in the field, returned to the village, and presented to local people for comment. This technique has been used with success by Boom (1987) and Alcorn (1984). Herbarium specimens (or a volume of photographs) can also be employed for this type of interview. Using the plant interview, researchers can involve more local people in the process, and their resulting conversations and discussions about the plant can help others recollect additional information that might otherwise be forgotten or withheld. Alexiades (1996) suggested that a disadvantage of this type of interview is the possible inability of people to recognize plant fragments, especially when pressed. However, we never found this to be the case in Belize; in fact, quite the opposite was true—people were able to recognize the smallest fragments of plants that were presented, not only by using cues from the appearance of the sample presented but also by observing taste, odor, and texture and by knowing its habitat. During the project, we rarely utilized this technique. It was employed only when a person could not, or would not, walk to the forest to collect plants with us or when we were querying a large group of people, such as at a community meeting, seeking to understand a plant's use in the context of "generalist" versus "specialist" information.

The **artifact interview** employs the use of artifacts that have been produced from plants. The plants, along with the methods by which they are obtained and processed into the final product, are discussed and can be gathered and identified for the researcher. This technique was utilized, for example, in understanding the manufacture of baskets from lianas growing in the forest. Baskets made by Isobel Xix were seen around his house and eventually offered for sale. Mr. Xix took us into the forest and provided basic lessons in harvesting and preparing the *Desmoncus* palm stems he was weaving into baskets. We used a video camera to capture details of working with the liana that would not adequately have been described by photographs and notes (Figs. 1.13 and 1.14). Another example was in the crafting of brooms from the leaves of *Cryosophila stauracantha* (Heynh.) R. Evans along with stems of sapling trees. In addition to his reputation as a powerful healer, Don Eligio Panti was able to make the local brooms commonly used for sweeping the mud floors of traditional Maya houses in his village. After discussing the brooms awaiting sale in the corner of his house, he shared with us his technique for crafting them and then took us into the forest to collect the plants used to make the brooms.

The **checklist interview** involves the preparation of a list of plant names or characteristics and discussion of their uses. It can also be used in the context of producing a

(above) Basket woven from the stems of *Desmoncus orthacanthos* Mart. [MB].

(left) Isobel Xix weaving a basket from the stems of the palm *Desmoncus orthacanthos* Mart. [MB].

Jars in Barbara Fernandez's herb shop in the old market of Belize City [MB].

list of plants employed, for example, to make a canoe or dwelling. The only time this technique was utilized during our research was to construct a list of 10 plants commonly known by rural and urban Belizeans from most ethnic and cultural groups, to develop a sense of the multiplicity of uses for a small set of individual plants. This was for the development of data to produce a graph for the "multiple use curve" concept (Balick, 1994). This graph compares the relationship between the number of different uses of a particular species and the number of healers interviewed and voucher collections obtained. The graph is used to obtain a sample size that offers a reasonable certainty the information gathered on a specific plant is relatively "complete," at least in the region sampled.

Finally, Alexiades (1996) discussed the **market survey**, in which local markets are the focus of the inventory and the interviewing—for example, herb sellers become the source of information on topics such as sourcing, habitat, harvest, costs, and utilization. The works of Robert Bye, Edelmira Linares, and Gary Martin in such settings are well-known examples of this type of research (e.g., Bye, 1986; Bye and Linares, 1983; Martin, 1992). In Belize, we were fortunate enough to have the cooperation of Barbara Fernandez, owner of an herb shop in the Belize City market and author of *Medicine Woman: The Herbal Traditions of Belize* (1990). "Aunt Barbara," as she is known by many in Belize, shared with us information about plants that she purchases from *yerbateros*, or herb gatherers, and dispenses for treating the illnesses of her patients, many of whom live in the urban environment.

Phillips (1996) discussed quantitative methods by which ethnobotanical data could be analyzed. He defined **quantitative ethnobotany** as "the direct application of quantitative techniques to the analysis of contemporary plant use data." In our research in Belize, we utilized quantitative techniques in forest inventory work in collaboration with Charles Peters, Rob Mendelsohn, and David Campbell, during which we examined the diversity of species in local forests and estimated the amount and value of harvestable non-timber forest products available in selected plots. Economic valuation studies of the medicinal plants in selected hectare forest plots are also one type of quantitative ethnobotany (Balick and Mendelsohn, 1992).

It is worthwhile citing two other concepts/techniques not discussed in Alexiades

(1996) but utilized during the work in Belize. The first we refer to as **salvage ethno-botany**, which is used when the plant/people relationship in the area under study is a tenuous one, with outside cultural forces causing the degradation of ancient traditional cultural uses of plants. While traditional knowledge is dynamic and constantly evolving as people embrace new skills and let go of others perceived to be of lesser value in contemporary times, its rate of evolution has increased during recent times, resulting in the loss of much ancient lore and practices. A parallel example is the background rate of species extinction, set at a certain number of species per unit of time, versus contemporary rates of species extinction, which have increased as the result of human activity. In salvage ethnobotany, the role of the ethnobotanist is to gather the remnants of whatever information currently exists, before those who possess it pass on, become too forgetful, or even are embarrassed to discuss the ancient ways. We have undertaken this work with elderly people who have a wealth of information but no apprentices to whom they can pass their knowledge. In one case in Belize, an elderly healer trained us in his family's knowledge of plant utilization up to a certain point, then suggested that if he were to pass away before teaching someone in his family his technique of snakebite treatment, the envelope tucked carefully in the center of his Bible was to be willed to our project. In another case, in a village in southern Belize, an elderly healer was prevented from practicing because he lived near a younger, more energetic man who demanded that *he* have a monopoly as the village healer. The elderly healer approached us out of the earshot of the younger man and requested that we work with him so that his knowledge become permanently recorded in the specimens, notes, and databases that were being developed by the project. Salvage ethnobotany takes its name from the concept of salvage botany, a term much utilized in the 1970s and 1980s during the Projeto Flora Amazônica expeditions, organized by Sir Ghillean T. Prance and Dr. Murca Pires, to endangered sites in the Brazilian Amazon (and elsewhere throughout the tropics) that were being destroyed as a result of road building or other human activities, to create a record of what once existed and provide data for future studies of plant diversity, distribution, and abundance. In a way, salvage ethnobotany is much like archeology, yielding a few clues to the use of plants that the ethnobotanical team carefully studies and then draws its conclusions. The archeologist may seek to

Gregory Shropshire, Pablo Cocom, and Rolando Cocom laying out a survey plot in the forest near the Chial Road in the Cayo District [MB].

Don Eligio Panti: "Messages from the Gods" [MB].

reconstruct a complete ceramic pot from a few potsherds; the ethnobotanist seeks to reconstruct the ancient modalities—plants and processes—for treating local diseases or for supporting human nutrition. Globally, more and more of our work is undertaken with people whose traditional cultural practices are in danger of disappearing. Data based on linguistic studies (Lewis, 2009) suggest that 473 of the nearly 7,000 spoken languages in the world are nearly extinct, with many others endangered—and when languages disappear, so, too, do the traditional belief systems and practices that are embedded within them (Balick and Cox, 1996).

The second technique is **immersion ethnobotany**, proposed in Balick and Cox (1996), which reduces the distance between the subject and observer, as, for example, when the observer "submits to being treated by an indigenous practitioner." In this way, the observer experiences firsthand the effects of the healer's treatments (while at the same time letting go of the feeling of control of the study). We have experienced numerous treatments by Belizean healers as well as healers in other areas of the world. Rosita Arvigo, as a resident of rural Belize and a healer in her own right, has been exposed to many hundreds of plants since moving to Belize with Gregory Shropshire in 1982. Over three decades of living in a study site and depending on the local natural and cultural resources for one's existence could certainly be considered immersion ethnobotany. This technique, when properly used, fosters a profound respect for the power and complexity of traditional healing as well as other traditional practices—for example, house construction, boat-making, and agriculture—while at the same time offering perspectives that differ from what might otherwise be a detached and impersonal study.

In concluding this introductory chapter, we want to acknowledge the inspiration for the title of this book. We often questioned Don Eligio Panti about the source of his knowledge—of a particular plant, or of how to treat a specific disease. Despite the fact that he apprenticed with an elderly traditional healer when he was young, he would always point to the heavens and then to his head and say that he received divine inspiration from the Maya gods and goddesses, as "Messages from the Gods." Indeed, many of the healers we worked with suggested that they were inspired or learned from dreams and visions, as we discuss in the next chapter.

2

TRADITIONAL HEALERS, BUSHMASTERS, & THEIR SACRED REALM

TRADITIONAL HEALING SYSTEMS in contemporary Belize are quite different than those of the past. This is due to the cultural richness of the nation, its colonial history, and the propensity of healers to experiment, exchange information, and continually adopt new practices. For example, ethnomedical systems of the Maya, whose civilization flourished up until the Conquest, subsequently have been influenced heavily by the cultures that immigrated to Belize, such as the East Indians, who brought along with their indenture the sacred and ancient tenets of Ayurveda (translated as "the science of life"). In addition, Spanish and English control over the indigenous people of Belize was sporadic and inconsistent. People moved around quite a bit in response to various kinds of pressure and oppression, and only a very few settlements in central and southern Belize have been continuously occupied over the last 500 years (Wilk, 2010). Thus, traditional healing in the region is now an amalgam of beliefs, therapies, and culturally specific illnesses that reflect the heterogeneity of contemporary Belizean society. Certainly, the purist would only find disappointment in searching for "true" Maya healing among today's practitioners. These were long ago entombed with the ancient *h'men*, or "healer priests," the result of the collapse of Maya civilization and subsequent persecution, beginning in the 1500s, by the *conquistadores*, who burned the once-vast indigenous libraries of medical and religious texts, destroyed the centers of medical learning, and prohibited traditional practices, all in the name of the advancement of civilization and spiritual beliefs.

To the ethnobotanist, Belize is a fascinating place to study traditional healing because of the cultural and ethnic mix that has brought so many elements from around the world into this very special ethnomedical system—actually, a series of systems. We have worked with local people who possess varying degrees of knowledge about traditional medical practices, from individuals with an understanding of the use of a few plants for common problems in daily life (generalists) to those with an acute awareness of the complexities of treating patients with chronic disease (specialists). In truth, each of the

Illustration of a traditional Yucatec Maya *primicia* ceremony. *From Gann (1924).*

culturally based ethnomedical systems we have investigated is deserving of its own separate volume, and thus we are able to present, even after more than a decade of fieldwork, only a small fraction of the information from some of the systems and cultures. In Chapter 4, which focuses on traditional plant uses, we attempt to identify the origin and culture of use when possible, but much of this is educated guesswork on our part as well as that of the healer who provided that information. Even such a statement as "I learned how to use this plant in a *chicle* camp in the Yucatan" cannot fully ascribe the origin of the utilization of this species to the Yucatec Maya. Laborers in these camps, we are told, came from very different backgrounds.

As pointed out in Chapter 1, traditional medicine and spiritual beliefs are intertwined in this region, evident by the large number of health conditions that are brought about, or treated with prayers, amulets and magic. The very nature of Don Eligio Panti's title—*h'men*, or "healer priest" in Mopan Maya—reflects the combination of his talents and training and is a credential for treating both the physical and spiritual ailments of his people. One important element of his work was the performance of the *primicias*, or "ceremonies of thanksgiving to the Gods for favors granted" (Pendergast, 1972). These ranged from small *primicias* in an individual's household or agricultural field to curing *primicias* by traditional healers to larger, public *primicias* undertaken to ensure that there would be sufficient rainfall for the village's crops growing in the cultivated fields, or *milpas*.

Thompson (1930), in discussing agriculture at Succotz, described a *primicia*, known in Maya as *u walikol*, for clearing the land to make a *milpa*. First, a small altar is set up on the edge of the area that will be cleared. The altar is composed of four forked corner posts, with thin sticks placed in the forks to form a table. A small cross is placed at the back of the altar. His description of the ceremony follows:

> The ceremony . . . to be described is attended by no one save the owner of the milpa that is to be made. On the morning on which he is to start work, he brings with him five calabashes (*Luts*) containing a special posol [maize drink] known as *sakqab* (white juice) because it contains no lime. These calabashes are placed one

at each corner of the altar, and one in the center of the front edge.

This latter, which is larger than the rest, and is known as the *Holtse*, represents *Quh*, who, my informant, after some thought, hazarded was the lord of the thunder. The corner cala-bashes represented *Qaanan Qasob* (the rain gods, meaning bountiful forests), *Halats winkob* (the big men), *Ahbehob* (those who pass before—the captains), and *Balam winkob* (the tiger men or big men). In Yucatan the *Yumbalamob* are the guardians of the towns. They are four in number, and sit one at each of the four entrances to the town, corresponding with the four cardinal points.

Don Eligio Panti performing a *primicia* ceremony with Rosita Arvigo on her farm, accepting her as his apprentice [GS].

When the altar is arranged the Hmen recites the following prayer . . . [the original Maya and a literal translation are provided, following which is the "Free Translation" presented here] In the name of God the Father, God the Son and God the Holy Ghost, Amen. Here before you I stand. Three times do I stand before you to worship God the Father, God the Son, and God the Holy Ghost. Behold my Lord how I stand in your presence now to venerate you. I stand in the presence of your holy name, Lord God and in the presence too of the Lords of the forests, who are mighty men. Forgive my sins because I am here to worship these Gods. That you may not forget me without cause I offer these five calabashes of posol in order that the mighty men, the lords of the forest, who live on the mountain tops, and who are the true lords, and are those who pass before to clear the road; in order that they, I repeat, may drink. Behold my Lord my good inten-tions in the presence of the Gods. I am preparing the drink-offering for my milpa. Forgive me O my great masters. Accept then but one cool draught of posol that the anger that lies in you hearts towards me may be cooled. In the name of God the Father, God the Son, and God the Holy Ghost. Amen.

When this prayer is completed the Hmen drinks one of the calabashes of posol. The others are left standing on the altar. If at the end of a fixed period the posol has turned sour, that is a sign that the lords of the forest are not satisfied for some reason or other and have refused to receive the *primicia*. In that case the *primicia* is repeated on some other occasion. Meanwhile, if the *primicia* is received the owner of the milpa is free to go ahead with his work. The ceremony is repeated before burning, sowing, when the ears begin to form, and at harvest. (pp. 115–7)

Rosita Arvigo, as Don Eligio Panti's apprentice, learned the *primicia* ceremony when Don Eligio performed several on her farm. The earliest was in 1985 to mark the occasion of accepting Arvigo as his apprentice. Don Eligio built a simple altar of vines and branches on an ancient Maya home site on Arvigo's farm in western Belize. The purpose of this *primicia* was to introduce his new apprentice to the Nine Maya Spirits to ask that they would give her every consideration they would him. The old Maya *h'men* laid out nine gourd bowls filled with corn *atole*, one for each of the Nine Maya Spirits, and placed four white candles at each of the four corners of the altar to honor the four *bacabs* who hold up the four corners of the earth as well as the four directions. Don Eligio chanted the *cantico*, or *primicia* prayer, in Maya nine times. Within the Maya *cantico* was the request that the Maya spirits would look favorably upon his apprentice and help her in all ways to achieve the goals of learning his system of medicine. Following the ceremony, as Arvigo (1994) noted, Don Eligio reported of the Maya spirits, "They're here right now. They've come. They have heard our *primicia* chant and drunk the spirit of the *atole*."

THE DISCOVERY OF THERAPEUTIC PLANTS

During this study, we queried many people for their opinions as to the way people first learn about a plant's use, outside of an exchange with other healers. As the result of these discussions, we offered several hypotheses for consideration (Arvigo and Balick, 1998). Some plants undoubtedly attracted attention due to their aroma. Certain taxa give off an odor when exposed to the elements and others when they are crushed, as in the leaf of allspice (*Pimenta dioica* (L.) Merrill) or cow foot (*Piper auritum* Kunth). These are both attractive odors, and perhaps were used originally to flavor foods. After consuming these plants, it is possible that observations were then made on their curative powers—the ability of allspice leaves to dispel gas and other digestive upsets, or how the pleasant, tingling taste of the cow foot leaf could be utilized externally to treat aches and pains. Foods and medicines are very closely linked in the cultural use of plants in Belize and will be discussed in greater detail later in this chapter.

Trial-and-error experimentation doubtlessly has played a significant role in the development of the Belizean ethnopharmacopoeia. As mentioned in Chapter 1, practitioners of traditional medicine are no less interested in learning new techniques and remedies and updating their skills and abilities than their counterparts in Western society. Ingestion of a plant known as a cure for stomach pains may have lead to an elimination of back pain in the same patient, and this was recorded—the dosages determined through trial-and-error testing on a group of patients, and the new therapy disseminated through exchange with other healers. There were probably numerous mistakes, and patients suffered or even died as the result of the ingestion of poisonous substances or toxic levels of therapeutic plants. Some cultures have identified a group of plants that, when ingested, will expel toxic substances from the body through increased sweating, urination, and vomiting. The properties of the plant product once kept in every Western household where young children resided—ipecac, derived from the roots of *Psychotria ipecacuanha* Stokes—to be administered in the case of poisoning were first observed by indigenous peoples in Brazil. Those cultures utilized it to induce vomiting in cases of suspected poisoning, along with other plants used to expel toxins from the body as needed. In Belize, purgative agents such as *Aloe vera* (L.) Burm. (known in Creole as *sink-am-bible*), *Ricinus communis* L. (known in Spanish as *iguerra,* or castor bean), *Carica papaya* L. (known in English as papaya or in Mopan Maya as *put*), and *Jatropha curcas* L. (known in English as physic nut and in Spanish as *piñon*) could all help expel toxic agents from the intestinal tract. Expertise in the use of these purgatives is necessary, as many of them can be highly toxic if not prepared and administered properly. Nine thorns of *Acacia cornigera* (L.) Willd., still containing the ants that inhabit this part of the plant, are said to be useful in treating poisonings when boiled in water and taken as a tea. There are many other plants that could have been used if and when the experiment went wrong.

Observations of domestic and wild animals certainly provided people with cues about the therapeutic value of plants. Curious persons, seeing an ill forest animal eat a certain plant species, might have collected this plant, tasted it, perhaps noted its bitterness, and then fed it to their own animals or used it on themselves. Thus, if successful, a new medicine was discovered and developed. Doubtlessly there were great failures, as some plants consumed by animals—for example, the berries of certain species in the Anacardiaceae (poison ivy family)—are toxic to humans. The study of the use of plants by animals for the treatment of illness is the field of zoopharmacognosy (Rodriguez and Wrangham, 1993).

The shape, color, appearance, or other physical characteristics of the plant is thought to provide guidance on its use. For example, *Sphagneticola trilobata* (L.) Pruski, locally called "rabbit's paw," is consumed as a tea to treat hepatitis. Perhaps the yellow color of the flower attracted the notice of an early healer seeking to cure a patient whose eyes had turned yellow as the result of this condition. The red tubers of *Smilax* species, locally called "China root," are often given to patients suffering from internal hemorrhage and as a tonic for the blood as well. This relationship between characteristic and cure is referred to as the "Doctrine of Signatures," where "like cures like" (or as Gann, 1918, noted in his observations of this phenomenon during his study of the Maya of northern British Honduras, "*similia similibus curantur*"), and has largely been ignored by Western science. The Doctrine of Signatures was proposed by Philippus Aureolus Theophrastus Bombastus von Hohenheim (1493–1541), a Swiss physician also known as Paracelsus, who developed his hypotheses based on more ancient practices. The notion that cures could be found in plants and other objects that looked similar to the disease or body part was developed in ancient medical practices such as those in China, India, Africa, and South America. Nyazema et al. (1994) discussed the efficacy of plants used traditionally in the treatment of urinary schistosomiasis in Zimbabwe based on the red coloring of the fruit or sap. Those authors investigated three plants that yield a red extract upon boiling, which proved to have appreciable antischistosoma activity as compared to the common and expensive Western treatment, praziquantel. The authors concluded that there is support for this traditional use based on modern scientific studies. Alexiades (1999) pointed out that it is difficult to know whether this concept serves more as a cue for the identification of medicinal plants or as an explanation of their efficacy once identified and incorporated into the native pharmacopoeia. Bennett (2007) suggested that most examples were "post hoc appellations rather than a priori clues" and that this concept "should be considered for what it primarily is—a way of disseminating information."

In Belize, there are many plants utilized in traditional medicine whose efficacy or discovery is credited by people to the concept of the Doctrine of Signatures. Alternatively, as both Alexiades and Bennett suggest above, the shape or color of the plant may also be a way to remember or teach its specific traditional medicinal uses. Following are four examples from our own studies:

- *Gumbolimbo* (*Bursera simaruba* (L.) Sarg.) has a red, peeling, shaggy bark that looks like some of the skin conditions it is used to treat—for example, sunburn, kitchen burn, diaper rash, heat rash, itch, and contact dermatitis from the black poisonwood tree sap (*Metopium brownei* (Jacq.) Urb.) that causes violent rashes that spread over the body. The inner bark of *gumbolimbo* is rubbed on the affected area of the skin and is extraordinarily effective in relieving itching as well as the appearance of the rash or dermatitis.

- The *zubin* or cockspur tree (*Acacia cornigera* (L.) Willd.) has large thorns up to eight centimeters long in which live ants that will sting the intruder upon any contact with the tree. The ants are said to protect the tree from being strangled by vines and other competing plants. The pain from their sting lasts for hours and often makes people quite ill. The bark and root of the *zubin* are used in traditional medicine for the treatment of snakebite, as a first-aid remedy. The victim is to cut a piece of bark roughly 2.5 × 15 centimeters and either boil it in water and drink it as a tea or chew the bark directly, swallowing the juice and using the leftover fiber as a poultice for the wound. Then, a section of the root is to be pulled up and chewed until the patient arrives at a hospital or snake doctor.

- The red-colored flower of the hibiscus (*Hibiscus rosa-sinensis* L.), and only the red-colored flower, among the many color varieties found in cultivation in Belize, is considered useful for the purpose of staunching the excessive flow of blood after childbirth, to prevent miscarriage, and as a tonic tea for anemia. The flowers and leaves are boiled into a tea and cooled, then taken in half-cup doses after delivery if the flow of blood seems excessive. To prevent miscarriage, the flowers are boiled with cinnamon sticks, and the tea is sipped throughout the day. As a tonic for anemia, the flowers are either eaten or made into drinks, as the taste is pleasant and soothing; the texture is somewhat mucilaginous.

- *Hamelia patens* Jacq. var. *patens* is known as polly red head, or *sanalo todo* ("cure all") in Spanish. The flowers and fruits are red, and the plant is used to heal stubborn wounds, sores, boils, carbuncles, and even leishmaniasis. The leaves, stems, and flowers are boiled into a tea-like liquid that is used as a hot bath for the wounds. Then, the leaves and flowers are toasted and passed through a sieve to make a fine powder that is applied directly to aid healing of the wound or sore.

Many healers report that dream visions are the source of their knowledge about the utility of plants. Don Eligio Panti considered God and the Nine Maya Spirits as his best teachers, and they would visit him in his dreams to help him deal with difficult cases. He noted that they would show him the plant in its forest habitat; instruct him on how to find, prepare, and administer it; and predict the patient's outcome.

Leopoldo Romero told of dealing with a difficult case of blindness resulting from diabetes. He retreated to the forest one day to search for a cure and, during what he describes as an evening "meditation" in the wilderness, had a dream about using fresh-water snails for the cure, a dream described in his own words in Chapter 3. He collected the snails, prepared them according to the directions in his dream, and used them to treat his patient, reporting success after several weeks of therapy. Interestingly, extracts from snails were used to treat vision problems in traditional Chinese medicine as well (Wang, 1998). We have found that dreams play an important role in the development of the therapeutic approach in Belize. From the Western scientific perspective, it is some-times difficult to accept that dreams can help solve scientific dilemmas; however, there are numerous anecdotes where scientific discoveries have been attributed to dreams. Notably among them is the resolution of the intricate structure of the benzene ring in 1865 by the chemist August Kekulé, who had pondered this problem without success until finally realizing, based on a dream he had about a snake biting its own tail, how the carbon atoms in the molecule could fit together. As the result of our studies, we have learned over the years to examine as thoroughly as possible the folk knowledge presented to us, with healthy skepticism combined with the willingness and patience to evaluate and understand it. One conclusion from our studies of traditional knowledge and beliefs is that there are many truths in this ancient wisdom.

GENERALIST AND SPECIALIST HEALERS IN BELIZE

As in Western medicine, traditional healing in Belize is divided into specialties, and healers range from the equivalent of the general practitioner to the more specialized ther-apist, such as the snakebite healer or bonesetter. Staiano (1981) considered four major types of health care choices in Belize: biomedicine (Western medicine), bush medicine, general lay knowledge, and spiritual healers. She offered fascinating case histories of individual patients in that study, which resulted from her fieldwork in southern Belize. Arvigo and Balick (1998) discussed seven categories of healers in Belize,[1] as follows:

The **doctor priest/priestess** is a man or woman mandated, through experience and power, to address both physical and spiritual ailments. Other societies refer to this type of person as a shaman, a term originating from the Tungus in Siberia (Eliade, 1964).

---

1. There are other categories of healers in Belize, such as, among the Q'eqchi', *ilonel*, who know both plants and prayers/ritual; *chinam*, who know prayers and ritual; and *ajk'e*, who are diviners, often using objects such as stones, bones, or corn (Wilk, 2010).

The doctor priest is referred to in Maya as *h'men* ("one who knows") and has the ability to contact the spirit world to ask its forces for assistance in the diagnosis and treatment of ailments as well as the lifting of evil spells, which can cause disease or death cast by other shamans or by the spirits acting on their own. The *h'men* draws some of his power from the *sastun*, a crystal or other such object that is an instrument of enchantment or divination.

The **village healer** is the primary health care provider for an entire village, having learned by raising many children, grandchildren, and even great-grandchildren. The village healer may be a man or a woman. A female village healer is usually a specialist in the care of women and children, including gynecology and obstetrics. A male village healer is more knowledgeable about general family practice and has often learned from his wife through collecting plants from the bush.

The **grannie healer**, despite the name, can be male or female. They are people who have raised many children using home remedies to treat common household ailments—diarrhea, painful menstruation, skin disease, constipation, anxiety, and other conditions. Grannie healers are experienced and very capable in the use of traditional medicines but practice, for the most part, on their immediate families. When confronted with a difficult case, they will call for the help of the village healer and work together on the patient.

The **midwife**, known also as a traditional birth attendant, usually has been trained by a family member to deliver babies in rural areas. This person is most often a woman, but among the Q'eqchi' Maya of southern Belize, the husband often attends the birth of his children. Young males are trained by their fathers for this task from childhood in some families. The midwife is also a provider of prenatal care, although in some villages, there are prenatal clinics. The midwife is expert in dealing with the ailments presented by women as well as children. It is traditional that the midwife visits the new mother's home each day for a nine-day period after the delivery to check on the condition of both mother and child and to administer teas, poultices, powders, and baths as needed. This reflects the sacredness of the number nine to the Maya, a number that appears in many ethnomedical recipes and healing rituals.

The **massage therapist**, known in Spanish as the *sobadera*, is expert in treating muscle spasms, backaches, sprains, stress due to overworked muscles, and general complaints of aches and pains. The massage therapist may be male or female and uses baths, poultices, teas, and oils to massage afflicted areas of the body. The training for this type of therapy is handed down from the previous generation, and some massage therapists often describe their skills as a divine gift.

The **bonesetter**, or traditional chiropractor, has been trained by his or her parent or grandparent to treat sprains, fractures, broken bones, and pulled ligaments. Through techniques of manipulation, the bonesetter often achieves good results. Some bonesetters also practice a form of traditional acupuncture, known as *pinchar*, that employs stingray spines placed at any of approximately 80 points for the treatment of a variety of conditions. This treatment can be combined with herbal remedies, such as teas, poultices, powders, and most frequently, hot baths of herbs. Gann (1918) described the work of the bonesetter as follows:

> Simple fractures of the long bones are set very neatly and skillfully in the following way: the fractured limb is pulled away from the body with considerable force in order to overcome the displacement; over the fractured bone is wound a thick layer of cotton wool, and over this are applied a number of small, round, straight sticks, completely surrounding the limb, their centers corresponding nearly to the seat of the fracture; these are kept in place by a firm binding of henequen cord. The limb, if an arm, is supported in a sling; if a leg, the patient is confined to his hammock till the fracture is firmly knit. Excellent results are secured by this method, the union being firm, and the limb nearly always uniting in good position. (p. 37)

The **snake doctor** is a specialist who is highly respected by the community for his or her ability to treat the toxic and venomous conditions caused by snakes, spiders, scorpions, worms, bats, rats, rabid dogs, and puncture wounds from rusty implements. In areas of the country where there is much sugar cane, abundant snake populations feed on the ever-present rodents in the cane fields. The snake doctor, male or female, is trained by his or her family to be the one who possesses the secret family formulas, which are carefully passed on to only one person per generation. Snake doctors note that they often receive patients released from hospitals as "hopeless cases"—too late to benefit from injections of antivenin—and treat these people with success in their homes or clinics.

Another type of healer not discussed in Arvigo and Balick (1998) is the *ensalmero* (if a man) or *ensalmera* (if a woman), who is a specialist in using prayer as a healing agent. These healers generally administer only prayers and are thought to be especially helpful in the ailments of infants, children, and those ailments that may be of a spiritual nature. They are respected members of the community and generally address their many prayers for a variety of illness to a pantheon that includes a combination of Christian saints and Maya spirits. There are many different prayers for a variety of ailments that must be committed to memory and are most often handed down orally from older family members. In recent times, the prayers have been passed on in written form. It is considered a great and generous spiritual gift to be given prayers by *ensalmeros*.

CULTURE-BOUND SYNDROME (CULTURALLY SPECIFIC ILLNESS) IN BELIZE

Culture-bound syndromes, culture-specific disorders, or folk illnesses are diseases that are characterized as being restricted to particular cultures, although their distribution range may vary considerably.[2] An example of a folk illness with a wide distribution range is *mal de ojo* (evil eye), which can be found throughout Latin America and

---

2. We thank Ina Vandebroek for contributing to this discussion of culturally specific conditions.

in transnational Latino migrant communities in the United States but also in the Mediterranean (Baer and Bustillo, 1993; Pieroni, 2000). Some of these conditions cause relatively minor health problems, while others can be life threatening. Often, a one-to-one correlation with discrete biomedical diseases does not exist. Sufferers from these diseases usually have a personalistic explanation for them, believing that they have a supernatural origin, as opposed to the Western school of thought that now mostly relies on a naturalistic explanation of illness. Such conditions are most frequently treated with local traditional health care, including herbal remedies, and spiritual and/or religious healing (Risser and Mazur, 1995).

There exists a rich medical anthropology literature on culture-bound syndromes in which these illnesses are usually characterized as ethnopsychiatric syndromes and thus as having a psychosomatic basis. Other perspectives in medical anthropology and ethnomedicine have studied these folk illnesses in relation to identity, gender roles, and the human body as well as the physical, social, political, and economic living environment of those suffering from these diseases (e.g., Thomas et al., 2009), sometimes depicting them as metaphors for people's representation of the self or their living conditions (Larme, 1998; Rebhun, 1994). On the other hand, Baer and Bustillo (1993) have shown that while folk illnesses may have psychological and social components and functions, the predominant physiological symptoms among sufferers of folk illnesses in their study are well recognized by biomedical health care providers (see also Vandebroek et al., 2008). The work of Baer and Bustillo (1993) demonstrates that folk illnesses may be linked to increased morbidity and mortality and are worthy of biomedical attention. Although clear social and psychological dimensions to these illnesses and their treatments exist, biomedicine can be used to treat the symptoms that are a component of these illnesses. Sometimes, even an etic (biomedical) diagnosis can be made, as in one case of *mal de ojo* that was linked to sepsis and another case of *susto* (fright illness) that was linked to colic (Baer and Bustillo, 1993). Furthermore, epidemiological studies have shown that similar to discrete and well-recognized biomedical diseases, epidemiological patterns in the distribution of folk illnesses can be demonstrated according to gender, social group, and age (Carey, 1993).

Regardless of the orientation of a particular field of anthropological study, depicting folk illnesses as a unique sociocultural behavior, a metaphor for the role of individuals within their society or environment, or a biomedical phenomenon modified by culture may prove to be too reductionist. Or, as Klein (1978) suggested, it is likely that they are the product of complex interactions between the individual's state of (physical) health, the social and cultural systems in which (s)he lives, and the individual's personality.

Recently, practitioners in the field of psychiatry have devoted a great deal of attention to these culture-bound diseases. The definition in the *Diagnostic and Statistical Manual of Mental Disorders, DSM-IV-TR* (American Psychiatric Association, 2000) is that they consist of

recurrent, locality-specific patterns of aberrant behavior and troubling experience . . . indigenously considered to be 'illnesses,' or at least afflictions, and most have local names . . . Culture-bound syndromes are generally limited to specific societies or culture areas and are localized, folk, diagnostic categories that frame coherent meanings for certain repetitive, patterned, and troubling sets of experiences and observations. (p. 898)

The book also contains an appendix on this topic that includes a glossary of culture-bound diseases that might be encountered in clinical practice in North America and attempts to link these to conventional categories of clinical diagnoses. Such linkage is extremely difficult, as the folk diseases can be a mixture of symptoms that do not correlate well with conventional categories of psychiatric illness. The glossary lists and defines 25 culture-bound syndromes, including conditions that might be found in Hispanic patients, such as *ataque de nervios, bilis, locura, mal de ojo,* and *susto.* Balick and Lee (2003) concluded:

> It is not surprising that in mental illness, ethnic variation will affect how a patient expresses distress. This is often referred to as somatization—a term describing symptoms caused by stress and experienced as bodily sensations. It is a condition that requires a thorough medical examination to evaluate the symptoms to exclude pathology as the source for the illness. (p. 108)

This topic was discussed in the Surgeon General's report on mental health (1999) and will undoubtedly receive increased attention as medical science and its practitioners begin to investigate the effect of cultural beliefs on health status.

There are many concepts and beliefs that underlie the treatment of culturally specific illness in Belize, varying with the region, ethnicity, calendar, and training the healer has received. These are described in some detail in Arvigo and Balick (1998), and several of these conditions are discussed in the following section.

An important task for the healer in Belize is to determine whether the disease is of physical or spiritual origin. Once this is ascertained, the course of action for each type of disease is quite different. Physical ailments can be addressed with the skills possessed by the village healer, grannie healer, or midwife, but supernatural ailments call for treatment by the doctor priest or priestess. The latter's skills are able to ascertain whether, for example, the cause of the illness rests in the unquieted soul or the presence of an evil spirit. Such illnesses are treated with prayer, incense, charms, amulets, special rituals, as well as herbal baths and teas. Examples of a few of the many spiritual ailments recognized in Belize include *susto, pesar, tristeza, envidia, mal de ojo, maldad,* and *mal vientos* (for more detail on these conditions, see Arvigo and Balick, 1998).[3]

Susto (fright illness) is primarily a condition of infants and children, although it can occur in those of all ages, both male and female. It can be caused by many types of situations—for example, seeing a fight erupt among adults in the household; witness-

---

3. Examples of Q'eqchi' culture-bound syndromes can be found in Boster (1973).

ing a frightening event of any kind; being awakened from a peaceful sleep by a loud, frightening noise or event; the sudden loss or death of a loved one; a suicide or murder in the household; or seeing a tragic accident in which a loved one is killed or injured. Symptoms of *susto* can include unhappiness, chronic indigestion, constipation or diarrhea, poor and interrupted sleep, nightmares, and cold sweats. In addition, the person appears disconnected from normal aspects of daily life. In infants, *susto* presents as crankiness, inability to sleep, nausea, and vomiting, with skin conditions and diaper rash that persist rather than heal. Each healer has his or her own ways of treating *susto*, which usually involve herbal baths, prayers, and teas.

Pesar, also known as grief, affects people of all ages. It can be caused by the loss of a loved one who has left the household or passed away. When caused by the loss of a romantic attachment, it is known in Belize as *macobe*. If children are mistreated, lose someone close to them, or a new baby is born into the household and another is displaced from the mother's breast, a child may experience *pesar*. This is thought to be a condition that can be fatal if not properly diagnosed, and it presents in children as behavioral changes, including tearfulness, crankiness, eating dirt, finger biting, loss of appetite, and in severe cases, development of black and blue marks on the body. Adults with this condition exhibit painful, shallow breathing, frequently taking deep breaths that are let out with a mournful sigh. *Pesar* is treated with prayers, herbal baths, and ingestion of various herbal teas.

Tristeza is sadness that can be caused by many different types of situations. The person most often feels a lack of interest in life, has no energy or interest in work or play, and prefers to stay at home in a darkened house. In Western medical terms, *tristeza* is most closely related to depression. The treatment involves specific prayers for *tristeza* and nine herbal baths, along with one other interesting treatment: The person with this condition is instructed to take a basket of freshly picked flowers to the riverside at noon and drop the flowers, one by one, into the river and watch as the current takes them away. In Belize, Thursdays and Fridays are said to be the most effective days for undertaking this act, which can be repeated as necessary.

Envidia is translated as envy, a common human trait that has devastating effects on the object of the envy. This disease occurs when one is subjected to a regular stream of negativity, projected by a person who feels envious or jealous. The result is said to include both physical and emotional problems. Victims of *envidia* feel a lack of interest in the daily affairs of life, sleep poorly, have no appetite, and experience nightmares, digestive problems, and a lack of self confidence that results in a breakdown in daily life. People suffering from *envidia* will feel like moving from their place of residence or work to escape the stream of negativity. Its treatment includes prayer, herbal baths, and teas. If necessary, the healer will prescribe added protection in the form of an amulet that might be blessed with special prayers to ward off the envy. Incense, most often made of dried copal resin with dried rosemary leaves and burned on Thursdays and Fridays, can also be prescribed. Don Eligio Panti used to say that *envidia* was the most serious and threatening disease among people in Belize.

Mal de ojo, known as evil eye, is said to be transmitted by the envious or hateful look of another person. Others are born with "hot eyes" that can cause sickness as the result of their gaze. Symptoms are similar to those of *envidia* as well as a sore, weepy eye exuding pus or phlegm, combined with the appearance of itching skin, rashes, and hot skin. Nightmares and interrupted sleep patterns with heart palpitations are common, as is a sudden disinterest in the usual affairs of daily life and the inability to get along with loved ones or family members. Treatment includes a series of nine traditional prayers said each day for nine days, with an herbal bath composed of nine different species of plants, along with copal resin burned in the home on Thursdays and Fridays.

Maldad is translated as evil and is thought to be an intentional act, as compared to envy, which is not always deliberate. *Maldad* can occur when a person is jealous and angry toward another and hires a practitioner of evil magic to cast a spell on the other person. The spell is aimed at ensuring the other person's failure in some or all aspects of life. Local people believe that this can lead to death if not treated properly. This disease is only diagnosed by a doctor priest or priestess, and it presents a long list of complicated and confusing symptoms, usually including frequent nightmares, chronic depression, itching skin, insomnia, indigestion, night terrors, anxiety, and a lack of interest in work, career, home or family. *Maldad* is diagnosed through communication with the spirit world. Treatment includes specific prayers, herbal teas, baths, and burning copal incense in the patient's home.

Mal vientos are characterized as bad or evil winds, or malevolent spirits under the control of practitioners of evil magic, whose goal is to do harm to individuals. Don Eligio Panti used to say that these evil spirits were persons who died in sudden, violent acts—drowning, murder, or suicide—and who wander aimlessly between the physical and spiritual planes in their desire to make contact with the physical world. They are willing to do harm to others due to their inherent state of confusion. The condition of *mal vientos* presents with poor sleeping patterns, nightmares, heart palpitations for no apparent reason, itchy skin, lack of appetite, indigestion, depression, and reports of hearing voices. Treatments include prayer, herbal teas, herbal baths, burning incense, and carrying protective amulets blessed by a *h'men*.

A traditional story about the origin of the disease caused by *mal vientos* in Succotz was related by Thompson (1930):

> Formerly the people of Socotz lived happily and free from those numerous maladies caused by evil winds. These good times came to an end when a certain sorcerer became enraged with the inhabitants of the village and plotted their destruction.
>
> In order to accomplish this, he made nine dolls of black wax (*qes*), which he buried under the ground, one close to each of the gullies that meander through the village. Luckily for the future of the community another sorcerer divined the wickedness that was contemplated. With the help of some of his friends he searched for and found seven out of the nine dolls. Search as they would, they could not find the remaining two images. Feverishly they searched all through the night, but in vain. They knew that at dawn the two dolls, if not found and destroyed, would come to life, and start their evil tricks. When the sun had risen, and the evil was

past remedy, they informed the elders of the village of the danger that threatened the community. The infuriated villagers sought out the evil sorcerer, and, dragging him and his family into the neighboring woods, murdered the lot. But the evil had been done, and to this day Socotz suffers from the two evil winds caused by the two wax images that escaped. These two evil winds are the source of much of the sickness in Socotz. (pp. 166–7)

Practitioners in Belize consider much information on the treatment of spiritual diseases with plants to be sacred or secret, and thus it is not discussed in great detail in our book. Some plants are considered very powerful, even dangerous in the hands of an inexperienced person. However, there are some common, less toxic plants used locally against spiritual ailments: basil (*Ocimum basilicum* L.), rosemary (*Rosmarinus officinalis* L.), marigold (*Tagetes patula* L.), garlic (*Allium sativum* L.), and lemon (*Citrus limon* (L.) Osbeck). These are frequently put in a cup of water, boiled, and consumed as a tea twice or three times per day. Species used as incense to treat spiritual disease include myrrh (*Commiphora myrrha* (Nees) Engl.), frankincense (*Boswellia carteri* Birdw.), copal (*Protium copal* (Schltdl. & Cham.) Engl.), and rosemary. The resins of the first two and/or leaves of the third are placed in a clay or metal container and burned, filling up the room with smoke, during which time curative prayers are recited.

PREPARATION OF PLANT MEDICINES

There are many hundreds of plants that are used in the treatment of illness in Belize, and Chapter 4 provides information on this topic that has been gathered during our study. Plants are harvested, prepared, and administered in a multitude of ways. Most of the species discussed traditionally are therapeutically administered as teas, poultices, herbal baths, powders, tinctures, salves, or as smoke. A number of detailed plant recipes are presented in Arvigo and Balick (1998), along with their modes of administration, and these are summarized in the following paragraphs.

A small handful of dried or fresh leaf or other material is added to three cups of water and made into an herbal tea. These can be consumed hot, warm, or at room temperature and, for the treatment of some diseases, go along with a special diet that prohibits certain foods that could negate the activity of the tea. For example, in the treatment of gastritis, the patient is prohibited from consuming cold drinks and cold food, acidic foods, sour fruits, beef, pork, and chili peppers. It is felt that such agents might shock the muscles of the gastrointestinal tract, intensifying the symptoms of the condition.

For the treatment of swellings, bruises, sprains, fractures, boils, sores, headaches, skin conditions, and other diseases, a poultice of fresh plant material is prepared. Local people believe that the plant should be collected after sunrise, when the dew on the leaves has dried. The plant is felt to be in its most powerful state at this time. Fresh leaves are mashed, a pinch of salt is added, and this poultice is applied to the affected area. Lightweight cotton gauze is wrapped on the area to hold the poultice, which is left on overnight. Poultices are changed daily and employed as needed.

One of the most important treatments in traditional medicine is the herbal bath. Nine species of leaves are employed in bath formulations, with choice of leaves depending on the complaint. Each condition requires a different type of bath formula. Two handfuls of leaves are boiled in the largest pot in the house for at least 20 minutes and then allowed to cool to around body temperature. Patients stand in a tub, shower, or other enclosure and pours the bath mixture from the container over themselves, using a cup or gourd. Drafts of wind are avoided during the bathing process. Herbal baths are felt to be an important and very effective treatment used by the traditional healer and are employed for skin conditions, nervousness, insomnia, headaches, rheumatism, arthritis, backaches, muscle spasms, swellings, sprains, and many diseases of childhood. Bath therapies are prescribed for a series of either three or nine days.

Many species of medicinal leaves are used by traditional healers to produce therapeutic powders. These are made from dried leaves and are used to treat stubborn skin sores, ulcerations, skin fungus, infected deep wounds, diabetic sores, and various skin conditions that are weeping, infected, inflamed, running, or open. For the treatment of an infected skin wound, a healer washes it with soap and water, dries it off, applies castor oil to the area, and sprinkles a finely sifted wound powder on the area. To keep the wound free of outside contamination, a clean cotton cloth is wrapped around it for the night, and the process repeated the next morning—up to twice a day. Traditional healers select bitter, strong-tasting leaves for these wound powders. *Neurolaena lobata* (L.) Cass, known in Belize as jackass bitters, is one species widely used after toasting on a clay cooking surface, called a *comal* in Spanish, until the leaves and plant parts are crispy to the touch. The leaves are then strained in a sieve and collected in a gourd or jar to be applied directly to sores, fungus, and infections.

Around the world, practitioners of herbal medicine commonly use tinctures of plant extracts as therapeutic agents. The plants—fresh leaves, roots, stems, dried barks, etc.— are soaked in alcohol for several weeks and then utilized. In Belize, an anise-flavored alcohol is used to produce the tinctures, as it is thought to help soothe the stomach and nerves and assist the body in absorbing the herbs.

Herbal therapies can be absorbed into the body by being applied directly to the skin, using the carrier medium of salves and ointments. In the past, the most valued fat utilized to produce salves was that rendered from an armadillo, in which the plants—fresh or dried—have been steeped for up to three days. In modern times, fat from beef, pork, coconut oil, or vegetable shortening is used, with a candle or wild beeswax added to solidify the mixture. Such salves are commonly used to treat itching skin conditions, skin fungus, ringworm, minor wounds, and burns.

For the treatment of acute muscle spasms, backache, or cramping of overworked muscles, the patient may be asked to smoke certain dried plant materials to deliver the healing substances directly into the body through the lungs. Leaves of the small but fascinating plant *Mimosa pudica* L. var. *unijuga*, locally called *dormilón* in Spanish (due to the way its leaves collapse, or "sleep," upon touch), are harvested, dried in an oven, powdered lightly and rolled into corn cob husks, and then smoked for the treatment of muscle spasm, backache, and nervous irritability. Two or three deeply inhaled puffs of smoke are said to greatly alleviate the pain. The powder can be sprinkled directly on food as well and is said to have sedative properties.

These are the basic methods of preparing plants for use as traditional remedies in Belize. Many plants are harvested from forest areas, while others are grown around the house. In one household, the mother proudly told us that she had raised her nine children entirely on the medicines found growing in a small area of her yard. There are protocols for harvesting the plants, depending on the time of day, moon phase, or other locally recognized variables. Local people insist that the plants are made more powerful when they are collected with a "clean heart," and prayers are said to appease the spirits and gods who are responsible for the plants. Many of these prayers are sacred and considered family secrets, and thus they are not discussed here at the request of the healers with whom we worked. Don Eligio Panti used a very simple prayer, a combination of Maya belief and Christianity, during his plant collecting activities. Arvigo asked and was granted permission to reproduce this herb collector's prayer. Each time he would cut a leaf or root or pull out a stem, he would say the following (translated from the Spanish):

> In the name of God the Father, God the Son, and God the Holy Spirit, I take the life of this plant and I give thanks to its spirit. I am Don Eligio, the one who walks in the mountains collecting medicine to heal the ailments of the people. I have faith with all my heart that this plant will heal the sickness of the people [or specific name of the patient].

Traditional healers and the *yerbateros*, or herb gatherers, who work for them say prayers of gratitude to the spirit of the plant so that it will accompany them home to help with the cure. Among herbalists throughout the world, there is the common belief that saying prayers will help the plants remain strong and add to their therapeutic power. It also reflects the strong commitment to nature that is one of the striking personality traits of the traditional healers that we have observed in Belize. No fragment of leaf, root, or stem is wasted, and extreme concern is expressed about the loss of forest habitats that is reducing the diversity and abundance of their pharmacopoeia.

THE VANISHING MEDICINE CHEST

The consequences of deforestation and the destruction of wilderness habitats are numerous and impact the local, regional, and global environment as well as its people. Rarely considered is the impact that wilderness conversion and species loss has on traditional cultural practices such as ethnomedicine. As Farnsworth et al. (1985) pointed out, up to 80 percent of the developing world's population uses plant medicines as an important component of primary health care. This is certainly the case in Belize, where both the rural and urban populations rely on traditional medicine for some aspect of their primary health care. Most adults now living in Belize were raised on traditional medicine to some degree, and the current percentage of people using traditional medicine in Belize is still significant. The reasons for this usage include low cost, efficacy, and cultural acceptance. In our observations of the fees charged by healers and local, Western-trained physicians, as well as the costs of plant medicines compared to the costs of pharmaceutical medicines, during the 1990s, we estimated that traditional medicine—including healer visits and remedies—cost about 20 percent of what is charged in the modern biomedical clinics. Traditional medicines are efficacious for many conditions, especially those common in this region. Finally, there are cultural reasons why Belizeans accept traditional medicines—as we have been told many times, "It was good for grandmother and grandfather, so it will be effective and safe for us as well."

Young (1981) analyzed the use of traditional medicine, which he referred to as "non-use of physicians," offering an interesting decision model for both the initial and subsequent choice of treatment in a rural Mexican community. He identified four ways of treating illness in that region: self or home treatment with traditional remedies; treatment by a folk curer, with traditional remedies; treatment by a *practicante*, a non-physician, unlicensed person who practices Western-style medicine; and treatment by a licensed physician. Factors in making the decision on seeking treatment included how serious the illness was, if the folk cure was known to be effective for the particular condition, the patient's judgement of how effective traditional cures might be and the cost of treatment as well as the availability of transportation to the Western physician. Young provided a most useful framework for understanding the preferences for and the choices made in seeking treatment of illness in the Belizean setting, with one exception. The *practicantes* referred to by Young are known as community health care workers, or *promotores de salud* in Belize. They are men and women trained by the government and a number of international health organizations to be first responders in remote villages. For example, they will diagnose patients with malaria or bad headaches, dispense common over-the-counter drugs such as aspirin and some antibiotics, and treat small wounds and other conditions.

One effect of deforestation is the dramatic reduction in supply of certain primary and secondary forest species used as medicinal plants available to the traditional healers. This of course does not include those weedy species that occur in forest and field clearings or the many plants in the ethnopharmacopoeia that can be grown in dooryard gardens (Voeks, 2004), but it does impact a large number of "bush" species widely utilized in Belize. The figure above gives an idea of the situation that faced Don Eligio Panti in the collection of plants used in his medical practice. In 1940, during the early stage of his practice, he only had to walk 10 minutes from

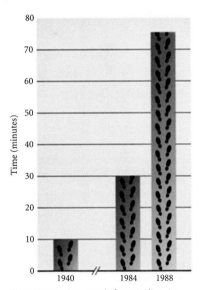

Time needed to reach forest sites to collect plant remedies as reported by Don Eligio Panti for 1940 and observed by the authors in 1984 and 1988. *From Balick and Cox (1996).*

his house to the secondary forest site where he collected medicinal plants used to treat his patients. In 1984, when he began working with Rosita Arvigo, it took them 30 minutes to reach a site that had sufficient quantities of these same species. In 1988, when we both began to work with Don Eligio, it took 75 minutes to reach an area where the plants for that day's patient load could be found. It was not the case, as some might suggest, that he was walking much slower, but that deforestation and the concomitant loss of economically important plants had reduced his effectiveness in offering primary health care services to the community.

It appears that this scenario is occurring in tens of thousands of habitats around the world each day, as traditional healers and herb gatherers struggle to find the medicinal plants on which they and their patients depend. The extirpation of plant populations of medicinally important species is also affecting pharmaceutical medicines that are dependent on wild-harvested raw materials for processing. For example, the wild harvest of *Pilocarpus* species in northeastern Brazil for the production of pilocarpine has virtually ceased due to overharvest and habitat destruction. To ensure a reliable supply of pilocarpine for the glaucoma patients who depend on it, the species has now been domesticated as a cultivated crop in that area, and this has resulted in a loss of cash income to the local indigenous peoples of the region who were once the primary harvesters (Pinheiro, 1997).

PROFILE OF A TRADITIONAL HEALER'S PRACTICE: DON ELIGIO PANTI

During February 1989, we observed a total of 361 patients consulted by Don Eligio Panti for a variety of conditions and were able to obtain an idea of their age, sex, and background as well as the most common complaints they asked him to treat. The patients we interviewed comprised a very diverse group; functioning as a general practitioner, Don Eligio treated people of all ages in his community as well as those who came to see him from other towns and countries. Sixty-one patients (16.9 percent) were from 0 to 5 years of age, 13 (3.6 percent) were from 6 to 15 years, 100 (27.7 percent) were from 16 to 30 years, 92 (25.5 percent) were from 31 to 50 years, and 95 (26.3 percent) were 51 years or older.

Of the total number of patients who were observed during that period, 129 (36 percent) were male, and 232 (64 percent) were female. Patients came from six different countries: 282 (78.1 percent) from Belize (158 [43.8 percent] from Cayo, 45 [12.5 percent] from Belize District, 28 [7.8 percent] from Orange Walk, 26 [7.2 percent] from Corozal, 14 [3.9 percent] from Toledo, and 11 [3 percent] from Stann Creek districts), 53 (14.7 percent) from Guatemala, 7 (1.9 percent) from Mexico, 5 (1.4 percent) from Honduras, 3 (0.8 percent) from El Salvador, and 2 (0.6 percent) from the United States (tourists who had heard of Don Eligio). We did not record data on the origin of nine patients (2.5 percent).

As noted, illness and disease in Belize are considered to have both natural and supernatural causes. In the group observed, 188 patients (52.1 percent) could be distinguished as having purely natural causes, 134 (37.1 percent) had conditions considered to be caused by supernatural forces, and 15 (4.2 percent) had conditions caused by a combination of both physical and supernatural forces. We were unable to make a clear distinction for 24 patients (6.6 percent). Of the patients who came for conditions distinguished as supernatural, a greater percentage were females—84 patients (62.7 percent) were female, and 50 patients (37.3 percent) were male. Of the patients who came for treatment of conditions diagnosed as having natural causes, a greater percentage again were females—122 patients (64.9 percent) were female, and 66 patients (35.1 percent) were male. Eleven females and four males were felt to have conditions with both supernatural and natural origins. Sixty-nine percent of the treatments administered included prayers, incense, protective amulets, and other forms of magic.

When examined as a total group, the most frequent complaints were stomach ache, muscular/skeletal problems, malaise, headaches, bad luck, fright, and crying/hysteria. The most common complaints, sorted by age groups, were crying/hysteria, preventative treatments, fright, and loss of appetite in those 0 to 5 years of age; malaise, stomach ache, muscular/skeletal problems in those 6 to 15 years; stomach ache, malaise, muscular/skeletal problems, headaches, weakness in the legs, complications of pregnancy, fainting, and bad luck in those 16 to 30 years; stomach ache, malaise, muscular/skeletal problems, romantic enchantments, and bad luck in those 31 to 50 years; and stomach ache, muscular/skeletal problems, malaise, headaches, weakness in the legs, and bad luck in those 51 and older.

FOODS AS THERAPIES

A popular health concept in Belize is the value of foods as medicines. Often, grandmothers, mothers, and healers are overheard advising each other to use certain foods in the diet to increase health and stamina and cure certain diseases. The foods most commonly used as medicines include the following:

Amaranth (*Amaranthus* spp.) is commonly known as *calaloo*. It is an annual that bears edible leaves and seeds. The mild-tasting, dark green leaves are considered a specific home remedy for treating anemia, exhaustion, and chronic tiredness and to act as a blood purifier. *Calaloo* is sold in markets throughout Belize and is both a wild and a cultivated crop. There are at least three varieties found in the country. It is prepared in soups and stews and is eaten boiled. The mashed leaves are applied to new cuts to staunch bleeding.

Papaya (*Carica papaya* L.) is eaten both ripe and green. When the unripe, green fruit is cooked, as one would prepare young squash, it is considered useful to rid the body of intestinal worms. The ripened fresh fruit is taken as a food by those who are "delicate of digestion" to aid that process, especially for the digestion of proteins. The ripe fruit is rubbed on the face to act as a cosmetic lotion, which is said to lessen wrinkles and hydrate the skin.

Cho cho or *chayote* (*Sechium edule* (Jacq.) Sw.) is boiled and eaten to rid the arteries of plaque, thereby reducing high blood pressure. One entire fruit with some leaves and a piece of the vine can be boiled together and eaten. It is often added to soups and stews as well as grated raw for inclusion in salads.

Limes or lemons (*Citrus* spp.) are commonly employed as medicinal foods. Limes are cut in half and rubbed over the forehead to alleviate headaches and to assist the drainage of the sinuses. Very young and green limes are pierced with straight pins in a cross pattern and carried as good luck charms to ward off evil. Consuming lemons or limes regularly is thought to be healing and tonic to the liver, stomach, and gallbladder. A combination of lime juice and brown sugar (one tablespoon of each) is used as an agent to rid the face of acne when applied twice daily.

Chaya (*Cnidoscolus chayamansa* McVaugh) is a popular and ancient food source of the Maya, found growing abundantly in the Yucatan. The bushy plant grows to three meters in height and bears dark green leaves that are troublesome to gather because of the stinging white sap that emits from the plant when the stems are plucked. When boiled and prepared in various ways, chaya leaves are tasty and sweet. They are eaten to build the blood in people with anemia and are considered tonic to people of all ages, but especially children and women of childbearing age. The raw leaves mixed with lemon juice are made into a nice tasting beverage.

Green plantains (*Musa* × *paradisiaca* L.) are grated, boiled, and eaten as a medicine for stomach ulcers and diarrhea. The leaves are used to wrap tamales for steaming and have been prized as topical applications for very serious skin burns.

Young, green baby coconuts (*Cocos nucifera* L.), when about three inches in length, are chopped, boiled, and given as a tea for childhood diarrhea.

A plant known as wild basil (*Ocimum campechianum* Mill.) is a popular food and medicine. Three days after childbirth, midwives will give a vaginal steam bath of basil leaves. This is said to ensure that the uterus returns to its proper position after childbirth and to prevent infections and swelling. The juice of fresh basil leaves is used to treat skin conditions caused by organisms such as worms or larvae that bore into the skin.

Oregano (*Lippia graveolens* Kunth) tea boiled with garlic cloves (*Allium sativum* L.) and stems is taken orally as hot as possible for the treatment of upper respiratory tract infections. Cooled oregano tea is also an excellent soak for sprains and bruises. When combined with the leaves of *susumba* (*Solanum rudepannum* Dunal), it is a most superior wash for burns that prevents blisters and scarring.

The golden hairs of young corncobs (*Zea mays* L.) are collected, dried, and boiled as a tea for urinary tract infections. The same tea is also considered by men to be good for virility because, as we have been told, "it cleans the lines."

POISONS THAT HEAL

Traditional healers and bushmasters in Belize recognize a number of poisonous plants that can be utilized for their medicinal value. The application of potentially toxic species in medicine requires special training to produce therapeutic results rather than death of the patient. *The Checklist of the Vascular Plants of Belize, with Common Names and Uses* (Balick, Nee, and Atha, 2000) lists 24 species as poisonous, which includes plants that are irritants. There are 531 species listed as medicinal; some share both categories. Following are examples of toxic plants also used in healing.

Chicoloro (*Strychnos panamensis* Seem. var. *panamensis*) is a common vine growing in the forest understory. It is easily recognized because the opposite secondary branches give the appearance of a cross. According to the traditional healers in Belize, the cross formation on a forest vine is a sign of both medicine and poison and, in effect, a warning that the person using it should be experienced and trained in the use of such plants. *Chicoloro* contains the compound strychnine, a toxic agent that in high enough doses can kill through its effects on the central nervous system. Despite its toxicity, the plant is very important to healing in Belize and elsewhere in the region where it grows. The tea made from the vine and root is given in specific doses for treating intestinal parasites. It is also used for severe gastritis, constipation, and gallbladder disease, but only with great precision. Externally, the leaves are used in herbal bathing formulas for a variety of ailments.

Another toxic plant is *Mikania guaco* Humb. & Bonpl., a vine with pale lavender flowers in the Asteraceae. It is common along roadsides and forest margins. Morris (1883) observed that it is

> used as a febrifuge and anthelmintic, but the chief interest connected with this plant is on account of its being supposed to be a powerful antidote for the bite of venomous serpents. So strong is the impression of the powerful medicinal virtue possessed by guaco that no Indian ever traverses the dark and dense forests without carrying a potion of it in his pouch. (p. 86)

The physic nut (*Jatropha curcas* L.) is administered both orally and topically. Orally, the mature seed is chopped and boiled to be consumed as a liquid purge. If the wrong dose is given, it can be fatal. The milky sap that exudes from the leaf stems is gathered on the fingertips and rubbed inside the mouth of babies suffering from thrush.

Flowers of the toxic species locally known as belladonna (*Datura* spp.) are gathered and soaked in alcohol to use as a topical application on painful joints, with good results reported. The heated leaves are pounded with a pinch of salt and used as a poultice for backache when wrapped in a thin cloth and applied directly over the painful area.

Bark of the highly toxic poisonwood (*Metopium brownei* (Jacq.) Urb.) tree was used by one aged traditional healer, Thomas Green, as a treatment for psoriasis. He noted that he places a one-inch-square piece of the bark in an eight-ounce glass of water until the water begins to turn black. At that point, the bark is removed from the glass, and the remaining water is administered as one teaspoon twice daily. Mr. Thomas, as he was known by all, was the only healer we spoke to who reported this treatment—and the only one who had the courage to try it.

Many other toxic species cited by Morris (1883) and other early travelers remain unknown and unvouchered at the present time, representing challenges for future ethnobotanical studies. One such species is a vine (*tie-tie*) identified as belonging to the Menispermaceae, and while it is not clear if there were any medicinal uses, it was reported to be used by the Indians to stupefy fish. They pound it in a mortar, dam up a pool and then throw the pounded mass into the water. In a short time any fish in the pool come to the surface in an unconscious state, and so are easily caught. It is much to be desired that good specimens, including flowers and fruits, of these and many other plants known to be used by the natives of Central America, be forwarded for identification, and that the specimens be accompanied by full and clear descriptions of the purposes to which they are applied. (p. 88)

PSYCHOACTIVE SPECIES

Although the region around Mexico was the center for the utilization of mushrooms to achieve a psychoactive state, we did not observe or hear of the ritual use of hallucinogenic mushrooms in contemporary Belize. One introduced species, *Cannabis sativa* L., is reputedly used for recreational purposes as well as in tinctures for medicinal use, according to some of the healers we spoke to. It was not seen or collected during the fieldwork.

The "alligator apple" (*Annona palustris* L.) was mentioned by Morris (1883) as a poison as well as a narcotic plant. He noted that the tree

bears a narcotic and probably a poisonous fruit, very similar to Sweetsop, which should be carefully avoided. It is said to be eaten by alligators when it drops into the water, and hence its name. The wood, which is known as corkwood, is used for lining boxes, stopping bottles, etc. (pp. 86–7)

We are unaware of the present-day use of this plant as a narcotic or poison, although consumption of the fruit of the closely related *Annona muricata* L. has been linked to atypical parkinsonism in Guadeloupe, in the Caribbean (Champy et al., 2005).

Morris (1883) also discussed the use of a hallucinogenic snuff:

Among the forest-trees one bearing acacia-like leaves, and rough, leathery moniliform pods, probably a species of *Piptadenia*, is of interest on account of the use to which the seeds are sometimes applied by the Indians. After being roasted they are pounded and mixed with powdered lime, and made into a kind of snuff which is said to produce "a peculiar kind of intoxication, almost amounting to frenzy." (p. 87)

This may relate to the psychoactive species *Anadenanthera peregrina* (L.) Speg., which is prepared in an identical way and found in the Caribbean. We have no record of this tree being present in Belize. However, it is widespread throughout the Orinoco Valley of South America and parts of the Caribbean, where it was once widely used for its psychoactive properties (Altschul, 1972). Morris' report is an interesting addition to the historical record of the use of this species and suggests a range extension of this narcotic species to Central America or evidence of its use in trading.

SOLVING A MYSTERY—CAPOCHE

Morris (1883) mentioned the travels of Henry Fowler, then the Colonial Secretary of British Honduras, who led an expedition that explored much uncharted territory in the colony. Morris quoted Fowler's observation on an interesting but enigmatic use of a plant:

> This tree, called by the Indians Capoche, appears to be the natural cinchona of the country, for it is used for fevers and has a bitter taste. The tree is very scarce. The Indians make cups from the wood for the purpose of water being steeped in them, which is given to the children for fever and also as an anthelmintic. (p. 88)

Morris further reported that as no fruits or flowers had been obtained of the *capoche*, it was impossible to identify it, but he surmised that it was "probably a member of the Lauraceae." During our fieldwork and subsequent herbarium studies, we found no mention of a febrifuge with this name, administered through preparation of a cup carved from its wood. Presumably, warm or tepid water was poured into the cup, allowed to steep until the bitterness of the wood was transfused into the water, and then consumed by the feverish patient. Some of the voucher collections of plants discussed in the Morris volume are to be found today in the economic botany holdings of the Royal Botanical Gardens, Kew, at the Centre for Economic Botany (CEB). Sarah Canham, then on the staff of The New York Botanical Garden Institute of Economic Botany, visited Kew in January 2000 to make an inventory of the Belizean artifacts catalogued at the CEB, to illustrate some of the historical plant products in this book. While there, she came across accession number 45307, comprised of bark and leaves of the "*capoche*" tree. In the entry book, dated February 13, 1882, it was noted that "the tree appears to be very scarce and much appreciated by the Indians; the bark is used in fevers and from the wood, cups are made in which water is put and afterwards drunk by children as a remedy." Still, we were at a loss to determine the taxon, due to the lack of adequate material. With a renewed interest in Lauraceae material from Belize, we discovered an unmounted specimen (collected by Romero and Mesh, s.n.) at The New York Botanical Garden which had remained unidentified for several years. The collection was determined by Daniel Atha to be *Ocotea veraguensis* (Meisn.) Mez, of the Lauraceae. In addition, it was discovered to be a new record for the flora of Belize. The tree is known to Leopoldo Romero as *copalchi*. The bark and leaves are used to treat snakebite and diabetes. This ethnobotanical information was passed down to Romero over the course of his studying and practicing bush medicine in the Cayo District. Upon comparison, there is little doubt that the material collected by Fowler and the material collected by Romero and Mesh are conspecific. We concluded that the tree known in Belize as *capoche* over 100 years ago is the same tree now known to very few Belizeans (perhaps only one) as *copalchi*. This single but fascinating example, detailed in Balick et al. (2002), illustrates the importance of combining studies of literature, museum and herbarium collections, and fieldwork to resurrect ethnopharmacological data that had previously been considered lost.

The sacred realm of healing in Belize is not only fascinating, but holds lessons for contemporary health care, both in-county and elsewhere. Integrative medicine, combining evidence-based traditional therapies and practices with allopathic care, is a particularly relevant model for widespread employment of traditional Belizean modalities. As Don Eligio often contended, there are many lessons from his people's ancient practices that have relevance today, and it was his dream that their use in contemporary biomedicine would result in greater respect for his ancestors, their knowledge, and the medical practices they have so carefully developed and used over centuries.

3

TRADITIONAL HEALERS, BUSHMASTERS, & THEIR BACKGROUNDS

THIS CHAPTER PRESENTS AND honors the individual traditional healers and bushmasters who worked with us during the project. As individual local experts, they were crucial to undertaking this work, and as a community of healers, they guided the trajectory of the research and educational activities that comprised this project. Each was interviewed with a tape recorder or video camera, and following the introduction, we offer excerpts from their interviews as well as our conversations with them over many years. While many of these experts have passed, we present some of the biographical sketches in the present tense.

JUANA AND ANTONIO CUC

Juana and Antonio Cuc were Q'eqchi' Maya from southern Belize, Toledo District, but lived in the village of San Antonio, Cayo District. They raised their 15 children without the assistance of medical clinics and were the grannie healers for an extended family that included great-great-grandchildren and numbered almost 300. The Cucs were both born in the 1920s, before the arrival of Western medicine in their village. Their parents were accomplished traditional healers, responsible for the health of their family and close neighbors. Doña Juana prepared a list of medicinal plants that were needed for the day, and Don

Antonio Cuc and Juana Cuc of San Antonio, Cayo District, in front of plants gathered for this project [MB].

Antonio went to work in his cornfield, gathering the plants on the list from the fields and forests. He returned with his sack filled with leaves, vines, and roots that she processed into the teas, salves, and powders that she used in her practice. As they were married for over 60 years, Don Antonio learned a great deal from his wife about midwifery, gynecology, and pediatrics. Their backyard garden was filled with many of the plant species used for medicines. Doña Juana passed away in 2001 and Don Antonio in 1994.

BARBARA FERNANDEZ

Barbara Fernandez is known as "Aunt Barbara" in the Belize City market, where she sells her herbs and books. She is of Creole descent, born in the 1930s in Belize and trained by an African healer. As a small child, she would intuitively know which plants to gather for which diseases. She tells the story of her first patient—when her mother suffered a fall and the wound on her leg became swollen, sore, and infected. Young Barbara went to gather some herbs, ground them with salt into a poultice, washed her mother's wound, and wrapped it with the plant poultice. Twice daily, she washed the wound and changed the bandages, using a fresh poultice each time. Within a week, the wound was healed, and Aunt Barbara had begun her career as a traditional healer in Belize. Following that experience, villagers came to their door seeking the child for advice—she reports that she "received guidance" at the right time to inform the patients what to do. At the age of 13, she was noticed by an African immigrant to Belize who offered to train her. She tells of how her teacher would take her blindfolded into the forest and hand her roots, vines, plants, and barks to taste and smell; she learned to identify them based on these char-

Barbara Fernandez in her shop in the old market of Belize City [MB].

acteristics alone. Aunt Barbara has a very successful market stall in Belize City where each day she sees patients from the community and helps them with both physical and spiritual ailments. She is deeply religious and has an abiding faith in her plants and their power to heal—a belief common to all of the healers we worked with for this study. She also is the author of the book *Medicine Woman: The Herbal Traditions of Belize* (1990).

THOMAS GREEN

Thomas Green was born in the Belize District on January 7, 1907. He lived on the Macal River for over 50 years and was an accomplished canoe (dory) maker who learned about medicinal plants initially from his aunt, Sofie Flowers. She would send him out to the forest and fields to gather plants for her use in midwifery. He also learned by "keeping with the older heads"—elderly people in his community who knew much about plants. He recalled that one of these people was his uncle, Joe Amson. When he was a teenager,

Mr. Green had an experience with a physician that caused him to lose faith in Western medicine. Prescribed pills for the treatment of fever contracted in the *chicle* camps, he had an adverse reaction to the bitter medicine—"My ears rang and I never can take more, I only had one dose!" He then tried a traditional cure, and the fever disappeared—"I do fine with the herbs," he noted. From then on, he began to treat people who came to him with their pains and fevers, and always with plant therapies that he would harvest and prepare. His wife, Theresa, was a noted midwife who traveled up and down the Macal River by dugout canoe to visit her patients. Well traveled, he spent a part of the Second World War in Scotland harvesting timber for the war effort. As one of Belize's best known dory makers, he went to the World's Fair in New Orleans in 1984, living there for months to demonstrate how to carve a canoe out of a single tree. He was a specialist in the plants that grow along the river, knowing their uses for medicine, construction, and many other purposes. He passed away in 2003.

WINSTON HARRIS

Winston Harris was born in San Ignacio, Belize, on May 10, 1942. His family was originally from Nicaragua, and his grandparents immigrated to Belize. He began to learn about herbs at a very young age, six or seven years old, and was taught by his grandmother, who was an herbalist and friend of Don Eligio Panti. As he grew older, he spent a great deal of time in the forest, hunting with his dogs, and also met Don Manuel Tzib, a well-known healer and uncle to Winston's wife. Don Manuel agreed to teach him about the use of herbs for curing many conditions. Today, Winston Harris is most well known for his abilities as a bushmaster and instructor for the British and

(top) Thomas Green, dory maker and traditional healer [MB].

(bottom) Winston Harris instructing a group of students on the use of bush medicine [MB].

Belizean armed forces in techniques of jungle survival. He is often sought out for that purpose by other groups as well. He is a snake doctor, bushmaster, and traditional healer, although he spends most of his time in the wilderness, training soldiers. He instructs them in ways to find water, food, and medicine and to construct bush shelters. His father was a well-known bushmaster as well.

DON ELIGIO PANTI

Don Eligio Panti was born in San Andreas, Petén, Guatemala, in 1893 to Gertrudes Cooh and Nicanor Panti. Don Eligio's father, Nicanor, was said to be an infamous practitioner of "black" magic, with a reputation for enchanting young women. Don Eligio

Don Eligio Panti collecting medicinal plants [MB].

recalls that his father was cruel to his young wife, Gertrudes, and indifferent to their son. When Nicanor murdered a man at a wedding feast, he had to gather up his family and flee on foot from the Petén to what was then British Honduras to escape the wrath of the local police and the family of his victim. Young Eligio, around two at the time, was carried in his mother's *rebozo* (a sling for children) as they slept by day in the forests and abandoned houses and traveled by night over pathways and roads until reaching the Mopan Maya village of Succotz in western British Honduras, where Gertrudes had a brother. They lived with her brother in Succotz until Nicanor built another house in the village and began their new life there. Young Eligio did not attend school, as there were none in that part of the country—he was never able to read. At age nine, he cut his first *milpa* (cultivated field), planting corn, beans, and pumpkins. When he was 15, he married Gomercinda Tzib, a 14-year-old Maya girl from San Antonio, Cayo District. Promptly after getting married, he signed up for work in a *chicle* camp to earn a living for his new family as well as to support his mother, recently widowed when Nicanor was shot to death by an angry man whose daughter Nicanor had enchanted. It was in the *chicle* camps of the Petén that Eligio met his mentor, Jeronimo Requena, who taught him what he knew about medicinal plants, prayers for healing, and the Maya spirits. They had only one season of *chicle* camp together, when the two, master and apprentice, were left behind in the final gathering season of 1933 as the other workers returned to their homes. The two men were charged with looking after the remaining provisions until a crew came to gather up the last of the items. Jeronimo taught Eligio about the uses of many plants and performed the sacred *primicia* to announce Eligio's intention to carry on with the healing work for both physical and spiritual ailments. Jeronimo was a powerful spiritual healer, and Don Eligio told us that he could turn himself into a jaguar and stalk the night forest. Jeronimo told Eligio that he was hiding from patients by living in the forest ("*chicle* bush") for six months a year, as he did not have the patience to deal each day with sick people. Before agreeing to pass his knowledge on to Eligio, Jeronimo questioned him:

> Do you have patience, boy? This is no easy road to follow. There is no rest for the healer. Night and day they will come to your hut with their sad stories, their sickness. Their troubles are plenty. People do not understand the healer and often mistrust us. When we cure what the doctor cannot, the doctors call us *brujos* (witches), and whisper lies about us. They say we work with the devil. It is a lonely life, I warn you!

Young Eligio, now intent on becoming a healer as his master had been, and perhaps as forgiveness for the sins of his father, continued the lessons with Jeronimo until they left the *chicle* camp. Once home in San Antonio, he began to treat his family and friends, and soon, word spread about his abilities and people came from other towns and villages to see him with their problems. For two years, he performed nine *primicias* each year to ask the Maya spirits to send him a *sastun* (a divining crystal) "so that I can do your work really well." Finally, he related, the *sastun* appeared at his home in a hammock on the day that he completed his 18th *primicia*. With the *sastun* (*sas*, meaning light or mirror, and *tun*, meaning stone or age), he could determine if a sickness was of a spiritual or physical nature. He could communicate with the Nine Maya Spirits and ask for help with healing. He could do divinations and enchantments—but vowed never to use it for personal gain or harm to others. He was now the *h'men* of his people and the bearer of great responsibility from the ancient ones.

Thus Panti continued to work, healing patients and collecting medicinal plants from the forests around his village, until he was in his late 90s. He passed away on February 4, 1996, and his funeral was a national event, attended by the Prime Minister of Belize and other dignitaries. His obituary was printed in *The New York Times*. Rosita Arvigo related the story of her apprenticeship with Don Eligio Panti in her book *Sastun: My Apprenticeship with a Maya Healer* (1994).

ANDREW RAMCHARAN

Andrew Ramcharan was born in Calcutta Village in Corozal District in northern British Honduras in 1903. His grandparents founded the village in the late 1800s and named it after their home in Calcutta, India. His family were indentured laborers who came to British Honduras from India to work off unpaid debts. They never returned to India and founded a large family in the New World. Andrew's father was a famous snake doctor, and when his young son asked him for training in this art, his first assignment was to cure a dog recently bitten by a fer-de-lance (*Bothrops asper*). His father guided him in this work, giving him the proper medicines until the dog recovered. Training of a snake doctor in Belize is different from that of other healers, as there is an element of mystery—secret formulas and treatments—that are rarely shared with anyone who is not a disciple of the snake doctor. "Mr. Andrew," as he was popularly known, began to treat patients on his own when he was in his early 20s, married, and had already started a family. He told

Andrew Ramcharan discussing his work with a class in Calcutta, Belize [MB].

us that his father warned him snake doctors were forbidden from having intimate relations with their spouse while caring for a snakebite victim because it would weaken his powers and cause his wife to become ill. He fathered 16 children and lived on the side of the main highway leading to Corozal. He noted that in his 70 years as a snake doctor, he had never lost a patient: "Many were those who were carried to me on a stretcher having been sent from the hospital or clinic to die. I cured them all. Every one!" He treated snakebite as well as the bites or stings of dogs, rats, bats, scorpions, bees, and hornets and contact with toxic plants. Because of the nature of the treatment, involving many secret prayers and powders, his patients stayed with him until they were better. He was a kind, humble, and humorous man, as is the case with many traditional healers. He passed away in 2003.

PERCIVAL HEZEKIAH REYNOLDS

Percival Hezekiah Reynolds—or "Sledge," as he was affectionately known—was a schoolteacher and plumber by trade until he retired and dedicated himself to the study of natural healing and medicinal plants. He liked to refer to himself as a "scientific herbalist" because of his extensive reading and experimentation with plant-based medicines. His clinic was filled with books about herbs as well as anatomical charts and other materials. Jeronimo Requena, the man who trained Don Eligio Panti, was Mr. Reynolds' grandfather. Percival Reynolds lived in Esperanza Village on the Western Highway in Cayo District for many decades. He had nine children, who grew up to be distinguished members of the community. For many years until his death in September 1997, he had a clinic just outside of his house where the hand-painted sign proudly proclaimed "Triple Moon Herbs: The Best Herbs for the Worst Sickness." He referred to his brand of botanical medicines as Triple Moon because of his belief that there were three different phases of the moon that were the optimal times for collecting medicinal plants: new moon, half moon, and full moon.

Percival Hezekiah Reynolds standing in the doorway of his Triple Moon brand herb shop [MB].

"Mr. Percy," as he was popularly known, was a gifted healer and active member of the Traditional Healers' Foundation. He believed in the power of faith and prayer in healing and enjoyed growing his own plants as well as new introductions, which were carefully tended in his garden. One request made repeatedly of Michael Balick was to bring seeds of *Taraxacum officinale* Weber, the common dandelion, a lawn weed of North America, to grow in Reynolds' garden for use in his practice. He considered this an important, but hard to get, medicine, and a few of these plants grew in his garden. His formulas were mixtures of garden plants, fruits such as grapefruit, and native herbs collected from the nearby forests. They were mixed in his clinic and sold in small, neatly marked packages. He never charged for treating children, and there was always a line of people outside his door. The Triple Moon sign, now weathered, still stands outside of his former clinic. He passed away in 2005.

HORTENSE ROBINSON

Hortense Robinson was born in Cozumel, Mexico, in 1928 while her parents were working in a *chicle* camp. Her mother and both grandmothers were midwives who worked in their village as well as with the women who traveled with the *chicleros*. Her cultural background is probably the most diverse of all the healers we have been so privileged to work with. Her family lineage includes Maya, East Indian, Afro-American, Scottish, and Belizean. When she was not working with her mother in the *chicle* camp, she was playing with the Maya children of Yucatan, Mexico, and learned to speak Yucatec Maya as well as Spanish. From the age of six, she began collecting herbs with the Maya children in the camp for her mother and grandmothers, delivering the herbs to the households of patients, and helping with childbirth when needed.

Hortense Robinson, midwife [MB].

"Miss Hortense,"[1] as she was popularly known, suffered from asthma as a child. When 13 years old, she had a serious attack and was taken from the *chicle* camp and flown to a hospital in Merida, Mexico. There, she had two experiences that were crucial in transforming her life and beliefs about healing. First, the medication she was given caused a bad reaction, and she nearly died. She recalls having a near-death experience in which she saw herself walking slowly up a mountain. At the top of the mountain was an old man with a long white beard and long white hair wearing a long robe. "Where are you going?" he said to her in a gruff, low voice. "I don't know," she answered, "I think I am lost." The old man had an enormous book made of stone tablets in front of him. He opened the book, and as he turned the tablets, Miss Hortense heard a large and frightening noise as if two boulders had crashed against one another. With a stick, he searched the stone tablets for her name, but he could not find it. He looked sternly at her and said, "Turn around and go back the way you came—you are not supposed to be here." As she turned around, she awoke to feel the nurses shaking her and patting her on the cheek. Miss Hortense said that she felt life was never the same after this experi-

---

1. In Belize, the title "Miss" is pronounced as it is spelled here but is usually written as "Ms."

ence, in the sense that she seemed to be blessed with good luck and succeeded at whatever she set her hand to. She also felt that this gave her a deep understanding of health, disease, and healing.

Just a few days later, as she was recuperating, the second defining moment of her life occurred. She left the hospital ward to use a washroom and, as she was leaving the toilet, she saw several *chicleros* bring a woman in labor to the hospital. They left her in the wrong corridor, and she remained there unattended. "Come here, child, and help me," the woman pleaded. Miss Hortense went to the woman and saw that the baby's head was crowning. Within minutes, she delivered the baby, and then she helped the mother to deliver the placenta by putting light pressure on her abdomen as she had seen her mother do many times. She wrapped the baby, umbilical cord, and placenta in her own hospital gown and chatted with the mother while the nurses came running from another direction. They relieved Miss Hortense of the newborn and wheeled the mother off into a room where they attended to her. One of the nurses bathed her and then gave her first cigarette to her, to calm her nerves. The nurse told her not to speak of this experience to anyone, especially "not to any of the doctors."

Miss Hortense knew then that she would be a midwife like her mother and grandmothers, and she recognized that she had a gift for taking care of people in need, especially the sick. She and her family returned to Belize when she was 15 years old. Because she was the only girl of seven children, she was never sent to school—her mother needed her to help with the housework. Miss Hortense adopted an unwanted baby delivered in her mother's house that same year and started working as a washwoman to support her new baby. Miss Hortense married at age 20 and had eight children. She eventually adopted 14 unwanted babies that she delivered and raised them all on her own, becoming a single mother while her own children were still young.

Hortense Robinson was probably the most respected herbal midwife in Belize. She received many referrals from hospitals and clinics. Many patients arrived at her doorstep saying, "They told me at the hospital to come here, that you are the only person who can help me." One of her specialties was infertility. She based her treatment of women on the concept of the prolapsed uterus, which she says 9 of every 10 women have and that "most do not even know it." She felt that the prolapsed uterus is the cause of most female complaints, and she taught her uterine massage technique with Rosita Arvigo for nearly a decade.

Hortense Robinson, nicknamed "*Mil Secretos*," or "a thousand secrets," lived in the town of Ladyville where she had her midwifery practice for over 50 years. She passed away in 2010.

LEOPOLDO ROMERO

Leopoldo Romero—or Polo, as he is known—was born in July 1944. At the age of 14, he began work in a mahogany logging camp in Belize, where he would be in the forest each day with a healer and co-worker nicknamed "Blockhouse." This man taught Polo to be a skilled bushmaster and traditional healer, giving him lessons during their workday. Polo learned to treat people injured on the job, to make teas from the roots of forest plants, and to collect water from local vines. Blockhouse taught Polo many skills, such as how to make a stretcher from bush plants to carry out a worker hit by a falling log or who had slipped out of a tree and how to splint bones using materials found in the forest. "I learned by doing," Polo said, "through emergency situations especially" (Arvigo, 1999). Over the 13-year period of his education in bush medicine, Polo apprenticed to four healers, and then he began his own practice, collecting and selling plant medicines for local use. He is deeply concerned about the loss of medicinal plants due to forest clearing, fires, and general disrespect

Leopoldo Romero next to a *gumbolimbo* tree (*Bursera simaruba* (L.) Sarg.) [MB].

for the environment. With this destruction, he laments, valuable medicinal plants are lost, such as Billy Webb, balsam, and *chicoloro*, which are getting harder and harder to find. One of his responses has been to transplant seedlings from endangered areas to protected areas where they can be grown for future generations. He noted, "In the 1970s, we used to go around the bush edges to gather material; now you have to travel for miles and still wander around to find the material" (Arvigo, 1999). Polo has been a long-term collaborator on our ethnobotanical studies, participating in many field trips, and has assisted many other biologists who come to Belize to carry out studies of the ecology or natural history of the country. He is always willing to go after an unusual plant we are seeking, determined to find it despite having to spend many days in the forest on this quest. He has become a noted lecturer and teacher and has been involved with numerous students and other groups that have visited Belize.

BEATRICE WAIGHT

Beatrice Torres was born in January 1948 in the village of Santa Familia in western Belize. She was educated there and worked for a short time as a nursing assistant before marrying Neri Waight, a laborer in the mahogany lumber mill, in the 1970s. Her mother had raised all her children using herbal remedies. Her grandparents were part of the wave of refugees fleeing war and strife during the Caste Wars of the Yucatan that began during the second half of the 19th century. Her mother's father was also from Yucatan. In every generation, as far back as anyone can remember, her family included healers, midwives, and snake doctors. As a child, Beatrice paid little attention to her mother's profession as a traditional healer, and she grew up knowing just enough to care for her own children. After her second child took ill, however, she realized that she could not afford to go to the health clinic for every ailment, so she began

Beatrice Waight in her backyard garden [MB].

to seek out her mother's aid and, in time, was able to care for all of the household ills with traditional remedies.

In 1973, when Beatrice was 25 years of age, a woman from her village came to her desperately seeking help. Beatrice noted that this woman had five children, four of whom had died, each a week or so apart. Her remaining child, a one-year-old baby. was quite ill, and the doctors had given up hope. Beatrice consoled the mother by telling her to pray and have faith in the plants that she would be using. Beatrice then administered a series of herbal baths, massages, and prayers and gave the baby a treatment that consisted of a tea made from marigold (*Tagetes patula* L.), vervain (*Stachytarpheta cayennensis* (Rich.) Vahl.), and the bark of guava (*Psidium guajava* L.). She instructed the mother to take the baby home, and if the baby's diarrhea stopped, it meant that there was hope for the child. Indeed, the diarrhea did stop, and the treatment continued for seven days. With the baby's recovery came the beginning of Beatrice Waight's career as a healer, as she decided to make this her life's work. She ascribed the power of healing to prayer, faith, and plant remedies.

With nine children, Beatrice Waight was unable to find much time to treat patients other than her own family. One night, she reports, she had a dream that her deceased father came to her and said, "When are you going to start working? We have waited for you long enough, now it is time to begin." Beatrice protested, saying she had no time. "Oh, then alright," said her father, "I will take you with me right now." She begged him not to leave her children without a mother— "Then I will take your husband!" he exclaimed. "Oh, no—this would be worse!" said Beatrice. "Then you must begin to do the healing work of your ancestors," said the father, who then bade her fall to her knees, gave her his blessing, and disappeared. Beatrice Waight passed away in 2010.

JUANA XIX

Juana Xix examining plants in her backyard garden [MB].

Doña Juana Xix was born around 1928 to a Mopan Maya family originally from Guatemala. She was orphaned at an early age and raised by her maternal aunt. She married at 14 and had 18 children. Her mother-in-law was the most noted and accomplished traditional healer of the day in Succotz, where she and her husband settled. When her father-in-law died, her husband's mother came to live with her. It was under the tutelage of her mother-in-law that Doña Juana learned how to care for a large family with traditional home remedies. Her mother-in-law delivered all of her children, and eventually, Doña Juana began to assist the elder woman with her midwifery work and to help with the many patients who came to their door seeking help with their illnesses. Juana Xix is an accomplished grannie healer, one who cares for their family members from children to grandchildren to great-grandchildren. Her progeny now number more than 250, and relatives living in Belize come to her first when there is a medical problem. She says she is able to treat most of their ailments and is happy to refer them to the local medical clinic when she does not understand their illness, but she says, shyly, "That rarely happens."

She and her husband, Don Isobel Xix, reside in Succotz in the Cayo District surrounded by several small houses inhabited by her sons and their families. She has several children living in the United States, as do many Belizean families. She runs a small restaurant and guesthouse and is happy to share her knowledge and her stories with her visitors as well.

IN THEIR OWN WORDS

During extensive conversations and interviews, recorded on video and audio, with some of the traditional healers we worked with, they were able to discuss a wide variety of topics relating to their lives and professions. Filmmaker Françoise Pierrot was the primary videographer and video editor for this project, working with the footage to produce videos for local television and classroom use as well as providing copies for the families and healers. In addition, National Geographic interviewed Don Eligio Panti and Rosita Arvigo, with the intention of producing a film that unfortunately was never completed. In this section, we present excerpts from the interviews with Thomas Green, Winston Harris, Don Eligio Panti, Percival Hezekiah Reynolds, Hortense Robinson, Leopoldo Romero, Beatrice Waight, and Juana Xix to provide insight as to their philosophies, concerns, and teachings. The abbreviated conversations are presented as transcripts of recorded conversations, with some editing for readability and flow. The bolded initials that precede the statements are the interviewers (**LR** = Leopoldo Romero, **MB** = Michael Balick, **NG** = National Geographic, **RA** = Rosita Arvigo). The complete interview transcripts were prepared by Rebekka Stone while she was a research assistant for this project.

THOMAS GREEN

MB: When did you begin to learn about the plants along this river?

TG: Well, they used to send me to show me some of the medicines, the herbs, and they used to send me for them. They used to send me for such herbs, and such herbs, the midwives and so, and finally, well, I started learning the plants. That was from my youth. I was 14 [or] 15 years old, then. And then I keep on, and they show me the use of some of the plants. Some of them, they use them. And maybe I start this work like this—"You see the plant there?" I grow [up] amongst a batch of girls. I used to keep amongst them, and some of the time they are no feeling good and they come and they tell me and I used to help them and, you know. Some this part of carelessness or lazy. [Some of their sickness was caused by their own carelessness or laziness about getting their own herbs.] But, one thing I never tried to learn or help them with. You have to look good in the face when they's coming, and they tell you they have pain here or pain there, because some times they are pregnant and they want to get rid of it. So you have to learn that. Well, that you could look in the eyes and see that they are pregnant, or look down on the throat and see that. You have too much time to know when they are pregnant.

MB: Who were the people that taught you what you know about these medicines?

TG: Well, there's so much of them. You see, there was one set of old men, and we used to keep with them. There were two or three of us that used to keep with them. Well, one of them was a snake doctor, and the other one use to practice herbs and things like that. And one like Benji Patnett. The next was a friend of mine, James Jones. We used to keep together and practice together. Each herbalist has his own plants. Not all of them work the same plant. Even for the same disease, same sickness. So, it's hard to tell like that. But some use the Waika [one of several plants locally known by that name], they use one medicine that you got. But the rest of the medicine that you have here is the African. They used to use the guinea pepper antidote and the snake head. They dry the snake head, the pepper, and the Tommy Goff [snake], and you roast the head. And when the head dries, you got to be careful not to make ants or a fly fly over it, so [you] couldn't lay [it] down. So what you do, you parch that properly till it dries. And you set it up and then you, you slow fire. And when you are not cooking it you got something like a tub, something that you can put it into until you get it [parched]. You rub all of them together. And that's what they used to use. But too—they have some herbs—the time of mixing the medicines—if there is a pregnant woman around, you ask she to come and just beat it a little, you know? And that helps. Because the pregnancy of the woman affects the man with the poison of the snakebite. They start bleeding. Just like this. Sometime ago up here at duPlooy [a lodge on the Macal River and the site of the Belize Botanic Garden], Peter, he was playing with a little snake. It's about that, about one foot, very little over. And we tell him that snake, that's trouble. That's a bad snake. Oh, he says, that's alright. And he [the snake] bite him. And in the evening he said the snake bit him. I gone home. I didn't pay him any mind. The night, they had to rush him to the hospital—so he had been in the hospital two or three days before he come out. But when he comes out [of the hospital], you look on the spot, the place the hole no open. I tell him this thing will get bad on you again, and if you want to start to bleed from here, only the dead can bring you back. So then I gone and I get some soft candle, lime, garlic, and I make my poultice and I put on him, wash him with this and a little bit of this *cordoncillo* [*Piper amalago* L.] bush. And I put that [on] that day. And the next morning, when I come back, I take off that and I put down my next poultice on him, [and] when I was coming over from the other side, the hole [was] open. I told him he is safe now. And I go back out to the place where I got the bush and I wash it. And the man alright. The Waika them got, some of them that the snake bites the man and they suck it. They have a next one [another treatment], that if the snake bites you on the right foot, you put the medicine on the left foot and it cure. I don't know how they do it. And you see that depends on the God they serve.

Because some people believe in Christ and the other one and the Holy Spirit. It is two different, difference between that and the other one.

MB: In your work, do you experiment with plants and learn about what new uses they might have?

TG: Yes. The one that somebody will come and tell you, "Look, this bush is good." And then you try to drink it. You drink any of them personally and you guarantee it. Because plenty of people come and they tell you that this is good and that is good. And finally they never tried it. So, they got no guarantee. The only one that I don't drink is the mauve leaf coco [an undetermined species], and I never try that. And I don't think they drink it. They use it for skin irritation. But it itch too much.

MB: How do you decide on the ones that you would like to experiment with if nobody tells you about them? Do you have other ways of thinking about them?

TG: Well, no. But, somebody will tell you about some of them. Most of them you get it from word of mouth we call it—90 percent. Some of them—you could get some of them by dreams also.

MB: Do you ever have those dreams?

TG: No. There were two boys in Santa Elena. One of them had stoppage of water and couldn't urine. So the mother, the mother-in-law, dreamed to him and she tell him what to use. The physic nut [*Jatropha curcas* L.]. It's good. Well, he gone, and he get the physic nut. And he use it. He said that the mother-in-law carry him to the tree and she tell him that it's the thing to use. So he use it. So he gone and he sit down and he tell the next one. And the man gone and the next one had the same complaint and that man charged him for it. And the boy nearly beat the other one. And he say, "I never tell you that for business." He say, "I tell you that because maybe that might happen to you or some of your friends." You know? And that will help them.

MB: When did you learn how to make boats, Mr. Green?

TG: Well, I was at Platón. And well, then you could have fishing and hunting for a living. But I wanted one small boat to go about in. And me and one boy, Polin, we bargained. We would clean the rice and I go with him in the morning and he go with me in the afternoon and work a little bit. Well, the first day I gone with him. And that evening he went to work. It's the first time that he's seen and he'd line the boat. The next day I gone and I help he clean the rice, and he never gone back with me in the evening. Well, I kept it, because it was me who wanted the boat. Well, when the boat done, I come up it had a fault, it was hard to steer, that was me gripe. So then I come to Larry. I tell him, it is hard to steer. He say there is no remedy to this because they cut off the piece already. But then I make my next one. And I keep on make [boats] until finally about four or five months before him dead, then he tell me he had explained everything to me. Every time I carry one, and I bring it, he explain it. But one of the [tips] was the flooring of the boat is the art of the bow, here. And if you don't set that [correctly], you can't bring [use] it. And then he explained the whole thing to me. But with that I keep on. But then you work, to work my dory, but with an axe. I did not want to work it. Too hard. You work it because you work with chainsaws, because the chainsaw is much easier.

MB: How do you pick a tree for a boat?

TG: Well, you try to get one, a straight one. That's all.

MB: What is your favorite wood?

TG: Tubroos. Tubroos [*Enterolobium cyclocarpum* (Jacq.) Griseb.] [is] the best. You see, they stand more strong, and they float up. The cedar [*Cedrela odorata* L.], they soak [up] water. If the sun strike it too much, they burst. Cedar and mahogany [*Swietenia macrophylla* King], they can't stand [the hot] sun.

MB: Have you taught anyone? How to make a boat, like you were taught?

TG: Here you have to pay somebody to learn.

MB: Why is that?

TG: You have to pay the man. And before he learn [teaches] you, he start to criticize you. Here and in other parts, everything, if you have an apprentice, he get pay for his work. In other parts, no, you have to pay for to learn. That's the one of the thing that you have to be careful with. Well, I used to, I used to, could have work the axe before. Because I used to make like trough and thing. I say, old man, I say, I could make one dory. But then I would make it. Rough, you know? But after I learn the trick of the bow and the stern, you know, I got no trouble again.

MB: What are the tricks of the bow and the stern?

TG: Well, in the flat bottom, you don't got no trouble because it sits down, floats across. But when it is round, no, as it touch [the bottom of the river]. And as it touch, it keels.

MB: Who invited you to go to the World's Fair in New Orleans to make a boat?

TG: Well, they had a Belize Pavilion. Somebody encourage them to tell different the things. And at the same time I was building, working one boat for Mick [Fleming]. And between him and Stewart [Krohn], they came together; they ask me if I will go. I tell them yes, but they settle to no effect. You'll be well treated. I tell you, I gone and I worked. They never find no fault on this or on that. That was it. I was there about five months.

MB: How many boats have you made?

TG: Some about 40, 41, 42, something like that.

MB: How many are still on the river?

TG: That one, [and] one or two down the line. Some down at Ontario, but not much left in them there. They are all dead out, just like this one here.

MB: How much would you sell a boat for when you would make it?

TG: It all depend on the size. One like this, 21 feet long, I would say about $1,200 Belize to get it.

MB: How many children do you have, Mr. Green?

TG: Two living. And I got about 11 grandchildren.

MB: Are any of your children or grandchildren interested in boat making?

TG: No. Well, it's—as you come to that it's something inborn. Because my grandfather by my mother's side, he was a shipwright. And them—one of the boys, Neal, he worked to repair, so like patch. But they don't have, them forget too.

MB: Do you think a lot of people will be making boats in the future or it's something that is going to be lost?

TG: Lost. Growing back no logs. They may make them with pieces, pieces, plank boat. You see you don't get much log now. And the little cedar that is—they are cutting them out. You don't find much big long one either. Pure short.

WINSTON HARRIS

MB: When you were young, were you raised with all this information about medicines and in this sort of way?

WH: Well, from when I was about maybe six or seven years of age, my grandmother used to be really smart with herbs. And she use to never take us to the doctor. She would always, herbs, anything, herbs. Miss—they use to call her Tila. But I don't know her name, how to say her name, because she died when I was maybe 12 years of age. But she use to know herbs. She and Don Eligio [Panti]. Don Eligio could always stay with her because they used to be herbalists, with herbs. My grandmother used to cure typhoid fever, malaria—any kind of fever with herbs.

MB: Where did she learn from?

WH: I could never say this. But I think my grandmother, according to my uncle. He told me that my grandmother and my grandfather, they are not from Belize. They are from Nicaragua.

MB: And did she teach the family those things?

WH: Well, she didn't have—well, I have to say no time, because we were little. And I am the only one that adopted herbs. I like it. I don't know why. I learned quite a bit from her. Like, we were sitting in a motel in Belize City and a plant is in that motel. The one with the yellow flower that you can see through the window? That's a really good plant. That *piss-a-bed* [*Senna alata* (L.) Roxb.]. That's a good plant, that's really good. That's good for yellow fever, it's good for typhoid fever, it's good for bilious, it's good for lots of different things. Just you have different ways of using it. It has a flower. You use the flower, you see like, when some people born and, kids them, lots of kids born like that—they would be yellow [from jaundice]. And you take that flower, and crush it, and bathe the kid with that. You would leave it in the sun for maybe, for a little while so that the sun would heal it? Then they bathe you with that and it would take out the yellowness of your body. That *piss-a-bed* is really, really good.

MB: So of all the people in the family, you are the only one that took, that learned a little bit from your grandmother.

WH: Well, I learned from my grandmother a little bit, not much because I was too little. And she use to show a lot of things, but still, you can't adopt it. You can't take it because you are too little. And then I, after a while, I start moving by myself. I go away to Black Rock, stay there for a week—just think about herbs, use different type of herbs, and things like that. Then I start growing up and come to the place that I married from Manuel Tzib's niece. After when I married to her, well, I use to like [to] hunt with dogs. I use to run around with dogs and the dogs that came out at his place. And [Manuel] start talking to me. He say, he say, "Winston." He say, "You are a good man." He say, "You walk all the time by yourself?" I tell him, "Yes." He said, "I will teach you something." He say, "I want you to come to my house for a week." So I went to his house for one week and he start telling me [about] different types of plants. This plant is for this, this plant is for that, this plant is for this. And I didn't write anything. I just stick it in my head and start using the plants. I would maybe use certain things like that rabbit's paw [*Sphagneticola trilobata* (L.) Pruski]. He tell me that he don't know the name of the herbs, but he know it's arthritis herb. And I tried [it] with many people in San Ignacio for arthritis and it definitely works.

RA: You just mash it?

WH: No, you boil and you drink it, like, just like how you would drink this skunk root [*Chiococca alba* (L.) Hitchc.] today. It's the same thing. That's really good for that. And, well, he start teaching me things, teach me things about your eyes. Things about different kind of sickness. And I see, when I come across it, when I start doing it, a lady came from Canada and she had a big wound on her wrist here. And he say, one day I was passing, he say, "Winston." He say, he said, "Tell me something," he say, "you don't know no one that know about herbs around here?" I knew Don Eligio know about herbs, but Don Manuel was teaching me. And I wanted to prove that the herb that he told me about work right. So I went to Don Manuel and I told him, "Mr. Manuel." I tell him, "You know what? I have a lady that," you know I told him in Spanish, I said, *una gringa*, "asked me if [I] have anything to cure a wound that she had because she went to many doctors and the doctors, they didn't cure her. And she spent a lot of money and she didn't cure." But some lady that went from here to Canada, she was Canadian but she use to live in Belize, and she had some kind of sinus in her nose and they cure her with herbs in Belize. So she went back and said the words to that lady, and the lady flew down here to Belize.

MB: And what did she have?

WH: Just a wound. I couldn't exactly tell you what kind of wound, but it was just something that was just eating her flesh, you know, and I went to Mr. Manuel, and I told him exactly the type of wound, the looks, and things like that. He said, "Alright," he said, "let's go, I will show you something." And we went and walked and walked and walked. He was showing me different types of herbs. [What to use], how to use it, and things like that. And then he came to the herb. He said, "Look at this." He say, "You can walk all the jungle and you can never find another herb like this one," he told me. He said, "This is the only herb that looks like this." He say, "And this is the anti-cancer herb." He told me like that. And I thought he would like give me a bag of leaves—six, seven leaves. He told me, "Two leaves for tea, two leaves to toast in fire and turn into dust, fill the wound with the leaf—nothing more than only the leaf." He say, "And you take three leaves for a bandage." "Seven leaves [total]?" I told him like that [asked]. He say, "Yes, seven leaves." I took it and went to the lady. I told him [asked] if she wanted to do it? She said, "Yes you can do it." If she want, you can do it. And I took and I start cleaning the wound. I took two leaves to boil, make a tea, and I start cleaning the wound with this same leaf. What he told me I did. I cleaned it with the same leaf. Slowly, so that it would not hurt her that much. I cleaned it with the same leaf. Then I took it and I dry it, then I toast the leaf on the stove and crush it and just fill it up. Fill, well, well. Fill it. Then I took two leaves, put it over it, and then I bind it with a cloth. In three days time I went and take off. He told me, "The third day," he said, "I want you to go." He say, "And you take off that bandage. And when you come back you tell me exactly how the wound looks." And so the third day in the morning, early, I went and took off the bandage. And you can see all the flesh start growing. All the flesh that was bad stick on to the leaf. And it [the flesh] start growing. And then I said, I told the lady, "What you should do, I want you to go and buy a bottle of baby oil." You know, we call it sweet oil. "And you take and put it in the wound and that would cure you." And it definitely cure her. So when she almost get well, the lady tell me, she say, "Winston," she say, "I owe you some money." She don't owe me a penny. She said, "But you are the one that did it for me." I tell her, "No, I will take you to the old guy that did it for you." So I took her to Mr. Manuel and I introduce Mr. Manuel to her. She started talking and she told Mr. Manuel, she said, "This wound that I have here," she said, "many times I went to the doctors. What they do they would just stick needle in my wound and give me lots of different shot and the wound still continue eating her flesh." And she said, "I thank you that you cure my wound."

MB: So, then Mr. Tzib was a *curandero*?

WH: Yes. But he never did use to show what he know. But you know, he was just like me. And maybe I know a lot but I don't say to no one. I keep it to myself and don't say anything to no one. And he told me like this, he say, "Winston." He say, one day [when] he was sitting in, you know, a little camp that they made in the field. He say, "Winston," and he start laughing. He tell me then, he said, "Mr. Eligio," he say, "he know a lot." I tell him, "Yes, Mr. Eligio is smart." He say, "But you know, I teach Mr. Eligio lots what I know?" I tell him, "But who you teach Don Eligio? Don Eligio *is* the Teacher!" He said, "No," he say, "I teach him a lot." He said, "But I am not like that, I don't them show what I have." He say, "I keep it to myself." He say, "You are the only one that I'm talking to and you are the only one that I am showing things." He told me like that. But it didn't went [wasn't] too long that [before] he got blind and couldn't show me no more. [Even when blind] you can take a leaf to him and he can just touch that leaf and he can almost tell you exactly which leaf it is.

MB: Your work with the soldiers, how does that function? Who had the idea of teaching the soldiers how to survive in the jungle?

WH: Well, the first time I went in the jungle with the soldiers, well, with Mr. Mick Fleming, I took him for a tour in the jungle. And he say, "Winston," he said, his sister, you remember there was that year that his sister came? And I took them. So he say, "Winston," he say, "you know what?" And he say, "I'm thirsty." I tell him, "But I told you no water in this area." And he say, "But I'm thirsty." And we start walking, and walking, and the lady came to me. He say, "Winston," he say, "I'm thirsty." He say, "You don't have no vine in this jungle that you can drink?" I tell him, "Yes, you have." I tell him, "You have the red vine, you have the water vine, grape vine." I told him like that. And he say, "Alright," he said, "go and cut the vine." So I went and cut a vine for the lady and she start drinking it and they took [a photo] that I am giving her the vine. And the brigadier went to Mr. Mick's place. And as he get there he start showing them the photo of the vine that I was giving her sister to drink. And the brigadier tell him, he said, "I want you to see if you can find Winston." Because the brigadier knew me [a] long time ago from when I was working with Little John. I used to work with a guide we knew as Little John, a tiger hunter. And then, when I saw that Mick came to my house, he say, "Winston," he say, "I want to take you a trip tomorrow. Up there to the international airport." I tell him, "Alright." And I went with him. So when I went in they took me to [the] Officer's mess. I went in there and then they took me to [the] Sergeant's mess. And they took me to different mess until I came to brigadier. When I came to brigadier he started asking me things. He said, "I want to work with Winston." He said, "And I would like you to try it. I don't want you to continue but if you don't like it, just quit. And if you like it, you continue." So they took me the first time in the jungle. And when I reach in the jungle, I work for the first week, with a group that they flew from England down to me. And I work with them. And then, all of them had poisonwood [*Metopium brownei* (Jacq.) Urb.] in the eye. Then I start using Mr. Tzib's same medicine that I told you, and they get well.

MB: Which one is that?

WH: The one that I call the wild ginger [*Renealmia aromatica* (Aubl.) Griseb.]. I start using that and then after that, they send another group and I did the same thing. And then another group, and another group. And by then I can't quit. So then, you know, I couldn't work at my farm. I can't hold my farm. Well, after that I went and I tell him, "I can't continue working with you." Because I know Carsewell, he is, I don't know his title, but me, I call him Carsewell. "I can't work with you no more," I said. He said, "Winston, why?" He said, "I will pay you more." After that well, I start working and then they start, they came up with jungle survival. And I started teaching jungle survival.

MB: When we talk to a lot of people in Belize, they say, you know, my grandmother knew a lot, and a lot of people knew a lot, a long time ago. And people today forget, you know, don't know as much.

WH: It's not that, it is because of the system of living, I think. Everybody think they start living too high, they don't look back. And you got to look back.

MB: Why do people forget? Why do they give it up so easily?

WH: It's not that. But I think many people know things. But that's the only thing I say. According to the living, system of living, Belizeans, they are not living like before. Most of them—if any little pimple, they will run to the doctor. They don't remember that herb. [They] want to go to that doctor. They accustom to pay that money to the doctor. And they would never go to the herbalist.

RA: Why do you think so few people want to learn about herbs?

WH: You know why they want to learn, mon? Because the time is coming that we got to learn things that was from before.

RA: You think a lot of people want to learn now?

WH: Yes. They definitely want to learn. Because they want to learn that because they know the time will tell. The time is coming back to the new ones [younger generation]. You have to [know] what they use to knew before.

MB: Why?

WH: Why? I couldn't ask. The thing is it's hard to say why. Either the people they will be diminished or something will happen. And that's why everybody is moving back to what used to happen before.

MB: We find it is a problem in our work because a lot of people are forgetting all these things. You know? How do you fix that? How do you turn that around to where people will get more respect for the herbs?

WH: The only way you can turn things like that around, you have to make something like a class, a lesson, and start teaching people. You know, you don't mind if they want to learn or they don't want to learn. You invite people and start teaching them. And then they will start. It is just like anything. They would start picking up bit by bit like when you are training on a machine. And then afterwards you will find lots of people will pay attention to it because when they find out what the root of it [is]—lots of people pay attention to it.

RA: A school, you think?

WH: That's the best thing. That's why, in the jungle, I have to talk to those soldiers for two hours, two complete hours, about herbs. The use of it, how to use it, what's the use of it, what for fresh cold, what for pain, what for this, and down the line. For two hours. That's my end. I have my end. My part of it. I have to give them a lecture about herbs. And then the medic, he would be along with me, and he would tell him about his medicine. But then sometimes I and he, his medicine, we clash because the things that he use, sometimes he let the patient [use] them, if it's the poison wood, and he use his medicine. [The wound] would always start running.

MB: It gets worse?

WH: Yes. Yes, it starts running and running and running. And the patient, they don't like it. And when I use the medicine for the poisonwood, "Oh," they say, "throw away their medicine and take Winston's medicine." That's the one that I told you, that you can use the jackass bitters. You can use the wild ginger. And we have another one that we use. You remember the one that we root? The leaf that I picked up with only one leaf that smell almost like the ginger? That one you can use. That one is really good. That one have a stronger smell than the other one. You crush the root and take out the shoot and rub it on the patient. And it don't burn or nothing. You just, well, you know poisonwood is hot. And you rub that, but you don't feel no pain. What it will do is cool down the area. And in the morning you look at that area and it is gone—so quick.

MB: So, who are you teaching in Belize?

WH: Right now, no one. The only one that know quite a bit, know almost everything I know is my son, George. He work with me in the army.

RA: How old is he?

WH: He is 19. They want to take him, I don't know, they will take him to the University in England. That's what the brigadier told me, asked me if I want him to go to the University and study and when he come back he could be maybe a doctor or something like that.

RA: Does he want to?

WH: He said that he would go—but then the soldiers, they start telling him that England is so cold and this and that. He's afraid.

RA: He goes with you to the jungle trips?

WH: Yes, mon. He is smarter than me in the jungle, he is. Yes, it's good.

MB: How would you get these chiggers [points to his leg]? You know those little spiders that bite you?

WH: The chiggers?

MB: Yes, they itch.

WH: Well, what you should do, you see that the *mano de lagarto* [*Neurolaena lobata* (L.) Cass.]? You just take it and crush it and rub it on that stuff, and you, and you would see how you feel. . . . Well I'm telling you I check that Captain Carsewell. He sit on a poisonwood and the whole of this was burned. And he was burn up for sure. Then I told him and I saw him start walking and his trousers was chafing his legs. And I told him, "Captain Carswell, is something wrong?" He say, "Winston," he say, "I sit on that log," he say, "and now I feel a heat that is burning my leg." And I told him, "When we are going to reach the area that we would set up our camp," I told him, "let me see." So he took down his trousers and he was burned from here to here. And I went and get a big amount of that stuff and I start squeezing it in a bowl. And no water, just a juice and I rub it on his leg. And in the morning Captain Carswell walked just good, and his leg was not that sore.

DON ELIGIO PANTI

NG: How did you learn to cure and who taught you?

EP: I learned from God, I don't have a teacher, I only studied my herbs and all my roots and all my prayers, I only studied, I don't have a teacher. But I guess I am all right since everyone comes looking for me. People must think I am doing something good, everyone comes to me for good health seeking my devotion, my herbs, my roots, my prayers.

NG: And when you cure, how do you know what ailment people have?

EP: I touch their pulse; from there I know the illnesses. From the pulse, when the pulse jumps, there is illness. The pulse is like a clock, when it is slow, O.K., but when it is fast, there is sickness. I then prescribe herb baths, medicine to drink, to put in their stomachs, they then can throw up. Sometimes they eat hair, cockroaches, mice, they put things in their stomachs, people do stupid things.

NG: When you cure someone, how do you know what disease the person has?

EP: I know by their blood, by their pulse, I know what ails them. I know which herb, which root, and which prayer I will pray. I know very well.

NG: We know that Maya medicine is mixed with the Maya faith and the spirits of the Maya, could you talk a little bit about the spirits of the Maya?

EP: It is the nine spirits. I pray to them to help me with my work. I work this way. But I don't converse with them as we are now; they converse with me when I am sound asleep. They speak to me and all they tell me, I remember, when I wake up I know what

I should do, if one of my patients is not healing, I ask the spirits during my sleep what I am lacking in doing, if it is an herb or a prayer. They tell me, [and] when dawn comes, I run and get the herb, if it is in an herb, and go to the sick person and the person gets well. This is how the Mayas do it.

NG: How do you use the herbs to cure?

EP: I mix them, nine mixtures [nine herbs per treatment]. If the person is not gravely ill, then I may use only three. But if the person is gravely ill, I must use all nine plants well mixed. I use all nine, I cook them, I bathe them, and they get well. Yes.

NG: When you gather your plants in the hills, do you say a prayer?

EP: Yes, in the name of God and in the name of the Mayas, those two names help me. I set my sack down, and my faith and all my heart, I never go gathering without imploring the name of God and those of the Mayas. They are my companions and my teachers.

NG: What is the importance of dreams in your work?

EP: Every time I dream, they call me to do such things. To look for more herbs. When I awake, I have more enthusiasm for looking for more herbs, more roots, and to study more prayers. Each day, God helps me and [along with] the [ancient spirits of the] Mayas, to do good work; they give me my prayers, blessings in order that I do more good work. The work then comes out very good, I am guaranteed with my herbs, wherever people will call me, "doctor," "good doctor"—not medicine man, but doctor.

NG: When you are working with the sick people, what is the most important thing in your work?

EP: The most important thing are the herbs. I have it guaranteed, when I see a gravely ill patient, I give them the herbs. Once they are more recovered, I give them some to stimulate their appetite, their speech, their everything. That is how I work.

NG: What does prayer do?

EP: The prayers give potential to the body, desire to talk, to converse, appetite to eat—because when they don't eat, don't sleep, they are on their way to death. Eating is for life, to sleep is to rest.

NG: How do Christians regard you?

EP: They can't see me. They say I work for the devil. But they are wrong. I told them so, I am not in collusion with the devil, I am not condemned, I am working, I am not condemned, I am not in collusion with the devil, I gather my herbs in the name of God, of God and of the Mayas. I tell them, that is why I cure.

NG: When you began your curing, where was the hill [pointing to San Antonio Village]?

EP: When I arrived, the hill was dead. But now it is alive, it is up there, there is more population, when I came there were only four houses.

NG: And when you had to go searching for your plants, did you have to go far?

EP: The mountains [where the plants grew] used to be nearby, but now, I have to go very far.

NG: What happened?

EP: People are moving in, they are making the mountains disappear, they are opening roads. Before, there were no roads, one had to go in as an animal, they used to get lost, because there were no roads. Now, everywhere there are highways. Now, you know where you are, where your house is. That is the way it got started.

NG: Don Eligio, the Mayas who were here a long time ago, the ones that built the temples and the ruins, what happened to them?

EP: They live here. They are here, from the time of the deluge, they stayed here. Now they are the gods of the earth. They stayed here to sprinkle the earth. They stayed here to help give everything one asks them for. They help with the harvesting, they are here.

NG: Don Eligio, what do you think today's world can learn from the [ancient] Mayas, what can today's Mayas teach today's world, to help it?

EP: They could [teach] if people believed, but they say there are no more Mayas, they say there are no more owners of the mountains . . . that is why the Mayas are not doing much for people, people don't think they are special. For those who are baptized and alive, they do not pay attention to them. Before they asked permission [through prayers to the ancients] every Friday, [but] now, they say it is foolishness, there is no master of mountain, we are the masters, and we do the same, we look after it, there is no Maya, there is nobody. They loathe the Mayas, they do not worry about anything. If you ask their permission, they will help you, just you, not all the world, just you, because you are the one asking—asking for your harvest, for your life, for everything. Whatever you ask them, they will give you.

NG: Don Eligio we are seated in the midst of this mountain which was given to us by God, to look for food, medicines, fresh air. When you see this mountain, which emotions do you feel?

EP: I get the emotion of working, of doing with my life something positive, so that God may help me, so that the Mayas may help me, to do the job, with all my heart, I am going to work with all my heart, and they will also put all their hearts into helping me. And I will put their *primicia*, and they will help me. What I ask of them, I get.

NG: You know that these days, people are poisoning the earth, cutting, burning, killing the earth, what do you think of this?

EP: That is natural, if you do not do it this way, there is no life, what will you eat, you must work the earth. From up above food or money does not come, one must work the earth. But God blesses us. We are burning, but God helps us. It is a sin to cut the twig—when the twig sees your axe, it cries because you are going to kill it. A lot of disasters occur because people work without the Mayas' blessings and without God's blessings, they don't ask for blessings, they just go about working without asking for blessings, they receive punishment, sometimes they get cut by an axe, sometimes they fall, sometimes trees fall on them, they are being punished because they did not ask permission. The permission is the *primicia*, they know you are going to work the earth and they help you.

NG: Don Eligio, is there something you want to tell the people of today, young and old, how to live in peace?

EP: Yes, I want to, but they do not obey, if I tell them they will hit me, these people do not have common sense, they are badly mannered, they do not respect older people.

If you do not stand aside, they will throw you down. They think they know, but that is one of their misfortunes.

PERCIVAL HEZEKIAH REYNOLDS

MB: How did you come to know about the use of herbs in healing?

PR: While we were back home, my mother used to give us little herbs and some herbs, and tea, like seeds, tea. And *piss-a-bed* [*Senna alata* (L.) Roxb.], she used to give us because it's a laxative. And you boil the root and the flowers. And you drink that and your apparatus pass a lot of bile and colon. We never used to get sick and go to the hospital. So we didn't know nothing about that, and we growed. The first time I went to a hospital was when a van hit me off a bicycle and knocked me down. So I went to the hospital. And after that I didn't went back for anything again up to now, 65 years, no. I had diabetes and I had arthritis. I had the diabetes first. And I went to find out what it was, because my head used to swing. And when I ride the bike I got tired. They claim when I went to the clinic, they claimed that I don't have [high blood] pressure. And so, I keep this swinging head, my head swinging. So I went back to the doctor, and I met a lady doctor. She said I must take my urine to the lab. And when I take my urine to the lab, the man tell me, he say, "Mr. Reynolds, I don't see how you are walking and you don't drop down." I ask him why. He says, "Your sugar is very, very high." And I went home. And there I started to treat myself. I never study nothing about the herbs, but I was reading between the lines. And I say, I don't know why it come to me, but I start to use fish. I eat the fish. That was my diet. I feed on a big fish [uses his hands to indicate about a foot in length]. And I don't put salt, I put honey over the fish. If I eat any flour, I would toast the flour until it's dark brown. And I start to feed on that. And the main thing that I drink for the diabetes was honey and lime, 50 percent. I squeeze the juice, and then I get the same amount of honey and I mix them together. And I take one tablespoon, three times a day. And the water that I drink is lime juice, sweetened with honey. That's [the] water I drink. I don't drink any hot tea. I don't drink any Coke. I don't drink soft drink. I drink that lime juice and honey entirely. I put it in the refrigerator. And it's cold, and I drink that. And I drink that about for a month and a half. And I [don't go] back to the doctor anymore. And I don't see any signs of diabetes, and things like that, anymore.

Years after that, after I start to study herbs, I meet one young lady, and they give me a book. And I see the book is marked *The Healing Powers of Herbs*. And I start to read this book. And when I read the book, it say, "This herb is good for this, and this herb [is] good for that and that." Then I start to get interested. And when I hear anybody that have that sickness, I take the herbs and I fix them together, and I give them, and they get help. So that encourages me, right? And I start to work just like that. Then finally, I start to get books from outside. Because I get one, you know, the same book said, "You can't be a herbalist without the herbalist book." So I order the book. The herbalist book has the picture and the plant's name. And it also describes the characteristic of the plants. And it also gives you the dose of each plant and herbs. And how you reap the herbs. And how you carry the herbs. How you dry it. And if the herbs must dry in the sun, or if the herbs must dry in the shade, because of the certain oil—all of this I read in the book. And I study the book.

MB: So you didn't learn when you were a child.

PR: No, no, no.

MB: Did you spend any time in the bush when you where a child?

PR: Well, yes. I learn a lot about the tree[s]. But I didn't learn about the herbs because, right now—this is one thing you have to know about herbs. You have people who know a *lot* of bush, a *lot* of herbs, trees—this is named this, this is named that. But when you go to talk about knowing about the herbs it's a different thing. Because when all the time we drink *piss-a-bed*, we drink *sorosi* [*Momordica charantia* L.], we drink *vervain* [*Stachytarpheta cayennensis* (Rich.) Vahl], we used to use *vervain*. My mother give us *vervain* for worms. But, my mother don't know anything about the medicinal properties about these things. So, if you don't know the medicinal properties about the herbs, you don't know the herbs. You don't know the herbs—you only know the name of the herbs. You know the name *vervain*. And you know the name *sorosi*. But, you don't know the herbs, you only know *about* the herbs.

MB: So did you apprentice with anyone in Belize?

PR: Well, no, I never gone as an apprentice, I practice on my own. Because I do a lot of scientific thing where I do practicing by myself, and so. The little pinto bottles, stones, liquids, and bring all things what I have in my mind. I watch them, how they behave, and I get a lot of idea from there too, see?

MB: So part of your work is experimenting with new ideas and new plants.

PR: Yes.

MB: Could you tell me what you do if a patient comes to you with a problem that you don't know what the cure is? How would you experiment? How would you find the answer?

PR: Well, when I first started, when I was young, I used to take the doctor's diagnosis for it, but I used to do a lot of wrong herbs, because the diagnosis was not right. So I give according to what the diagnosis [was]. But later on, when they come back, the doctor said it's not that, and they come with something else. Then I have to go to something else. And sometimes I hit it and sometimes no. But now I don't do that. I do my own diagnosis.

MB: Could you give me an example of something that you discovered in this work?

PR: Yes. You'll find if somebody come to you, and it's all over the place, and the doctor can tell what's wrong with them, and you want to help, and the doctor can't tell them what it is, and they are worrying and things like that. You start a conversation. Now from the knowledge that I have, because I know of the symptoms of so many diseases that I start to question you along that line, how you feel. If your pain might be a broad pain or if it's a sharp pain or things like that. Those things tell me exactly where you stand. Then from there I pick up and could start and come up with the right diagnosis. And I say, so-and-so, you going to get this formula. For years, I haven't missed one up until now. Take the example like pleurisy and asthma, you know? Sometimes peoples are going to the doctor for years and they treat them for asthma. And sometimes the people, they come here, and when they come here, I would find out that they don't have asthma. Because something with asthma—asthma will always affect you on the changes of the weather—sudden changes of temperature, your body reaction—that's because asthma is cold, but it comes from a really dirty system. The system is really dirty, and so it produce that mucus, and them around the lungs. But pleurisy is an inflammation around the lungs. So the changes of the weather don't have nothing to do with pleurisy. It makes a difference. So right away, I know, when I question you, and so, you would say, "I go out in the rain—and just one time—and it doesn't seem to bother me, for that it come on." So I will treat you for pleurisy and they will get better. And then the next thing, too, the next sickness is tumor. Sometimes the doctor that knows that a woman have tumor in the ovary or in the womb. And sometimes when I give them the medicine, because the formula, I usually do for stone and tumor, and when the people are urine [urinating], they urine [pass] a stone instead of tumor. But sometimes, it's [a] tumor because, you see, they start urine [urinate] out the little pieces of meat, until they will come clean. The system come clean, and then I send them back to the doctor to diagnosis this thing, and they usually use this one [diagnostic tool] in Belize [City], this one where they see the inside of you. And they usually report that it is clean, and they don't have to operate on them. And so I know, with confidence, that it's a job well done.

MB: What sort of plants do they use for that condition?

PR: Well, we use things like *bu kút* [*Cassia grandis* L.f.]. *Bu kút* is a plant that not any-body could use because it is a dangerous plant. You know, it's something like the fern. The fern is very dangerous.

MB: Which fern?

PR: Most of the fern family. They are very dangerous herbs to work with, to somebody who [is] practicing [healing].

MB: What could they do to you?

PR: They are very poisonous. You see, you must have the correct dose, according to what is wrong with the patient. Because you can take a stronger dose for a certain sickness than [for] another sickness, you see? So you have to know what you are doing for those type. It's the same thing for this *procute* [an unspecified plant]. *Procute* is very good. It's something that kills pain. It dissolves tumors. You know? It's proven over the years. You can put your head on the block, you know? Once it's tumor, you can try it. It's also good for goiter. But, with goiter, one herb does not always cure goiter on everybody. So, also, I do a formula that never misses. Because I use willow bark [*Thevetia gaumeri* Hemsl.] along with *bu kút*. And, because the willow bark has the niacin, and also [has] the vitamin B, because the niacin is a type of vitamin B, that's what makes it go well. Because it's not necessarily iodine that is the only shortage when you have goiter. In some people, yes, you give them iodine and they start to recover. But some people have goiter and you could feed them on iodine, too, and it wouldn't help. Because they have a underlying cause of vitamin C, too, see?

MB: How many people do you see in your clinic every week on average?

PR: Well, maybe some weeks, I'll get six people. And maybe some days alone, I will get more [than] six people. On average maybe three person a day, maybe.

MB: What are the conditions that people come to you for?

PR: Well, they mostly come to me when they are under condition where they all over the place [regarding their health]. [From] different countries and they seek their health and they never get it and they're still suffering and their health becomes less. Near their home they have a drug store, right? And so they hear about me and they will come to me. And I find what to do with them, once they get here, I find what to do.

MB: So you have a very high success rate?

PR: Yes. I think one person in all my clinic, that I ever knew, he come to my clinic and died afterwards. A fellow who they give over from the hospital to die and they wrapped it [the area of the tumor] right here [indicates the place on his arm from elbow to wrist]. He had the cancer and he get here. And he couldn't live over 14 days. But if I get him before that, before I had 14 days working with him before he died, he would not have died.

MB: Do you see a lot of spiritual illness as well as physical illness?

PR: Yes a lot of that thing. This lady she come to me with like something standing up in the belly. And I check it out for gas, for gastritis. But, I say I never seen gastritis that look like this. But, so I started to check it and I found it wasn't gas. But I still try the herbs for this gas. It was a Spanish lady. And for this, I had a woman who was interpreting for me, and the husband told the lady that it's harmed his wife, and she was sick for years. And he said he has been taking her all over the place. So, when he tell me that, I change the medicine to spiritual healing. Because you have the herbs that have magical power. So I tried them. So I follow the [herb] book and I tried them. And they work. So I put that thing, you know, I wash that lady's belly when it was swelling up, and as I put it on the lady's belly, the belly got went right down and stop [makes the sound of a deflating balloon]. And when that stop, the lady get up off the bed [and] she said, "How much, Sir?" Then the husband walk over to her. He was waiting to hear we announce that his wife died. When he bring that wife here, that he bring, then he take them to Mundal, and then he take them to Mr. Garcia. And Garcia did a very good examination and didn't find anything wrong the lady and the lady's belly is up there and just foaming. He say, in just Spanish, "*Ya, me voy. Ya, me voy.*" The lady toss and say, "*Ya, me voy. Ya, me voy.*" And the lady over there tells me that the husband say that, "You have an evil thing." So I go into [the forest] for the bush that [is] good for evil thing and wash it and it get [she gets] better immediately.

And the next one, come with some—she had varicose vein, but I never see that varicose vein look in that manner. You know the varicose vein poked out like you have a depression, like it was sinking down in the flesh. And it bleed here and there. So she had three place on the foot there. And the sister brought her here said that they harm her. I said alright, and I went for the same bush, and I fix it up, and a little roach jump out [of the foot]. And the lady went [away]. I do not see the lady anymore. I don't know if the lady get better or so, but, something come out of that foot. It was a roach, a funny roach. The first time I see a roach look like that.

MB: What sort of personality do you think it takes to be a traditional healer or a spiritual healer? Do you think anyone can learn this?

PR: Anyone, anyone. But it depends on how dedicated you are. Because if you are like me, if you come here, and you have a problem, something I have never seen yet. And I tell you, "Come back Wednesday." And I meditate on that. And when you come back on Wednesday I have a cure for it. Somebody come right in my sleep and tell me, "Use this as a herb on that person." And so, I put the herb on. I give you that herb, and you get better.

MB: When you say someone will go away and you meditate on that, could you describe what that means?

PR: Yes, like, you are thinking about this thing, because you don't know what it was. This is first time you see something like this or hear something like this, you no read it in no book. I pray in the night, before day in the morning. And then, when I pray and go and sleep on it and my mind on that. And tonight, some—I don't know if it's a spirit or what—come and it tell you exactly what herbs to use for that sickness. And then I get up, same night, and I write it down. The next day, I'll get it, and when you come back and I say, "Drink this, and you'll get better." But not everyone get that gift, not everybody get that kind of inspiration. Because you have a lot of people [who] are gifted and dream a lot of herbs, but they no put it into practice, you know. We have a lot of people who dream.

MB: When you meditate, how do you do it?

PR: You can lay flat on your back in your bed while the silence pass in the night and you can drop asleep right there. And it come to you. Sometimes somebody come and say let me show you something. And they take you, and you don't know where they carry you. They show you this and this and that and that. Just like that.

MB: How do Western doctors feel about traditional medicine?

PR: They resent it, but if they would study it, then the unknown would be known.

HORTENSE ROBINSON

MB: What was life like growing up as a child in a *chicle* camp?

HR: Sometimes it would be rough, sometimes it's fun. Like, when the weather is good and they don't have to change from one camp to the next camp. Maybe they'd settle down for a month or two in one camp. You're spending a good time, because you get to know all the people in the camp and all the kids come together and play. And, like this time [today] they all separate and go home and get washed and go to bed and things like that and go home to their parents because in camp they were separated, you know, in a home. Then, you find that the people, they come in from the bush. All the *chicleros*. They're playing guitar, and all, some playing cards, playing games. That was the life in the *chicle* bush. Singing and playing guitars until it was time to go to bed.

Who drink, drink *cerveza, tequila*. All the adults, they go to bed. And all the kids, they sleeping. Then at times you find some of the parents come together and tell stories. And we enjoy it. We starts getting sleepy and we go to bed. Yes, there was always many, many children were at the camp. And you leave from the mahogany camp, and from the *chicle* camp to the mahogany camp. In those days, some people come down from the *chicle*, they do a little bit of *milpa*, then they go to the mahogany camp. And they fall the mahogany. Some people work at the back of the tree with the two hand saw. And the children they are collecting the sawdust. For the tomato, to make the bed, you know, the sawdust to mix with the soil and plant. And we collect the bark from the mahogany. And we burn it and make the ashes. From the ashes we get like bleach to take stain from all the working clothes. Like you make this Clorox, so we could take the stain from all the clothes. And we collect them and pack them in containers, so that all parents have it. And we collect firewood. It was big fun to collect wood. We have a bet who could get more wood for the day. And the parents they were very fond of if because when the father come home they do not have to cut firewood—because we collect it now. And we think that that was great for us.

We learned a lot about *chicle*. We learned how to cook it. When it was finished, and when to take it out and put it in the water to let it cool and mold it out, and put it on to burn [to brand it]. You've got to burn it with whose name you'd burn into. And, to make the *chicle blanco*, you've got to put it in the fire. You have to wash your hands, properly clean. Trim your nails, then properly clean. And then just do it like that. And from the heat of your hand start making a little lump like that. And keep collecting it. Wash. And get all that juice out. Like chalk. And then, the *chicle* is ready to mold. And to put into the mold a little design, whatever you want. A little girl, a little horse, a little chicken with eggs, whatever you want. When they cool down, we have them to sell. And each little design, you know, costs more than the regular blocks. We used to have fun doing that and put different colors and flavor to the *chicle*. Not sugar coated, just different flavor from mint or mint oil and spice. That used to be fun. We also work to carry firewood for the kitchen. And when we do it for the *chicleros'* kitchen, we earned some money to, kind of pay us for something. Whatever amount would bring, they give us something for it. That used to be great fun for us. Because we used to be able to go to the commissary and buy candies, things that we like. From the *chicle* camp, when we're going out, that used to be like, like big fun for us. We're having a farewell, you know. And then they have a lot of noise that night because tomorrow morning with everybody leaving camp—all the horses them knew they going to take the load out, all the camps them, to amount of mules you going to use to take out the cargo, you have one to ride, until you reach at the camp where the plane field and pick up the loads. Sometimes five miles from camp. Then we get on board the plane and come to Chetumal. That was when we were at Laguna del Carmen or Laguna Ohn. And that used to be our daily things. To prepare for the camp and if we were leaving camp, everybody had to keep up. You have to pack up everything, you can't leave anything because the next day you did not know if you would be at the same camp. You had to take everything. The ladies they pack the kitchen things. And now, going out, collecting all the monies from the *chicle*. And the men, they go locate a place to put the *milpa* and start cutting until the mahogany season is right. Start and then go over. Then just the ladies, they stay to look after the plants. Like you have corn, rice, tomato, beans, cabbage, carrot, were all in section. And the men, they go to work. As soon as the first of May they start to break off from the mahogany camp: In June started the *chicle*. And that used to be like from June to December. You know. And then, from January they start changing from one camp to the next camp.

MB: How did you begin your interest in herbs?

HR: Well, from the age of three. The first one I tried on my mother. She had a headache and I went to collect the herbs. And when I put it on her forehead, she said, "What you doing?" I said, "I'm doing some medication here," I say. For the relief of this headache. And I [bathe] her head with the water. Put the leaf them on her forehead, and I wrapped her head in a towel. And in a couple of seconds, she went off to sleep. And when she wake up, she feels better. And when she said to me, "What kind of herbs have you been collecting?" And then I started to show her. She said, "I didn't know that there was herbs that were useful." And she realize that I [was] interested in herbs. But she still wouldn't give me the freedom to go off with the Indians into the bush to collect herbs. You know, she was afraid. Because they always have that fear about the Indians, and how they weren't quite civilized in those days. Well, they didn't do any harm to me; I learned a lot from them. And I'm questioning them about the herbs. Picking up the herbs, I might show them and ask them, "What is this for?" And they would tell me, "Well, this one, you use this one and that one together. And you make a tea, or make a bath, or a steam bath or whatever. And I retained them all. I do it at home when my brothers are sick. And I help them. They all work out fine. And my mother used to do herbs too. A lot. Because all her patients she attended. She give them steam bath, with different herbs and things like that and I learned them all. She make bottles of medicine—preserve them, and have them for when the patient come; they're ready—quarts by pint. And we have the cohune [*Attalea cohune* Mart.], a wine that they used to make to preserve the plants. They leave the strongest, the second strongest drawing from the cohune head, and they leave it to ripe the cohune head, they call it. And they become like alcohol. And that's what they use to preserve the tonic from the plants. Well, we don't have no cohune palm to do that. It's so far, so, we do it the next way—we preserve it, we draw it, and put it in the bottle. And, leave it open until it cool. So that it could last some days.

MB: When you first started learning about herbs in the *chicle* camps, who were your teachers?

HR: The Indians. We had Miss Maria Chi. We had a lady the name of Philomena Koh and Celestino Mai. They were old people. But they used to go as cooks, to stay in the camps for the *chicleros*. Those are the people that were at the camp. The elderly people. When we were living in San Francisco Botes, we had two ladies—Cecilia, and one Maria, those are the people that used to take me out to the bush and pick herbs. We used to walk like three miles in the woods to collect herbs, all different kind of herbs. Whenever they would come back with the herbs and they would select them and say,

"Ok, this is for such a person, and this is for such a person, and this is for fever. And this is for cough. This is for pain of backs, or for belly ache." And you know, they explained to me, as to suit my age. And then later on, after I become a girl of 13 years, they explain to me when a woman is sick in the womb, what to use, how to attend to them, and how many days treatment, and like that. How many steam baths they should have. How many drinks they have for the day. They have a small little calabash they use for making herb tea. They say, "You give them before meals, one of this little cup." And that's the way I learned.

I am a midwife. I love to save the babies. The first baby I delivered in a hospital in Morelos, Mexico, I was the age of 13 years. I was a patient in the hospital. And the people who brought the lady in did not know where to take her. Coming from the *chicle* bush also. And they set the stretcher down on the walkway, and they went to locate the areas where the nurses were, where they were working. As soon as they come back with the stretcher from the hospital, the baby was already born. I had the baby in my arm like that, in a big face towel, with the placenta and everything, inside the towel. And, that was my first experience. When I went back home and told my mother what took place and said, "Oh." She said, "That's terrible. [Going in] as the patient, the girl [Hortense] deliver a baby!" And, right from there I got interested. I say, whenever I hear a person in labor to have a baby, I could imagine the help that they need. You know, and I put myself to do it. And I continue on helping people, until this time.

I'm satisfied when I am doing my work, the progress that I am making with that, and I am so fortunate, that I have not got any dead baby from carelessness, or I would say neglectfulness. Everything was splendid up to now. For the year I am slowing down because my daughter [is] doing it. For now I tend them [mothers] until the time that they give birth to the baby, and let [my daughter] take over. So that she can get used to it and do it well as I do. And I do it on the bed, on the ground, in the bush, wherever. And, I never have a dead baby. I remember when I were at Mountain Pine Ridge sometimes I have five patients for the morning. Sometimes for the week I have 15 patients. For one week. And I still used to do my household and tend to my kids. I'm just running around and do all my work, tend my patients, take tea to their bedside and give them herbs to drink. And give them a steam bath, until everybody get up and feeling well. I have babies in Benque, Cayo, Corozol, Belize [City], all over the country. There is no part of the country that don't have babies who I delivered. Right here, some of those boys that passed, just now, were babies who I delivered when I first came here. I have year 32 going on 33 since I am here living in Ladyville. And all the teenagers, them [points outside], I have delivered them.

MB: Do a lot of people see you for spiritual problems, spiritual diseases, as well as the physical problems?

HR: That's right. I have a lady right now come to me and says that she can't sleep in the night. She's saying, people come into the room and look at her. Turn out and [then go] make noise in the kitchen. And she took me there to spend a night in her home. And while I was there I didn't see anything but I heard it. And all the children were sick in the home. And the next day I get up early in the morning and I make a one bucket boil of some herbs, and I let all of them take a bath, and I come home. And, that's it. They haven't heard anything. They didn't be molest [harmed] by anything. And the next family come down, just about a week ago, and they have the same problem. And I did the same to them. I gave them some bath in a gallon, and she phoned me, yesterday morning, she tells me that she is sleeping well and the family is happy. The problem is just from the envy of neighbors. You know, they are envious people. I gave them baths, and I gave them teas to drink. And they're ok. They're fighting it. Prayers are very helpful. "Our Father." That's the powerfullest prayer for me. "Our Father who art in heaven, hallowed be thy name. Thy kingdom come, thy will be done on earth as it is in heaven. And give us this day our daily bread and forgive us our trespasses as we forgive those who trespass against us. And lead us not into temptation but deliver us from all evil. Amen."

Spiritual healing is like—you could learn, if you believe. And it's much easier for you to learn if you are a person that is gifted to hear voices calling on you. And dreams also. Sometimes, you have scary dreams, and you are not afraid. You do not wake up frightened. You wake up and wondering, "Why do I dream such dreams?" You know. It sometimes is the spirit that comes and tries to give you a gift on how to go about approaching a person who is [in] need of help. You know. And then, you study it and work along with people who know about it. You know. And people—it's so simple. Some people do it, but—like let it be—like it's something great or it's something, you know. But, I don't try it that way. When it's something that could be done simple. Some people let you believe that it is something very hard to do, or it is scary to do. And then that person can't advance. In learning, you know, because when you reach the point when there is scariness, or too many things, or too many words to say, you know, that person gets confused. And then they say, "Better not to go forward with it, because I could be hurt." It's not like that. I don't say it like that. The little that I do, I don't find any difficultness in it. Not for once. And it's simple. It's very simple. Because you have a time that you collect your herbs. And if you have a—like for instance, here in Belize, or in this country, or like in Mexico where they have grave space there. They are sure that the person's body is there, they are buried. You know. You could go and ask for help. You know. And collect whatever plant that is on that grave. You could use it as medicine to heal one of those sick. And you use a bit of alcohol. You use holy water. And a water that they sell in Mexico, we call it, "*Agua de los sietes poderes*"—"Water from the seven powers." Just godly things; nothing evil. Well, now on the evil side, I don't interfere with it; I don't. But things like—that you have to mention the word of God, or use the power of the Almighty—I do like it. And it makes me feel great whenever I heal a person.

If the person must have faith and believe that he will be saving by the power of the Almighty, Our Father—he must have faith in whatever he or her is taking [to] heal himself. And have faith and trust to the Lord that nothing overcome His power—that is the cure. You have to have faith, and if you have faith, you can do it. Sometimes I have to walk nine miles to see my patient. Coming back, collect plants. I sacrifice myself to the recovery of my patients, you know. I have that belief that, by walking that nine miles, my patient will be recovered. I feel so great doing it.

MB: Can anyone be a spiritual healer in your opinion? Are there any special qualities a person must have?

HR: Well, if a person believes in seeing the Virgin or the Spirit or whatever, I think it's easier for that person to have that gift. You know. Like, sometimes you hear people calling you. And you recognize the voice, and you go and open the door and that person is not there. And, so many times it could happen. It happened to me. Sometimes I hear Miss Clara hailing me. Shout loud. And I'm the only one that heard it, only me. And I go out and say, "You were calling me?" "No." And we understand that it's a Spirit calling, in her voice, using her voice, you know, maybe to distract or something. Sometimes, maybe sometimes you going to do something that is not right, or the time is not right. There is a good spirit who comes and distracts you. Until you straighten out and get back to it. That's the time that it should be. I do more physical healing than spiritual. Because I don't introduce myself to everybody as, you know, a spiritual healer.

MB: Do you use the pulse to tell you things about the person?

HR: That's right. You take your hands to the person like that [puts fingers 1 and 2 on the wrist]. You know. And this pulse here [moves fingers up to the arm just below the valley of the elbow area]. When I'm working at the patient, I put them to sit down. Maybe I give them some water to drink. Maybe that person is a smoker, or a coffee drinker. I let him rest for a while, until I think that coffee digests. And I give them a glass of water to drink and converse them a little so that they can, you know, just keep calm just for the moment. And then I go right up to them and take their pulse again [indicates on the wrist]. Just like when they just smoke a cigarette or drink a cup of coffee, a drink of beer or whatsoever. And then, I take this, you know [indicates the temple area around the eyes, then the area of the neck below the ears]. And you push this behind here [referring to the area under and behind the ears]. And that person will explain to you what he felt. What he is feeling. And you tell him, "Just don't be afraid to tell me what you are feeling." You could know if the [blood] pressure is high or low. And then sometimes people say that their head swings. You know. It could be from low or it could be from high [blood pressure]. But there is something different [between the two types, high and low], that you look at the veins then [indicates the upper chest area and along the interior of the arm]. If the veins are soft [pressure is low]. And when the pressure is high, you find people with veins that are very lumpy, and very, very stiff. As soon as they get one drink of this medicine, the person himself feels from here [indicates the head above the eyes] to here [indicates the breast] in an hour and a half, a difference.

Where [did] these plants [go] that we [were] collecting in the past?

MB: Well, a lot of them have gone to the National Cancer Institute where they have been screened against 40 different types of human cancers and two types of AIDS, HIV-1 and HIV-2. The scientists there are extracting them in water and other solvents, and then putting them into test tubes with a living virus or a living cancer cell. And then, those that are effective—that is, if they kill the virus, without poisoning the human cells—then they go to a next stage, where the scientists at the National Cancer Institute are trying to take out the chemical that is responsible for that activity, identify it, and then try it out on animals, and eventually people. It's a very long process.

HR: Let's pray that something will come up, to save so many lives, so many young people who are dying from this malignant disease. I pray every night that the scientists will come up with one good use.

MB: Do you know a lot of people around here that have malignancies?

HR: Oh. Yes. But they keep hiding. You know. Like if you know that they are around the vicinities, they go to the next places and keep hiding. You hear about people [who] will say that so many people will have AIDS at such a place, and such a place, but you don't know them. Until, when you hear that that person died, then you know that, well, he died from AIDS. Or they died from cancer, they keep hiding. They won't expose themselves [so] that you could try and help them with whatever, you know. And I keep telling them, "We are not looking for money. We're just looking for the cure." Soon as the medicine cures somebody then we have guarantee that it's good. Well, then, the money is behind that. I said, "But right now we are just trying." But they don't give up themselves. A lot of people in Belize are dying from AIDS. And you don't know that they have it until they die. They just say, "You heard such one night? That person died from AIDS." Where before, you didn't hear one thing. They died? And you knew that that person was sick.

MB: Where do you get the plants you use in your work?

HR: It's getting harder and harder to gather these plants because each person has so many acres of plants, and for so many years they are not using it, until now, and everybody is pushing it down with the bulldozer. They are taking down all those great trees. So, places that we used to go and collect, they are no longer there. They are all gone. So, that's the reason we had asked the government to preserve that piece of land at Terra Nova, so that we could be assured that there are some good plants there—medicinal plants. And what is not there we could take from the next land that they are clearing to plant there, so that in years to come they could have it, the young generation, the coming generation could find a use of it, the plant. That's the reason that we are trying to leave these things of them, and learn the use of them—to provide it for them and teach them about it.

MB: If all the forest plants disappear, what will happen to your work?

HR: Well, there will be lost dreams.

MB: How do you feel about that?

HR: Very sad. And I believe that [is] what makes many people want to go away, just to forget everything that is there or longer there. You know, so it's best to leave and live the next life. Different from—you've been here; you're depending on those things; and they are not there to use any more. The little that is there, you can find maybe in your friend's backyard or a relative of yours' backyard. It's not enough to help other people that need help. While in the forest you could fetch enough to go and help other people, all over the world, all in the whole country. Sometimes coming from so many miles sometimes looking for just one quart of medicine, and you can't give it to them because there is none. You have to wait until next week before you can spend about five miles to reach out to make just that one quart amount of medicine. It's very hard on us. But like you know, the government does not see it the way we see it. It needs someone to go and explain to them and show them. They may not have the use of it. And that's just one or two person that may not have the use of this plant. But there are so many hundreds of people who need it. You see? I could go right now to the government now and say, "Well, you know. This plant is a good plant. It's so useable. It's so good for you; it's good for everybody." And he say, "Whenever I feel sick I just catch a plane and go to Merida or go to the United States or go to Guatemala." When the poorer class of people, they are hundreds, thousands. They can't do it. So they rely on the local medicine. And they get healed. And these people rely on the local medicine and they are not one or two person but it's thousands and thousands of people. Just from yesterday to today you see how many people come and ask for medicine. Not all of them can pay, you see? Even when its local medicine, not everyone can pay. Do you see what I am telling you? It's necessary to save this plant. Because there are some people that don't have money to go to the doctor. Some come and have the local medicine, and still can't have money to pay for it. If they pay for it, they won't be able to buy bread for the table this evening. So all they come and want to see you private, and can't pay you for that medicine, what it costs. If the cost of medicine is 10 dollars for the quart. Because you say that, I tell you, I'll make it five for you. "Ok. But I still can't pay you until next week." "Ok. You come for it. I'll give it to you." You know, and if he need a gallon, you keep on giving that person that quart weekly until he come back and say, "I'm feeling fine." "Ok. Go to your doctor and take a check-up. And see what the doctor have to say to you. If the doctor tells you your complaint has disappeared, you're feeling better, I say well then, let your conscience be your guide. One of these days you may come across something you want to give me as a gift—it's ok. And if not, forget about it." But I feel like I have sent a treasure up to the Almighty by helping someone who can't afford it. So, that's the problem that we are facing here now. Everywhere's clearing down, and there are more poor people suffering from wants of medicinal plants.

MB: What's the biggest concern for the traditional healers in Belize today?

HR: To preserve the plants. Yes, that's the biggest concern with the traditional healers, because without the plants, it doesn't make sense. You can't work without the plants. It's from the bark or from the leaf, or the root, or from whatever you want to use. If it's not there how can you work? It's just like a mechanic without their tools. So, we have to get the plants [to] them, to do the healing, to do the medicine, to heal the sick. Without the medicine, the *name* of the medicine does not help. You've got to use the extract, from the bark or the root or the leaves or the branches, to give the people, to prove to them that they can be healed by these plants . . . Years to come maybe the doctors that will go and put their lab right there and work hands in hands with them [the healers] . . . Nothing is impossible. Because, you know, we could know the use of one herb. And maybe when the doctor would take it and break it down and do it their way, they would find a better use than we know about it. That's what is in herbs . . . I think it's great what the scientists are doing. You know, to get a view from each person, and ideas from each healer, and put them all together and see what they can get from it. Because it's just like you can't prove [something] with one person's word. But we want to hear what this one person, and this next person, and the next person have to say about this plant—if it's good, or it's right, or the amount is correct, you know. I think that's what they looking for—more guarantee on these plants.

LEOPOLDO ROMERO

MB: What type of healer do you consider yourself—do you do spiritual healing or more traditional healing?

LR: When it comes to spiritual healing, I don't deal with the spiritual. I deal with healing. Because healing and spiritual healing is two different things. A spiritual healer is a person who deals mostly with prayers. It has to do with invisible spirits, which thing that most people do not believe in invisible spirits. But, still there are. And they have to be in touch with this type of spirit, with certain rules of prayers. And then, for instance, like when you have a patient that is bothered with some fits. You know what fits, like when they go into a coma or they get unconscious, you know? Or maybe, there occurs a breeze—what we call [a] bad breeze—hit that person, and they affect that person. Now if I am the guy who [is] curing this person, under the spiritual system, I have to got my system in proper order. When I say in proper order—like, you don't eat meat, and you don't try to be upset with the neighborhood, you have to be in good harmony. And you have to do some offerings for cure this person. Like, if you go to the jungle, you carry certain things, you make an offering.

MB: What would you carry?

LR: You carry a wax candle. You harvest the honey in the jungle. And you harvest the wax from the jungle, from the hives. Then they are the white hives. You extract the wax, and you have to make your candles out of that wax. You go in the jungle, you clean a little spot, you make a little table, with round sticks, you know? And you light your candles, and you got the calabash, the *jicara* [*Crescentia cujete* L.], and you carry what we call corn lob.[2] Corn lob—hot one. And you put it there on that table, and you take nine grain [kernels] of corn, and you put those nine grain of corn in a circle, and you set that calabash with the knob there in the center, and you see that the knob is smoking. And that's what the spiritual healer claims to say, that when this knob is—or this porridge is—smoking, that's what the spirits observe, you're feeding the spirit, so that the spirit will do such healing. And, if you need to do a bath to this patient, you have to go to the jungle, and you make your prayers. And then, you've got a little guy by the name of *Duende*.

   Duende is an invisible spirit, but he also appears to you like a human being. He's a short little man, with a long beard up to here [motions to the upper belly range], and

---

2. A type of corn porridge consumed as a hot drink and pronounced "corn lob."

he use a big white hat. The hat is as much or bigger than him. Because he is a small little guy, small, and he only has four fingers. He doesn't have five like we—he has four. His toes—four, too. And when you shake hands with him, never shake hands with him with five fingers, because he wrench the next one. You have to shake hands with four fingers. You hide your finger [motions to the thumb], and then you shake hands with him with four fingers. And those are the guys that tell you what type of leaf or what type of herbs you collect to cure your patient. He is the guy who give you the information for your healing [of] this patient. And then, when you're done with that, maybe he tells you that you have to pay a Mass, or you have to visit so much churches, like if you visit five churches, or seven churches, or nine churches, you have to do a prayer in each church. And that's thanking the spirit. You give thanks to the spirit that you heal this guy, or you help your neighbor. And, one next very important thing, you could know right away, if you have the experience, you could know right away, if this guy is a good spiritual healer, because a *good* spiritual healer *never* give you a price, he never interested in money. When he done with you or finished with one in your family, you say, "How much I have for you?"—he say, "Give me what you like to give me." And if you give him—that person—maybe you so happy that you get cured, or your family gets cured, that maybe you spend so much money with qualified doctors—these specialists—you spend so much money with specialists, and maybe doesn't give you proper satisfaction of your illness or your family's illness, you know. But this guy comes from nowhere and do a spiritual healing to your system or to your family, and you're so happy because this guy make a good miracle to you, you know, which a qualified doctor couldn't do. So when you come to him and say, "Hey, how much I owe you?" he answers "Give me what you want. You don't owe me nothing." Maybe you're so quite satisfied that you want to give heart and soul to this guy for [to] show him appreciation at [for] what he has done. So maybe you give him too much money, he never expect that amount of money, but you give it to him. He can't use all that money, he'll have to donate that to widows or children without parents, or help a patient that is really in time of need, that's where he goes and spend that money. If he do take, he might take a quarter of what you give him. But he no have access for using all that money. That's the way, how you can know if he's a good spiritual healer.

MB: If I were looking for *Duende*, to ask him some information about someone, where would I find him?

LR: *Duende*, you could find him where a lot of cattles are. Good cattle, he like being where good animals are. He doesn't deal with common animals. Like if they have Swiss Cattle, or Red Bull, or good quantity of animals. Mostly you'll find him at night, when [the] moon lights. He drives all those cattle in one place and he look of them—that's his curiosity of doing that. And you can know when the *Duende* always be around because you look at the horses, they have their hair platted [braided]. Have you ever seen horses with their hair platted? Well platted, and they give you trouble if you loose their plat. You're going to have to cut that horse's platting. Whenever you see horses like that, *Duende* always be around. Those are the places where you could find him. But for talk to him, or have information with him, you have to know his prayers, how for bringing [*Duende* to you], and you have to have good, good, good nerves. Because they have a powerful, more powerful spirit than we. Because those are invisible spirits from way back. It claims to say that they were music mans in heaven. When Jesus chased down Lucifer from heaven, he chased Lucifer and he chased that group of *Duendes*, because those were along with Lucifer. Those are the ones that make the disorder, and those are the ones that remains here in earth. We do have them.

I know a guy from Benque Viejo, that when he was a small guy—he is in the States right now—when he was a small guy, the *Duende* used to come around and pick him up and carry him into the jungle and play with him. He carried him one time for three days. He had it [the small guy] with him. And he said that where he was at was so beautiful. He said that he was in a beautiful garden. And he had game, all type of game, food. Any type of food that you would like to have, he bring it to you. And there is a woman in San Ignacio, and she also used to be persecuted by the *Duende*. If you would like to talk to her and get information from her, I could introduce you to her. She lives right in Santa Elena. She's a good friend of mine. She's a married woman, now have children. But when she was a little girl, the *Duende* used to follow. He followed her in a stage that, at that time, you never have water facilities like now—you go to faucet and connect your water. They used to go to the river to do laundry. And when she was in the river, and she never used to walk by herself, and she's around the other females doing laundry, and she's seeing this guy standing there, over there, and the rest of people can't see him, can't see the guy, only she could see him. So one old man give her the advice, "Try to get a puppy. But make sure it's a female, a female puppy." And she raised that puppy. And wherever she go, [she] carry that puppy. And that puppy is the one that used to save her, because anytime *Duende* comes around, the puppy sees the *Duende*. And she sees too, but the puppy bark after the *Duende*, and he get vexed and he goes.

MB: Have you ever seen *Duende*?

LR: I see it one time, but I never talked to him. I see him. He always be in full whites. He scared me. I feel my hair stand up [motions what his hair looks like by pulling his out slightly, horizontally]. I get cold feet. I feel myself heavy. But at that time I never know so much prayers like how I know right now, you know. I was about 18 years old. I see him on a road, coming home from a dance about three o'clock in the morning. I was passing through one pasture, a cattle pasture. And he whistled. He has a high whistling. Just like a cattleman. And I heard this whistle. I heard it on this side first [indicates one side]. The second time I heard it on this side [indicates the other side]. The third time I heard it in front of me. In a couple seconds, I see this guy, coming in [on] a horse, riding his horse. But he doesn't ride where we normally ride. He always ride on the neck of the horses. And where he plat the horse—that's where he put his little foot. He's got short, short little feet, like that. And that's where he put his feet. He holds himself from the horse head. And he traveling, and he's a good jockey. A very good jockey, yes. And those are the spiritual healers.

MB: So, who would you go to if you had a spiritual problem that would go talk to the *Duende* for you?

LR: Well, I would talk to him if I have a problem, for sure. I'm no scared. I'm no scared of him no more.

MB: So do you find that some herbs have toxicity or danger? Can anybody use them? Are there skills you have to have to know what is poisonous and what is safe?

LR: Yes, like, for instance, when you come to that test, like *chicoloro* [*Strychnos panamensis* Seem.] is one of the vines that is very toxic. So, you really have to need this type of herbs for make you to be giving to you [a very strong herb; a person must really need it to use it]. Because if you doesn't need this herb, and you drink it, it could make you feel dizzy, or drunken, or your eyes could go dark, or it could hamper something in your cells, you know. Because it's a strong dosage. Now, when it come to the test if you have a snakebite you could use as much—it wouldn't hamper. Because the snakebite is a poison, and this root, it also contains poison, you know. For instance, like, *sorosi* [*Momordica charantia* L.]—people here in Belize use *sorosi* for many purpose, for many uses. But you have to know how for use *sorosi*. Because if you use too much *sorosi*, then your body, or your cells in your body has a defect. It would give you some kind of bad reaction, somewhere about, you know. Maybe in your kidneys, maybe in your cells, maybe in your bladder, maybe in your stomach. Due to the potential of these herbs, you have a potential—very strong—and that could hamper you. But, when it comes to the other side of the story, like, if you are anemic, or you have malaria, you know, and you drink it—you could drink as much, and it wouldn't hamper yourself. It's preventing you from malaria or other types of diseases, like diabetes, you know. If you have diabetes you could drink *sorosi* tea with jackass bitters [*Neurolaena lobata* (L.) Cass.]—that's very good. But let's just say that you drink this herb for curiosity. Now then, you are trying to do something bad to your body, because then, your body doesn't need it. It no have to drink it. And, there is other medicines, barks and trees, that you could drink it. And, even if your body doesn't need it, you could drink it, and it is always welcome to your system. It won't hamper your system. Like, for instance, I could mention a few of them, like *gumbolimbo* [*Bursera simaruba* (L.) Sarg.]. You could drink *gumbolimbo* and wild yam [*Dioscorea* sp.], and that is very good for your nights, very good for your bladder system. It's good for your eyes, you know, it's very good. And you could drink him anytime. And it wouldn't hamper your system. Ok, we've got another one like a balsam. You've got wild balsam [*Myroxylon balsamum* (L.) Harms]. You got the tame one—you could drink that like tea, instead of coffee. And that's very good for your system. That prevents you from having ulcers. It prevent you from having gastritis. It prevent you from having upset stomach, like when you eat food and the food goes rancid in your stomach. And when you belch it, you belch it fully. But if you drink this balsam often, you wouldn't get that problem. It prevents you from that type of problem.

Then you've got another one like, *epazote* [*Chenopodium ambrosioides* L.]. You could drink *epazote* every morning. You could eat it in food. It's an eatable fruit. It's an eatable herb. You could cook it with beans. You could cook it with egg soup. You could cook it with fish soup, and so. And it wouldn't do you nothing, you know. Like the cow foot leaf [*Piper auritum* Kunth]—you could combine the cow foot leaf, and it's very good. You could drink it instead of coffee. You could make tea. It's a drinkable tea, you know. It's from the cow foot leaf. It keeps your organ running good and it prevent you from rheumatism. It prevent you from having asthma, and things like that. And it's no harmful plant. The other ones that I would say are harmful in one way that if you drink too much of the other one that is toxic, you know, those are the ones who hamper your system inside. But those ones like balsam and the ones that I mention before, they are ok. Billy Webb [*Acosmium panamense* (Benth.) Yakovlev]—I use Billy Webb when I [am] working in the jungle, working with the machete, with an axe, doing hard work, and I have to sweat a lot. I cut the Billy Webb and I put it in my water bottle. And I drink it all day. Just that you don't make it that strong. You make it, but make it weak. And that keeps running your system. Prevent you from many diseases. Those are things that you could drink. But, like, *sorosi*, *chicoloro*, and so, I would not advise you, say to drink it like the other ordinary plants. Because it hampers you, yes.

In 1992, the early part of the '92, there was a patient in the local hospital, by the name of Therese [family name omitted]. And she was blind. She get blind—she wasn't blind—but she get blind. And she was diabetic. She lost her vision by diabetes. She used to be alcoholic, and through that medium she get blind and get the diabetes. She was in the hospital for many days, weeks. And the hospital dismiss her, you know, give up. Can't do no more for her in the hospital. So, when the hospital release her, the alarm already went out that she had AIDS. Now, when she send call to her mother that the hospital can't admit her no more, she tell her mother, her mother didn't came. She sent for her dad, her dad didn't came. And her two sisters, none of them came. And she had three brothers—none of the brothers came for her. And so she sleep two nights at the hospital veranda. And when people pass, and people who know her, and say, "Hey, Therese." And she say, "Please, give me a piece of bread. I'm hungry. I'm here from yesterday. Nobody want take care of me. I'm blind. I can't help myself." People saw her and bring her a piece of bread or buy something for her to eat, you know. So, she knew me, before she get blind. She knew me, and she was a friend, you know. So she sends a person for call me. But before the message came to me, I already know that she had AIDS. That's what people said, that she had AIDS. And when the message come to me, I said, my goodness, but why her parents don't take care of her? She has her mother, brothers, sisters, and father, you know? It was a surprise to me. So, anyhow, it's a message, and I took the message, and I went. I was just arrived with two bottles of balsam for Rosita [Arvigo]. And I remember quite good that. And as I reached to my home with the two bottles of balsam, there was a guy waiting for me. So I just put it down, I drink some water, and I went to the hospital with my working clothes, straight up, everything.

So when I got there she was on a long bench, laying down. I said, "Therese?" She say, "Yes?" "Do you recognize my voice?" She say, "Polo, is you?" I say, "Yes, Therese." She said, "Lord help me." I said, "What's wrong?" She say, "I'm here for two days. Nobody want take me out from here. The hospital don't want me in there." She say, "And I don't know where for go. I can't see. I'm blind." She say, "I can't help myself. I'm helpless. And nobody want to be responsible for me." She say, "You wouldn't mind and take me home and do something for me? I know you will cure me." It shocked my heart, you know? It shocked my heart. And I couldn't afford to just turn my back to her. So I took her from [by] her hand and she quickly stand up. I say, "Therese, sit down." She say, "I want to go. I don't want to stay here." I said, "Before I carry you from here I have to talk to the hospital—the doctors. Because I don't, you know, want no problem with the doctors." So I went into the hospital, and I talked to the doctors. They said, "Well, we sent for her people, the whole family. She is too serious, and nobody want to be responsible. But if you want to be responsible, no problem, we give you a paper right now." I said, "Give me the paper." So he take a paper and he filled out a form. And he sign it, and he make her sign, and the nurse in charge of duty, the head nurse—the head nurse who take care of the whole squad—she signed too.

And I took her by her hand, and a little bag of her clothes, her dirty clothes, because she can't wash nothing. And I took her home. Everybody on the street just watching me. "My God. This man carrying that girl with AIDS, she was pale. That girl got AIDS." People were just whispering to one another. "See, Terry got AIDS, and this man just bring her home." I bring her—at that time I used to rent a little small house. So I put her in there. I fix something up to eat. And I say, "Therese, open your mouth." When she open her mouth—I see it's all eaten out—white, white, white. I say, "Oh, oh, it's bad in there." I say, "Therese, let me open your eye." And I opened her eye. You see the yellow of an egg? The yellow eggs—and I saw her eyes look inside like the yellow egg. And I say, my goodness, what could I do? Anyhow, I fix something. She says she is hungry. So I fix something for she to eat. I buy a cabbage. I buy those small things that give a purple water, beets. I buy one of them. Cabbage, beets, carrots, and tomato. I chop it up good, and I give her to eat. And she eat everything. She was hungry. I say, "Therese, right now I no got nothing for give you, but anyhow, I am coming back right now." I took my flashlight and I went behind those pastures and I find the *sorosi*, I find jackass bitters, in the night. And I boil it, and I give her to drink. The next day morning, I get up early in the morning. I make breakfast. I eat. And I give her to eat. Even take her and I bathe her, because she can't do nothing for herself. So I make a bath. So I bathe her and everything, and I pulled her out, and I left her on that bed. I took all her clothes, and I went to the river and I wash all her clothes. She only had three suit of clothes, and I left her naked in the room. And I wash her clothes and I hang it. When I done with that, I took my machete, and my bag, and I went to the jungle. I find Billy Webb, I find *chicoloro* and I find this same one, *zorillo* [*Chiococca alba* (L.) Hitchc.]—like this right there—I find the root and I dig it up. And I bring it, I clean it, I wash it, and I cook one gallon—I make one gallon of liquid out of that. And I tell her, "Therese, you want to get better?" And she say, "Sure. Nobody wants me. Look, see what stage I am in? I'm just ready for the vultures." I say, "If you want get better, drink this every time you feel thirsty. It is possible. You have to consume this gallon of water for the day." She said, "Sure," and she took it. "Man, it's bitter." "It's good for you. Drink it." And she starts to feel—make a funny face, but she drink it. And she drink it. That the first day. The second day, I took her to there [visited her] in the evening. Tonight I'll get the *sorosi*. And the next day, I'll cook the medicine. That was the second day. The third day, I went to the bush. So, I had a neighbor, and I tell the neighbor, "Listen, I going to the bush. I have this girl. See this my key, here? Well, Therese is in my room. Please give her some food, I pay you. I'll buy you what to cook for her. No rice. Nothing with corns and rice, no white flour, no white bread or anything like that. No beef, no pork, no chicken. Vegetables. I buy *calaloo* [*Amaranthus dubius* Mart. ex Thell], I buy spinach, and things like that to give her for eat. For a gallon of the water; Billy Webb, *chicoloro*, and *zorillo*. I buy it, and I left it there. And I tell the lady, "Cook this for her." And then I went. I didn't think. I never gone to stay there overnight. So, I went to the bush, and when I reached to the bush, my mind give me, for meditate in the jungle. And so I stayed overnight and I meditate the whole night. Around two o'clock—between two and two-thirty—I thought, something coming to my mind, and it tell me, "You know what good for that girl? What could bring that girl's vision back? *Jute*. You know what is *jute*? They call it snails. It something with shells, like this long [indicates about 1.5 with his fingers]. He grows in sweet waters, creek. It's round. It's in a shell. He long, like this. And it's in a shell. You harvest that, and you collect it from the water, and you cut the point, the real point [of the shell], and put it for cook. You make like one soup. So you cook it. And you cook it with *juanacka*—it's a cow foot leaf. You put the cow foot leaf in there. And you give that for eat. And you take the raw snail—take the shell and you break it—and take the snail, and you mash it, mash it up properly good. And you soak it in water. And left it in the dew for overnight, and the next day morning, you wash her eyes with it. That's what comes into my mind. And it stay in me. And I do that. And I went. The next day morning I move from there and I went straight to the creek, because I knew where the creek is where I could get that. And I crossed three miles around the jungle. Up, down, up, down hills, and I reach to that creek. And I collect the *jutes*. I went back, and she was afraid. She said, "When are you get home?" She say, "Ah, you come already?" I say, "Yes." "I thought something happen to you and I was wondering and you didn't tell me and that you would stay too long and you gone from yesterday and today you come." And I never eat nothing from that for the rest of the night. So I came hungry, you know? But I do that medicine before [I] eat anything. And I give her that soup to drink, and as quick as she finish that plate, she went to sleep. It's a hard, hard, heavy vitamin. Just dropped right there [makes snoring noises], sleep. Then I mash the rest. And I put it under the water. And every day, she washed her eyes. And here is the people in San Ignacio who could tell you. In two weeks' time, the girl has her vision back. And she comes back to life, and she has good eye vision. And she get fat and strong. Everything was ok. And when she get better, she cured, she used to wash and cook for me. And anybody who come from San Ignacio, you ask for Terry and they will tell you that she was a blind woman and she came back to vision. And everybody say, "That's the man who bring that girl back." And so that's part of what I call concentrate the mind in the jungle and meditation. That's part of my meditation. That's what I did. Because I never knew about [how] this snail could have been good, and so, but the meditation gave me that point of view. And I prove it.

BEATRICE WAIGHT

MB: Was your mother or father a traditional healer?

BW: My father, he was born in it, you know? But in the olden time they wouldn't say nothing much, you know? What he knew he knew just for his family. And from there, well, I learned a little from him. And my grandmother. My grandmother, she too, she knew a little bit. So, I kind of pick [what I know] from the both of them. My father's specialty was fixing bones, like when it is ruptured, and dislocated, and these [bad] pains that you suffer in your back. And he would give you this thing called *ventosa* [cupping] and baths. That was his specialty. There's a lot of people, they knew that at any time of the hour of the day or the night, if something went wrong with their bones, they rush to him right away.

MB: Did he use a lot of plants to set the bones?

BW: No. Just massaging. Massaging with a little Vicks or oil. And that will do it. He had a special plant that was called *suelda con suelda* [*Anredera vesicaria* (Lam.) C.F. Gaertn.]. If it was ruptured, he would put like a plaster on it like this. And by the end of maybe three months or so, it is already healed. He used *suelda con suelda* because they say, when you have a ruptured bone or a bone that is dislocated, this *suelda con suelda* would do something like draw everything together, and make it heal quicker. That's the reason that he used *suelda con suelda*. I learned a lot from my father. Learned most of the massaging that he does, and the *ventosa*, and the *suelda con suelda*—when to put it, when to not, and how often and things like that. The baths—especially the baths. He would say, if you have a bad wind sometimes that gives you cramps in your back or you have too much headache, or fever—he always would recommend a bath.

MB: What was your grandmother's specialty?

BW: She was a midwife. I learned from her about curing children and the evil eye. When the mole [fontanel] has sunken, hold the baby; and why the baby bothers so much at night, what to do with the baby and things like that. She comes from the Yucatan also. She moved down here along with my mother. For some reason, I don't know—I think it was that the land was bad for some companies and that some companies did not want them there anymore, or something like that. So they had to leave the land and come and live here.

MB: And so, is your father still alive?

BW: No. He passed away six years ago. He was 83 when he died.

MB: And did he teach other people about bone setting?

BW: No. Only one of my brothers knows a little bit about it. Just a small amount. Like, he didn't encourage us then, like now, and well—if I learned, it was because I was always curious, inquisitive, wanting to know this and that, but it is not really that he wanted to teach anybody. I don't know if it was that he didn't have the time, or maybe that he was tired, you know. But, I couldn't say that. But it was somehow that he didn't want anyone like him, saying that I am going to teach this one, and know.

MB: Do you think that it is important that the parents teach the children?

BW: I think that it is very important. Like in my case, I will educate my children in whatever I know about medicines, about everything. Because I see that this is very important. Because when I pass away, somebody could take over and just continue, you know, the good work. That's how I see it.

MB: How many of your children are interested in this?

BW: Well, right now I could say that I have two of them that are very interested—Bertha and Nery. Bertha is 15, and Nery, he is 18. And even at her young age, Bertha would question me about a lot of things that sometimes, I sit down and think that I should not tell her, not at her young age. But sometimes I tell her. Because she will say to me, "Suppose that you die tonight. And you have not told me something that I have wanted to know?" And I tell her, "You are correct." It's just like, I say that she's not old enough to know certain things, but then she will catch me in that weakness and just like that I would tell it to her.

I think that she [Bertha] is very sensible. She is very caring for other people. If she would hear that somebody was sick, she will come and tell to me, "You don't know what is good for this kind of sickness that my friend has?" I tell her, "Yes I know." She say, "You wouldn't tell me so I could go and tell my friend?" And I tell her, "I don't have tell you. I think that your friend should come and ask me." But, she say, "If I am her friend, shouldn't I try to help her?" And so I tell her, "Yes." And I try to tell her some points. So just by that I tell her.

MB: Do you ever dream about a problem?

BW: I think I was born with something special. But, I don't know. Because my mother said that when I was born I had a little veil [caul] from my head to my umbilicus, and a tooth. The veil was cut open when I was born, and the tooth disappeared after three days. And she thought that I was somebody that would die at a very early age. So, she wouldn't tell nobody, not even my father. So when I was married for about two or three years, she came one day and she told me, "I think I have to tell you a secret, that you [were] born with something special apart from all the rest of my children." So I was frightened and I said, "But why is it that you haven't told me?" And she mentioned it to me, "I think you have something special that nobody else has, none of the other children." And what I notice is that when something bad or something good is going to happen to me, I have a revelation of it. But then, when I notice when I pray, I would do continuous praying, it is more like I am very sure of this dream I had tonight, because of that. But when I am not in that forces, I would not say it. You know? And when I have a sick somebody maybe that is going to come to me and somehow this type of plant is revealed to me. You know, that I'll use. Definitely from the night I dream someone will come to me, the next morning, and for sure, that person will come. And so, I've never been surprised. I know ahead of time anything that will happen. Every time I have a medicine prepared, you know, like, the most difficult things that I know will come, I have them maybe two or three days prepared [in advance], you know, so at anytime when someone would come, well, I have something prepared. So, that will never take me, like, by surprise.

The most I deal with is babies. Some people will come for baths and spiritual healing. That I will do the most. But the massage and things, not that much. Because the one thing that I don't like, like some men—or even some women—don't like other women to deal with their man. And so I avoid that. I don't want to be looked upon like, she just want to massage my husband, like that. So mostly, I don't deal with men, rather than my husband. I don't treat men that much. Rather I deal with babies and women.

MB: What do you think about the future of traditional healing? Will it grow or die off?

BW: Well, I don't think in my case it will die. Because in my case, I am trying to tell my children and they too are seeing that traditional healing is a very good way of healing. Because all of my children would say, that money that we have to go and spend with that doctor—we won't say that that doctor is not good, you know, because like, in surgery and thing, we have to use that doctor—but for little common problems like cough and headache or diarrhea in children, or—which, in my case, I don't see that a problem [for me to heal]. But unless someone doesn't know that, they need them [doctors]. To me, I think it's very good to train your children so these children could train others, you know. And in the future that [traditional healing] will not die. But then, if you don't train your children, there's a possibility it could [die].

MB: And does it worry you that more and more forests are being cut down everyday, and the medicines are disappearing?

BW: It worries me a lot . . . In finding medicine in the next few years, it'll be hard for us to get a place we need to collect the medicine we need. Because, especially like the ones that cut down *milpas*. If they see a tree that they know, because most of the people that lives here knows about medicine—they won't take that into consideration, they will just chop down everything, you know, and burn it, too, and that's it. I think that there should be a law, you know, that would say that the endangered [medicinal plant] species should be left or should be preserved or something like that, you know. That would help us. But, if nothing is not been done, you know, then, we are at risk.

MB: Do you have anything that you would like say about your life? About your work? About the future?

BW: Well, I am very pleased about my work, and when I help others I feel [it is] more like helping myself. You know, 'cause I always remember the first commandment, the golden rule, love your neighbor as yourself. And if your neighbor is sick and he or she asks you for help, I think you should do it. If you know it, you should do it. If you don't know, well, just you should seek help for her. And try to do something for yourself. And for my future, I think that, probably as time go by I would like to train anybody that, you know, would want to learn about what I know. I would be willing to do it. You know, because I see a need for it. If I teach others what I know, I feel a blessing that God will put on me, for that reason.

MB: How often do you work with your children in teaching them about traditional healing?

BW: When I am mixing medicines or collecting. Sometimes they go out by themselves and collect for me. They know a lot about the medicines that I use. So when I am boiling medicines or for a bath or things like that, they will help me with the skillet or, you know, something like that.

MB: Bertha, what are your impressions of your mother's work?

B³: My mother knows a lot of herbs, because whenever my sisters or brothers are sick, she just tells me to go and—or my brother to go—and find the leaves, and just say the name. And I ask her to describe it to me and then I learn it, after a while I learn it. I go and find the leaves and herbs by myself. I'm very interested in what my mother knows, because when I grow up I want to be like her and know a lot of medicinal plants. Sometimes I ask her to tell me some secrets she has for medicinal plants, but she tells me I'm too young. So I tell her that I will wait until I am big enough so she can tell me those secrets that she can't tell me now.

JUANA XIX

MB: The last time we spoke, you explained to me that all of your 18 children were brought up—all of these children were brought up on herbs.

JX: Yes. For all of my children that I had. I never went to the hospital. I had a midwife that was my mother-in-law. Only with herbs.

MB: Yes. And when they got sick, what happened?

JX: This—she taught me some leaves. And with these leaves, I [soaked] them. And they were heated. I soaked them in tree oil, cooking oil, almond, and a little bit of Vicks, and I soaked them with the *eremuil* [*Malmea depressa* (Baill.) R.E.Fr. subsp. *depressa*] leaf. With this, my children got better. I did not have to go to the clinic.

MB: Where did you learn?

JX: My mother-in-law. It was my mother-in-law—she was number one in these herbs. She was like a *comadrona*—she was a caring midwife. She took care of the women when they were birthing their children. I learned everything that I know about herbs from her. When she massaged, I learned how to massage and turn around the fetuses when they were backwards. Yes, and after, when the women had their baby, she gave them the avocado [*Persea americana* Mill.] leaf, with the oregano [*Lippia graveolens* Kunth] leaf, and a pinch of salt, for three mornings after delivery—this is in order to clean the womb.

MB: And, where do the herbs come from that you use?

JX: Right here [pointing to her backyard and nearby environs].

This [holding up a leaf] is the avocado leaf. You take this, three leaves of this with the oregano leaf. And you put it in saltwater and bring it to a boil. And in about five minutes, after it has cooled down to warm, and then the woman drinks it. This is very good. Yes, the other day when the woman delivered her baby, this is what she drank.

---

3. **B** refers to Bertha Waight, daughter of Beatrice Waight, who was present for the conversation and was featured in an educational video made by the project entitled *Diary of a Belizean Girl: Learning Herbal Wisdom from Our Elders*.

MB: Would you say that the majority of the species you use come from your garden?

JX: Yes. This was how they used to cure it. Our ancestors never used any of the doctor's medicines.

MB: Are you teaching anyone in your family how to use these herbs?

JX: I have a daughter that is learning the herbs that I use. And my other son also wants to learn. He tells me that he wants me to teach him my herbs. I tell that this is good.

MB: Then, we are not going to see the loss of your knowledge?

JX: It is not going to get lost.

LR[4]**:** In your manner of thinking, of these traditional medicines—what is it that you think about the healing of the past and the healing of this generation, with the boys and girls that are growing up today? What do you think about them? Is it necessary that they continue learning? Or that the traditional medicine dies?

JX: It is better with the herbs—that they learn about the herbs. Because, this, the herbs, they heal a person more rapidly. And when they heal, it is not like the pills. With the pills, when something hurts, you take one, and it calms you down. But in a short while, you have another pain. It's not like that with the herbs. The herbs heal you. If you live in a far away place where it takes you hours to get anywhere, it is better to search diligently for the herbs than to leave and go out away from that place.

MB: And the disappearance of the forest here with the healing plants, are you worried about this?

LR: What she is referring to [is] that, for example, they are cutting down the mountain, they are cutting down the forest, are they not? And these wonderful plants that grow in the forest are being destroyed?

JX: Yes.

LR: Does this worry you?

JX: Yes. Because the medicine is gone. One cannot get ahold of it anymore. And perhaps, in order to get it, you will have to walk very far.

LR: And before, they used to get ahold of it really close?

JX: Yes.

LR: And now?

JX: And now the medicines are ending. Now there are few of them.

THE UNBROKEN CHAIN

As is clear from these conversations, the training of a new traditional healer or bush-master often begins at an early age, following a traumatic experience or one that points the person in this direction. This is not an unusual story; it is heard quite often among herbalists as well as those in the healing arts. In ancient times, the training was formalized, based on texts and schools of learning. Following the Conquest and the collapse of the Maya civilization, training became based on oral tradition, with information

---

4. **LR** refers to Leopoldo Romero, who was present at this interview.

passed from master to student over a long period of apprenticeship. In contemporary times, there are many career choices for young people and little incentive to earn their income on the products of an ever-declining rainforest resource base. As a result, as we have seen time and time again, in Belize and elsewhere in the world, elderly traditional healers often have no apprentices to carry on their work following their retirement or death. Each year, the body of information about the ancient use of plants for healing and as foods, construction materials, and traditional culture becomes smaller, as the knowledgeable people pass on. In Belize, the traditional healers have recognized this loss, and most consider it to be a tragedy for their culture. Part of the rebirth of interest in traditional ways has included renewed interest in ethnomedicine and its potential role in the health care delivery system. Another force propelling interest in traditional knowledge is the wave of ecotourists visiting Belize each year, now totalling hundreds of thousands, and the need for local guides expert in natural history, including ethnobotany and ethnomedicine. Many of these guides are young people, and their sources of information are their relatives as well as the published literature. We are frequently acknowledged by the guides we meet and told that the ethnobotany project—and its publications, videos, and other educational materials—is an essential resource for them in learning about the traditional uses of plants in their country. Finally, the Bush Medicine Camp, run by Rosita Arvigo and Dr. Patricia Gildea each year, pairs its young participants with knowledgeable elders who teach them how to use local plants for medicine, food, and shelter and to appreciate nature. As was recognized long ago, the youth of Belize hold the fate of the wilderness in their hands.

4

PLANTS OF BELIZE & THEIR USES IN ANCIENT & CONTEMPORARY TIMES

An Ethnobotanical Compendium

THIS CHAPTER IS A compilation of the useful plants of Belize, citing local uses and local names when known or hypothesized along with specimen data and sources of information where available. Chapter 4 is organized into two sections: Non-Flowering Plants and Flowering Plants. Within those two sections, there are four subsections: Ferns and Fern Allies, Gymnosperms, Monocotyledons, and Dicotyledons. Within each of those subsections, plant families are arranged alphabetically, with a brief description of that family presented in bold.

The descriptions were for the most part derived from *Flowering Plants of the Neotropics* (Smith et al., 2004), with information on some of the cultivated species based on descriptions in *Hortus Third* (Bailey and Bailey, 1976), guidance from Dr. Robbin Moran (ferns), as well as our own observations. Family descriptions are followed by commentary on the number of genera and species within that family known from Belize and the number of taxa having uses that are listed in this book. The species identification and plant families follow the structure published in Balick, Nee, and Atha (2000) with some exceptions. The Angiosperm Phylogeny website (http://www.mobot.org/mobot/research/apweb/welcome.html) contains significant changes in the arrangement of some of the plant families found in Belize as compared to our earlier work. These are as follows (family names in bold are those accepted by the Angiosperm Phylogeny website as of April 20, 2010): Agavaceae within **Asparagaceae**, Asclepiadaceae within **Apocynaceae**, Asphodelaceae within **Xanthorrhoeaceae**, Bombacaceae within **Malvaceae**, Cecropiaceae within **Urticaceae**, Chenopodiaceae within **Amaranthaceae**, Dracaenaceae within **Asparagaceae**, Flacourtiaceae within **Salicaceae**, Hippocrateaceae within **Celastraceae**, Hydrophyllaceae within **Boraginaceae**, Myrsinaceae within **Primulaceae**, Punicaceae within **Lythraceae**, Sterculiaceae within **Malvaceae**, Theophrastaceae within **Primulaceae**, Tiliaceae within **Malvaceae**, Turneraceae within **Passifloraceae**, and Viscaceae within **Santalaceae**. In this volume, we have used the new accepted family names based on the findings of the Angiosperm Phylogeny Group (III) and presented the older family names in brackets as appropriate for these specific families. In these cases, the family descriptions have been merged to some degree to reflect the wider taxonomic circumscription. The reader should note that the region's existing historical, ethnobotanical, chemical, pharmacological, medical, and floristic literature primarily is organized by the older family names; however, this will no doubt change in future publications. In the case of botanical nomenclature at the species level, we consulted The Flora Mesoamericana website (http://www.mobot.org/mobot/fm/intro.html), Tropicos (http://www.tropicos.org), and the International Plant Names Index (http://www.ipni.org) to obtain the most current names and authors. In some cases, we had to make a taxonomic judgement based on a review of these sources when they were not in agreement. Again, virtually all of the historical, ethnobotanical, chemical, pharmacological, medical, and to some extent, published floristic literature must be accessed by using the older names, which, as in the case of family names, will change with time. The most recent contemporary divisions within the Angiosperms are as follows: Basal Angiosperms, Monocots, and Eudicots. However, in this chapter, we are maintaining the more traditional concepts of Moncotyledons and Dicotyledons used in our previously cited floristic checklist.

Our goal has been to include at least one plant image per genus. These include photographs by the individuals listed in the photography credits at the beginning of this book, identified by their initials in the photo captions. There are also illustrations from the collections of the LuEsther T. Mertz Library of The New York Botanical Garden, indicated by [NYBG] in the caption. A series of original illustrations was kindly prepared for this volume by the noted botanical artist Francesca Anderson; these are credited as [FA] in the figure captions. We were unable to locate images for a small fraction of the 551 genera discussed in this chapter and would welcome contributions for a future edition.

The common names that follow the species names were recorded by us during the fieldwork or otherwise taken from published literature. Following the names is the language of their origin, when known. We transcribed the spelling of the names as we understood them or as they were spelled to us by local guides. Thus, the names that are recorded in this chapter are limited by our level of understanding of the specific language. Languages in which names were recorded include Creole, English, Garifuna, Spanish, Mopan, Q'eqchi', Yucateca, and Mayan. Mopan, Q'eqchi', and Yucateca are the 3 distinct languages spoken by the indigenous Maya living in Belize; the category of Mayan was used when the specific Mayan language was not known to us. When there was no indication of the origin of a language, we included the name with the qualifying statement "Additional name(s) recorded." We expect that additional linguistic data and clarifications will be gathered by others and trust that what we present here can be updated in the future. We offer our most sincere thanks to the linguists who worked with us, in particular Don Victor Cal, who spent a great deal of time reviewing the common names; the full list of linguistic specialists who helped in this project is provided at the beginning of the book. However, we, the authors, accept full responsibility for any linguistic errors that appear in this publication.

Specifics of plant uses are emphasized in bold (e.g., **food**, **cough**, **cordage**, **skin ailments**). Ethnobotanical information is derived from three sources: 1) first hand field observations vouchered by herbarium specimens are cited following the mention of the use [e.g., B2656, referring to Michael Balick et al. collection number 2656]—the list of collector's initials is presented below; 2) personal commentary by a knowledgeable individual we either interviewed or who wanted to contribute additional uses to those listed in the preliminary manuscript that was circulated to our local experts [e.g., HR, indicating that this information was provided by Hortense Robinson]—the list of local experts and their initials is also presented below; and 3) from the literature, in which case a reference citation follows the specific use [e.g., Flores and Ricalde, 1996]. In the case of species used medicinally, details on preparation of the remedy and the amount of plant part used, along with dosage, are provided when known to record this information for the future. Conditions that are specific to a culture or region—for example, *susto*—are described in great detail in Arvigo and Balick (1998), and that reference should be consulted if further information on these syndromes is desired. However, as pointed out in the disclaimer, this book is not to be viewed as a pharmacopoeia or guide for treatment of medical conditions using local plants. Numerous species, although used traditionally as therapeutics, contain compounds that are potentially toxic or otherwise injurious to humans and domestic animals in small doses or during long-term use. In many cases, we do not report dosage, as this information was unknown by the person discussing the use of that particular plant. Some reports of mixtures are incomplete (e.g., only a few of the plant components are known, again, based on the knowledge of the one to several people interviewed), while others may be incorrect.

Collectors whose herbarium specimens are cited in the text include Rosita Arvigo, indicated as [A], as well as Michael Balick [B], Julie Chinnock [C], Daniel Atha [DA], Noah Goldstein [G], John Brown [JB], Peter E. Meleady [M], Michael Nee [N], Mary Palmer [P], Andrew Reed [R], and Wilfred Warrior [W]. As already mentioned, the collection number follows the uppercase letter (e.g., [B2656]).

In addition to the ethnobotanical collections and interviews, we discussed some of the information that was recorded over the years with a small group of individuals for clarification, and for some of the plants, these individuals contributed additional uses. The local experts who provided information in this way are acknowledged in the chapter and include Thomas Green [TG], Andrew Ramcharan [AR], Hortense Robinson [HR], Leopoldo Romero [LR], and Beatrice Waight [BW]. Additional commentary on the traditional or contemporary use of plants in Belize and elsewhere is from Rosita Arvigo [RA] and Michael Balick [MB], as well as from the literature, in which case a conventional citation is given.

Due to the nature and limitations of our study, no medical evaluations of patients or diagnoses of their conditions were made during our fieldwork. Thus, in many cases, we are unable to confirm the recommendations for use with specific diagnoses. One particular term that resulted in confusion during the ethnomedical interviews was "cancer." Several healers mentioned "cancer" in their discussions of medicinal plant properties. In Belize, the term "cancer" is often used to describe external skin ulcers or open sores as well as internal growths or tumors. When external or internal "cancer" was not specified, we concluded that the remedies taken internally were more likely to be used for internal growths and that the remedies applied externally were more likely used for skin ulcers or oozing sores.

We have compiled this chapter for the Belizean people interested in learning about and teaching the traditional uses of plants. While recording such information is important, and we have attempted to accomplish this in some measure through this book, it is only by *keeping cultural traditions in practice* that the wisdom of the Belizean elders who worked on this project, many of whom have now passed away, can be saved for the benefit of future generations. Based on our research in Belize as well as in other places around the world, it is clear that traditional knowledge can help guide those who choose to use it on a path toward a more sustainable, more healthful, and more meaningful lifestyle. This was the spirit in which this information was provided to us by the elders who expressed great concern that the plant-based traditions of Belize—their roots—would be discarded and forgotten. Those who worked on this book know that it is up to future generations to ensure that this cultural legacy is secure.

NON-FLOWERING PLANTS

CYATHEACEAE

Plants terrestrial. Stems arborescent, rarely short-erect, scaly. Leaves monomorphic, 3.25–13 feet long. Petioles exuding copious mucilage when cut. Blades 1- to 4-pinnate, but most commonly 2-pinnate-pinnatifid. Sori abaxial, round, usually borne at the fork of a vein. Indusium present or absent. In Belize, consisting of 3 genera and 9 species. Uses for 1 genus and 1 species are reported here.

Cyathea myosuroides (Liebm.) Domin

To **promote sweat** to treat **fever**, 1 handful of leaves and stems is boiled in 1 gallon of water for 10 minutes and used hot as a bath once daily, at bedtime or naptime, until better [B2601; HR]. If a bath is not desirable, the leaves are placed on the bed, and the person rests upon them [B2601].

DENNSTAEDTIACEAE

Plants terrestrial. Stems long-creeping, hairy. Fronds 3–20 feet long, monomorphic or slightly dimorphic. Laminae often large and decompound, sometimes with indeterminate growth, the apex resting while the pair of pinnae below it develop. Sori round and in marginal cups or elongate to linear and covered by false indusia. In Belize, consisting of 6 genera and 12 species. Uses for 1 genus and 1 species are reported here.

Pteridium caudatum (L.) Maxon
helecho [Spanish]

To ease **cold sweats**, usually occurring with asthma, exhaustion, excessive coughing, or with anemic babies, 9 leaves along with 9 lemon tree leaves (*Citrus limon*) are steeped in 1 gallon of water and used tepid as a bath [B2142; BW].

Pteridium caudatum [RM].

GLEICHENIACEAE

Plants terrestrial. Rhizomes long-creeping, frequently dichotomously branched, bristly or scaly. Fronds 3–20 feet long, monomorphic, indeterminate, often resting on the surrounding vegetation. Pinnae opposite, typically forked repeatedly with a resting bud between the forks, the ultimate divisions pectinate or pinnatifid. Sori round, abaxial, non-indusiate, with relatively few (2–10) sporangia. Spores yellow. In Belize, consisting of 2 genera and 3 species. Uses for 1 genus and 1 species are reported here.

Dicranopteris pectinata [RM].

Dicranopteris pectinata (Willd.) Underw.
  helecho [Spanish]
  For **fever sores** and other **skin ailments**, 1 handful of leaves is boiled in 1 gallon of water for 10 minutes and used as a bath [B2610]. To treat **leishmaniasis (baysore)**, the leaves are toasted, crushed into a powder, and applied directly to the affected area once daily after bathing with the same leaf decoction as described above for fever sores [B2610].

LYCOPODIACEAE

Plants terrestrial, epiphytic, or rarely rupestral, habit erect or pendulous. Stems dichotomously branched, occasionally with lateral branching. Leaves 0.2–0.8 inch long, simple, entire or rarely denticulate, one-veined, homophyllous (leaves all alike) or anisophyllous (with reduced leaves, usually spore-bearing, in the terminal divisions). Sporangia solitary in the leaf axils or on the upper side of the leaf, reniform or nearly globose. In Belize, consisting of 2 genera and 7 species. Uses for 1 genus and 1 species are reported here.

Lycopodiella cernua (L.) Pic. Serm.
  fern [English]
  To **improve blood circulation**, 1 entire plant is boiled in 1 gallon of water for 10 minutes and used warm as a bath for 30 minutes twice daily, at noon and bedtime, for 3 days [B2609]. Alternatively, the leaves and a pinch of salt are mashed and applied as a poultice over the affected area [B2609].

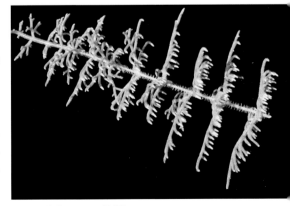

Lycopodiella cernua [MB].

LYGODIACEAE

Plants terrestrial or epipetric. Rhizomes short-creeping, hairy, the hairs multicellular. Fronds up to 10 feet long, twining. Petioles containing a C-shaped vascular bundle. Pinnae short-stalked, once-forked with an inconspicuous resting bud in the fork. Sori born in marginal finger-like lobes about 0.25 inch long. Sporangia solitary on a vein tip, each covered by a flap of green tissue. In Belize, consisting of 4 genera and 13 species. Uses for 1 genus and 3 species are reported here.

Lygodium heterodoxum Kunze

To treat a **fever** caused by evil magic (*obeah*), 9 leaves are mashed, steeped in 1 gallon of cold water, and used cold as a face wash for 9 Fridays [B2481].

The stems of the leaves are used as **cordage** to weave baskets [B2481].

Lygodium venustum Sw.

wire wis [English]; *alambre, bejuco de alambre, corremiento* [Spanish]; *xix el bá* [Mopan]

To relieve **headache**, 9 leaves are mashed in water and applied as a poultice to the head once or twice daily as needed [B1780; Arvigo and Balick, 1998].

To treat **foot or skin fungus**, 2 cups of leaves are boiled in 1 quart of water for 10 minutes and used as hot as can be tolerated as a bath over the affected area [A935, B1780, B2446; Arvigo and Balick, 1998]. This anti-fungal decoction is never to be used internally [A935].

To treat **sores**, **rash**, or other **skin ailments**, fresh plant juice is applied directly to the affected area [Arvigo and Balick, 1998]. To relieve **itchy skin**, 2 cups of leaves are boiled in 1 quart of water and rubbed directly on the skin 2–3 times daily as needed [B2110]. Alternatively, 1 entire plant along with the leaves and young stems of polly red head (*Hamelia patens* var. *patens*), the leaves and young stems of pomegranate

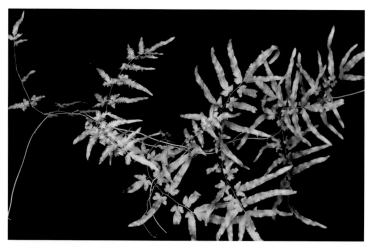

Lygodium venustum [MB].

(*Punica granatum*), and the leaves and young stems of arnica (*Montanoa speciosa*), are mashed in water and used as a wash over the affected area [B2190]. To treat **rash**, refer to stinkin' bush (*Cornutia pyramidata*) [B2104].

To relieve the pain of **wind (*viento*)**, either 9 small vines, 6 medium vines, or 3 large vines are boiled in 2 gallons of water with a pinch of salt for 10 minutes and used as a bath once daily, at bedtime, until better [B2292].

To **cleanse the urinary tract** when the urine turns a dark, reddish color and smells strong and bad, 1 entire plant is boiled in 2 quarts of water and sipped all day long until finished. This treatment is repeated the following day, if necessary [B3688].

As a **calmative**, for when a person experiences fear during a thunder and lightning storm, 1 entire plant is burned and the smoke inhaled [B3541]. To calm babies suffering from **fright (*susto*)**, 2 quarts of the vine and leaves are boiled in 1 gallon of water for 20 minutes and used warm as a bath once daily, at bedtime, for 3 days. This treatment is repeated as needed [A935].

This plant was **sacred** to the ancient Maya and used as an altar decoration during **ceremonies** [Arvigo and Balick, 1998].

Lygodium volubile Sw.

fever fern [English]

To treat **rash**, 1 large handful of leaves is boiled in 1 gallon of water for 20 minutes, 1 cup consumed cold, and the remainder used cool as a bath over the affected area [B2624]. For **fever**, 1 cup of this same decoction is consumed cool 3 times daily [B2624]. To treat **skin infection**, 1 large handful of leaves is boiled in 2 quarts of water for 20 minutes and poured over the affected area twice daily, in the morning and in the evening [B2624].

NEPHROLEPIDIACEAE

Plants terrestrial, epiphytic or rupestral. Rhizomes erect to suberect, short, generally producing numerous wiry stolons, scaly, the scales peltate. Leaves monomorphic, 1.5–10 feet long. Petioles with several vascular bundles arranged in a U-shape. Blades 1-pinnate, typically with a small fiddlehead remaining at the apex. Rachises grooved adaxially, the old dead ones persistent, bearing circular scars at the point of pinna attachment. Pinnae numerous. Veins free. Sori discrete, round, indusiate. In Belize, consisting of 2 genera and 5 species. Uses for 1 genus and 2 species are reported here.

Nephrolepis biserrata (Sw.) Schott

quash tail [English]; *cola de pisote* [Spanish]

To treat **urine retention (stoppage of water)** caused by renal obstructions, 1 rhizome (the larger "root") is scraped clean of its scales, boiled, and mashed with a stone to release a liquid, and 2 tablespoons are consumed once daily for 1 day [B2597]. To treat "**cancer**" (this reflects the local use of the word to describe a type of **open sore**, as in **ulcer**, or an **internal growth**), the liquid prepared above is consumed daily [B2597].

Nephrolepis brownii (Desv.) Hovenkamp & Miyam.

> For **urine retention (stoppage of water)**, though not as highly regarded as the maidenhair fern treatment (*Adiantum tenerum*), the roots and leaves of 3 medium-sized plants are boiled in 2 quarts of water for 10 minutes and consumed throughout the day [B2681].

POLYPODIACEAE

Plants epiphytic, less often terrestrial or epipetric. Rhizomes creeping, scaly. Leaves 3–28 inches long, monomorphic or (less commonly) dimorphic. Laminae simple to 1-pinnate, rarely more divided. Sori round or rarely elongate to linear, non-indusiate. Spores yellow. In Belize, consisting of 9 genera and 28 species. Uses for 3 genera and 3 species are reported here.

Nephrolepis brownii [RM].

Microgramma percussa (Cav.) de la Sota

> *xox pim* [Q'eqchi']
>
> To treat **deep sores**, the leaves are mashed with a stone and applied as a poultice to the affected area [B2494].

Microgramma percussa with close-up of sori (right) [RM].

Niphidium crassifolium (L.) Lellinger

To treat **swollen, pus-filled skin ailments**, 1 leaf is mashed and applied as a poultice to the affected area as often as necessary until better [B2499]. This is 1 of 12 plants used in an unspecified antidote to treat **snake-bite** from the Tommy Goff, a group of pit vipers in the genus *Bothrops* [B3551].

Phlebodium decumanum (Willd.) J. *Sm*.

bear paw fern [English]; *cola de mico* [Spanish]; *calawalla, canawana, tallawalla* [Creole]; *sic k'en* [Q'eqchi']

To treat **stomach ulcer**, **stomach pain**, **gastritis (*ciro*)**, **chronic indigestion**, **high blood pressure**, or to reduce a **"cancerous" (diagnosis uncertain) tumor** at an early stage, one 3-inch piece of root is boiled in 2 cups of water for 10 minutes and ½ cup consumed 4 times daily for 6 weeks [Arvigo and Balick, 1998]. To treat **systemic "cancer" (internal growth)**, 1 air-dried root is powdered, 1 teaspoon of the powder is mixed with ½ cup of tepid water, with some healers adding 1 teaspoon of dried boa snake meat (an undetermined species), and this beverage is consumed 3 times daily until better [B1844]. For **internal tumors** (growths or perhaps what is locally called "cancer"), refer to the God Almighty bush (*Struthanthus cassythoides*) [LR].

(top) *Niphidium crassifolium* [RM].

(bottom) *Phlebodium decumanum* [MB].

To treat **high blood pressure**, six 4-inch pieces of root are boiled in 1 quart of water for 5 minutes and ½ cup consumed cool twice daily, in the morning and in the evening, every other day for 4 days. After a 4-day treatment, the blood pressure is checked; treatment is stopped if the pressure is down but repeated if the pressure has not decreased [LR].

PTERIDACEAE

Plants terrestrial, epipetric, or epiphytic. Rhizomes erect or creeping, scaly or hairy. Leaves 0.3–10 feet long, monomorphic or dimorphic. Blades 1- to 6-pinnate. Sori borne on the lower surface of the lamina, often following the veins. In Belize, consisting of 11 genera and 40 species. Uses for 3 genera and 6 species are reported here.

Acrostichum aureum L.

tiger bush [English]; *helecho* [Spanish]

To treat **craziness**, the rhizome is scraped and cleansed to get at the "slimy part," which is then eaten [A352]. Alternatively, the rhizome scrapings are boiled with the rootlets in 1 quart of water for 10 minutes and 1 cup consumed twice daily [A352].

Adiantum tenerum Sw.

black stick, maidenhair [English]; *palo negro* [Spanish]; *roq ch'ikwan* [Q'eqchi'];
oc en su cun [Mayan]

For **cough**, as an **expectorant**, as an aid to **detoxify an alcoholic**, to **increase lactation** in a nursing mother, to aid in **kidney function**, or to treat **intestinal parasites**, 3 stems with leaves are steeped in 3 cups of boiling water for 20 minutes and sipped all day [Arvigo and Balick, 1998].

To treat **headache** caused by evil magic (*obeah*), 1 handful of leaves is boiled in 1 quart of water for 10 minutes, 1 cup consumed hot once daily, and the remainder sprinkled over the body as a bath [B2560].

For **urine retention (stoppage of water)**, 1 entire large plant with roots is washed well, boiled in 1 quart of water for 10 minutes, and ½ cup consumed every hour until the urine passes normally [LR].

To help break a woman's **amniotic fluid** during childbirth, 1 leaf is boiled in 1 cup of water for 5 minutes and consumed during labor [B2356].

For babies who **cry** too much when the cause may be the **evil eye (*mal ojo*)**, 9 leaves with stems are boiled in 2 quarts of water for 10 minutes and used as a bath once daily, at bedtime, for 3 days [BW]. Alternatively, 1 root is mashed in olive oil and used to massage the baby [BW].

To treat **dandruff**, the leaves are macerated and applied directly to the scalp [Arvigo and Balick, 1998].

The leaves are used as a **decoration** on ceremonial altars [A134].

Adiantum villosum L.

roq ch'ikwan [Q'eqchi']

To treat **fits** or **convulsions**, 1 entire plant is mashed in water and used as a tea and a bath [B3653]. For **kidney infection**, an unspecified treatment is prepared and used as a cold tea and a bath [B2706].

Adiantum wilesianum Hook.

roq ch'ikwan [Q'eqchi']

To treat **influenza** or **fright (*susto*)**, 1 entire fresh plant is mashed in warm water and used as a bath once daily, at bedtime, for 5 days [C40].

Adiantum wilsonii Hook.

Santa María [Spanish]; *rah li ch'och, ru'uj raq ajtza* [Q'eqchi']

To revitalize a person who is **tired** or **weak**, 1 handful of leaves along with 1 handful of *Santa María* leaves (*Aphelandra aurantiaca*, another plant with the same common name) are mashed in 1 quart of cold water, soaked in the sun for 1 hour, and consumed cool as a beverage as needed [R28, R23]. To clear **congestion** and quell a **cough** caused by "the morning mist," 1 handful of leaves is boiled in 1 gallon of water for 10 minutes and used as a bath over the head and face once daily, in the morning, as needed [B2477].

Pityrogramma calomelanos (L.) Link var. *calomelanos*

silver leaf, silver leaf fern [English]

To relieve **anxiety** or **nervousness**, the leaves are used to prepare a bath [B2647]. As local healer Hortense Robinson noted, "All ferns are good for nerves" [B2647].

SELAGINELLACEAE

Plants terrestrial, on rocks, or rarely epiphytic. Stems creeping or erect, branched regularly or irregularly, the branches flattened with 4 rows of leaves. Leaves 0.1–0.4 inch long, one-veined. Strobili (clusters of overlapping sporophylls) compact, cylindrical or quadrangular or flattened, usually simple, entire, borne at the branch tips. In Belize, consisting of 1 genus and 14 species. Uses for 1 genus and 3 species are reported here.

Selaginella sertata Spring

To **improve blood circulation**, 1 entire plant is boiled in water and used as a tea and a bath [B2707].

Selaginella umbrosa Lem. ex Hieron.

roq ak'ach [Q'eqchi']

For a variety of **bladder ailments**, including a **bladder infection**, ½ quart of stems, leaves, and roots is boiled in 1 quart of water for 10

(top) *Pityrogramma calomelanos* RM].

(bottom) *Selaginella sertata* [RM].

minutes, steeped for 30 minutes, and ½ cup consumed hot all day until finished; this treatment is repeated for 3 days [B2705]. Additionally, this same decoction is used cool as a bath, in combination with the tea [B2705].

To treat **skin fungus**, the leaves are mashed and applied as a poultice to the affected area several times daily as needed [B2512]. To treat **craziness**, 1 entire plant is dried, burned, and the smoke inhaled. This treatment is repeated until better [B3540].

Selaginella sp.

To reduce **swollen feet**, 1 handful of leaves is boiled in 1 gallon of water, cooled slightly, and used warm to soak the affected area [R26].

TECTARIACEAE

Plants mostly terrestrial. Rhizomes short-creeping or decumbent, less commonly erect or long-creeping. Leaves 8–40 inches long. Laminae simple to 4-pinnate-pinnatifid, often pubescent with short (<1 mm) and jointed hairs, especially on the upper surfaces of the axes. Veins netted. Sori typically round, indusia present or absent. In Belize, consisting of 6 genera and 17 species. Uses for 1 genus and 2 species are reported here.

Tectaria panamensis (Hook.) R.M. Tryon and A.F. Tryon
rah li ch'och [Q'eqchi']

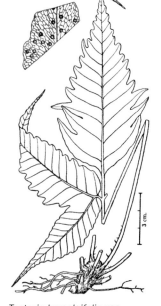

To treat **arthritis ("burning bones")** or **bone pain**, 9 leaves along with 9 *sák e pú chuch* leaves (*Piper tuerckheimii*) and 9 Spanish elder leaves (*Piper schiedeanum*) are boiled in 1 gallon of water for 30 minutes until the liquid is dark, and 1 cup is consumed hot 3 times daily, before meals, for 9 days, followed by rest for 9 days, until better [R30, R31, R34]. Additionally, this same decoction is used as hot as can be tolerated as a bath to soak the affected areas once daily, at bedtime, for 9 days, followed by rest for 9 days, until better [R30, R34].

To treat a terrible **itch** or **swelling** on the body, caused by stepping where snakes have been fighting or defecating at night on the path to one's *milpa* when one is asleep, the leaves are mashed, steeped in cool water, 1 cup consumed, and the remainder used as a bath over the affected area [B2478, B2478a].

Tectaria heracleifolia (Willd.) Underw. var. *heracleifolia*
murciélago [Spanish]; *q'eq curuz i pim* [Q'eqchi']

The leaves are a component of a Maya bath formula [B2243]. Generally, as a **panacea** to treat many ailments, and specifically to treat **swelling**, **pain**, or **sores**, 1 handful of the vine and leaves is boiled in 1 gallon of water for 20 minutes and used as a bath once daily as needed [B1895, B2243; RA]. To

Tectaria heracleifolia var. *heracleifolia* [NYBG].

treat **biliousness**, resulting from a congested gallbladder and liver, 1 handful of the root is boiled in 3 cups of water for 10 minutes, cooled, and 1 cup consumed warm as needed [B1895].

To treat **epileptic seizures**, 1 handful of fresh leaves along with 1 handful of fresh *sák e pú chuch* leaves (*Piper tuerckheimii*) are boiled in 1¼ cups of water for 8 minutes and consumed warm 2–4 times daily for 5 days [C33, C34]. For this treatment, the leaves are collected only on Mondays and Tuesdays [C33].

PINACEAE

Large trees, sometimes releasing strong fragrance from bark and/or leaves. Leaves needle-like, persistent, resinous, arranged in spirals, solitary, or in bunches, simple. Flowers usually monecious. Fruits a woody cone containing seeds, usually winged along 1 margin. In Belize, consisting of 1 genus and 2 species. Uses for 1 genus and 1 species are reported here.

Pinus caribaea Morelet var. *hondurensis* (Sénéclauze) W.H. Barrett & Golfari
  pine [English]; *pino, ocote* [Spanish]; *chaj* [Q'eqchi']
  The needles are used to prepare an energizing bath [A401].

    For **back pain** or **urinary ailments**, three 1-inch pieces of young branches with needles are boiled in 1 quart of water for 20 minutes and 1 cup consumed warm 3 times daily [A401]. To treat **cough**, the decoction described above for back pain is combined with lemon juice and consumed warm 3 times daily as needed [A401]. Alternatively, ½ cup of fresh pine needles is boiled in 3 cups of water for 10 minutes, juice from ½ lemon is added, and 1 cup is consumed warm 3 times daily as needed [A401]. As a third reported treatment, one 1-inch × 6-inch, resin-filled piece of wood from the center of a fallen tree is boiled in 1 quart of water for 10 minutes and ¼ cup sipped warm at bedtime until the mucus loosens and comes up with the cough [BW]. This

Pinus caribaea var. *hondurensis* [MB] with inset of dried pine resin [RBGKew].

resin-filled wood is found when the tree is felled or has been lying on the ground for some time. It is not recommended to cut down a tree to prepare this treatment [RA].

To treat **parasites**, equal parts of trunk resin and garlic are mixed together to make pills about the size of a vitamin pill, and 1 pill is consumed once daily after dinner for 3 days when the moon is young [A401]. As a treatment for "**leukemia**" (unspecified blood disorder), one 4-inch strip of wood and 3 garlic cloves are boiled in 3 cups of water for 10 minutes and 1 cup consumed warm 3 times daily for 9 days, followed by 9 days of no treatment. This cycle is repeated as needed [A401]. To remove excess prostate and uterine **mucus**, refer to *guaco* (*Aristolochia grandiflora*) [LR]. For non-insulin-dependent **diabetes** mellitus or **venereal diseases** such as **gonorrhea**, refer to "male" chicken weed (*Chamaesyce hypericifolia*) [A315] or chicken weed (*Euphorbia armourii*) [A322]. To prepare "**old man's tonic**," refer to broomweed (*Sida rhombifolia*) [Arvigo and Balick, 1998].

The green wood is burned to produce "a thick smoke used to **repel insects**" [Horwich and Lyon, 1990]. Lundell (1938) noted that the Petén Maya burned the wood of this species for use as **torches**. According to Thompson (1927), "the pine was the principle source of illumination in pre-conquest Central America; indeed, it is so used in many of the remoter Indian villages to this day." This is an important **timber** tree in Belize [MB]. Standley and Record (1936) reported that the wood was used in **construction** and **carpentry** for making pilings and railway ties.

PODOCARPACEAE

Usually tall trees, slightly resinous. Leaves persistent, usually spirally arranged, rarely opposite, leathery, simple. Inflorescences terminal, axillary, or solitary. Flowers dioecious. Fruits a cone, drupe-like. In Belize, consisting of 1 genus and 1 species. Uses for 1 genus and 1 species are reported here.

Podocarpus guatemalensis
Standl.

cypress [English]
Standley and Record (1936) reported that the wood of this species was used in **construction** for making house posts, sills, railway ties, and boats.

Podocarpus guatemalensis [IA].

ZAMIACEAE

Palm-like plants, sometimes with bulb-like or tuber-like stems or columnar and elongated, with few to numerous leaves at the apex. Leaves spirally arranged, leathery in texture, pinnate or bi-pinnate; leaflets usually linear or lanceolate. Inflorescences borne at the apex of the caudex among the leaves. Flowers dioecious. Fruits fleshy. In Belize, consisting of 2 genera and 4 species. Uses for 1 genus and 1 species are reported here.

Zamia polymorpha D.W. Stev., A. Moretti, and Vázq. Torres
camotillo, *mata ratón*, *mata ratónes* [Spanish]

Lundell (1938) reported that the roots were eaten as a **food** when cooked, but we have no further information on this claim.

To **poison** mice and rats, the tuber is grated, mixed with ground corn (*Zea mays*), and placed in or near the infested area, where it will be eaten by the rodents. When the rodents return to their burrows, they die of internal bleeding [B1803, B2058]. This mixture is said to poison to death whatever eats it [B2058].

Zamia polymorpha, showing tuber (bottom) [MB].

FLOWERING
PLANTS

ALISMATACEAE

Herbs, found in aquatic or wetland habitats; stems with milky sap. Leaves alternate, basal, simple, 1.5 feet or longer. Inflorescences terminal, in racemes. Flowers white. Fruits a flattened achene. In Belize, consisting of 2 genera and 4 species. Uses for 1 genus and 1 species are reported here.

Sagittaria lancifolia L. subsp. *media* (Micheli) Bogin
To treat **diarrhea**, one 3-inch piece of the chopped vine along with one 3-inch piece of *nanci* bark (*Byrsonima crassifolia*) are boiled in 1 quart of water for 20 minutes and ½ cup consumed cool every hour, up to 6 times daily, until better [B2171]. For bloody **diarrhea**, one 3-inch piece of the chopped vine along with one 3-inch piece of *nanci* bark, one 3-inch piece of cinnamon bark (*Cinnamomum verum*), one 3-inch piece of *sapodilla* bark (*Manilkara zapota*), and the contents of 3 young baby coconuts (*Cocos nucifera*) are boiled in 1 quart of water for 20 minutes. The color should be pale pink-red, and ½ cup is consumed cool every hour, up to 6 times daily, until better [B2171].

To treat **heavy menstruation**, 1 handful of the chopped vine is boiled in 1 quart of water for 5 minutes and ½ cup consumed cool every 2 hours [LR]. For **postpartum bleeding**, 1 cup of this same decoction is consumed every 30 minutes [LR].

AMARYLLIDACEAE

Perennial herbs, arising from bulbs. Leaves alternate, simple, linear or strap-shaped. Inflorescences umbellate and borne on a stalk or of solitary flowers and terminal. Flowers variously colored, including white, red, and yellow. Fruits a capsule. In Belize, consisting of 4 genera and 5 species. A use for 1 genus and 1 species is reported here.

Hippeastrum puniceum (Lam.) Urb.
This red-flowered plant is cultivated and used as an **ornamental** [MB].

ARACEAE

Herbs, terrestrial or epiphytic with creeping, subterranean, or aerial stems, often covered with thickened roots that frequently attach to stems of other plants. Leaves alternate, simple, sometimes perforated or lobed. Inflorescences terminal or axillary with a spadix subtended by a single spathe containing many small, sessile flowers. Fruits a berry, often juicy and colorful. In Belize, consisting of 12 genera and 49 species. Uses for 8 genera and 15 species are reported here.

Hippeastrum puniceum [RH].

Alocasia macrorrhizos (L.) Schott

Carib gal, Creole gal, wild coco [English]

This plant is cultivated and used as an **ornamental** around homes and gardens [B2313; Arvigo and Balick, 1998].

Bailey and Bailey (1976) noted that the corms were used as a **food**. As they contain injurious calcium oxalate crystals, the corms must be properly prepared before consumption.

To treat **skin ailments**, especially **itching** or new **burns**, 1 large leaf and stem is chopped, boiled in 2 gallons of water for 20 minutes, and poured as a bath over the body twice daily as needed [A144; Arvigo and Balick, 1998]. To treat **wounds**, the leaves are toasted, powdered, and applied to the affected area [Arvigo and Balick, 1998]. For expelling mucus from **sinus congestion**, 1 fresh leaf is steamed, placed over the forehead to which "Vicks" (Vicks VapoRub® topical ointment) has been applied, a cloth wrapped around the head, and left on overnight; this is repeated for 3 consecutive nights [Arvigo and Balick, 1998].

Alocasia macrorrhizos [IA].

To treat **arthritis**, 1 leaf is warmed over a flame, rubbed in oil, and applied as a poultice to the affected area. A cloth is wrapped lightly around the leaf, and this poultice is changed every 15 minutes [BW]. To treat **varicose veins**, especially to **improve blood circulation**, **prevent bursting**, and **reduce pain**, fresh leaves are mashed and applied as a poultice directly to the affected area [Arvigo and Balick, 1998]. To ease **rheumatism**, 1 fresh leaf is lightly coated with cooking oil, steamed over a heat source, wrapped around the affected area, and left on overnight [Arvigo and Balick, 1998].

This plant is said to bring good **luck** to the household where it is planted [BW]. When a person is brought to court on a false accusation, the red part of a heart-shaped leaf is cut out and wrapped around a coin in a square pattern. The coin is wrapped with the Queen's head up if the accuser is female and with the King's head up if the accuser is male. The accused person, wearing the wrapped coin placed in the left shoe under the heel, must arrive to the courtroom before the accuser. By virtue of the charm, the accuser must tell the truth in court [BW].

Caution is used when using this plant directly on the skin or mucous membranes due to the oxalates found in the fresh sap of this genus [Arvigo and Balick, 1998]. Contact with the leaf juice, unless first heated, causes burning and itching [BW].

Anthurium pentaphyllum (Aubl.) G. Don. var. *bombacifolium* (Schott) Madison

ruq' li maus [Q'eqchi']; *uèx um, uíx um* [Mayan]

To treat **skin sores**, especially deep infected **wounds**, the leaves are mashed and applied as a poultice directly to the affected area [B2498]. To relieve **mastitis**, the leaves are warmed on a clay oven and placed on the affected area [B3538].

Anthurium schlechtendalii Kunth subsp. *schlechtendalii*

pheasant's tail [English]; *cola de faisán* [Spanish]; *xiv ak tun ich, xiv yak tun ich* [Mopan]; *tye i pú* [Q'eqchi']

For **muscle or joint pain**, **sprains**, **back pain**, **arthritis**, or **rheumatism**, 3 large leaves are chopped, boiled in 2 gallons of water for 10 minutes, and used as a warm bath [Arvigo and Balick, 1998]. To treat **severe cramps**, **muscle spasms**, or **paralysis**, the limb or affected area is covered with a blanket and placed over the steam of the boiling decoction described above for muscle pain. **Caution** is used to avoid burning the skin [Arvigo and Balick, 1998]. To treat serious cases of **rheumatism**, **arthritis**, and even **paralysis**, 1 double handful of fresh, chopped leaves is boiled in 2 gallons of water for 10 minutes and used warm as a bath. If a bathtub is not available, the decoction is used as a steam bath [B1794].

Anthurium schlechtendalii subsp. *schlechtendalii* leaves and inflorescences [MB].

To reduce **swelling** and **inflammation**, especially from swollen breasts, boils, back pain, or joint pain, the leaves are mashed and applied as a poultice to the affected area. The poultice "will stick to the affected part and can be worn all day" [B3507; Arvigo and Balick, 1998].

To treat **back pain** or **muscle spasms**, the center vein of the leaf is mashed, the leaf applied directly to the affected area, and a hot water bottle placed over this for 1 hour [Arvigo and Balick, 1998].

To treat **urinary infection**, 1 handful of roots is boiled in 3 cups of water for 10 minutes and sipped all day until symptoms improve [Arvigo and Balick, 1998].

For *chilillo* (a dog acting out of control), 1 handful of roots is boiled in 1 quart of water for 10 minutes and ½ pint of liquid force-fed to the dog 3 times daily for 1 day only. The dog is monitored for 3 days and the treatment repeated if necessary [BW].

Anthurium verapazense Engl.

To treat red and itchy **skin irritations**, 1–3 leaves are boiled in 2 gallons of water for 10 minutes, 1 cup consumed cool, and the remainder poured cool over the affected area as a bath [B2546].

Dieffenbachia maculata (Lodd.) G. Don

This plant is cultivated as an **ornamental** [B2359].

(top left) *Anthurium verapazense* [MB].

(top right) *Dieffenbachia maculata* [MB].

(bottom left) *Monstera siltepecana*, fruit (left) and leaf (right) [TC].

(bottom right) *Philodendron aurantiifolium* subsp. *aurantiifolium* with inflorescence [MB].

Monstera lechleriana Schott
embroidery plant [English]
To treat **skin sores**, 9 leaves are boiled in 2 gallons of water for 10 minutes and used as a bath over the affected area [B2718].

Monstera siltepecana Matuda
As a **protectant** and **calmative** for someone who has passed through a place "where the devil was," causing that person to shake and tremble, 9 leaves are chopped, boiled in 2 gallons of water for 20 minutes, 1 cup consumed warm, and the remainder used cool as a bath twice daily, in the morning and in the evening [B2550].

Philodendron aurantiifolium Schott subsp. *aurantiifolium*
To ease **joint pain**, 1 quart of leaves and stems is boiled in water for 5 minutes, mashed, placed on a cloth, and applied as a poultice to the affected area. When boiling the plant, just enough water is used to cover the plant material in the pot [B2679].

Philodendron hederaceum [MB].   Philodendron radiatum var. radiatum [MB].

Philodendron hederaceum (Jacq.) Schott
heart vine [English]

For **skin sores**, especially stubborn ones, the leaves are rubbed directly on the affected area [B2603]. To treat **varicose veins**, 9 leaves are chopped, boiled in 2 gallons of water for 5 minutes, and used as a bath over the affected area [B2603].

Philodendron radiatum Schott var. *radiatum*
belly full tie tie [Creole]; *uxb'* [Q'eqchi']

For **erysipelas**, 3 leaves are boiled in 1 gallon of water for 10 minutes and used cool as a wash over the affected area. Following this, fresh leaves are mashed and applied as a poultice to the affected area, wrapped in a cloth, and changed every 10 minutes. Treatment continues until better, usually about 2 days [BW].

To treat **back pain**, a piece of stem is tied around the waist and back and replaced daily until better [BW].

For excessive **obesity**, or "when a person is too fat to lose weight," 9 pieces of stem are braided together and worn as a belt. This belt is replaced daily [BW].

To treat **cough** in a dog, 9 pieces of the vine are braided together, 3 limes (*Citrus aurantiifolia*) strung on this, and worn by the dog as a necklace until better [BW].

The aerial roots are used as **cordage** to tie poles used for making frames of temporary shelters in camps. An example is the sleeping shelter built in cornfields when a person is working on the *milpa*, cultivating crops. This plant is considered a substitute for the *macapal* plant (*Trichospermum grewiifolium*), which is more commonly used as cordage for these ties [B3539].

Philodendron sagittifolium Liebm.

sake chol châl, sake chol chól [Mayan]

The aerial roots of the vine are cut and used fresh as **cordage** for tying roofing material, house frames, and other things around the house and farm [B3615].

Spathiphyllum blandum Schott

plantain grove, wild plantain [English]; *platanillo* [Spanish]; *tèic, túic, tyuk, tz'uql* [Q'eqchi']; *tsabal* [Additional name recorded]

The young spadix (spike that bears the flowers) is boiled, coated with black pepper (*Piper nigrum*), and eaten as a **food** [B3506].

To alleviate **difficult urination** or counteract **food poisoning** when "bad" (rotten or spoiled) food is eaten, 1 teaspoon of leaf ash is mixed in 1 cup of cool water and consumed 4 times daily [B2602].

Spathiphyllum phryniifolium Schott

Standley and Record (1936) noted that the young inflorescences "of plants in this genus are cooked and eaten as a vegetable in certain parts of Central America." Standley and Steyermark (1958) noted that the cooked inflorescences can be found in the markets of Guatemala. They related the following anecdote:

(left) *Philodendron sagittifolium* [TC].

(top right) *Spathiphyllum blandum* [TC].

(bottom right) *Spathiphyllum phryniifolium* with inflorescence [TC].

Some few years ago a North American visiting Guatemala was served in some hotel with soup in which he discovered what he took to be diminutive ears of corn. Upon return to the United States he reported the matter to the United States Department of Agriculture. Because of the hope of some day discovering the wild ancestor of corn, the Department was interested and sent two men to Guatemala to investigate the plant. Investigation revealed that the supposed corn consisted of the spadices of *Spathiphyllum*, which bear a remote resemblance to much-reduced ears of maize. (pp. 352–3)

Syngonium podophyllum Schott

consuelda, contra hierba, hinchazón [Spanish]; *kwán chi* [Q'eqchi']; *hop ya* [Mopan] This plant is commonly found growing in pastures as an epiphyte on bay cedar trees (*Guazuma ulmifolia*) [A119]. The leaves are a component of a 9-plant Maya bath formula [B1779]. The fruit is eaten as a **food** [Robinson and Furley, 1983].

Syngonium podophyllum [MB].

To reduce **swelling** or to treat **general skin ailments**, such as **sores**, **boils**, **dry skin**, **fungus**, **itching**, **rash**, and **bruises**, 9 leaves are boiled in 1 gallon of water for 10 minutes and used warm as a wash over the affected area [B1779; Arvigo and Balick, 1998]. Alternatively, to reduce **swelling**, 9 leaves are mashed, soaked in 1 bottle of rubbing alcohol for 3 days, and applied to the affected area [B1881]. To treat **swelling** or **rheumatism**, 9 leaves are crushed, soaked in 1 cup of rum for 3 days, and applied as a poultice to the affected area [A119]. In another reported treatment for **general swelling and pain**, **rheumatism**, or **arthritis**, 3 large leaves and 1 quart of alcohol are soaked in the sun for 7 days and applied to the affected area 3 times daily with cotton [Arvigo and Balick, 1998]. To help open an **abscess**, 5 fresh leaves are heated in a hot pan and applied as a poultice directly to the affected area 3 times daily for 3 days [C46].

To reduce **pain following tooth extraction**, 1 leaf is squeezed into 1 quart of water and used as a bath over the affected area [B1779].

To treat rectal **bleeding** in dogs, the leaves are mashed and the juice applied to the affected area [Arvigo and Balick, 1998].

Xanthosoma violaceum Schott

The tuber is eaten as a **food**—for example, boiled in water and eaten like a potato (*Solanum tuberosum*) [B2212]. Thompson (1927) reported that the Maya of San Antonio (Toledo District) and Succotz (Cayo District) ate the boiled roots as a **food** and also boiled the leaves and young shoots and ate these "as we would serve sprouts."

ARECACEAE

Palms, understory vines and shrubs to large trees, solitary or clustered stems, sometimes covered with spines. Leaves spiral, palmate, pinnate or simple, mostly clustering near the apex of the stem. Inflorescences of solitary flowers to much branched, subtended by one to many enclosing bracts. Flowers usually numerous and whitish. Fruits a drupe, usually with a hard endocarp. In Belize, consisting of 27 genera and 44 species. Uses for 19 genera and 24 species are reported here.

Xanthosoma violaceum [MB].

Acoelorraphe wrightii (Griseb. & H. Wendl.) H. Wendl. ex Becc.
Hairy Tom, Hairy Tom palmetto, Honduras pimenta, palmetto, pimento palm [English]; *palma, prementa, primenta* [Spanish]; *chi it, papta, pimient* [Q'eqchi']; *taciste, tasiste* [Yucateca]
The palm heart (the apical growing point of the stem) is roasted on coals, mixed in oil, sautéed with onion and garlic, and eaten as a **food**, although it is very bitter [A371]. The boiled seeds are also eaten as a **food** [A371].

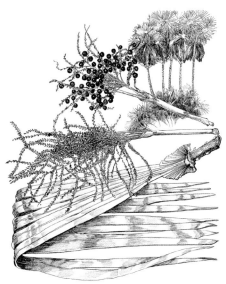

(above) Fence made of *Acoelorraphe wrightii* stems [MB].

(left) *Acoelorraphe wrightii* [FA].

To treat **sores** and **wounds**, the palm hearts and roots are boiled either independently or together and the decoction used as a wash over the affected area [A371]. To prevent **sores** from becoming infected and to stop wounds from **bleeding**, the inner fibrous material surrounding the meristem is scraped, dried, and dusted into the affected area. The scrapings must be dried, as they might otherwise burn the skin when applied [A371].

The stems are used in **construction** to make house walls and small barnyard structures such as chicken coops [B3614]. Split or entire stems are lashed together in a row and used for walls, and whole stems are used as support posts for small structures, including fence posts [A285]. The stems are also used to make broom handles [A285]. The trunks are used in **construction** to make saltwater posts [A490].

To make **white lime**, the stems are burned in a fire hearth with limestone until very hot and reduced to ash. Water is sprinkled over the ash to complete the process [A490].

The entire palm with stem and leaves is tied to bridges and lamp posts as street **decorations** for special events and parades [BW].

Acrocomia aculeata (Jacq.) Lodd. ex Mart.
grugru palm, suppa palm [English]; *cocoyal, cocoyol, cocoyul, sipa* [Spanish]; *map* [Q'eqchi']; *tuc* [Yucateca]

Acrocomia aculeata [FA].

The fruits are made into a conserve and eaten as a **food**, although children will also chew on mature fruits [A383, B3681]. The kernels of the seeds are eaten either raw or boiled with sugar [Horwich and Lyon, 1990]. Sap from the stem is processed to make a **fermented beverage**. The palm is cut down, the meristem partially cut open, and a trough made in the area containing the youngest developing leaves, through which sap flows and fills the stem. This sap is collected and fermented into a beverage locally known as "wine" [B2051].

For non-insulin-dependent **diabetes** mellitus, 3 fruits are boiled in 3 cups of water for 10 minutes and 1 cup consumed 3 times daily [A383].

This plant is said to be the "Devil's Cup," out of which the devil will give a drink of water at the end of the world, when everybody lines up for the Judgment Day. The devil will offer to give water to people in exchange for a child, so he can gain souls for himself. The *cocoyol* holds only a tiny bit of water, however, so it is said to be a devil's trick [BW].

The hard outer surface of the seed has been used to make **jewelry**, including rings and rosary beads [Horwich and Lyon, 1990]. According to Steggerda (1943), Maya in the Yucatan used to put inlays into **rings** made from these seeds, which were considered to "bring good luck." Additionally, he noted that the bark is sometimes used to make a **container** for holding soil in which vegetables are grown.

Asterogyne martiana (H. Wendl.) H. Wendl.
ex Hemsl.

> monkey tail [English]
> Standley and Record (1936) noted that the leaves were
> sometimes used for **thatch**, but because they are small
> (up to 1 meter long), they were not the preferred spe-
> cies for this purpose.

Astrocaryum mexicanum Liebm. ex Mart.

> cohune, warree cohune, warri cohune [English];
> *güiscoyol*, *lancetilla*, *pacaya* [Spanish]; *akté* [Q'eqchi'];
> *chapay* [Additional name recorded]
> Parts of the stem, flower, and fruit are eaten as a **food**
> [B2721]. The palm heart is cooked and eaten [B3516].
> When the bract is still closed, the entire flower can be
> roasted in the fire like corncobs and eaten [B2721].
> The fruit is eaten raw, once the spiny outer covering is
> peeled off [B3516].
>
> The stem, when cut and sharpened at one end, is
> used as a **planting stick** [B3516].
>
> The common name refers to the sharp, dangerous
> spines that cover the stem. They are said to "resemble
> the bristles of the white-lipped peccary (*Tayassu pecari
> ringens*), or 'Warree'" [Marsh, Matola, and Pickard,
> 1995].

Attalea cohune Mart.

> cohune [English]; *tutz* [Q'eqchi']; *chunciey*, *corozo*
> [Additional names recorded]
> Local people have always recognized the importance
> of this palm as a source of **shade** for their farm animals
> and to provide a variety of **economic products**; thus,
> since early Maya times, people have left these palms
> standing when clearing the forests for agriculture. The
> trunk is also very hard and thus difficult to cut down.
> It has become one of the most prominent trees in the
> forests of Belize, especially when disturbance—either
> induced by humans or by nature—has taken place
> [MB]. Stevenson (1932) wrote of its importance:
>
>> The Cohune plays an important role in the life
>> of the forest laborer in the south of the Colony.
>> He uses the leaves (fronds) for thatch, and the leaf

(top) *Asterogyne martiana* [FA].

(middle) *Astrocaryum mexicanum* [FA].

(bottom) *Attalea cohune* [FA].

stems for the sides of his house, the top of his table, and his bed. He obtains oil from the nuts and food from the heart of the 'cabbage.'" (pp. 3–5)

The palm heart is eaten as a **food** either fresh, pickled in brine and vinegar, or cooked in stews and soups [RA]. To extract a wine-like **beverage**, a hole is made into the heart of the trunk and the liquid collected. Each tree yields about 1 gallon, and consumed as is, it is said to be like wine [RA]. The endosperm, after oil extraction, is used as **fodder** for pigs and the mesocarp as **feed** for chickens [McSweeney, 1993]. The crushed endocarp and epicarp are used as **fuel** for cooking, either directly or when converted to charcoal [McSweeney, 1993]. A great deal of additional information on the use of this palm was provided by this author in a later paper [McSweeney, 1995].

The fruits are made into **butter**. The ripe, fallen fruits are collected, cracked opened, and 5 pounds of the inner fruits ground and boiled in 2 quarts of water. Ground-up coconut meat (the endosperm of *Cocos nucifera*) is added and the mixture left to sit overnight. In the morning, the palm fat is scooped off the top, salted, and used as butter [B2709; LR]. **Oil** extracted from the fruit is used as a **food** and for cooking. To extract the oil, the fruits are collected, placed on a stone, and pounded with another stone to crack the hard outer shell (endocarp), which releases the oil-rich seed (endosperm). This is removed from the hard shell, pressed, and boiled in water for 1 hour. Very dry seeds that have been on the ground for 3 months yield more oil than fresh seeds. To make 1 pint of oil, 3 quarts of pressed endosperm are boiled in 2 gallons of water. The oil rises to the top, is scraped off, strained, and bottled for use. It was commonly seen for sale in Saturday markets for about US$5.00 per quart [B1862, B2407, B2709; RA]. During the First World War, the British army obtained many tons of these nuts from Belize to make charcoal **filters** for use in gas masks to protect their soldiers from the gas attacks involved in trench warfare at that time [Horwich and Lyon, 1990].

To treat **skin ailments**, the oil is used as a carrier oil for a salve with various herbs [RA]. To treat **open sores**, rancid oil is applied to the affected area [RA]. Fresh oil is applied to the **hair** to improve the general health of the hair and to the **skin** as a soothing lotion [RA]. To make a **soap** or **shampoo**, refer to mamey apple (*Pouteria sapota*) [B3753].

To staunch **hemorrhaging**, the young and unopened leaf rachis is scraped with a knife, the scrapings steeped in water for 30 minutes, and the decoction sipped [B2709, RA].

To treat **internal infections**, the liquid contained in the stem of the leaf blade is consumed [B2709; RA].

The leaf is used as roof **thatch**, but it is considered only temporary roofing, lasting up to 3 years before falling apart [RA]. The leaf is also used as a **fiber** to weave fans, bedrolls for sleeping, and the walls of bush houses [RA; Horwich and Lyon, 1990].

The oil has been and still is used as **fuel** in lamps [RA; Lundell, 1938].

To **repel mosquitoes**, 1 pint of oil is mixed with 1 teaspoon of *Carbo Negro* (an over-the-counter product) [RA]. Old inflorescences, when their fruit is removed, are dried and pounded to produce a **fly brush** that looks similar to a horse's tail [Horwich and Lyon, 1990].

To **attract edible grubs**, a hole is made into the palm heart on the new moon, and in 7 weeks' time, numerous palm weevil grubs (*Rhynchophorus palmarum*) can be collected, fried, and eaten like shrimp. These grubs are said to be tasty, oily, and nutritious [RA].

Bactris major Jacq. var. *major*

pork and doughboy [English]; *biscoyol, hones, jauacte, palma de espina* [Spanish]; *cocano boy, pokenoboy, warrie cohune* [Creole]; *k'ixk'ib'* [Q'eqchi']

The mesocarp (the part of the fruit just under the outer skin) and endocarp (the nut-like seed) of the fruit are eaten as a **food** [B3299]. The mesocarp has a sweet-acidic juice obtained by sucking on the flesh. The endocarp is chewed and eaten like a piece of coconut (*Cocos nucifera*) [A330]. The taste of the mature fruit is said to be similar to "black cherries" [Horwich and Lyon, 1990]. A fermented **beverage** or "**wine**" is made from the ripe fruits [A330].

To treat **uterine "cancer" (sores, abscesses, or ulcers on or within the uterus)**, one 12-inch piece of the root along with the skin of 1 provision bark tree fruit (*Pachira aquatica*) are boiled in 1 quart of water for 20 minutes and 3 cups consumed daily for at least 9 days. If symptoms do not disappear, the treatment is continued following a 9-day break from taking the tea. This 18-day cycle is repeated until better or the symptoms disappear [A330].

To make cooking **tongs**, locally known as *kis-kis* in Creole, the split stem of this palm is rubbed with oil, gradually bent, and the starchy pith inside removed. The stem is heated over a flame and the ends bent until

(top) *Bactris major* var. *major* [FA].

(bottom) *Bactris mexicana* [FA].

they are nearly touching and tied. The *kis-kis* is used to catch crabs, hold hot coals, spread hot coals for baking, as well as other purposes that protect the hands [RA]. It is said that the use of *kis-kis* to remove meat and biscuits from the cooking fire resulted in the older common name for this plant, "pork and doughboy," which has been abbreviated through usage to "pokenoboy" palm [Horwich and Lyon, 1990].

Bactris mexicana Mart.

warrie cohune [English]; *palma de espina* [Spanish]; *k'ixk'ib', halawte'* [Q'eqchi']

The seeds of the fruit are eaten raw as a **food** following removal of the outer covering

[B3260]. The fruit provides **food** for many bird species; hunters know this and seek their quarry near these palms [B2011].

The stems are used in **construction** to make roofing crossbars, and other palm leaves are attached to them for roof thatch [B3618].

Calyptrogyne ghiesbreghtiana (Linden & H. Wendl.) H. Wendl.

San Miguel palm [English]; *pum èk, pum úk, se'min* [Mayan]; *capuche* [Additional name recorded]
The leaves are used as **thatch** for making huts [B2555].

Chamaedorea ernesti-augustii H. Wendl.

The leaves are used to **wrap** and **protect** copal resin (*Protium copal*) after collection [B3496]. A number of round pieces of copal are placed in a kind of tube formed by the leaves [MB].

Chamaedorea oblongata Mart.

xate macho [Spanish]; *xate maach* [Q'eqchi']
The seeds provide **food** for many bird species; hunters know this and seek their quarry near these palms [B2006].

Chamaedorea pinnatifrons (Jacq.) Oerst.

monkey tail [English]; *chem chem, x te, xal a cam, xáte* [Additional names recorded]
The inflorescence is eaten as a **food**. It is roasted or boiled while still enclosed within its bracts. Then, the bracts are peeled away, and the inflorescence is fried with other foods such as eggs and tortillas [B3327]. The immature seeds are usually eaten with tortillas as a **food**, but they have the unfortunate tendency to make one's mouth itch after eating [B2474].

The leaves are used to **wrap** copal resin (*Protium copal*) after collection, in a manner similar to that described above for *Chamaedorea ernesti-augustii* [B2474].

Stirring the *chicle* (*Manilkara zapota*) as it is cooking with the stems and roots of this palm helps the *chicle* latex to congeal; use of the palm in this way serves as a **thickening agent** [B3272].

(top) *Calyptrogyne ghiesbreghtiana* [FA].

(middle) *Chamaedorea ernesti-augustii* with close-up of fruit and inflorescence [FA].

(bottom) *Chamaedorea ernesti-augustii* [IA].

(top left)
Chamaedorea oblongata [FA].

(top right)
Chamaedorea pinnatifrons [FA].

(bottom left)
Chamaedorea tepejilote [FA].

(bottom right)
Chamaedorea tepejilote with close-up of leaf [IA].

Chamaedorea tepejilote Liebm.

pacaya [Spanish]; *k'ib' i xul* [Q'eqchi']; *sak eh kém, sak eh km* [Mayan]

The inflorescence is eaten as a **food**. It is cooked while still enclosed in its bracts; the inflorescence is then extracted and eaten [B3348, B3498].

To relieve **pain** from spinal and back ailments, 1 leaf sheath is twisted and stretched, crushed to release the juice by rolling a bottle over it, plastered on the back from the neck to the waist, and secured in place with a cloth [B2703].

The fronds are used as church **decorations** [B2703].

Cocos nucifera L.

coconut [English]; *coco, cocotera* [Spanish]; *kook* [Q'eqchi']

This palm is widely planted in the lowland areas of Belize. Standley and Record (1936) reported annual production of approximately 12 million nuts, "of which from 9–10 million are exported either as whole nuts or copra." Today, the industry is but a shadow of its past.

Cocos nucifera [FA].

The fruit is eaten as a **food**, both when the tender, immature endosperm (the inside of the fruit) is clear-white and gelatinous as well as when the fruit is mature and the center has hardened [B2271]. The tender water of an immature fruit is highly **nutritious**, being a rich source of potassium and other minerals [Arvigo and Balick, 1998]. "It is a good food for **diabetics** because it contains no fat or starch" [Arvigo and Balick, 1998]. The mature endosperm is a source for extracting **cooking oil** [B2271]. To make **butter**, refer to cohune (*Attalea cohune*) [B2709; LR].

As a **first food** for children and to treat **indigestion, gastric ulcers, colitis, hepatitis, diarrhea**, and people in **weakened conditions**, the very young endosperm, which is free from fiber and easily digested, is mashed along with banana (*Musa acuminata*) and consumed [Arvigo and Balick, 1998]. For **infant diarrhea**, the endosperm of 2 immature fruits (about 3–4 inches long) is boiled in 1 cup of water for 10 minutes and 1 teaspoon consumed several times daily [Arvigo and Balick, 1998]. To treat bloody **diarrhea**, refer to *Sagittaria lancifolia* subsp. *media* [B2171].

To treat **constipation, indigestion**, or **vomiting**, the water is consumed [Arvigo and Balick, 1998]. For **constipation, bronchitis, piles**, and as a good **nerve tonic**, the water and 1 tablespoon of honey are mixed and consumed [Arvigo and Balick, 1998]. To **prevent curdling** of milk in the stomach of an infant, the water is mixed with infant formula [Arvigo and Balick, 1998]. For **stomach pain** due to food poisoning, refer to scorpion tail (*Heliotropium angiospermum*) [B1939]. For **tapeworm**, the mature endosperm is grated, squeezed, and 1 cup of this milk is consumed once daily in the morning and 1 ounce of castor oil (*Ricinus communis*) a few hours later [Arvigo and Balick, 1998].

To treat **gonorrhea, toxemia** of pregnancy, **high urine acidity**, or **heart, liver, and kidney disorders**, the water from an immature fruit is consumed. For **kidney failure**, it is advised that this water only be given by a physician [Arvigo and Balick, 1998]. To treat **cough** and **lung trouble**, the oil is said to be good and is used in an unspecified way [Fernandez, 1982].

For **dehydration**, the water along with lime juice (*Citrus aurantiifolia*) are mixed and consumed as a beverage [Arvigo and Balick, 1998]. As a "cooling" medicine to treat **excessive body heat**, milk from the mature, grated, and squeezed endosperm is mixed along with lemon juice (*Citrus limon*) and consumed [Arvigo and Balick, 1998].

To **prevent pimples and wrinkles**, the tender flesh (endosperm) of an immature fruit is said to be an excellent **face wash** [Arvigo and Balick, 1998].

Cryosophila stauracantha (Heynhold) R. Evans

give-and-take [English]; *escoba* [Spanish]; *mis* [Mopan]; *akte', mesb'eel* [Q'eqchi']; *coxan, akuum* [Mayan]

The palm heart is eaten as a **food** when cooked on the coals of an open fire. It is very bitter and should not be eaten by pregnant women in the first trimester, as it may have **abortifacient effects** [BW].

The common name "give-and-take" indicates that the spines can "give" a bad, stinging cut and other parts can "take away" pain, bleeding, or infection [Arvigo and Balick, 1998]. To alleviate **pain**, staunch **hemorrhaging**,

Cryosophila stauracantha [FA].

or **prevent infection**, the base of the leaf sheath (pink, cotton-like, and sticky) is peeled back, and the inside of the fiber is scraped off with the fingernails or a knife and applied directly to the fresh wound as a sterile gauze [B1765, B3236; Arvigo and Balick, 1998]. The leaf is used to treat **bone pain**, **convulsions**, and **giddiness** [Fernandez, 1982].

To treat **snakebite**, the palm heart is boiled in 1 quart of water and consumed slowly until finished. This is said to prevent the poison from spreading throughout the body too quickly [RA].

As a **fish poison**, the leaves are mashed, placed on top of the water in a small pond, and the stunned fish are harvested and eaten. Fish not collected do not die, however. The paralyzing effect wears off, and they swim away [RA].

The young, dried leaves are woven together using basket tie tie (*Desmoncus orthacanthos*) on a slender stick and used as a **broom** [B1765, B2024; Arvigo and Balick, 1998]. The stems are used in **construction** to build house walls said to "last 100 years" [B1765].

Desmoncus orthacanthos Mart.

basket tie tie, basket whisk [English]; *b'ayl* [Q'eqchi']

The sweet tasting palm heart is eaten as a **food** [B3569].

As a **fiber** used to weave baskets, mats, handicrafts, lampshades, and wall coverings, the stems are harvested from the forest, and while still green, the spines are scraped off. The stems are then split into 4 equal strips and, after the soft center pith is removed, are used for weaving [B2201, B2670, B3275, B3569]. At least one rattan factory in Belize has experimented with this fiber in making large chairs and sofas [MB]. To make a broom, refer to give-and-take (*Cryosophila stauracantha*) [B1765, B2024; Arvigo and Balick, 1998].

Desmoncus orthacanthos [FA].

To **poison vampire bats**, locally known as "rat bats" (*Desmodus rotundus*), the fruits are poisoned and placed in chicken coops and pigpens, where the bats eat the fruits and die [B2670].

Euterpe precatoria Mart. var. *longevaginata* (Mart.) Henderson

cabbage palm, mountain cabbage palm [English]; *mokooch* [Q'eqchi']; *jal au t, mal au té* [Additional names recorded]
The palm heart is eaten as a **food**, usually cooked [B2689, B3610].

The stem is said to be strong, water resistant, and buoyant and is used in **construction** to make pilings on the docks [Horwich and Lyon, 1990]. In the past, the stem was lashed together in **construction** for making rafts, which were used to transport heavy timber and supplies along local rivers [Horwich and Lyon, 1990].

Gaussia maya (O.F. Cook) Quero & Read

la palma, palmasito [Spanish]
Unlike some other palms, including species of *Attalea*, *Euterpe*, and *Sabal*, the heart of this species is *not* edible. In fact, sap from this palm causes severe itching [B3249].

The durable "wood" is used in **construction** to make house frames and walls said to last 20 years [A508].

Geonoma interrupta (Ruiz & Pav.) Mart. var. *interrupta*

monkey tail palm [English]; *pacaya* [Spanish]; *k'ib' i tzuul* [Q'eqchi']; *cambuc* [Additional name recorded]
The palm heart is roasted over hot coals until the inside is hot, and eaten as a **food** [B3334].

Manicaria saccifera Gaertn.

Standley and Record (1936) reported the use of the leaves for **thatch** said to last for many years. They also noted the importance of the **fiber** that covers the inflorescences in making **handicrafts**.

(top) *Euterpe precatoria* var. *longevaginata* [FA].

(upper middle) *Gaussia maya* [FA].

(lower middle) *Geonoma interrupta* var. *interrupta* [FA].

(bottom) *Manicaria saccifera* [FA].

Roystonea regia (Kunth) O.F. Cook.
royal palm [English]
This tree is often cultivated as an **ornamental** [Marsh, Matola, and Pickard, 1995].

The palm heart is relatively sweet and eaten as a **food** [McSweeney, 1993].

The leaves and trunk are used in **construction** [Marsh, Matola, and Pickard, 1995].

Sabal mauritiiformis (H. Karst.) Griseb. ex H. Wendl.
bay leaf [English]; *botán* [Spanish]
Standley and Record (1936) reported that in addition to the leaves being used for **thatch**, the stem, being very strong and durable, was employed in **construction** for making house posts and pilings.

To make **white lime**, this trunk along with a similar-sized *gumbolimbo* trunk (*Bursera simaruba*) are burned into ashes with a white limestone resting on the fire. The white lime powder is collected, passed through a strainer, and soaked for 3 hours with corn (*Zea mays*) that will be used to make *tortillas* [BW].

Lundell (1938) reported that the young leaves were used as a **fiber** to weave hats and mats by the ancient Maya of the Petén. The leaves are used as roof **thatch** [B2024]. The wood is split and used in **construction** to make walls [B2042]. The wood is also used in **construction** for making saltwater rafts and pilings for piers [RA].

Sabal yapa C.H. Wright ex Becc.
bay leaf palm, big thatch, thatch palm [English]; *xan* [Q'eqchi']; *botán*, *huano* [Additional names recorded]
The palm heart is eaten as a **food** prepared fresh, cooked in soups, or pickled in vinegar and salt [B3280]. This palm, along with cabbage palm (*Euterpe precatoria* var. *longevaginata*), is said to have a more bitter palm heart than the more commonly used and sweeter cohune (*Attalea cohune*) and royal palm (*Roystonea regia*); however, the former 2 species are said to be preferred for this bitter taste by some people in Belize [McSweeney, 1993].

The leaves are used for **thatch** [B3280]. The highly durable and water-resistant palm trunks are used in **con-**

(top) *Roystonea regia* [FA].

(middle) *Sabal mauritiiformis* [FA].

(bottom) *Sabal yapa* [MB].

struction for making outer walls of buildings and posts for piers said to last 50 years or more. Walls are typically made from split trunks and posts from whole trunks [B3280]. Lundell (1938) reported that the ancient Maya in the Petén and Yucatan areas used this palm for **thatch** and left the tree standing when clearing their agricultural fields. Some people in Belize are beginning to plant this palm as an **economic crop** for the harvest of its leaves [MB].

Schippia concolor Burret

mountain palmetto, mountain pimento, silver palmetto, silver pimento, silver thatch [English]

This palm is only found in Belize [Balick, Nee, and Atha, 2000]. It is cultivated as an **ornamental** [Balick and Johnson, 1994].

The pollen and nectar of the young flowers provide **food** for small, black, stingless bees (*Melipona* sp.) known locally as "drunken baymen" [B2065].

Thrinax radiata Lodd. ex Schult. & Schult. f.

fan palm, saltwater palmetto, saltwater pimenta, saltwater pimento [English]; *chit* [Additional name recorded]

The ripe, white fruit provides **food** for wild birds; hunters know this and seek their quarry near these palms [A412]. The fruit is also eaten as a **food** by humans [Horwich and Lyon, 1990].

To treat **diarrhea**, 1 handful of chopped roots is boiled in 1 quart of water for 10 minutes and ½ cup consumed every 2 hours until better [B1956].

As a **fish poison**, the leaves and stems are mashed and placed in a stream during times of low water for 1 day, after which the fish float to the surface, where they are easily harvested [B2141].

The trunk fibers are used as a **stuffing** for pillows and mattresses [Horwich and Lyon, 1990]. The leaves are used as **thatch** and may last as long as 15 years [A412].

Schippia concolor [FA].

Schippia concolor inflorescence [MB].

Thrinax radiata [FA].

Thrinax radiata with panicle of white fruit [MB].

The durable stem is used in **construction** for making pier posts and stakes, especially in saltwater [B1956].

ASPARAGACEAE (INCLUDING AGAVACEAE AND DRACAENACEAE)

Herbs to woody vines or shrubs or trees with multiple branches or not, sometimes with subterranean or columnar rhizomes. Leaves simple, alternate, occasionally opposite or whorled, sometimes succulent. Inflorescences axillary or terminal, in umbels, racemes, or spikes. Flowers whitish, yellowish, or greenish. Fruits a fleshy berry or capsule, reddish to purple-black. In Belize, consisting of 8 genera and 11 species. Uses for 6 genera and 7 species are reported here.

Agave sisalana Perrine

Thompson (1930) reported that the southern Maya in the San Antonio area (Toledo District) made **fishing nets** from the leaf fiber to use with fish poisons. Four pieces of vine from the liana *barbasco* (*Serjania lundellii*) were thrown into the water, and the net was stretched across a river, enclosing it at the top and bottom. Shortly thereafter, all of the fish within this area floated to the surface, where they were collected. Thompson also reported that these Maya made **hammocks** from the leaf fiber. The leaves were cut and dried in the sun or put quickly through a large fire. After 3 days, the fibers were scraped from the leaf with a machete and rolled between the hands to make long **cordage** for weaving.

Agave sisalana [MB].

Asparagus plumosus [DA].

Lundell (1938) noted that this plant and henequen (*Agave fourcroydes*) were both "probably cultivated by the Maya prior to the conquest." He reported that it was said to be exported from the Yucatan region as **cordage** to make binder twine as well as "used locally for making hammocks, nets, bags, ropes, cord, and other things."

Asparagus plumosus Baker
This plant is cultivated as an **ornamental** [B2318].

Cordyline fruticosa (L.) A. Chev.
This cultivated shrub is used as an **ornamental** [B2276].

Dracaena americana Donn. Sm.
candlewood [English]; *isote del monte* [Spanish]; *ixote* (male), *tuét*, *tuút* [Additional names recorded]
This plant is collected from the wild for use as a **houseplant** [B3270].
The leaves are stripped to obtain fiber used as **cordage** in the bush to tie things together [B3553].

Sansevieria hyacinthoides (L.) Druce
snake plant [English]; *culebrilla*, *curarina*, *lengua de suegra*, *lengua de vaca* [Spanish]
As a first aid treatment for **snakebite**, 3 leaves are boiled in 3 cups of water for 5–10 minutes and 1 cup consumed hot every hour. If no fire or stove is available, the leaf is chewed and the juices swallowed [B1972; RA]. After the leaves are chewed and the juices swallowed, leftover leaf fibers are used as a poultice applied directly to the bite [LR].
The stem fibers are prepared in a manner similar to henequen (*Agave fourcroydes*),

Cordyline fruticosa [MB].

Dracaena americana [GS].

producing a silk-like **fiber** [B1972]. To prepare the fiber, the leaves are pounded gently with a piece of wood, combed out straight, washed, rinsed in water, and hung in the sun to dry. It is said that the fiber is rather rough but becomes smooth and soft after washing [LR].

Standley and Steyermark (1952) reported that this plant was widely used in traditional **medicine** and was "one of the most esteemed of local plants."

Sansevieria trifasciata Prain

sansevieria, snake plant [English]; *culebrilla, lengua de vaca* [Spanish]

To treat non-insulin-dependent **diabetes** mellitus, 1 leaf is mashed in 2 cups of water and 1 cup consumed twice daily, in the early morning before breakfast and again before dinner [A91; Arvigo and Balick, 1998]. Alternatively, fresh leaves are chewed [A91; Arvigo and Balick, 1998].

Sansevieria trifasciata [IA].

To treat **skin sores**, the leaves are mashed and applied as a poultice over the affected area [A91]. For **skin sores** and **rash**, the leaves are boiled in water and used as a bath over the affected area [Arvigo and Balick, 1998].

Fresh leaves are used in 3 ways to treat **snakebite**. They are chewed all day long and the juices swallowed, used to prepare a decoction that is consumed, or mashed and applied directly as a poultice to the bite until the symptoms improve [A91; RA; Arvigo and Balick, 1998]. Alternatively, refer to hog bush (*Psychotria pubescens*) [A488].

To **keep chickens healthy**, the leaves are mashed and the juice placed in their drinking water [Arvigo and Balick, 1998]. To **maintain health**, **prevent illness**, and to treat **skin sores of chickens**—for example, **weakness**, **colds**, and **skin fungus**—the leaves are mashed in water and applied as a poultice to the affected area [A91].

To make **cordage**, the leaf is pounded and the fibers extracted for use as rope [Arvigo and Balick, 1998].

Yucca guatemalensis Baker
izote, isote, yucca [Spanish]
Standley and Record (1936) noted that the young flowers were cooked and eaten as a **food**. Lundell (1938) reported that the young flowers could also be mixed with meats and cooked.

Yucca guatemalensis with a collector sorting plants [MB].

During our fieldwork, we noted people harvesting thousands of plant stems from the forest; the plants ranged in size from 18 inches to several feet long. In speaking to some of the people working at an "*isote* camp," they told us that these stems were being shipped to Europe for use in the **horticultural trade** [MB].

BROMELIACEAE

Small to large herbs, terrestrial or epiphytic, solitary to numerous, sometimes on vine-like stems. Leaves simple, stiff to soft, alternate, arranged in a rosette, sometimes forming a cup, sometimes with spines on the margins. Inflorescenses sessile or formed on a scape, often showy with bracts subtending the main axis. Flowers variously colored. Fruits a capsule, berry, or aggregation of berries forming a multiple fruit. In Belize, consisting of 10 genera and 51 species. Uses for 5 genera and 8 species are reported here.

Aechmea bracteata (Sw.) Griseb.
water orchid, wild pineapple [English]
This plant is cultivated as an **ornamental** [A476; LR].

Bushmasters know that the basal rosette of leaves serves as a reservoir, providing a valuable source of **water** for people working or traveling in the bush [A476; LR].

Aechmea magdalenae (André) André ex Baker
silk grass [English]
According to Standley and Record (1936), "the leaves of this plant furnish one of the best fibers known, remarkable for its fineness, strength, and length." Schultes (1941) provided a very detailed description of the use of this plant as a **fiber** source in southeastern Mexico, Central America, and northwestern South America. He

concluded that "although fibres of most of the Bromeliaceae have not, on the whole, been commercially promising, that of *Aechmea magdalenae* is of superior quality. It has been shown to possess great powers of resistance to the effects of saltwater." This fiber is made into "cordage or even cloth" [Horwich and Lyon, 1990].

Aechmea tillandsioides (Mart. ex Schult. & Schult. f.) Baker

gallinasco [Spanish]; *ik'l, paajch'* [Q'eqchi']; *chub* [Yucateca]

The roots of the mature plant are made into a **brush** used for household chores [B3567]. It takes about 10 years for the plant to grow roots strong and large enough to make a brush [BW].

Ananas comosus (L.) Merr.

pineapple, pine [English]; *piña* [Spanish]

This plant is cultivated for its fruit, which is eaten as a **food** [B2284, DA1064]. According to Thompson (1927), the Maya did not know about this plant until it was introduced after the Spanish colonization.

To treat **kidney stones**, refer to wild okra (*Parmentiera aculeata*) [LR].

For **intestinal worms**, 1 large unripe but nearly mature fruit is grated, the juice squeezed out, 3 tablespoons consumed by children once daily, and 6 tablespoons consumed by adults once daily, in the morning before eating on the new moon. If no worm passes in 24 hours, the same formula is mixed along with ½ cup of milk and 2 tablespoons

(top) *Aechmea bracteata* showing inflorescence [IA].

(upper middle) *Aechmea magdalenae* silk grass fiber [RBGKew].

(lower middle) *Aechmea tillandsioides* showing inflorescence [MB].

(bottom) *Ananas comosus* [IA].

of sugar and the dosage repeated. This treatment is only done within a 7-day period of the new moon because of a belief that worms are more vulnerable to the medicine during that time, as they are outside of a protective "sack" that they have during other phases of the moon [RA].

Androlepis skinneri Brongn. ex Houllet

paajch' [Q'eqchi']; *tchu vc, tchu véc* [Mayan]

The dried inflorescence was once used as a scrub **brush** to do household chores [B1807].

Bromelia pinguin L.

wild pine, wild pineapple [English]; *piña de puerco, piñuela, pitapuerco* [Spanish]; *ix tot* [Mopan]

(top) *Androlepis skinneri* leaf and inflorescence [MB].

(bottom) *Bromelia pinguin* [IA] with inset showing fruit and leaves [MB].

The ripe fruit is eaten as a **food** [B1808, B2048; Arvigo and Balick, 1998]. The young flowers are also eaten, coated with flour, and fried in oil like the pacaya blossom (*Chamaedorea tepejilote*) [B1808, B2048]. The "heart" (*cogollo*) is boiled, fried with eggs, and eaten. To make a **beverage**, 6–9 ripe fruits are mashed in 1 quart of water, strained, and consumed [B1808, B2048].

For **sprains**, **fractures**, or **broken bones**, the leaves are pounded with salt and applied as a poultice directly to the affected area [Arvigo and Balick, 1998].

The leaf is lightly beaten with a stick and hung to dry, after which the fibers are easily separated and used as **cordage** for making rope, baskets, and hammocks [B1808, B2048; Arvigo and Balick, 1998].

It is said that this is possibly the species whose ripe fruits were hung in structures where animals were kept to **repel vampire bats**, locally known as "rat bats" (*Desmodus rotundus*) [Robinson and Furley, 1983].

Bromelia plumieri (E. Morren) L.B. Sm.

pim i winq, chac chom [Q'eqchi']

To treat children with **parasites**, "several" seeds of this plant are boiled in water and eaten 2–3 times daily [Mallory, 1991]. Steggerda and Korsch (1943) noted that the Yucatan Maya took the "dusty portions" of this plant and used them to cover wounds to staunch **bleeding**.

Vriesea gladioliflora (H. Wendl.) Antoine
ek' [Q'eqchi']
The leaves are cut from the plant, and the remaining basal section, including roots, is dried to make **brushes** for cleaning tables and doing other household chores [B3616].

CANNACEAE

Herbs 3–8 feet tall with horizontally branched rhizomes. Leaves alternate and simple. Inflorescences terminal, usually branched, sometimes simple, with green bracts. Flowers primarily red, yellow, or white. Fruits a capsule. In Belize, consisting of 1 genus and 2 species. Uses for 1 genus and 1 species are reported here.

Canna indica L.
canna [English]; *huevo de gato*, *platanillo* [Spanish]; *naq'rit li mis* [Q'eqchi']; *x'chi qui laba* [Mayan]
This plant is used as an **ornamental**. It is said to be occasionally, if rarely, cultivated [B2312].
Some people report that the starchy root is eaten as a **food** [RA]. The leaves are used to **wrap** tamales for cooking [B2393].

COMMELINACEAE

Herbs, succulent and usually terrestrial with erect to spreading stems. Leaves alternate, simple, sometimes purple on the undersides. Inflorescences terminal or axillary. Flowers green, white, or purple. Fruits a capsule. In Belize, consisting of 6 genera and 21 species. Uses for 3 genera and 5 species are reported here.

(top) *Vriesea gladioliflora* with inset showing fruit [HW].

(bottom) *Canna indica* with close-up of fruit [MB].

Callisia cordifolia (Sw.) E.S. Anderson & Woodson
wandering Jew [English]; *la cucaracha* [Spanish]
To stop the pain of a **bee sting**, a 1-inch piece of 1 leaf is mashed with oil and applied to the affected area [BW]. To treat **headache**, 1 entire plant is soaked in a small amount of water, squeezed, the juice strained onto a cloth, and the cloth applied as a plaster to the head [A331]. To treat **earache**, 3 leaves are passed over a flame and 3 drops of leaf juice squeezed into the ear, following which the area behind the ears is massaged. To treat **hemorrhaging** associated with menstruation, cysts, or fibroids, 1 handful of leaves and stems is washed, mashed with a small amount of water to

Commelina erecta [MB].

Tradescantia spathacea [MB].

extract the juice, and 1 teaspoon consumed each hour. Hemorrhaging is said to stop on the third dose [LR].

Commelina diffusa Burm. f.

sacatinto morado [Spanish]

To reduce **inflammation** due to conjunctivitis (pinkeye), the flowers are crushed, and 3 drops of the sap are squeezed into the eye 3 times daily until better [B2291].

Commelina erecta L.

little bamboo [English]

For **clearing vision** from an eye infection or substrate formed on the eye, the juice of the stem is dripped into the eye, drop by drop, until the person sees more clearly [B2118]. For **sinus headache**, refer to warrie wood (*Diphysa americana*) [B2155]. To treat **erysipelas**, **sores**, or **swelling**, refer to *coqueta* (*Rivina humilis*) [B2123].

To treat **kidney stones**, **urine retention (stoppage of water)**, **burning urine**, as well as **constipation**, the fresh or dried leaves are boiled in water and consumed as a tea [B2118]. For **kidney stones** or **urine retention (stoppage of water)**, ½ quart of fresh leaves and stems along with ½ quart of fresh "female" chicken weed leaves and stems (*Chamaesyce mendezii*), ½ quart of fresh "male" chicken weed leaves and stems (*Chamaesyce hypericifolia*), and ½ quart of fresh *tan chi* leaves and stems (*Capraria biflora*) are boiled in 2 gallons of water and used as a steam bath. The person sits over the boiling pot of water while holding a blanket over their body [B2118]. To treat **constipation**, **urine retention**, or **kidney stones**, refer to *tan chi* (*Capraria biflora*) [B2116]. For **urine retention (stoppage of water)** where there is a blocked urinary flow, refer to bay cedar (*Guazuma ulmifolia*) [B2132].

Tradescantia spathacea Sw.

maguey sylvestre, *pabana* [Spanish]

To treat **high blood pressure**, 2 leaves are mashed in 2 cups of water and 1 cup consumed every second day until better [A518]. To treat **sprains**, **fractures**, and

swelling, the leaves are mashed with a pinch of salt and applied as a poultice to the affected area [A518]. To stop **hemorrhaging**, 3 leaves along with 9 *tulipan del monte* leaves (*Malvaviscus arboreus* var. *mexicanus*) are crushed, boiled in 1 quart of water for 20 minutes, and consumed cool as needed until hemorrhaging stops. This is said to stop the hemorrhage within minutes [B2327]. Alternatively, refer to *tulipan del monte* (*Malvaviscus arboreus* var. *mexicanus*) [B2357].

Tradescantia zebrina Hort. ex Bosse var. *zebrina*

club moss, wandering Jew [English]
To treat **earache**, juice is squeezed from 1 entire plant, the liquid strained, and 4 drops placed in the ear [A324]. For **pain** and **internal infections**, especially relating to the ovaries, uterus, and other organs, 1 entire plant is boiled in 1 quart of water for 20 minutes and 1 cup consumed 3 times daily for 9 days [A324].

COSTACEAE

Herbs to 8–9 feet tall with horizontally branched rhizomes. Leaves alternate, simple. Inflorescences terminal or axillary, in spikes or of solitary flowers. Flowers white, yellow, orange or red, and showy. Fruits a capsule. In Belize, consisting of 1 genus and 5 species. Uses for 1 genus and 2 species are reported here.

(top) *Tradescantia zebrina* var. *zebrina* [MB].

(bottom) *Costus pulverulentus* with inflorescence [IA].

Costus guanaiensis Rusby var. *macrostrobilus* (K. Schum.) Maas

w'eh te [Mayan]
The seeds are used as **bait** to trap birds [A623].
To **stimulate hair growth**, the young shoots are mashed in water and used as a hair rinse [A623].

Costus pulverulentus C. Presl

spiral ginger, wild ginger [English]; *caña de Cristo* [Spanish]; *ch' uun* [Q'eqchi']
The root is **edible** and used like ginger (*Zingiber officinale*) in **flavoring** foods [B3228]. As a **beverage**, the stem is peeled, and the sap is squeezed out, strained

into water, and consumed fresh. When mixed with corn (*Zea mays*) and allowed to ferment, the beverage, known as *chicha*, is a traditional drink [B2702]. The flowers are popular with hummingbirds because of their sweet **nectar** [N46823]. Local people will suck the **nectar** from the young, small flowers, which is said to provide **energy** [R41]. The entire plant is used as animal **fodder** [A223].

To treat **conjunctivitis (pinkeye)**, 3–4 drops of juice are squeezed from the stem directly into the affected eye as needed several times daily until better [B3228]. For **burning feet** (often in non-insulin-dependent **diabetes** mellitus), 9 leaves are mashed in water, set in the sun for 1 hour, and used to soak the feet for 20 minutes. This is repeated 7–9 times for relief [BW]. For **urine retention (stoppage of water)**, 3 roots and one 8-inch piece of an aboveground stem are boiled in 2 quarts of water for 10 minutes and ½ cup consumed every 2 hours [LR]. To treat **urinary infection**, the stem is eaten [B3544].

CYCLANTHACEAE

Herbs or shrubs, with subterranean rhizomes or epiphytic lianas. Leaves usually alternate, bifid to palmate in shape. Inflorescences terminal or lateral with several bracts subtending a spadix. Flowers white, fragrant. Fruits a fleshy syncarp of berries. In Belize, consisting of 3 genera and 5 species. Uses for 1 genus and 1 species are reported here.

Carludovica palmata Ruiz & Pavón
Panama hat palm [English]; *jippy jappa* [Creole]; *k'ula* [Mopan]; *kaláh* [Q'eqchi']; *hu unco* [Additional name recorded]
The young inflorescences are boiled, cooled, fried, and eaten as a **food** [B3534].

Small, white baskets are tightly woven using washed, sun-bleached, dried **fibers** harvested from the young, unrolled leaves emerging from the center of the plant.

Carludovica palmata with inset showing inflorescence [IA].

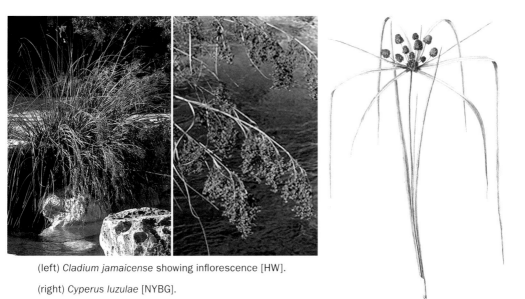

(left) *Cladium jamaicense* showing inflorescence [HW].

(right) *Cyperus luzulae* [NYBG].

These baskets last a long time and are a **handicraft** commonly sold by people in Belize [B3534].

CYPERACEAE

Grass-like herbs, often with rhizomes. Leaves narrow and grass-like. Inflorescences terminal or axillary, in panicles, corymbs, clusters, anthelae, heads, spikes, or spikelets. Flowers very small. Fruits an achene. In Belize, consisting of 21 genera and 151 species. Uses for 6 genera and 7 species are reported here.

Cladium jamaicense Crantz
curring grass, paint brush grass [English]
The lower stems (the thicker parts closest to the ground) are mashed and used as household paint **brushes** [A431].

Cyperus laxus Lam.
el sueño [Spanish]
To treat **gonorrhea**, 3 whole plants are boiled in 2 quarts of water until reduced to 1 quart, and 1 cup is consumed hot 3 times daily as needed [B2717]. To **induce sleep** in babies, the entire plant is used as a pillow stuffing [A145].

Cyperus luzulae (L.) Retz.
This plant is used as **fodder** for horses [G28].
For **urine retention (stoppage of water)** or **bad urine (*mal de urina*)**, 2 leaves and 2 root clumps are steeped in 3 cups of hot water, and one 6-inch piece of sugarcane root (*Saccharum officinarum*) is chopped and added when warm, after which the infusion is sipped warm all day long [B2662].

Eleocharis geniculata (L.) Roem. & Schult.

This is a seaside lawn grass that can be trimmed and used as an **ornamental** [A419].

Fimbristylis cymosa R. Br.

This plant is used for **ornamental** purposes, such as a ground cover [A423].

Rhynchospora cephalotes (L.) Vahl

For **urine retention (stoppage of water)** or **bad urine (*mal de urina*)**, 2 leaves and 2 root clumps are steeped in 3 cups of hot water, and one 6-inch piece of sugarcane root (*Saccharum officinarum*) is chopped and added when warm, after which the infusion sipped warm all day long [B2625].

Scleria secans (L.) Urb.

old lady scissors [English]

To relieve pain of a **sinus headache**, the apical meristems are boiled in 1 cup of water for 10 minutes, strained, cooled, and used as nose drops [B2606].

DIOSCOREACEAE

Vines or lianas arising from large, tuberous roots. Leaves alternate or opposite, entire or lobed. Inflorescences axillary or rarely terminal, in spikes, racemes, panicles, or cymes. Flowers inconspicuous, usually numerous. Fruits a capsule or samara. In Belize, consisting of 1 genus and 12 species. Uses for 1 genus and 1 species are reported here.

(top) *Eleocharis geniculata* [CI].

(middle) *Fimbristylis cymosa* with inset showing fruit [ST].

(bottom) *Rhynchospora cephalotes* [MB].

Dioscorea bartlettii C.V. Morton

old man's beard, white China root, white cocolmecca, wild yam [English]; *barba del viejo, cocolmeca blanca, marcador* [Spanish]

For **arthritis**, **back pain**, **rheumatism**, **fever**, the onset of non-insulin-dependent **diabetes** mellitus, **bilious colic**, **urethral inflammation**, **bladder infection**, **urine retention (stoppage of water)**, **kidney ailments** (sluggishness and malfunction),

(top) *Scleria secans* showing fruit (left) [RF].

(bottom) *Dioscorea bartlettii* vine with Rosita Arvigo and Leopoldo Romero (left) [GS]; mature tubers (right) [MB].

blood circulation, **blood cleansing**, **anemia** resulting from internal hemorrhaging, to **build the blood**, and to **loosen mucus** in coughs and colds, 1 small handful of the chopped tuber with the attached side roots is boiled in 3 cups of water for 10 minutes and ½ cup consumed 6 times daily, before and in between meals, as needed until better [A527, A758, A761, B1787; Arvigo and Balick, 1998]. To treat **internal hemorrhaging**, 9 small pieces of root are boiled in 2 quarts of water until reduced to 1 quart, after which ½ cup is consumed every 30 minutes until the bleeding stops [Arvigo and Balick, 1998]. To **improve blood circulation**, refer to *golondrina* (*Alternanthera flavogrisea*) [LR] and *dormilón* (*Mimosa pudica* var. *unijuga*) [LR]. For a **bladder infection**, refer to yellow malva (*Malvastrum corchorifolium*) [LR].

To **prevent miscarriage** when a pregnant woman suspects she may not be able to carry to full term, 1 small handful of the chopped tuber along with 1 tablespoon of chopped ginger root (*Zingiber officinale*) is boiled in 3 cups of water for 10 minutes and 1 cup consumed 3 times daily, before meals, for an unspecified amount of time [B1787]. Alternatively, 1 small handful of chopped root along with 1 ounce of cinnamon stick (*Cinnamomun verum*) are boiled in 3 cups of water for 10 minutes and sipped cool all day long for 3–7 days [Arvigo and Balick, 1998].

In the past, to **stamp** *chicle* blocks (*Manilkara zapota*) for identification and payment, a solid block of the tuber was cut out, the *chicleros* carved their initials into the block, and then used it as a stamp [LR].

As a **female contraceptive**, 1 young tuber about the size of a potato (*Solanum tuberosum*) is chewed, the juice swallowed, and the tuber pieces spit out. This is done 2 days after menstruation stops [LR]. Alternatively, 1 tuber may be boiled in water and consumed as a tea, although it has been reported that this tea causes constipation [LR]. Interestingly, for the treatment of male **impotence** and female **infertility**, a different preparation was reported—1 cup of the chopped tuber is soaked in 1 quart of alcohol (gin for men and *anisado* [anise liquor, *Pimpinella anisum*] for women) for 10 days and 1 tablespoon consumed 3 times daily.[1] During this treatment, cold drinks must be avoided [Arvigo and Balick, 1998]. To remove excess **mucus** in the prostate and uterus, refer to *guaco* (*Aristolochia grandiflora*) [LR]. To quickly clear **mucus** from the male urinary tract and to treat **prostate ailments**, refer to *flor azul* (*Miconia albicans*) [LR].

HELICONIACEAE

Herbs to several meters tall with creeping, branched rhizomes. Leaves alternate, distichous, simple. Inflorescences terminal, large, either erect or pendent. Flowers with large and often brightly colored bracts. Fruits a drupe. In Belize, consisting of 1 genus and 11 species. Uses for 1 genus and 7 species are reported here.

---

1. Note that the two reported traditional uses, as a contraceptive and to treat infertility, have different extraction methods (chewing and swallowing the juice vs. soaking in alcohol), which would result in a different set of compounds being ingested. The different compounds present in these extracts and their effects might be an interesting topic for future investigation.

Heliconia bourgaeana [MB].

(top) Heliconia latispatha [MB].

(bottom) Leaf and inflorescence of Heliconia librata [MB].

Heliconia aurantiaca Ghiesbr.

wild plantain [English]; *plantanillo* [Spanish]; *tz'ukl* [Q'eqchi']

As a **calmative** for babies who tremble, cry, and are hysterical, 1 leaf is boiled in 2 quarts of water for 10 minutes, ¼ cup consumed cool, and the remainder used as a bath. This treatment may be repeated a second time if necessary [B2485].

Heliconia bourgaeana Petersen

waha leaf, wild plantain [English]; *mox* [Q'eqchi']

To reduce **swelling**, 1 entire plant is chopped, mashed with salt, and applied as a poultice [B2713]. Alternatively, 1 entire plant is steeped in 2 gallons of hot water for 30 minutes and used warm to bathe the affected area [B2713].

Heliconia latispatha Benth.

platanillo [Spanish]

To treat **burns** on the body and **prevent scabbing**, young leaves are applied to the affected area. If the burns are widespread, the person is placed on a bed of the leaves. The mucilage in the leaves is said to help [B2654].

Heliconia librata Griggs

platanillo [Spanish]

The fruit provides **food** for birds, and the leaves provide **fodder** for other animals. Hunters know this and seek their quarry near this plant [B2012].

To draw out **splinters** and **thorns**, the root is macerated and applied as a poultice directly to the affected area [B2012].

Heliconia mariae Hook.f.
wild plantain [English]
In times of famine or need, the "heart" of the stem is peeled, cooked, mixed with other vegetables, or roasted alone on a fire and then eaten as a **food**. It is often boiled and sometimes refried [A595].

Heliconia vaginalis Benth. subsp.
mathiasiae (Daniels & Stiles) L. Anderss.
waha leaf [English]; *mox* [Q'eqchi'];
wah, *wahá*, *wahe* [Additional names recorded]
To treat **swelling**, 1 entire plant is chopped and applied as a poultice [B2697]. Alternatively, 1 entire plant is steeped in 1 gallon of warm water for 3 hours in the sun and used to bathe the affected area [B2697].

Heliconia wagneriana Petersen
tz'ukl [Q'eqchi']
The young, inrolled, emergent leaves are eaten as a **food** either raw or chopped and cooked like noodles [B3517].

MARANTACEAE

Herbs with horizontally branched rhizomes. Leaves alternate, distichous, simple. Inflorescences terminal. Flowers yellow to white to purple. Fruits a capsule. In Belize, consisting of 5 genera and 11 species. Uses for 3 genera and 5 species are reported here.

(top) *Heliconia mariae* with inset showing old inflorescence [MB].

(middle) Inflorescence of *Heliconia vaginalis* subsp. *mathiasiae* [MB].

(bottom) *Heliconia wagneriana* [MB].

Calathea crotalifera S. Watson

waha leaf [English]; *mox* [Q'eqchi']

To expel a **retained placenta**, 1 leaf with stem is chopped and boiled in 1 quart of water for 10 minutes and ½ cup consumed cool every 10 minutes until the bleeding stops [B2700].

The leaves are used to **wrap** tamales for cooking and serving [B2700].

Calathea lutea (Aubl.) Schultes

waha leaf [English]; *hoja de sal* [Spanish]; *mox* [Q'eqchi']; *mu shan* [Additional name recorded]

The uses for *Calathea lutea* and *C. crotalifera* are essentially the same, and they share the local names of waha leaf and *wahá*, said to be African names [B2016, B2701, B3237].

To expel a **retained placenta**, ½ cup of chopped leaves and ½ cup of chopped stems are boiled in 3 cups of water for 10 minutes and sipped slowly while hot until the placenta passes [B2701]. For **bloody urine**, especially in men, refer to black cockspur (*Acacia collinsii*) [A342; RA].

The leaves are used to **wrap** non-edible items for storage [B2016, B2701, B3237]. The leaves are also used to **wrap** tamales and chicken for cooking, baking and serving [B2016, B2701, B3237, M223].

Calathea lutea leaves (left), waxy underside of leaf (center), and inflorescence (right) [IA].

Maranta arundinacea L.

wild arrowroot [English]; *camotillo, sagú* [Spanish]; *xe' tz'in* [Q'eqchi']; *chaac* [Yucateca]

Standley and Record (1936) reported that this species was cultivated and also grew in the wild. They noted that people produced **starch** from the roots of the cultivated plant.

Maranta gibba Smith

mountain arrowroot, tie tie, wild arrowroot [English]

The stems have been used on occasion as a **fiber** to weave baskets [A37].

Thalia geniculata L.

purple platanillo [English]; *platanar* [Spanish]

The leaves are used to **wrap** tamales for cooking and serving [A367].

MUSACEAE

Large to giant herbs with corm-like rhizomes. Leaves alternate, arranged in spirals and simple. Inflorescences terminal and large, in cymes. Fruits a berry with a yellow-to-red, leathery pericarp. In Belize, consisting of 1 genus and 2 species. Uses for 1 genus and 2 species are reported here.

Musa acuminata Colla

banana [English]; *guineo* [Spanish]; *box, b'oosh haaz* [Mopan]; *q'eq i tul* [Q'eqchi']

The mature fruit is eaten as a **food** [Arvigo and Balick, 1998]. The fruit is easily digested and used as a **first food** for infants and invalids [Arvigo and Balick, 1998]. As a **first food** for children and to treat **indigestion, gastric ulcers, colitis, hepatitis, diarrhea,** and people in **weakened conditions,** refer to coconut (*Cocos nucifera*) [Arvigo and Balick, 1998]. The

(top) *Maranta arundinacea* [MB].

(middle) *Maranta gibba* showing old flowers fruit [MB].

(bottom) *Thalia geniculata* [RH].

green, immature fruit is eaten as a **food**. It is boiled in water in the skin until tender, after which the skin is removed and the fruits mashed, formed into cakes, and cooked like a *tortilla* or pancake [Arvigo and Balick, 1998]. The juice of the stem is rich in potassium [Arvigo and Balick, 1998].

To treat a **blister**, **burn**, or an **abscess**, fresh stem juice is applied directly to the affected area [Arvigo and Balick, 1998]. Alternatively, for a **blister** or **burn**, the young, green leaves are wrapped as a plaster directly on the affected area [Arvigo and Balick, 1998]. To treat "chiclero **sore**" or **chronic sores**, the sap is mixed with freshly ground black pepper (*Piper nigrum*), placed on a clean cloth, and applied as a poultice to the affected area 3 times daily [Arvigo and Balick, 1998].

As a **diuretic**, fresh stem juice is consumed [Arvigo and Balick, 1998]. For

Musa acuminata [IA].

gastritis (*ciro*) or **ulcers**, the fruit is eaten [Arvigo and Balick, 1998]. To treat **diarrhea**, the well-ripened fruit is consumed exclusively for 24 hours [Arvigo and Balick, 1998]. To staunch excessive **hemorrhaging** in childbirth, the root (solid white corm below the ground) is washed well, grated, squeezed through a porous cloth to obtain liquid, and 1 tablespoon consumed every 5 minutes for a maximum of 3 doses [Arvigo and Balick, 1998].

The leaf is used to **wrap** tamales for an added, pleasant taste [Arvigo and Balick, 1998].

To treat **head lice** (*Pediculus humanus*), refer to wild squash (*Cionosicyos excisus*) [HR].

Musa spp.[2]

plantain [English]; *banana de marano, platano* [Spanish]; *box, b'o'osh haaz* [Mopan]; *q'eq i tul* [Q'eqchi']

This fruit is eaten as a **food** and used widely in Latin American cooking [B2398, B2408].

For **headache**, fresh, green, or ripe fruits are grated and applied to the forehead as a poultice [HR]. When a **fever** accompanies the headache, a second poultice is

---

2. Also referred to as *Musa* × *paradisiaca* as well as by designations that more specifically reflect the genetic origin of the cultivated plant.

Musa spp. (plantain) with inset of fruit [MB].

prepared and applied to the feet; this poultice is left on until the fever decreases [HR]. To treat **vomiting** and **diarrhea**, 1 skin of the green fruit is burned to ash, powdered, 1 teaspoon added to ½ cup of warm water, stirred until the ash is fully dissolved, and ½ cup consumed every 30–60 minutes depending on the severity of conditions. The dosage is lessened as the person improves [LR]. To treat **foot fungus**, the stem is cut down to 3 inches above the ground after harvesting the fruits. A hole is dug in the center of the stem that produces a liquid in 12–24 hours. This liquid is collected and applied to the feet once daily. It takes about 3 days of treatment to heal. After the 3-day treatment, the fungus is treated for 3 days with the powdered skin from the green fruit, described above for vomiting [LR]. To treat **tuberculosis**, the leaf sap is used in Maya traditional medicine, but no further details were provided [Flores and Ricalde, 1996].

The green leaves are used to **wrap** tamales for cooking and serving [LR].

ORCHIDACEAE

Herbs, terrestrial or epiphytic with fleshy roots. Leaves usually alternate, simple, thin to leathery. Inflorescences axillary, basal or terminal of solitary flowers, racemes, or panicles. Flowers small to large and showy. Fruits a capsule. In Belize, consisting of 93 genera and 279 species. Uses for 3 genera and 3 species are reported here.

(top left) *Lockhartia pittieri* [BA].

(top right) *Sobralia fragrans* [©2007 Greg Allikas www. orchidworks.com].

(bottom) *Vanilla planifolia* vine (left) [IA] with inset of flower [BBG]; fruit (right) [MB].

Lockhartia pittieri Schltr.

As an antidote to **snakebite**, the leaves are mashed with rocks and applied directly to the bite. This is said to break down the proteins contained within the venom and has the ability to "pull" out the venom from the person's flesh.

Sobralia fragrans Lindl.

teelom pim [Q'eqchi']

To produce **sterility** in women of childbearing age who should not or do not want to have children, 1 handful of leaves is boiled in 3 cups of water for 10 minutes, strained, and 1 cup consumed lukewarm once daily for 3 days.

Vanilla planifolia Jacks.

vanilla [English]

The fruiting pods are placed in boiling water, fermented, dried, and used as a **flavoring** [Horwich and Lyon, 1990]. Lundell (1938) noted that the Petén Maya used this as flavoring in cacao beverages (*Theobroma cacao*). Today, the pod is added to cacao beverages as flavoring.

As a **love charm**, a person might make the following potion to give to their wayward lover to keep that lover from leaving. Nine seedpods are split open, and the inner part, which is moist and gel-like, is mixed with 2 drops of skunk urine (*Conepatus* spp.) and 3 drops of the *Siete Macho* perfume (a popular perfume sold in stores). After 9 days of steeping, this mixture is used as a perfume for 9 consecutive Fridays and a prayer said each Friday, during which the lover's name is mentioned. To get the skunk urine, the skunk must be caught in a trap, held, and massaged over the bladder to pass the urine [A545; LR]. To attract a lover, the liquid described above as a love charm is poured onto a handkerchief, which is later taken out of the pocket and held downwind toward the person desired to be a lover [LR].

POACEAE

Small herbs to woody bamboos. Leaves alternate, distichous or spiral. Inflorescences compound, usually in spikes or panicles. Flowers usually small. Fruits a caryopsis. In Belize, consisting of 78 genera and 258 species. Uses for 11 genera and 11 species are reported here.

Cymbopogon citratus (DC.) Stapf.
lemon grass, fever grass [English]; *zacate limón* [Spanish]

To make a **stimulant beverage**, pleasant **tonic**, **tea substitute**, or **coffee substitute**, the leaves are used [Arvigo and Balick, 1998]. **Caution** is used while drinking this tea, as it is not recommended for daily consumption by those with heart trouble, who have extreme weakness, or are aged [Arvigo and Balick, 1998].

To treat **cough**, **colds**, **stomach cramps**, **promote sweat**, or excrete **phlegm**, 9 leaves are lightly chopped, simmered in boiling water for 5 minutes, and consumed as

Cymbopogon citratus
[IA].

needed. This tea is especially useful for children and infants [RA; Arvigo and Balick, 1998]. To **cleanse the system**, an unspecified tea is prepared and consumed [Mallory, 1991].

To treat **fever** in adults, 10 leaves and 1 mashed root are boiled in 3 cups of water for 10 minutes and then consumed very hot, following which the person is wrapped warmly in blankets and put to bed [Arvigo and Balick, 1998]. To treat **fever** in children, 10 leaves are boiled in 3 cups of water for 10 minutes and ½ cup consumed 6 times daily while the child is kept warm [Arvigo and Balick, 1998]. Alternatively, refer to John Charles (*Hyptis verticillata*) [Arvigo and Balick, 1998].

To relieve **back pain**, **headache**, and **muscle spasms**, the root is mashed, soaked in oil, and rubbed on the affected area [Arvigo and Balick, 1998]. For **headache**, refer to *rosa* (*Rosa chinensis*) [B2333].

Cynodon dactylon (L.) Pers.

bahama grass, bermuda grass [English]; *grama* [Spanish]; *pach'aya'* [Q'eqchi']; *canzuuc* [Additional name recorded]

To **build the blood** for people who feel tired and run down, the fresh or dried root of 1 plant is chopped, boiled in 1 quart of water for 10 minutes, and 3 cups of the decoction consumed daily for 9 days [B2089].

Eragrostis excelsa Griseb.

Known as "a grass that cuts," this plant is considered a **hazard** to people who walk through it unprotected, as the sharp edges of the leaves will cut bare legs and feet [A413].

Gynerium sagittatum (Aubl.) P. Beauv.

caña brava [Spanish]

To treat **gonorrhea** and **difficult urination**, the leaf is chewed [Arnason et al., 1980].

Imperata contracta (H.B.K.) Hitchc.

fox tail grass [English]

Lundell (1938) reported that the leaves were used by the Petén Maya for **thatch** and **filling** pack saddles.

(top) *Cynodon dactylon* [FA].

(bottom) *Gynerium sagittatum* [DA].

Lasiacis ruscifolia (H.B.K.) Hitchc.

zit [Yucateca]

Steggerda (1943) noted that the Yucatan Maya women used the stems as a **fiber** to weave carpets and the young boys used the hollow reeds for **whistles** and to **drink water** from shallow water holes.

Oryza sativa L.

rice [English]; *arroz* [Spanish]

This is an important, essential **food** crop in Belize [MB]. To give rice a red color, refer to *achiote* (*Bixa orellana*) [Mallory, 1991]. It is cultivated in lower, wetter areas and planted from mid-May through June during heavy rainfall [Horwich and Lyon, 1990]. To prepare canoes for hulling rice, refer to *guanacaste* (*Enterolobium cyclocarpum*) [Horwich and Lyon, 1990].

For a **fish poison** tradition, refer to *barbasco* (*Serjania lundellii*) [Thompson, 1930].

Panicum maximum Jacq. var. *maximum*

guinea grass [English]; *nim laj pach'aya'* [Q'eqchi']; *sacaton* [Additional name recorded]

This plant is used as **forage** for cattle and horses [DA1083].

Saccharum officinarum L.

sugarcane [English]; *caña de azucar* [Spanish]; *utz'a'ajl* [Q'eqchi']; *tó* [Additional name recorded]

This plant is a very important **commercial crop** in Belize, especially in the north [MB]. It is cultivated and processed to make table **sugar** [MB]. To make **sugar**, refer to Pine Ridge bay cedar (*Trema micrantha* var. *floridana*) [B2441]. To make sure the **sugar** comes out well, refer to Pine Ridge bay cedar (*Trema micrantha* var. *floridana*) [B2441]. For a **sugar substitute**, refer to *patito* (*Amphilophium paniculatum*) [B3207].

As a snack, the sweet sap is extracted from the raw cane by chewing on the stem, taking care not to swallow too many of the fibers along with the juice [B2392]. To

(top) *Oryza sativa* [MB].

(bottom) *Panicum maximum* var. *maximum* [ST].

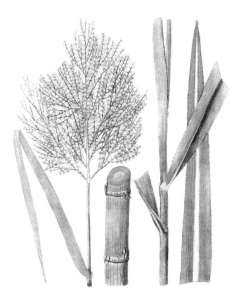

Saccharum officinarum [NYBG].

Saccharum officinarum on a sugarcane press on Chaa Creek [MB].

prepare as a **food**, refer to *cocoyal* (*Acrocomia aculeata*) [Horwich and Lyon, 1990] and craboo (*Byrsonima bucidaefolia*) [Steggerda, 1943]. To make **jam**, refer to coco plum (*Chrysobalanus icaco*) [A414]. To make **wine**, refer to wild cherry (*Pseudolmedia spuria*) [B3733; RA] and sea grape (*Coccoloba uvifera*) [A411]. Thompson (1927) noted that this plant was mixed with corn (*Zea mays*) in water and left to ferment for several days, producing a mildly intoxicating **beverage**. *Chicha*, as this drink is known, is still produced today [MB]. To make a **cacao beverage**, refer to cacao (*Theobroma cacao*) [MB, RA].

To treat a raspy, **sore throat** in babies, refer to periwinkle (*Catharanthus roseus*) [BW]. To treat **cough** and loosen **mucus**, refer to mango (*Mangifera indica*) [BW] and *annona* (*Annona reticulata*) [Arvigo and Balick, 1998]. To treat **intestinal worms**, refer to pineapple (*Ananas comosus*) [RA]. For **cough**, refer to *guanábana* (*Annona muricata*) [B2107, B2108, B2400, B2403, B2411, B2458], trumpet tree (*Cecropia peltata*) [A938], *ciricote* (*Cordia sebestena*) [A427], *jícara* (*Crescentia cujete*) [A376], *ix chel cher* (*Cornutia grandifolia*) [A792], *baston de vieja* (*Lippia myriocephala*) [A750], and *rabo de mico* (*Tournefortia hirsutissima*) [A677]. For **dry cough**, refer to cow foot (*Bauhinia divaricata*) [Steggerda, 1943] and licorice plant (*Lippia dulcis*) [Arvigo and Balick, 1998]. To treat a **baby's cough**, refer to *dama de noche* (*Cestrum nocturnum*) [B1971] and Spanish thyme (*Lippia graveolens*) [BW]. To treat a **cough** or **cold**, refer to wormwood (*Piscidia piscipula*) [Steggerda, 1943]. For **internal gas** causing stomach pain, refer to cancer herb (*Acalypha arvensis*) [Arvigo and Balick, 1993]. For **urine retention (stoppage of water)** or **bad urine (*mal de urina*)**, refer to *Cyperus luzulae* [B2662] and *Rhynchospora cephalotes* [B2625].

To **clarify sugar**, refer to bay cedar (*Guazuma ulmifolia*) [Standley and Record, 1936] and bur (*Triumfetta lappula*) [Standley and Record, 1936]. To **bleach sugar**, refer to wild cotton (*Urena lobata*) [R91]. To **attract and poison mosquitoes**, refer to horse's balls (*Stemmadenia donnell-smithii*) [BW].

Vetiveria zizanioides (L.) Nash
valeriana [Spanish]

To ease **"heart" pain** due to indigestion, 1 handful of chopped roots along with ½ cup of fresh mint leaves (*Mentha* sp.) are boiled in 1 quart of water for 15 minutes, steeped for 15 minutes, strained, and 1 cup consumed 3 times daily, before meals, for 9 days or until better [B2090].

To treat **heart ailments**, 1 bundle of roots (approximately 6 inches long and 4 inches in diameter) is boiled in 2 quarts of water for 10 minutes and ¼ cup consumed warm 3 times daily, generally for 3–4 days or until better [LR].

Zea mays L.
corn, maize [English]; *maíz* [Spanish]; *ixim* [Q'eqchi']

This is one of the most important **food** crops in Belize [MB]. These crops are planted in Belize at 3 different times: March, mid-April through May, and September [Horwich and Lyon, 1990]. To make **tortillas**, refer to *botán* (*Sabal mauritiiformis*) [BW] and breadnut (*Brosimum alicastrum* subsp. *alicastrum*) [LR, RA]. To make an excellent **coffee-like beverage**, the kernels are dried, roasted, ground, steeped in boiling water, and consumed [Arvigo and

(top) *Vetiveria zizanioides* showing roots [MB].

(bottom) *Zea mays* showing "silk" [IA].

Balick, 1998]. To make a **cacao beverage**, refer to cacao (*Theobroma cacao*) [MB, RA]. To make *chichi*, refer to wild ginger (*Costus pulverulentus*) [B2702] and sugarcane (*Saccharum officinarum*) [Thompson, 1927]. For a **container** to hold *espacha* (the local corn-wine) that imparts a flavor to it, refer to *jícara* (*Crescentia cujete*) [Robinson and Furley, 1983].

To treat **measles**, 1 handful of dried kernels is boiled in 1 quart of water for 20 minutes and consumed [Arvigo and Balick, 1998].

To **increase lactation** of nursing mothers, refer to sesame (*Sesamum indicum*) [Steggerda, 1943].

To treat **urinary ailments** such as **urine retention (stoppage of water)**, **burning urine**, **kidney stones**, **bladder infections**, and **gonorrhea** or as a **lymphatic system cleanser**, the "silk" (fine hairs at the end of the ear) from 3 ears of corn is boiled in 3 cups of water for 5 minutes and then sipped all day long [Arvigo and Balick, 1998]. To treat **bloody urination**, 1 handful of fresh kernels is cut off of the cob, steeped in 1 quart of water all day in the sun, and sipped all throughout the following day [Arvigo and Balick, 1998]. To treat **gonorrhea** and **difficult urination**, the "silk of black corn" is said to be consumed cool, but no further details were provided [Arnason et al., 1980]. To treat **gonorrhea** and **kidney** or **bladder ailments**, refer to wild coffee (*Malmea depressa*) [Steggerda, 1943].

According to Thompson (1927):

> A frequent method of curing illness at San Antonio [Toledo District] is to pass tiny tortillas made of the blood of a hen, maize, and copal [*Protium copal*], seven or nine in number, over and under the body of the sick man . . . These tortillas are kept for three days, at the end of which period the sorcerer takes them far out into the forest and, lighting copal, either burns them or throws them away. The tortillas are said to draw the evil out of the sick man's body. (p. 72)

One belief of many indigenous people of Central America is that this plant is the "food of the Gods," given to humankind as a benediction, while another is that the Gods created man from *maize* [Arvigo and Balick, 1998].

To prepare **mulch** for the cornfields, refer to Spanish needle (*Melanthera nivea*) [A579]. To make **bins** for corn, refer to *tah* (*Viguiera dentata*) [Steggerda, 1943].

To **poison** mice and rats, refer to *mata ratónes* (*Zamia polymorpha*) [B1803, B2058], butterfly weed (*Asclepias curassavica*) [A146], and *madre de cacao* (*Gliricidia sepium*) [Arvigo and Balick, 1998].

PONTEDERIACEAE

Aquatic herbs with rhizomatous stems. Leaves alternate, simple and glabrous. Inflorescences terminal, of 1–2 flowers. Flowers blue, purple to white, or yellow. Fruits a capsule or nutlet. In Belize, consisting of 2 genera and 3 species. Uses for 2 genera and 2 species are reported here.

Eichhornia crassipes (Mart.) Solms

water hyacinth, water lily [English]
The root is boiled and eaten as a nutritious **food**. It contains good, simple carbohydrates as well as iron and minerals [A400].

To treat **anxiety**, accompanied with **heart palpitations** and **tightness around the chest**, 1 entire plant is chopped, placed in a glass bottle containing 2 quarts of water, set in the sun for 1 day, strained, and 1 cup consumed daily for 3–9

Eichhornia crassipes [MB].

days. This treatment is said to have a calming effect [A400]. For **severe hysteria**, 1 cup of the decoction described for anxiety is consumed daily until better [A400].

Pontederia cordata L. var. *cordata*

To treat non-insulin-dependent **diabetes** mellitus (**sweet water** or **sugar in the urine**), 1 root is boiled in 3 cups of water for 10 minutes and 1 cup consumed 3 times daily before meals [G18].

Pontederia cordata [KN].

SMILACACEAE

Lianas or vines, often with thorns and tendrils. Leaves alternate, simple. Inflorescences in umbels or racemes. Flowers green, white, or cream. Fruits a berry. In Belize, consisting of 1 genus and 8 species. Uses for 1 genus and 4 species are reported here.

Smilax luculenta Killip & C.V. Morton

dog tongue sarsaparilla, dog tooth sarsaparilla, sarsaparilla [English]; *corona de cristo, sarsa* [Spanish]

To ease **joint pain**, 1 cup of chopped root is boiled in 2 quarts of water for 10 minutes and 1 cup consumed warm 3 times daily until better [B2154, B2165]. To **cleanse the liver** or to treat **skin discoloration**—for example, from white, brown, or pink patches—9 leaves are boiled in 2 quarts of water for 20 minutes and 1 cup consumed daily for 7–14 days [LR].

Smilax mollis Humb. & Bonpl. ex Willd.

sarsaparilla, wild sarsa [English]; *sarsaparillo* [Spanish]; *pu ja* [Mopan]

The fruit is eaten as a **food** [A211].

To treat **spasms** in babies with colic, 3 leaves are boiled in 1 cup of water for 5 minutes. The amount consumed depends on the baby's age and varies from 3 teaspoons to 1 cup at a time. Often, 1 day of treatment is sufficient, although this depends on the case and severity of symptoms [A211].

Standley and Record (1936) reported that the roots were used in Honduras as a **fish poison**.

Smilax ornata Lem.

red China root, red cocolmecca, red wild yam [English]

As a **blood tonic**, ½ cup of the dried, chopped root is boiled in 1 quart of water for 15 minutes, then steeped for 15 minutes, strained, and 1 cup consumed 3 times daily for 9 days [W1864]. As a **blood tonic** and a **male tonic** for energy, vitality, and sexual potency, ½ cup of the dried, chopped roots along with ½ cup of the dried, chopped provision bark roots (*Pachira aquatica*) are boiled in 2 quarts of water for 20 minutes, then steeped for 20 minutes and ½ cup doses, 6 times daily, consumed for 9 days [W1846; LR]. Additionally, following either of these 9-day treatments, either of the decoctions is consumed as a beverage to maintain the iron level in the blood [W1864].

To treat **dysentery**, 1 small handful of the fresh,

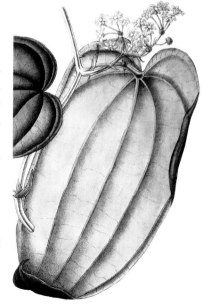

Smilax ornata [NYBG].

chopped tuber is boiled in 1 quart of water for 10 minutes, then steeped until luke-warm, strained, and ½ cup consumed every hour until better [LR]. Consumption of a tea made from the fresh roots of either this plant or the provision bark tree causes **constipation** [W 1864].

Smilax sp.

China root, red China root, wild sarsa [English]; *pu ja* [Mopan]

To treat **fatigue**, **anemia**, **acidity**, **toxicity**, **rheumatism**, **skin ailments**, and as a **blood tonic**, 1 small handful of chopped roots is boiled in 3 cups of water and con-sumed as a pleasant beverage [Arvigo and Balick, 1998]. To **strengthen** and **prolifer-ate red blood cells**, the decoction described above is combined with milk, cinnamon (*Cinnamomum verum*), and nutmeg (*Myristica fragrans*) and consumed as a beverage [Arvigo and Balick, 1998]. To stop **internal hemorrhaging**, for example after child-birth or during menstruation, 1 handful of the root along with 3 garlic cloves and 3 unopened red hibiscus flowers (*Hibiscus rosa-sinensis* var. *rosa-sinensis*) are boiled in water for 10 minutes, cooled, and sipped all day [Arvigo and Balick, 1998]. For male **impotence**, 1 handful of the root along with 3 tablespoons (or a 3-inch piece) of *guinweo* root (an unidentified plant species) are soaked in rum and 1 shot consumed twice daily, in the morning and in the evening [Arvigo and Balick, 1998].

TYPHACEAE

Aquatic herbs with subterranean rhizomes. Leaves alternate and simple, blades linear and spongy. Inflorescences terminal, spike-like, and cylindrical. Flowers small. Fruits a follicle. In Belize, consisting of 1 genus and 1 species. Uses for 1 genus and 1 species are reported here.

Typha domingensis Pers.

elephant grass [English]; *sacate ignea* [Spanish]

The root is boiled and eaten as a **food** [A357].

The flower's feathery seeds were once used as a **fiber** for pillow stuffing [A357, B1977]. Standley and Record (1936) noted that the leaves were sometimes used in Central America as a **fiber** to weave coarse mats and that the feathery seeds were used for pillow and cushion **stuffing**.

Inflorescence of *Typha domingensis* [SC].

XANTHORRHOEACEAE (INCLUDING ASPHODELACEAE)

Succulent plants, the stem short to tall. Leaves thick, formed in a rosette, often succulent with spines on the margins. Inflorescences a scape growing to 3.25 feet, in racemes. Flowers yellow. Fruits a capsule. In Belize, consisting of 1 genus and 1 species. Uses for 1 genus and 1 species are reported here.

Aloe vera (L.) Burm. f.

aloe vera [English]; *sink-am-bible* [Creole]

To treat **weakness** or a **sedentary** habit and as a **tonic** to treat the **liver**, **pancreas**, **kidney**, and **stomach**, 1 leaf about 6 inches long is mashed, soaked in 1 cup of water for 20 minutes, strained, and consumed as a purgative. **Caution** is used with this treatment, as the juice may be toxic when consumed excessively [Arvigo and Balick, 1998].

To treat **burns**, **sunburn**, **rash**, **stubborn sores**, **black skin spots**, **bed ulcers**, **diaper rash**, **boils**, and **fungus**, or to reduce **scarring**, fresh juice from the leaf is applied directly and liberally to the affected area [Arvigo and Balick, 1998]. To **prevent hair loss**, 1 fresh leaf is mashed and applied directly to the scalp for 7–10 days [Arvigo and Balick, 1998]. To **extract** deeply embedded **thorns**, **stones**, or **fish scales**, 1 leaf is sliced in half, applied over the affected area, secured with a band or cloth, and changed once daily for 3–5 days until the object comes out [Arvigo and Balick, 1998].

Steggerda (1943) reported that the Yucatan Maya used this plant to make **shampoo** and noted that it could be purchased in the markets of Merida, Mexico. In addition, the leaves can be macerated and applied to the head to treat **headache** and **neuralgia** [Steggerda, 1943].

Standley and Steyermark (1952) suggested that this plant was likely introduced into Central America shortly after the Spanish colonization, after its long history of use as a bitter **purgative**. They also noted that in Guatemala, the crushed leaves were applied as a poultice to treat **sores** and **boils**. They further reported that women used the bitter sap to **wean** nursing children by rubbing it on their nipples.

Aloe vera [IA].

Curcuma longa with inset showing rhizomes [MB].

ZINGIBERACEAE

Herbs to several meters tall with horizontally branched rhizomes. Leaves alternate, distichous, simple. Inflorescences terminal, usually on a separate, leafless scape. Flowers white, yellow, or red. Fruits a capsule. In Belize, consisting of 4 genera and 6 species. Uses for 3 genera and 4 species are reported here.

Curcuma longa L.

ginger, turmeric, yellow ginger [English]

The root is powdered and used as a **spice** [B2113]. It is added to the ripe *sorosi* fruit relish (*Momordica charantia*) that accompanies the curries eaten by the East Indians of Belize [B2113].

To **increase blood circulation**, 2 roots the size of the index finger are chopped and boiled in 1 quart of water for 20 minutes, strained, and sipped throughout the day [B2113]. For **bruises**, 1 root is mashed, heated in water, wrapped in a cloth, and applied hot as a poultice to the affected area [B2113]. To treat **adult colic** and **stomach pain**, the outer portion of the dried stem is scraped, and three 3-inch pieces of the dried stem, along with 1 garlic clove and 3 sweet orange leaves (*Citrus sinensis*), are boiled in 1 quart of water for 5 minutes and ½ cup sipped warm every hour until better. This formula is not to be used to treat children [LR].

Renealmia cf. *alpinia* with close-up of inflorescence (right) [RF].

Renealmia cf. *alpinia* (Rottb.) Maas
shma buí [Mayan]
The fruit eaten as a **food** when squeezed into water, releasing a yellow juice that is consumed warm as a soup [B2386].

Renealmia aromatica (Aubl.) Griseb.
rat plantain, wild ginger [English]; *k'isk'im tz'i'* [Q'eqchi']
To treat **urine retention (stoppage of water)**, 3 small leaves are boiled in 3 cups of water for 10 minutes and 1 cup consumed 3 times daily until better [A364]. To expel a **retained placenta**, 2 cups of the decoction described above for urine retention are consumed after childbirth [A364]. For **abdominal pain**, the leaves are warmed over a clay oven on a *comal* (baking tray) and then pressed against the abdomen [B3565]. To reduce **swelling** anywhere on the body, the root is mashed and applied as a poultice to the affected area [B3302].

Zingiber officinale Roscoe
ginger [English]; *gingibre, jengibre* [Spanish]
To keep the body warm on cold days, one 5-inch piece of root is boiled in 6 cups of water for 10 minutes and consumed hot [LR]. The root is grated and used to make a

Zingiber officinale with inset showing rhizomes [MB].

stimulant **beverage** consumed in place of coffee (*Coffea arabica*) or tea (*Camellia sinensis*) [Arvigo and Balick, 1998].

For a **high fever**, 2 ounces of grated root are boiled in 2 cups of water for 10 minutes, consumed as hot as possible, and repeated every 2 hours if necessary [HR]. To treat **broken bones**, the rhizome is boiled, macerated, wrapped in leaves, and applied as a poultice to the affected area [Mallory, 1991]. To treat **bruises**, the same poultice described for broken bones is applied immediately to the affected area [Mallory, 1991]. To treat **chest colds** or **muscle pain**, the root is chopped and boiled in water, after which a cotton or flannel cloth soaked in the hot decoction, applied hot over the affected area, covered with towels, and repeated 6 times [Arvigo and Balick, 1998]. To treat **bronchitis**, the root is grated, mixed with honey, and applied as a poultice to the chest [Arvigo and Balick, 1998].

For relief of **stomach pain**, **gas pain**, **indigestion**, and **colds**, 1 teaspoon of fresh root is grated into 1 cup of water, boiled for 5 minutes, and consumed [B2321; Arvigo and Balick, 1998]. For relief of **stomach ache**, **nausea**, **vomiting**, **car sickness**, or **air sickness**, 1 piece of fresh root is chewed [B2321; Arvigo and Balick, 1998]. For **gastritis** (ciro), refer to *guaco* (*Aristolochia grandiflora*) [B3669] and allspice (*Pimenta dioica*) [A818]. To treat **colic** pain in babies, refer to *artemisia* (*Egletes viscosa*) [LR].

To treat **painful menstruation**, the root is grated, placed on a warm towel, and applied as a poultice to the abdomen for 1 hour [Arvigo and Balick, 1998]. To ease the pain of **childbirth**, 1 ounce of the root is mashed, boiled in 1 cup of water for 10 minutes, and sipped warm until finished; only 1 cup is consumed [HR; Arvigo and Balick, 1998]. For very **delayed labor**, 1 ounce of the root is mashed and boiled in 1 cup of water for 10 minutes, after which a cotton cloth soaked in the decoction, applied warm to the woman's belly, and covered with a blanket for 20 minutes. If the pain eases, the baby is not ready to be born; otherwise, this poultice will increase the speed of the delivery [HR; Arvigo and Balick, 1998]. To **increase lactation** in a nursing mother who has lost her milk due to fright (*susto*), the root is mashed and soaked for 1 day in *anisado* (anise liquor [*Pimpinella anisum*]), after which 2 teaspoons consumed and the woman is combed from head to foot while a prayer asking for the return of her milk is recited [Arvigo and Balick, 1998]. To **prevent miscarriage**, refer to white China root (*Dioscorea bartlettii*) [B1787].

To treat **chronic and loud snoring**, 2 teaspoons of grated root is added to 1 cup of cinnamon tea (*Cinnamomum verum*) with honey or milk and 1 cup consumed once daily, at bedtime, until better [Arvigo and Balick, 1998].

ACANTHACEAE

Herbs, shrubs, rarely trees or climbing vines, mostly terrestrial or rarely aquatic. Leaves usually opposite, simple. Inflorescences terminal or axillary, in spikes, racemes, cymes, thyrses, or of solitary flowers, often densely clustered, frequently with brightly colored bracts. Fruits a capsule or drupe. In Belize, consisting of 19 genera and 50 species. Uses for 7 genera and 13 species are reported here.

Aphelandra aurantiaca (Scheidw.) Lindl.

Santa María [Spanish]; *pot k'unbul* [Mopan]; *Santa María Q'ehem* [Q'eqchi']
To treat **headache** in any part of the head or **bone pain**, 1 double handful of fresh leaves is mashed in cold water and used as a bath over the affected area twice daily for 7 days [C69, C70]. To revitalize a person who is **tired** or **weak**, refer to *Santa María* (*Adiantum wilsonii*) [R23, R28]. To stop **vaginal hemorrhaging** in a **pregnant** woman, the plant is mashed in cool water and consumed; this infusion is considered to be only "for women" [B2486].

To treat a person **vomiting fresh blood** or to treat **snakebite**, 1 handful of fresh leaves is mashed in cold water and consumed 4 times daily for 2 days [C27]. Alternatively, for **snakebite**, refer to Indian head (*Aphelandra scabra*) [C31].

Aphelandra scabra (Vahl) Sm.

Indian head, red anal [English]; *anal, anal grande, cabeza de Indio, pito* [Spanish]; *tz'ojl* [Q'eqchi']; *anal ché, chacanal* [Additional names recorded]
The leaves are a component of a Maya bath mixture used for both adults and children. One handful of the leaf mixture is boiled in 2 quarts of water for 10 minutes and used tepid as a bath once daily at bedtime [A61, A635, A768].

To reduce **fever**, **promote sweat** in babies with fever or malaria, treat **skin ailments** with a **red rash**, or as a **panacea** for whatever ails children or adults, the mixture

Aphelandra scabra [MB].

described above is used as a bath [B2004; RA]. For **swollen wounds**, this same bath formula is used to wash the affected area 3 times daily for 8 days [LR]. If the wound is from **snakebite**, 1 handful of chopped root is added to the bath formula [LR]. To treat **headache**, 9 leaves are crushed with a bit of cooking oil, placed on a cloth, applied as a poultice to the forehead, and changed every 20 minutes until better [LR]. For **snakebite**, 1 handful of fresh leaves along with 1 handful of *Santa María* leaves (*Aphelandra aurantiaca*) are mashed in cold water and consumed 3 times daily for 7 days [C31].

The entire plant is used as a table **decoration** and is a favorite of the elders for the *primicia* altar [A537].

Blechum pyramidatum (Lam.) Urb.
contra hierba, frondosa, tomatillo [Spanish]; *pixpil pim* [Q'eqchi']; *sahm ak ách* [Additional name recorded]

To treat **herpes simplex** and **herpes zoster**, 1 handful of fresh leaves and fruits is boiled in water for 5 minutes and consumed warm 3 times daily for 3 days [C48]. Steggerda (1943) noted that he found three "herb doctors" who boiled the leaves and used the water for bathing children who had **heavy night sweats**. When the sweats were so intense that the skin seemed to contain salt on it, boiled or fresh leaves were applied as a poultice to the body.

Blechum pyramidatum [MB].

Steggerda (1943) also reported that in colonial times, this plant was used by Yucatan Maya "for chills, snakebites, coughs, and fever." As a treatment for **snakebite**, the leaves and roots along with 1 blue vervain root (*Stachytarpheta cayennensis*) are mashed and applied as a plaster. This is said to remove the "scales of snake teeth" from the person who would otherwise die [A495]. Alternatively, 1 leaf and 1 handful of the vine and stem along with 1 entire plant of blue vervain including roots are chopped and mashed with enough water to moisten the mixture, after which a pinch of salt is added and the mash applied as a poultice to the bite and changed twice daily. The healer looks for the snake's "tooth scale" while treating the person [LR].

For **vaginal sores on a dog**, 1 leaf is boiled in 2 quarts of water for 10 minutes and used lukewarm as a wash over the affected area. This wash is followed by the application of a powder that comes from the dried vine. This treatment is repeated once daily if necessary [LR].

To increase the healing power of **prayer**, Maya healers hold the plant against a person's wrist while prayers are said into the pulses [B1904; RA].

Elytraria imbricata (Vahl) Pers.
Steggerda (1943) reported that the Yucatan Maya boiled 1 entire plant of this common weed in ½ quart of water, which was then used to treat women with **venereal diseases**.

Elytraria imbricata [SH].

Justicia bartlettii (Leonard) D.N. Gibson
anal blanco [Spanish]
The leaves and tender stems are used as part of a Maya bath formula containing 9 different plants [A203]. To treat **skin ailments**, **swelling**, **insomnia**, **burns**, and other ailments, 1 double handful of leaves is boiled in 2–3 quarts of water for 10 minutes, steeped until the temperature is reduced to warm, and poured over the body a little at a time using a bowl [RA]. To treat babies with **insomnia**, the decoction described for skin ailments is poured over the entire body as a bath at bedtime from the head down and then across the shoulders, making a cross pattern [LR].

Justicia breviflora (Nees) Rusby
zorillo macho [Spanish]; *au'kop* [Mayan]
To treat **lockjaw**, 1 handful of leaves is boiled in a bucket of water for 20 minutes and used as a bath 3 times daily until better [B2524]. To soothe and reduce the inflammation of **swollen testicles**, the leaves and roots are mashed in water and massaged over the affected area [B3672].

Justicia sp.
To treat **herpes simplex**, 1 large handful of fresh leaves is mashed in cold water and an unspecified amount consumed 3 times daily for 5 days [C74].

Odontonema albiflorum Leonard
For people who experience nighttime "**burning inside the body**," **esophageal reflux**, or other **gastrointestinal ailments** resulting from excess gastric acid, 1 double handful of leaves is boiled in 2 gallons of water for 10 minutes, 1 cup consumed warm, and the remainder used as a bath [B2547].

Ruellia brittoniana Leonard
The plant is cultivated as an **ornamental** for its flowers [B2330].

Justicia breviflora [GD].

Ruellia nudiflora (Engelm. & A. Gray) Urb.

wild radish [English]; *maravilla segunda* [Spanish]

To relieve the pain of **menstrual cramps**, 1 entire plant with roots is boiled in 3 cups of water for 10 minutes and ½ cup consumed hot every hour until better [A56, B1797].

Ruellia pereducta Standl. ex Lundell

maravilla del monte [Spanish]

To treat **headache** in a woman following childbirth, 1 entire plant is boiled in 1 cup of water for 10 minutes and used tepid as a bath over the head only [B1849].

Thunbergia erecta (Benth.) T. Anderson

Rigoberto [Spanish]

This shrub is cultivated as an **ornamental** for its blue-purple flowers [B2278].

(top) *Ruellia nudiflora* [MB].

(bottom) *Thunbergia grandiflora* [MB].

Thunbergia grandiflora Roxb.

This introduced tropical vine is cultivated as an **ornamental** and valued for its constant flowering [Marsh, Matola, and Pickard, 1995].

ACTINIDIACEAE

Trees or shrubs, often with pubescent stems when young. Leaves alternate, simple, blades variable in shape and size, coriaceous to papery, frequently pubescent. Inflorescences axillary, in thyrses, with few to numerous flowers. Flowers often white. Fruits usually a berry or capsule. In Belize, consisting of 1 genus and 1 species. Uses for 1 genus and 1 species are reported here.

Saurauia yasicae Loes.

wild orange [English]; *aguacatillo, jaboncillo* [Spanish]

Standley and Record (1936) reported that the fruit pulp, said to be transparent and appear somewhat like an egg white, was good to eat as a **food**.

AMARANTHACEAE (INCLUDING CHENOPODIACEAE)

Herbs, sometimes a semi-woody vine, rarely trees, sometimes succulent. Leaves alternate or opposite, simple. Inflorescences axillary or terminal, in cymes, sometimes compound, often in panicles, thyrses, or dense heads or spikes. Flowers small. Fruits often a capsule, nutlet, achene, or berry-like. In Belize, consisting of 15 genera and 32 species. Uses for 6 genera and 8 species are reported here.

Alternanthera bettzickiana (Regel) G. Nicholson

la coqueta [Spanish]

Standley and Record (1936) reported that this cultivated plant was thought to be a "**natural repellent**" of the *wee-wee* (leaf cutter) ant (*Atta mexicana*). "It is claimed that the ants will not pass through, under, or over the plant."

Alternanthera brasiliana (L.) Kuntze

The plant is cultivated as an **ornamental** [B2323].

Alternanthera flavogrisea Urb.

colondrina, golondrina [Spanish]; *ix can lol* [Mopan]

This plant is a component of a 9-leaf Maya bath mixture used as a **panacea** [Arvigo and Balick, 1998].

To treat **influenza**, **colds**, **urinary ailments**, **tiredness**, **postpartum complications**, **granulated eyes**, **headache**, non-insulin-dependent **diabetes** mellitus, **fever**, **internal infections** (especially of the reproductive organs, e.g., **ovarian inflammation**), and those who "eat too much salt," three 12-inch pieces of stems with leaves are boiled in 3 cups of water for 2 minutes and 1 cup consumed 3 times daily before meals [Arvigo and Balick, 1998]. Alternatively, 1 handful of leaves is steeped in 3 cups of tepid, room temperature water for 20 minutes and 3 cups consumed [A117].

For **swelling**, 1 quart of fresh leaves, roots, and stems is steeped in 1 gallon of water for 1 hour and used as a bath over the affected area [B1903]. To staunch **bleeding**, the fresh leaves are mashed and the juice applied as a poultice to the affected area [Arvigo and Balick, 1998]. For **mouth sores** and **thrush**, the fresh leaves are mashed and the juice used as a mouthwash [Arvigo and Balick, 1998]. To reduce **infection** of a **tooth abscess**, nine 6-inch stem ends along with 1 handful of Spanish thyme leaves (*Lippia graveolens*) are boiled in 3 cups of water for 5 minutes and used as a mouthwash 6–10 times daily for 2 days [LR].

For **poor blood circulation**, characterized by **leg cramps**, **numbness**, **bone pain**, and **low energy**, 1 double handful, made of equal parts of the leaves and stems from this plant along with the leaves and stems

Alternanthera flavogrisea [MB].

Amaranthus dubius [MB].

of *dormilón* (*Mimosa pudica* var. *unijuga*) and fresh wild yam tubers (*Dioscorea bartlettii*), is boiled in 1 gallon of water for 20 minutes, strained, and ½ cup consumed cool twice daily, in the morning and in the evening, until the entire gallon is consumed. This is repeated 6 months later [LR].

To stabilize the blood sugar level for treating non-insulin-dependent **diabetes** mellitus, 1 cup of fresh leaves and stems is boiled in 1 quart of water for 10 minutes and 3 cups consumed daily for as long as needed, usually 9 days [B1903]. To alleviate **ovarian inflammation**, 3 cups of the decoction described for diabetes are consumed daily for 3 days [B1903]. To help **postpartum recuperation**, 9 branches are placed in a cross formation on the *primicia* altars as an offering [BW].

Amaranthus dubius Mart ex Thell.

amaranth, pig weed [English]; *quelite, kelite* [Spanish]; *calaloo, calalu, callalu* [Creole]; *maak'uy, ichaaj* [Q'eqchi']; *kel tosh* [Mayan]

This plant, which is high in vitamins and minerals, is used as a **food** and a **medicine** [Arvigo and Balick, 1998]. The seeds are ground into flour and used as a **food** [B2380, R94]. Seed heads are gathered, winnowed, roasted, ground into a powder, and then mixed with salt, garlic, and chili (*Capsicum frutescens*) for use as a **seasoning** powder [Arvigo and Balick, 1998].

To treat **anemia**, the iron-rich leaves are eaten raw as a green salad, boiled and eaten as a "pot herb" much like spinach (*Spinacia oleracea*), or cooked and eaten in a soup [A92, B2209, B2380, R94; Arvigo and Balick, 1998]. To treat advanced stages of **anemia**, ½ cup of leaf juice is consumed 3 times daily until better [Arvigo and Balick, 1998]. To treat non-insulin-dependent **diabetes** mellitus, 1 handful of leaves and young stems is washed, placed in a pot with no water, and slightly heated to extract the juices, after which the leaves removed and ¼ cup of the juice consumed 3 times daily for 1–2 weeks while the blood sugar levels are monitored daily [LR].

For **insect bites**, the leaves are mashed and the juice applied to the affected area [A92]. To **cleanse wounds and sores**, 1 entire plant is boiled in water and used as a wash over the affected area [Arvigo and Balick, 1998]. As a **postpartum treatment**, this plant along with guava (*Psidium guajava*) are 2 of the 9 plants used as a tepid bath once daily, at noon or bedtime, on the eighth, ninth, and tenth day following childbirth [BW].

Arvigo and Balick (1998) reported that

[t]he leaves and seeds of amaranth were a **food** source for the Aztecs and Mayas. The Spanish conquerors were horrified to learn that the Aztecs mixed popped

Chamissoa altissima var. *altissima* [GS].

Chenopodium ambrosioides with close-up of plant (right) [MB].

amaranth seed with human blood to form into a ceremonial cake as an offering to their gods. For this reason, it was illegal to cultivate amaranth in New Spain for many generations. (p. 371)

Chamissoa altissima (Jacq.) H.B.K. var. *altissima*

To **increase lactation** in nursing pigs, the leaves are eaten as fodder [A328].

To treat various **skin ailments**, especially **new or open sores**, the leaves are boiled in water and used as a bath [A328].

Chenopodium ambrosioides L.

Mexican weed, worm bush, wormseed [English]; *epasote, epazote* [Spanish]; *wormwood* [Creole]; *apasoot* [Q'eqchi']

The leaves are used to **flavor** food when cooked with beans, tamales, and soup [BW; Arvigo and Balick, 1998].

To **stimulate the appetite** of sick people, the leaves are added to soup [BW]. To **stimulate the appetite** in crying children who will not eat, the leaves are chopped and mashed, the juice extracted, and 1–2 spoonfuls consumed daily for 3 days [B2338]. For **erysipelas**, refer to *coqueta* (*Rivina humilis*) [B2092, B2123]. To disinfect **cuts, infections**, and reduce **swelling (*hinchazón*)** in the body, refer to *ruda* (*Ruta chalepensis*) [B2316]. Standley and Steyermark (1946) noted:

> The plant finds still further use in local **medicine**. There came to the attention of the senior author a case in which fomentations of the plant and hot poultices were applied to an inflamed and supposedly an infected foot by one of the best-known North American doctors practicing in Guatemala. (pp. 140–1)

For an **upset stomach** or **indigestion**, 1 handful of fresh leaves is mashed in 2 drops of olive oil or sweet almond oil and ¼ cup of fresh leaf juice consumed once daily in the morning, before breakfast, or when experiencing indigestion [LR]. To reduce **intestinal gas** and to aid in **digestion**, the leaves are eaten raw or cooked in soups, beans, or tamales [Arvigo and Balick, 1998]. To treat **"bad wind in the belly"** caused by bewitching, refer to bay cedar (*Guazuma ulmifolia*) [Arvigo and Balick, 1998].

For **cranky children** or **babies who cry and are restless**, 1 large double handful of leaves is mashed in a 5-gallon bucket of water, set in the sun for 2–3 hours, used warm as a bath and a tea at bedtime, and repeated a second time if necessary [HR]. When a baby's navel is oozing blood, which is called *puhu* in Mayan and is said to be caused by a menstruating woman who visits without holding the baby, 1 cup of leaves is heated over the fire until dry, powdered, and sprinkled over the baby's navel. On top of this powder, 1 *hoja de la preñada* leaf (*Cissampelos tropaeolifolia*) is applied, and this poultice is changed twice daily [HR, LR].

To remove **intestinal worms**, the leaves are crushed, after which 1 ounce of leaf juice is mixed in ¼ cup of milk and consumed once daily for 3 days, in the mornings during the new moon phase of the month, followed by a purge from consuming 1 tablespoon of castor oil (*Ricinus communis*) on the third day [A153]. Similarly, to treat **intestinal parasites** in adults or children, the leaves are mashed and 1 teaspoon of leaf juice consumed alone or in ½ cup warm milk once daily for 3 consecutive mornings before eating. On the fourth day, 1 tablespoon of castor oil (*Ricinus communis*) is consumed as a purge and the stool observed to see if the parasites have passed [Arvigo and Balick, 1998]. Standley and Record (1936) noted that "the seeds are used widely as an agent for expelling **intestinal parasites** of man."

As a **sedative**, 3 small branches are boiled in 1 cup of water and consumed hot [Arvigo and Balick, 1998]. Standley and Steyermark (1946) reported "it is said that about Cobán the plant is employed as a 'narcotic,' the plant being placed beneath the pillow to **induce sleep**. Considering how unpleasant the odor is, one would expect the effect to be quite the opposite." To treat a **hangover (*crudo*)**, 1 root from 1 large plant is boiled in 2 cups of water for 10 minutes and consumed hot [Arvigo and Balick, 1998]. To treat the **spiritual sickness** that follows **shock**, refer to Bethlehem star (*Hippobroma longiflora*) [BW].

This powerful medicinal plant should not be given to pregnant women, the weak, or the aged [Arvigo and Balick, 1998].

Gomphrena globosa L.
amor seco [Spanish]

This plant is cultivated for its beautiful flowers as an **ornamental** [W1842].

To staunch **hemorrhaging** in women, 1 small handful of leaves and stems is boiled in 3 cups of water for 10 minutes, steeped until cool, and ½ cup consumed every 10 minutes until better [W1842].

Gomphrena globosa [KN].   *Iresine diffusa* [NYBG].

Iresine diffusa Humb. & Bonpl. Ex Willd.
   azrozillo, *yerba de caballo* [Spanish]; *bare tát* [Mopan]
   The entire plant is used as a **fodder** for horses [B3238].
      For **cataracts**, 1 top crown of leaves and stem is boiled in 1 cup of water for 10 minutes, strained, and 3 drops placed into each eye every 2 hours for 1 day, then no treatment is given for a day, and on the next day, the treatment is given again (e.g., 1 day of rest between treatments until the cataract comes off). If this plant is collected after 6 p.m., then according to local beliefs, the roots must be added to the decoction [A314]. To treat non-insulin-dependent **diabetes** mellitus, 1 large handful of fresh leaves is boiled in water for 15 minutes and consumed warm 3 times daily for 30 days [C67].

ANACARDIACEAE

Trees, shrubs, sometimes vines or lianas. Leaves alternate, often clustered at branch tips, usually pinnate, sometimes simple. Inflorescences in panicles, thyrses, spikes, racemes, clustered, or of solitary flowers. Flowers small, green, white, or yellow to deep reddish-purple. Fruits usually a drupe, sometimes a samara. In Belize, consisting of 8 genera and 10 species. Uses for 6 genera and 8 species are reported here.

Anacardium occidentale L.

cashew [English]; *marañon* [Spanish]

This is the tree that yields the well-known cashew "nut," which is actually the seed of the fruit. As a **food**, the seed is first roasted, driving off the effects of the toxic resin containing anacardic acid [B3660]. The fleshy fruit is eaten as a **food**, but eating too much causes a burning sensation in the mouth [BW]. Standley and Record (1936) reported that the roasted seeds were eaten as a **food** and that a **wine** was made from the fruits. Lundell (1938) reported that the fruits were used to produce "an agreeable fermented beverage." The fruit is used today to produce a **juice** that is either fermented into wine or mixed with rum [Horwich and Lyon, 1990].

Anacardium occidentale with fruit [MB].

To **promote sweat** to treat **fever** in children over 6 years of age, 3 young leaves are crushed, steeped in ½ cup of hot water, strained, and ½ cup consumed 3 times daily until better [BW]. To treat **dry cough** and **warts**, the fruit juice and leaf secretions are used in an unspecified way by Maya healers of the Yucatan [Flores and Ricalde, 1996]. To treat **diarrhea** in children, one 3-inch × 3-inch piece of bark along with 1 garlic clove are boiled in 1 cup of water for 10 minutes and sipped cool throughout the day until better [BW]. To treat **diarrhea** in babies, one 3-inch × 3-inch piece of bark along with the skin of 1 garlic clove are boiled in 1 cup of water for 10 minutes and sipped cool throughout the day until better [BW].

Standley and Record (1936) noted that the seeds produced an oil that was used as a leather and wood **preservative** to prevent damage from termites and other insects. Those authors also noted that the gum exuded from the bark of the tree was utilized as a **preservative** for the same purpose but that the wood, while easy to work, was not very resistant to decay.

Astronium graveolens Jacq.

glassy wood [English]; *palo mulatto* [Spanish]

Standley and Record (1936) reported that the wood from this tree was used in **construction** to make cabinets. Standley and Steyermark (1949) noted that the wood was excellent for **construction** to make fine furniture and "reported to be used in Veracruz for gunstocks, furniture, house posts, railroad ties, and bridge timbers. It is said to be susceptible to the attacks of termites."

Mangifera indica L.

mango [English]; *mango* [Spanish]

This tree is frequently cultivated around homes in Belize and commonly found growing in pastures [B2267, R47]. The delicious fruit is eaten as a **food** in all its stages: green (sliced and dipped into salt and *chili* [*Capsicum frutescens*]), semi-mature, and mature [RA].

To treat **warts** and **corns**, the leaves and fruit latex are used in an unspecified way by Maya healers of the Yucatan [Flores and Ricalde, 1996]. To treat **bruises**, 1 double handful of leaves is boiled in 1 gallon of water for 10 minutes, ½ cup consumed, and the remainder used tepid as a bath over the affected area once daily [BW]. To treat **cough** and **loosen mucus**, 3 leaves are boiled in 1 quart of water for 10 minutes, 3 tablespoons of sugar added, and 1 tablespoon consumed

Mangifera indica [DA] with inset showing close-up of inflorescence [IA].

every 30 minutes until better [BW]. For **cough** and **bruises**, refer to avocado (*Persea americana*) [B3324]. For **menstrual cramps**, 3 leaves are boiled in 1 quart of water for 10 minutes and ½ cup consumed quite warm 6 times daily. This usually helps after the first dose, and blood clots will also be passed [BW].

The timber of this species is used in **construction** to make boats and dugout canoes [Marsh, Matola, and Pickard, 1995].

Metopium brownei (Jacq.) Urb.

black poisonwood, Honduras walnut [English]; *chechem negro, che-chen* [Spanish]; *tuul che'* [Q'eqchi']

The red fruit is eaten as a **food** by a variety of birds [Marsh, Matola, and Pickard, 1995]. Steggerda (1943) reported that bees produced a black-combed **honey** when visiting the flowers of this species.

Like many other plants in this family, the sap from black poisonwood causes severe itching and rash, hence its common name [JB83]. For the accepted antidote to treat **skin ailments** caused by contact with this plant, refer to *gumbolimbo* (*Bursera simaruba*) [A745c, A909, A927, B1815, B2412, B2352; RA; Arvigo and Balick, 1998], *dama de noche* (*Cestrum nocturnum*) [B1970], and *bejuco de agua* (*Cissus microcarpa*) [B3584]. In another treatment for **skin ailments**, Steggerda (1943) reported that the Yucatan Maya applied urine to skin that had come in contact with the sap of this tree to counteract its poisonous activity.

Fruiting *Metopium brownei* with inset showing bark and black sap [MB].

In the past, for treatment of **erysipelas, psoriasis, general itching**, or **scaly skin ailments**, one 1-inch × 1-inch piece of bark was placed in 1 cup of water until the water started to turn dark, the bark removed, and 1 teaspoon consumed every 2 hours during the day and early evening until better [TG]. To treat **warts** and **unspecified skin "affections,"** the resin from the bark is used in an unspecified way by Maya healers of the Yucatan [Flores and Ricalde, 1996]. In view of the extremely toxic nature of this plant, it would seem imprudent to utilize the resin from the bark for any therapeutic purpose [MB].

As a **fish poison**, one 6-inch × 6-inch piece of bark is mashed and placed in a pond to stun fish [LR]. To **prevent diseases in barnyard fowl**, two 6-inch × 2-inch pieces of bark are cut and placed in ½ gallon of water, which is consumed by the birds. This will treat about 25 hens. The water is filled daily and the bark changed every 2 weeks [LR].

To release "**bad vibes**" after contact with a bad person, the receiver passes by or walks under a tree 9 times to give the "bad vibes" to the tree [LR].

Lamb (1946) reported that the wood was used in **construction** to make house posts, but to avoid the terrible itching that occurred from handling the wood, workers waited until the felled trees were dead and dried. He also noted that the strength and hardness of the wood that made it suitable for construction also made it good for **turnery**. It is still used today in **construction** to make local furniture [MB]. Standley and Record (1936) referred to the common name of this plant as "Honduras walnut" and suggested it was "the source of some beautiful furniture woods." The wood was said to be "harmless" once the caustic sap had drained out. Those authors noted that the heartwood was variegated in color, appearing

> brown and reddish brown with a greenish tinge and a golden subluster so characteristic of cabinet woods. It is hard and heavy, of rather fine texture, often wavy-grained, finishes very smoothly, and takes a lustrous polish. It is more attractive in small sizes than in large panels and is suitable for articles of turnery and handles of cutlery. It is practically unknown to the trade. While the supply of the timber is not very large, it is said to exceed that of rosewood, which is regularly exported from the Toledo District. (p. 37)

Those authors also listed this species as a timber used in **construction** to make house posts and railway ties and noted that the gum [sap] was used for "blistering."

Mosquitoxylum jamaicense Krug & Urb.

bastard mahogany, wild mahogany [English]; *palo de danto* [Spanish]; *chichimeca, nikte'* [Q'eqchi']

For treatment of non-insulin-dependent **diabetes** mellitus, ½ cup of chopped bark is boiled in 3 cups of water for 10 minutes, strained, consumed in ½ cup doses throughout the day, and continued for about 9 days, until the blood sugar normalizes. If a headache occurs, the dosage is limited to 1 cup daily [A760]. To treat **anemia** in children or adults, 9 leaves are boiled in 3 cups of water for 10 minutes and ½ cup consumed 3 times daily for 9 days [A760]. To treat **anemia** in adults or **internal infections**, ½ cup of bark is boiled in 3 cups of water for 10 minutes and ½ cup consumed 3 times daily for 9 days [A760].

As a first-aid response to **snakebite** resulting in significant swelling and soreness and to draw out the snake's "tooth scale," 4 ounces of fresh bark and 1 entire plant (if large) or 3 entire plants (if small) along with *dormilón* flowers (*Mimosa pudica* var. *unijuga*) and 6 entire *corrimiento* plants (*Cuphea calophylla*) are mashed together until moist, applied as a poultice to the bite, and changed twice daily until the wound is dry and healed. The "tooth scale" is searched for each time the poultice is removed, usually coming out after the second application. There should be no sign of blood at all to assume complete healing has occurred [LR].

Mosquitoxylum jamaicense [MB].

To treat **foot fungus**, the bark is mashed in water and salt, placed between the toes, and left on overnight as needed until better [LR].

This tree is a principal secondary **hardwood** species in Belize [Munro, 1989]. Standley and Record (1936) reported that the wood of this tree was hard, heavy, and strong, with a fine texture, and was pale reddish-brown in color with a yellow tinge and "moderately durable." They also noted that it was said to finish extremely smoothly.

Spondias mombin L.

hog plum [English]; *ciruela cochino, jobo* [Spanish]; *k'anab'al* [Q'eqchi']; *puk, pok* [Mayan]

Standley and Steyermark (1949) reported that the fruit was eaten as a **food** but was of a much lesser quality than *Spondias purpurea*. They also noted that the sap from the roots was consumed as a **beverage** when water was unavailable.

To treat a **blister** that forms around the mouth following a fever or excess sun exposure, the bark is cut and the clear resin applied directly to the affected area twice daily for 2–3 days [C55]. To treat **painful, white spots** or **sores surrounded by redness** on the tongue, the young leaves are mashed and fresh leaf juice applied directly to the affected area twice daily for 2–3 days [C55].

This tree is a useful secondary **hardwood** species and showed promise in Belize Forest Department experiments during the 1960s [Munro, 1989]. The branches are used to make **living fences** [C55].

Spondias purpurea L.

plum [English]; *jocote* [Spanish]; *rum* [Q'eqchi']; *kening* [Yucateca]; *ab ièl, ab úl* [Mayan]

The fresh fruit is eaten green or mature as a **food**. It is **astringent** and may cause constipation if too many are eaten [B2467, B3364]. Steggerda (1943) reported that there were several local varieties of this species and that the Yucatan Maya used the fruit as a **food**, boiling it with meat in their stews.

To treat **gum infections**, leaf secretions are used in an unspecified way in traditional medical practices by Maya healers of the Yucatan [Flores and Ricalde, 1996]. For **skin rash**, refer to polly red head (*Hamelia patens* var. *patens*) [B2099]. To treat **gonorrhea**, the plum-like fruit is eaten [Arnason et al., 1980].

Spondias radlkoferi Donn. Sm.

hog plum, plum, wild plum [English]; *jobo, jocote* [Spanish]; *pook, puk* [Mopan]; *rum pook'* [Q'eqchi']; *ho bo, hu hu* [Additional names recorded]

The fruit is eaten as a **food**, particularly by the howler monkey (*Alouatta pigra*) and by humans, but can cause constipation [A627, A825, B2270, B2439; Marsh, Matola, and Pickard, 1995].

The **astringent** properties of this plant make many of its parts highly useful in the treatment of a variety of medical conditions. To treat **headache** and **fever**, the leaves are mashed in a little alcohol, placed on the forehead, wrapped up for the night, and repeated for 3 consecutive nights [Arvigo and Balick, 1998]. As a treatment for **gum disease**, 1 cup of the chopped bark is boiled in 3 cups of water for 20 minutes, cooled, and used as a mouthwash at least 6 times daily [B2175]. To

Spondias radlkoferi [MB].

soothe the pain of a **sore throat**, 1 cup of leaves from the branch tips is boiled in 1 quart of water for 10 minutes, steeped until cooled, strained, and used as a gargle every hour for 1 day [B1824]. To treat **diarrhea**, 1 cup of the decoction described above for gum disease is consumed 3 times daily, before meals, until better [A800]. For a **bladder infection**, the decoction described above for sore throat is prepared and 1 cup consumed 3 times daily, before meals, for 3 days [A825]. To treat **diarrhea**, **gonorrhea**, or **sore throat**, 1 handful of flower buds and bark is boiled in 3 cups of water for 10 minutes and 1 cup consumed 3 times daily before meals. If this treatment is for gonorrhea, the dosage is continued for 10 days, after which the person is tested again [Arvigo and Balick, 1998]. For **gonorrhea**, 1 cup of chopped bark and 1 cup of leaves and young stems are boiled in 2 quarts of water for 10 minutes, steeped until warm, strained, and 1 cup consumed 3 times daily, before meals, for 9 days [B1824].

To treat an **internal bruise**, 1 cup of bark and 1 cup of leaves are boiled in 2 quarts of water for 10 minutes, steeped until cool, strained, and ½ cup consumed 6 times daily [A825].

To treat **skin rash**, 1 quart of chopped bark is boiled in 2 gallons of water for 20 minutes, steeped until cool, strained, and used to bathe the affected area twice daily [A800]. To treat **skin sores**, 1 quart of leaves is boiled in 2 gallons of water for 20 minutes, steeped until cool, strained, and about ¾ gallon poured over the affected area 2–3 times daily [B1824]. For **infected sores**, 1 small handful of leaves is mashed in a bit of water, applied as a poultice to the affected area, and changed 3 times daily until better [BW]. For **stubborn sores**, **rash**, **painful insect stings**, and pregnant women who are **tired** or **weak** beyond the first trimester, 1 large double handful of leaves and one 1-inch × 8-inch piece of bark are boiled in 2 gallons of water for 10 minutes and used as a bath [Arvigo and Balick, 1998].

For women feeling **weak** due to pregnancy, 1 double handful of leaves or bark is boiled in 2 gallons of water for 20 minutes, steeped until warm, and used as a bath once daily for 3 days [B1824].

For **nervousness**, 4 ounces of bark along with 4 ounces of the trumpet tree bark (*Cecropia peltata*) are steeped in 1½ quarts of boiling water for 25 minutes, strained, cooled, and ½ cup consumed twice daily, in the morning and at 4 p.m. This treatment should not exceed 5 consecutive days; however, it can be repeated after a 3-day rest from the tea [LR].

The trunk is used in **construction** to make house posts [A627]. The tree is cultivated and used as **living fences** [Marsh, Matola, and Pickard, 1995].

ANNONACEAE

Usually shrubs to medium-sized trees, rarely large trees or lianas. Leaves alternate, distichous, simple. Inflorescences axillary or terminal. Flowers white, yellow, rarely red or purple. Fruits a cluster of berries or syncarp. In Belize, consisting of 13 genera and 19 species. Uses for 6 genera and 10 species are reported here.

Annona glabra L.
alligator apple, bobwood, corkwood [English]; *yobapple* [Creole]
The wood is considered an important secondary **hardwood** [Munro, 1989]. Standley and Record (1936) noted that the wood was used in **carpentry** to make bottle stoppers but that the fruit was "scarcely **edible**." The wood was still in use for bottle stoppers some 55 years later [Horwich and Lyon, 1990].

Annona muricata L.
soursop [English]; *annona, guanábana* [Spanish]; *anaab', pak* [Q'eqchi']; *takob* [Additional name recorded]

Annona glabra [MB].   *Annona muricata* showing fruit [MB].

The fresh green fruit is eaten when soft and mature as a **food** [B2268]. The fruit is often made into **ice cream** [Arvigo and Epstein, 1995].

This is a component of a 9-leaf Maya bath formula used for various ailments but said to be especially good for **fever** [LR]. To draw the heat out of a **high fever**, 3 leaves are placed in a cross formation (3 at the left pulse, 3 at the right pulse, and 3 over the forehead) and 9 prayers said once daily until better. After praying, the same leaves are crushed with a bit of salt and water, applied as a poultice to the head, and changed every 20 minutes until better [BW].

For **cough**, 9 leaves along with 9 cotton leaves (*Gossypium hirsutum*) and 9 avocado leaves (*Persea americana*) are boiled in 1 quart of water until reduced to ½ quart, strained, ½ quart of sugar added, the decoction boiled again for 20 minutes, and 1 tablespoon consumed every 30 minutes until better [B2107, B2108, B2400, B2403, B2411, B2458]. To treat **cough** and **colds**, refer to *maravilla* (*Mirabilis jalapa*) [B2283]. To treat **rash**, the leaves are used to make a hot-water extract, but no further details were provided [Marsh, Matola, and Pickard, 1995].

To treat a **womb infection** following childbirth, 9 leaves along with 1 handful of fresh or dried Spanish thyme leaves (*Lippia graveolens*) are boiled in 2 quarts of water for 5–10 minutes, 1 cup consumed warm, and the remainder used as a vaginal steam bath. This is given 3 days after childbirth if there is no excess bleeding and continued until better [LR].

Annona primigenia Standl. & Steyerm.
wild annona, wild custard apple [English]; *anona del monte, anonillo, anonio* [Spanish]
The green, immature fruits are boiled until soft and eaten as a **survival food** [B3164]. Mature fruits are sweet and pleasant tasting [BW].

To treat a **high fever**, the leaves are placed on a hot *comal* (baking tray) or held over a fire for 5–10 seconds until heated, applied as a poultice to the forehead, changed twice daily, and prayers said in the same manner as described for a fever using *guanábana* (*Annona muricata*) [B3281]. To treat **sprains**, one 12-inch piece of fresh bark and a bit of salt are beaten on a rock or wooden surface, applied as a poultice to the affected area, wrapped in a cloth, and changed twice daily [B3281].

For **head lice** (*Pediculus humanus*), 1 handful of seeds is mashed with enough petroleum jelly to make a paste, applied to the scalp, left on for 2 days, shampooed out, and reapplied for 2 days—for a total of 4 times during an 8-day treatment [BW]. As an **insecticide**, 1 double handful of seeds along with 1 double handful of laurel seeds (an undetermined species), 1 double handful of John Charles tree leaves (*Hyptis verticillata*), and 2 heads of chopped garlic cloves are ground together either in a corn mill, on a stone, or in a blender with enough water to make a mash. The mash is soaked in 1 gallon of water to ferment, placed in the sun during the day, and brought in and covered during the night. After 1 week, the plant material is strained out, 5 ounces of liquid diluted into 1 gallon of water, and the water used as a spray on plants [LR].

Annona reticulata L.

bullock's heart, custard apple, wild custard apple [English]; *annona, annona del monte, anona blanca, anona colorado* [Spanish]; *pak li tzuul* [Q'eqchi']; *oop* [Additional name recorded]

The mature fruit is eaten as a **food** [A947, B2025, B2355; Arvigo and Balick. 1998].

To ease **headache**, 3 leaves along with 3 buttonwood leaves (*Piper amalago*) are crushed and applied as a poultice to the head [B2355, W1832]. To treat **headache** or **fever**, the leaves are warmed, crushed, and applied directly to the temples [A906, A947, B1817; Arvigo and Balick, 1998]. To draw the heat out of a **fever**, the leaves are soaked in alcohol and used to massage the body [Arvigo and Balick, 1998]. Alternatively, fresh leaves are tied to the soles of the feet [Arvigo and Balick, 1998]. To treat **fever** in babies, 9 leaves along with one 3-inch piece of *gumbolimbo* bark (*Bursera simaruba*) are soaked in 1 cup of wintergreen alcohol for 3 days and rubbed on the feet [B2106]. For **yellow fever**, the leaves are boiled in water and used as a steam bath [A906].

For **cough**, fresh leaves along with sugar are boiled to make a cough syrup [Arvigo and Balick, 1998]. To treat **asthma**, **bronchitis**, or a very bad **cold**, 3 leaves along with 1 garlic clove are boiled in 1½ cups of water for 5 minutes, honey added, and ½ cup consumed 2–3 times daily, including ½ cup at bedtime, for 3 days [A947]. For **mouth sores**, the leaves are prepared into a tea and used to wash the mouth [Arvigo and Balick, 1998].

To treat **skin ailments**, fresh leaves are boiled in water to prepare a bath [Arvigo and Balick, 1998]. To hasten pus formation on **boils**, raw fruit pulp is applied as a poultice to the affected area [Arvigo and Balick, 1998]. To remove **warts**, the leaves or fresh leaf juice are rubbed on the affected area twice daily, in the morning and the evening, over a 9-day treatment [B1817; Arvigo and Balick, 1998].

For **swelling** and **bruises**, the leaves are cooked in oil for a few minutes and applied as a poultice to the affected area [Arvigo and Balick, 1998]. To treat a **fracture** or **broken bones**, the bark is sliced, pounded into powder, and applied to the affected

Annona reticulata [MB].

Guatteria diospyroides [FM].

area [B1817]. To treat a **sprain**, **strain**, or **fracture**, the bark is mashed with a pinch of salt and used as a poultice, wrapped securely with a cloth, and changed daily [Arvigo and Balick, 1998].

To treat **head lice** (*Pediculus humanus*) and dry up their eggs, 1 handful of crushed seeds along with 1 handful of *ruda* leaves (*Ruta graveolens*) are mixed and applied as a poultice to the head, wrapped with a cloth, left in place overnight, and rinsed out the following day [A947, B1817]. For **head lice** and **dandruff**, the seeds are pounded, applied as a poultice to the head, wrapped in a scarf, and left on overnight [Arvigo and Balick, 1998]. As an **insecticide**, the seeds are ground to a powder, mixed with water, and sprayed in gardens [W1832].

To prepare a blue-black **dye**, the ancient Maya used the leaves and small branches in an unspecified way for use on cloth and thread [RA]. To make **cordage**, the bark is stripped and the fibers made into rope [A906].

Annona squamosa L.
saramuya [Spanish]
To reduce **fever** in children, 1 double handful of leaves is crushed, combined with copaiba oil (*Copaifera* sp.), and wiped over the body [B2305]. The leaves, along with *ruda* leaves (*Ruta graveolens*), are boiled in water and used as a facial wash [Arnason et al., 1980]. This plant is used to treat the **evil eye (*mal ojo*)** [Arnason et al., 1980]. Steggerda (1943) reported that the Yucatan Maya ground the black seeds into powder to treat **head lice** (*Pediculus humanus*).

Cymbopetalum mayanum Lundell
guanábano, *marta boni* [Spanish]
According to Standley and Record (1936), the dried, fleshy flower petals were used by the ancient Aztecs as a **flavoring** for cacao (*Theobroma cacao*) and other food items. Those authors reported that the dried petals were still sold as a **flavoring** for beverages and food in markets of El Salvador and Guatemala. Lundell (1938) reported that the Petén Maya probably knew about the use of this plant, as it was common to the region.

Guatteria diospyroides Baill.
To treat **pain**, 1 quart of fresh leaves is boiled in 2 gallons of water for 20 minutes, steeped until warm, and used as a bath over the affected area [B2691]. Alternatively, 9 fresh leaves and a pinch of salt are mashed and applied as a poultice to the affected area [B2691].

Malmea depressa (Baill.) R.E. Fr. subsp. *depressa*
chief of herbs, lancewood, wild coffee, wild soursop [English]; *suprecayo* [Spanish]; *eremuil* [Mopan]; *che che xiv*, *itz imul*, *marana* [Additional names recorded]
The mature, red fruit is eaten as a **food** [Arvigo and Balick, 1998].

In Belize, this is one of the most important components of bath and tea mixtures used as a **panacea** or to treat a variety of specific physical and emotional ailments. As a **panacea**, 1 quart of leaves is boiled in 2 gallons of water for 20 minutes and used as a bath [A49, A751, B1874, JB26; Arvigo and Balick, 1998].

To treat **anxiety**, **insomnia**, **evil spells**, **insanity**, **back pain**, **nightmares**, **depression**, **hysteria**, **nervousness**, **menstrual cramps**, and **headache**, 9 leaves are boiled in 3 cups of water for 10 minutes and 1 cup consumed, before meals or at bedtime, for 9 days [A49, A751, B1874, JB26; Arvigo and Balick, 1998]. To treat **headache** or **wind (*viento*)**, 1 quart of leaves is boiled in 2 gallons of water for 10 minutes, cooled, and used to bathe the head once daily for 9 days, usually at bedtime, following which the person must stay in bed [B2437; BW]. To treat **fever**, **insomnia**, **malaise**, **back pain**, **arthritis**, **rheumatism**, **nervousness**, or **skin ailments**, the decoction described above for headache is used warm, not hot, as a bath [Arvigo and Balick, 1998]. For **paralysis**, **severe pain or swelling**, and **nerve pain**, as in **sciatica**, the boiling decoction described above for headache is used as a steam bath for 20 minutes [RA].

Steggerda (1943) reported that the root and corn silk (*Zea mays*) were boiled in water and consumed to treat **gonorrhea** and **kidney or bladder ailments**.

To **cleanse the blood**, 6–9 leaves are boiled in 2 cups of water for 10 minutes, strained, and 2 cups consumed warm twice daily, in the morning and in the evening [JB26; RA].

The wood is used in **construction** for making house frames and **carpentry** for making hammers or axe handles [A823]. The bark is stripped from the tree and used as **cordage** to tie various objects [B3131].

This plant is considered the "male" counterpart of timber sweet (*Nectandra salicifolia*) [B1896].

Sapranthus campechianus (Kunth) Standl.

sufricaya, tabaquio, tuspi del monte [Spanish]; *bobtob, elemuy* [Additional names recorded]

The fruit provides **food** for birds; hunters know this and seek their quarry near these plants [A263].

Xylopia frutescens Aubl.

polewood [English]; *palata* [Spanish]

The fruit provides **food** for wild animals; hunters know this and seek their quarry near these trees [A221].

The wood of mature trees is strong, long-lasting, and used in **construction** for making house poles [B3251]. The pliable younger stems can be sharpened and

(top) *Malmea depressa* subsp. *depressa* [MB].

(bottom) *Xylopia frutescens* [GS].

used for such things as **fishing spears** or **bows** [B3251]. Standley and Record (1936) reported that the slender, straight trunks were employed as **poles** for maneuvering boats and canoes, which is also the case at present [Horwich and Lyon, 1990].

APIACEAE

Herbs, sometimes shrubs, rarely trees. Leaves alternate, sometimes opposite, simple or compound. Inflorescences axillary or terminal, usually in umbels, sometimes in heads. Flowers small, white, yellow, blue, or reddish-purple. Fruits a schizocarp. In Belize, consisting of 9 genera and 10 species. Uses for 2 genera and 2 species are reported here.

Eryngium foetidum L.

coriander [English]; *culantro, cimaron*, *kolentro* [Spanish]

The leaves and seeds are eaten as a **food** and used to season salads, soups, stews, and other dishes [A794, B2317, B3686; Arvigo and Balick, 1998]. Standley and Williams (1966) commented that "the fresh plant has a very strong and most nauseous odor, but when boiled in soups or stews it imparts to them a delicious flavor, such as could never be expected from the living plant."

To **dispel flatulence** in adults, the leaves are eaten [Arvigo and Balick, 1998]. To treat **diarrhea** in infants and **stomach gas**, **indigestion**, or **vomiting** in older children and adults, 6 leaves are chopped, steeped in boiling water for 15 minutes, and

Eryngium foetidum [MB] with inset showing inflorescences [IA].

¼ cup consumed repeatedly throughout the day [Arvigo and Balick, 1998]. To treat **dysentery**, 3 entire plants along with 1 green, immature lime (*Citrus aurantiifolia*) are boiled in 1 quart of water for 10 minutes and ⅓ cup consumed every hour for 3 full days, or longer as needed, until better [BW]. To calm **spasms** in babies with **colic**, 3 leaves are boiled in 1 quart of water for 10 minutes and ½ cup consumed 3 times daily until better [A794, B3686]. For **high cholesterol**, 36 leaves are boiled in 1 quart of water for 5 minutes, cooled, and ½ cup consumed 3 times daily, before meals, for 1 week. Cholesterol counts are then checked in a lab. This treatment is said to remove fat deposits from the heart [LR].

To treat **snakebite** and to remove the "tooth scale," 9 leaves along with 3 garlic cloves are chopped, fried in olive oil for 1 minute, placed in gauze, and applied as a poultice over the bite for 30 minutes. This is repeated continually with a fresh poultice. During the treatment, no oily foods are eaten, including avocado (*Persea americana*), grease, or fried foods [BW]. Alternatively, 3 leaves, one 1-inch piece of root, and 3 seeds or flower heads are chewed or mashed into a fine paste, applied to the wound, and repeated 4–5 times in a 24-hour period [Arvigo and Balick, 1998].

To treat **wind (*viento*)**, an unspecified amount of the roots along with garlic are boiled in water and used, but no further details were provided [Robinson and Furley, 1983].

Foeniculum vulgare Mill.
fennel [English]
Standley and Record (1936) reported that the leaves and seeds of this cultivated plant were used as a **flavoring** for foods.

APOCYNACEAE (INCLUDING ASCLEPIADACEAE)

Herbs, shrubs, trees, vines, or lianas, sometimes woody or succulent, containing milky latex in the stems and leaves. Leaves opposite, sometimes whorled or alternate, simple, sometimes reduced and deciduous. Inflorescences axillary, determinate, terminal, or nodal, in panicles, racemes, corymbs, umbels, or rarely of solitary flowers. Fruits usually a follicle, capsule, berry, or drupe. In Belize, consisting of 34 genera and 69 species. Uses for 22 genera and 25 species are reported here.

Allamanda cathartica L.
alamanda, campanitas amarillas [Spanish]
These shrubs are cultivated as an **ornamental** for their yellow flowers [B2281].

To treat **uncertainty** and **indecisiveness**, the leaves and flowers along with 7 white roses (*Rosa chinensis*), 9 small branches of *ruda* (*Ruta graveolens*), 9 flowers with 6-inch stems and leaves of yellow wild daisy (an undetermined species), and 9 flowers with 6-inch stems and leaves of red hibiscus (*Hibiscus rosa-sinensis* var. *rosa-sinensis*) are combined, crushed in water, steeped in the sun for 1 hour, and used as a bath [B2651].

(top left)
Foeniculum vulgare [NYBG].

(top right)
Foeniculum vulgare with inset showing close-up of seeds [IA].

(bottom)
Allamanda cathartica [MB].

To improve **luck**, **fortune**, and **love**, 9 flowers along with 9 white roses (*Rosa chinensis*), 9 red roses (*Rosa chinensis*), one 6-inch piece of cinnamon bark (*Cinnamomum verum*), ½ cup of milk, 1 cup of champagne, and 9 drops of "Florida water" (a local over-the-counter product) are combined with 2–3 gallons of water, steeped in the sun for 2 hours (or, if there is no sun, boiled together for 5 minutes), and used as a bath once daily for 9 days, from December 31st until January 8th [BW].

Caution is advised when using this plant, as the white sap it contains may cause rash or illness if a person is sensitive to the toxin [Marsh, Matola, and Pickard, 1995].

Asclepias curassavica L.

butterfly weed, polly red head, red head polly [English]; *chuchita, gato, hoja de veneno, ratón* [Spanish]; *kaq i atz'um* [Q'eqchi']; *cho, chushu yu shi, cuchillo xiv, mis* [Additional names recorded]

Asclepias curassavica [MB].

To treat **toothache** when the tooth must be extracted, white latex is gathered from the base of the leaves when pulled from the stems or from the stems when broken apart, placed on a piece of cotton, applied to the problem tooth, and left there for an entire day. It is said that after this treatment, the tooth will break off in about a month's time [B2435]. To alleviate the pain of **toothache** and **earache**, 1 entire plant is boiled in 2 quarts of water for 10 minutes, cooled, and used to wash the face [B2032].

To clear **sinus congestion** and break up **nasal congestion**, 1 drop of the leaf juice is placed into each nostril, the head tilted back for 10 minutes, and the sinuses drained [A727, B2241, B3694]. To treat **colds**, the latex is placed on cotton, dried in the sun, and inhaled [Robinson and Furley, 1983]. Standley and Williams (1969) reported this same use in the area around Cobán, Guatemala.

To stop the **itching** of small sores and rash, latex is applied directly to the affected area [B3646]. To treat a **wart** or **wound**, the leaf latex is used in an unspecified way [Flores and Ricalde, 1996]. The leaves of this species are 1 of 9 plants used in an unspecified bath formula for the **skin** [B2241]. As a treatment for **mastitis**, 1 fruit is heated lightly in ashes, then rubbed downward over the breast 9 times while still warm; this process repeated 3 times [BW].

This plant is called *hierba de veneno*, suggesting its toxic nature [Robinson and Furley, 1983]. It is said that if a single plant is eaten by a horse or cow, it is **toxic** enough to kill that animal [Robinson and Furley, 1983]. To **poison** rats, the leaves are ground, mixed with corn, and placed in the infested area [A146].

This species is considered the "male" counterpart of polly red head (*Hamelia patens* var. *patens*) [B3694].

Aspidosperma cruentum Woodson

milady, my lady, red malady [English]

To soothe **sores** caused by thrush, 1 handful of chopped bark is boiled in 4 cups of water until reduced to 3 cups and then used as a mouthwash [A489]. For **skin sores** and **rash**, 10 ounces of bark are boiled in 1 gallon of water for 15 minutes and used cool to soak or pour over the affected area twice daily [LR].

To **reduce heat** or **fever** in the body, ½ cup of bark along with ½ cup of *gumbolimbo* bark (*Bursera simaruba*) are boiled in 2 quarts of water for 5 minutes, cooled, strained, and consumed as a "refrigerant." This may also be consumed on a hot day to cool a person [A489; LR]. To help "**flush out the kidneys**," the decoction described for fever is consumed twice daily, in the morning and in the evening [LR].

The young trees grow straight and sturdy and are used in **construction** to make house posts [A489, R80].

Aspidosperma megalocarpon
Müll. Arg.

my lady [English]; *malady blanco*, *malereo* [Spanish]; *peech maax* [Yucateca]

To treat "**bad bile**," the inner bark of the tree is cooked and consumed, but no further details were provided [Arnason et al., 1980].

This secondary **hardwood** species was regularly harvested for timber in areas of Belize

(top) *Aspidosperma cruentum* [GS].

(bottom) Fruit of *Blepharodon mucronatum* closed (left) and open showing seeds (right) [HW].

[Munro, 1989]. Standley and Record (1936) noted that the "fairly durable" wood was used in **construction** to make "railway ties, house frames, rafting poles, and scaffolding." Lamb (1946) reported that the wood was said to be used in **construction** to make house beams, rafters, sills, rafts, boat poles, and railroad ties.

Blepharodon mucronatum (Schltdl.) Decne.
tie tie [English]

To reduce **swelling**, 1 quart of fresh leaves with young green stems is boiled in 1 gallon of water for 10 minutes and used tepid as a bath [B2685]. Alternatively, the leaves are mashed with a pinch of salt, applied as a poultice, and wrapped with a clean cloth [B2685].

Steggerda (1943) reported that the Yucatan Maya crushed the leaves and used them as an **antiseptic**, to treat **snakebite**, and to reduce **swelling**.

Cameraria latifolia [MB].

Cameraria latifolia L.

savanna white, savanna white poisonwood, white poisonwood [English]; *chechem de caballo* [Spanish]

Standley and Record (1936) reported that this plant was highly **toxic**, causing swelling and inflammation when in contact with the body.

Catharanthus roseus (L.) G. Don

all day flower, periwinkle, ram goat [English]; *chata, picaria* [Spanish]; *chula* [Additional name recorded]

The pretty white and pink flowers are widely cultivated as an **ornamental** [B2358].

To treat a raspy, **sore throat** in babies, three 3-inch branches with leaves and flowers are boiled in 1 quart of water for 10 minutes, 1 tablespoon of sugar added, and 2 tablespoons consumed every 10 minutes [BW]. To treat a raspy, **sore throat** in adults, nine 3-inch branches are boiled in 1 quart of water for 10 minutes and ½ cup consumed every hour [BW]. Additionally, for either of these raspy, **sore throat** treatments, "Vicks" (Vicks VapoRub® topical ointment) is rubbed on the chest and mint leaves (*Mentha* sp.) rubbed down the neck, behind the ears, and on the wrists [BW]. For **sore throat** or a **cold**, 9 pink flowers are soaked in 1 pint of water for 3 hours in the sun and sipped all day long [Arvigo and Balick, 1998].

To treat non-insulin-dependent **diabetes** mellitus, 9 branches with leaves and flowers along with one 12-inch piece of the woody *chicoloro* vine (*Strychnos panamensis* var. *panamensis*) are boiled in 2 quarts of water for 10 minutes and 1 cup consumed

Catharanthus roseus [MB].

Couma macrocarpa [RF].

Cryptostegia grandiflora [RL].

warm 3 times daily for 30 days. The blood sugar level is checked every 2 weeks and treatment discontinued when the sugar level is normal. This dosage is repeated in 30 days to maintain a proper sugar balance [B2358]. Alternatively, in the Toledo District, the root is ground into a coarse powder, boiled in 2 cups of water for 5 minutes, and consumed twice daily as needed [Mallory, 1991]. To treat non-insulin-dependent **diabetes** mellitus, **menopause**, **high blood pressure**, and slow the growth of **tumors**, two 12-inch branches with leaves and flowers are boiled in 3 cups of water for 2 minutes, steeped for 15 minutes, strained, and 1 cup consumed 3 times daily before meals [Arvigo and Balick, 1998].

Couma macrocarpa Barb. Rodr.

cow tree [English]; *barca* [Spanish]

Standley and Record (1936) noted use of this species as a **food**. According to those authors, this was

> one of the most interesting of Central American trees, and one that has received much publicity in periodical literature. When the bark of the cow tree is cut or broken, there issues from it a rich creamy latex that is sweet and palatable. It is not very sticky, and may be drunk like cow's milk. (p. 324)

Cryptostegia grandiflora R. Br.

The purple flower of this shrub makes it a popular **ornamental**, and it is cultivated in many Belizean yards [B2329].

Echites yucatanensis Millsp. ex Standl.

To treat **warts** and **corns**, the leaf latex is used in an unspecified way by Maya healers of the Yucatan [Flores and Ricalde, 1996].

Gonolobus fraternus showing flowers and fruit [GS].

Mandevilla hirsuta [MB].

Gonolobus fraternus Schltdl.

vine of the dead [English]; *behuco de muerto* [Spanish]

To **induce sleep** in adults and as a **calmative** for babies who are crying or suffering from fever or malaise, 1 quart of fresh leaves and chopped vine is boiled in 2 gallons of water for 20 minutes, steeped until very warm, and used as a bath once daily at bedtime [A529].

Lacmellea standleyi (Woodson) Monach.

milk tree, prickley vaca, vaca tree [English]; *palo de vaca*, *vaca* [Spanish]

Standley and Record (1936) noted that the fruit of this species was sweet and eaten as a **food**. Standley and Williams (1969) noted that people cut incisions into the trunk to obtain white latex that is "drunk sometimes, although not altogether agreeable in consistency." Those authors suggested that the fruit was "probably" **edible**, with the odor of mangos.

Mandevilla hirsuta (Rich.) K. Schum.

This is 1 of 9 different unspecified herbs used to **test the blood** [B2649]. For pasmo **(congested blood; more than the usual amount in an area of the body)** and other **bruising or clotting ailments**, 1 entire plant is boiled in 2 gallons of water for 20 minutes, steeped until very warm, and used as a bath. Additionally, a tea made from fresh limes (*Citrus aurantiifolia*) is consumed while the veins are bled simultaneously [B2649].

Metastelma schlechtendalii Decne.

Steggerda (1943) reported that the Yucatan Maya boiled the root in water and used the extract to "rinse the mouth in case of **canker sores**." He noted that the rinse could be repeated several times but that the liquid should never be swallowed.

Nerium oleander L.

narcisso [Additional name recorded]

Valued for its red flowers, the shrub is cultivated as an **ornamental** near homes and in other areas [B2331]. It is known to be **toxic** [B2331].

(top) *Metastelma schlechtendalii* showing close-up of flower [JC].

(bottom) *Nerium oleander* [IA].

Pentalinon andrieuxii (Müll. Arg.) B.F. Hansen & Wunderlin

snake root [English]; *salsa, tulipan del monte, waica contribo* [Spanish]; *ciro ak selsa, sak sa yumb'* [Additional names recorded]

As a **panacea** for many illnesses, including the treatment of **head-ache**, **swelling**, and feelings of **grief** (*pesar*), 1 handful of the chopped plant is boiled in 3 cups of water for 10 minutes and consumed or used as a bath until better [B1886]. To treat non-insulin-dependent **diabetes** mellitus, 1 handful of root along with ½ cup of Billy Webb bark (*Acosmium panamense*) are boiled in 3 cups of water for 10 minutes and 1 cup consumed at room temperature 3 times daily before meals [B2128].

To treat **snakebite**, any other **poison**, or a **hernia**, the root is roasted to a powder, ¼ teaspoon mixed with ½ cup of warm water, and consumed 3 times daily [A655]. For a temporary **snakebite** anti-dote, when walking or working

(top) *Pentalinon andrieuxii* [MB].

(bottom) *Plumeria rubra* [MB].

alone in the bush (forest), the root is chewed until the person can reach a doctor [B2128]. Alternatively, the root is chewed and applied as a poultice directly over the bite [A772]. For another reported **snakebite** treatment, refer to white breadnut (*Trophis racemosa*) [B3262]. To reduce pain and swelling from **snakebite**, 1 entire plant is boiled in 1 gallon of water for 20 minutes and used hot as a bath over the bite twice daily [A655]. For **snakebite**, the roots along with *dormilón* roots (*Mimosa pudica* var. *unijuga*) are mashed and applied as a poultice directly to the bite [P2]. Alternatively, ½ cup of this root infusion is consumed every 30 minutes [P2].

As broadly useful as this plant is, it is not without side effects. When taken orally, it is said to sometimes cause **vomiting**, though this is thought to mean that the treatment is "working" [B1886].

Plumeria rubra L.

frangipani, mayflower, Spanish jasmine [English]; *flor de mayo* [Spanish]

Said to be a **laxative** or a **purgative** when a person is suffering from a buildup of "harmful substances" in the internal organs, one 1-inch × 6-inch piece of fresh bark is boiled in 2 cups of water for 10 minutes, fermented overnight, and ½ cup doses consumed until vomiting occurs [B1928]. To treat skin **warts**, **toothache**, and **gum infection**, the latex from the stem and leaf is used in an unspecified way by Maya healers of the Yucatan [Flores and Ricalde, 1996]. While no further information on therapeutic use was provided, in general, this plant is quite **toxic** and must be used with great care if at all; it is particularly **toxic** when ingested [MB].

Rauvolfia tetraphylla [PA].

To extract **beefworm**, the white latex is collected on a piece of brown paper, applied to the affected area for 10 minutes to "stifle" the worm, and the worm squeezed out of the host's body [LR].

There is a local **legend** that the spirit of Princess Zak Nik Te lives in this tree. She was supposed to marry Uxmal, a prince who was an ugly dwarf, but drowned herself instead [LR].

Caution is advised, as this plant is highly toxic [B1928].

Rauvolfia tetraphylla L.

tèk ta men, *túk ta men* [Mayan]

The red fruit provides **food** for small birds; hunters know this and seek their quarry near this plant [B3192]. Standley and Record (1936) reported that the fruit was **toxic** to humans.

Standley and Williams (1969) noted that this plant was commonly employed in traditional medicine for the treatment of **malaria** and was also a reputed remedy for **snakebite**. Steggerda (1943) reported that the bark was "used by tobacco farmers to give odor and color to tobacco (*Nicotiana tabacum*)" in the Yucatan, Mexico, and the juice of the plant was used to treat **sore eyes**. He wrote:

Care must be taken, however, that only a small amount be used, for too much causes blindness. The directions call for only one application a day. One *yerbatero* adds that the root of this shrub is ground to a powder and applied to open wounds in which fly maggots have already appeared. (p. 215)

To treat **toothache** and **cavities**, the latex from the stem and leaf is used in an unspecified way by Maya healers of the Yucatan [Flores and Ricalde, 1996].

Rhabdadenia biflora (Jacq.) Müll. Arg.

mangrove vine [English]

To reduce **swelling**, 1 handful of leaves is boiled in 2 quarts of water, steeped until tepid, and used as a bath over the affected area. Additionally, the leaves are mashed and applied as a poultice to the affected area. The latex within the leaves is noted to be useful for reducing the swelling [A365].

Rhabdadenia biflora [MB].

Sarcostemma bilobum Hook. & Arn.

This plant is said to be **toxic** [B2134].

Stemmadenia donnell-smithii (Rose) Woodson

cojeton large, horse's balls [English]; *cojotón, comulyote* [Spanish]; *chac lik'in, chaklakin* [Additional names recorded]

The sticky exudate from this plant is used to remove **parasitic larvae** from people and animals. To extract **beefworm**, especially on the head, the latex is mixed with tobacco (*Nicotiana tabacum*), applied to the affected site for 30 minutes, and the larvae squeezed out [B1931]. To suffocate **botfly larvae** that develop subcutaneously, the latex is applied directly to the skin and the botfly extracted [B3117]. To treat **ringworm**, fresh latex from the tree is applied to the affected area, left on all day, and washed off at bedtime; it is said to be a one-treatment cure [RA]. To **attract and poison mosquitoes**, the latex is mixed with sugar and left out—for example, on a table. The mosquitoes are attracted to it, get stuck to the gum, and die [BW].

The latex is used to make household **glue** in Belize and to **fasten** cigarette papers in El Salvador [Horwich and Lyon, 1990].

Tabernaemontana alba Mill.

dog balls [English]; *cogetón, cojetón, cojón de perro, cojotón, huevo de chuchu, huevos de chucho* [Spanish]; *naq rit li tz'i'* [Q'eqchi']; *ton chà, ton chí, ton sàmin, ton símin* [Mayan]

The white, milky latex is the most valued part of this plant. The green fruits are boiled in water to release a latex used like **chewing gum** [A581, B3602]. Children like to chew on latex from the seeds, both mature and immature, and they get at it by splitting it open with their fingernails [B3201; BW].

To extract **beefworm**, **screwworm** and **botfly** larvae, the white latex from the leaves is mixed with tobacco (*Nicotiana tabacum*) and placed on the affected area, suffocating the larvae and bringing them to the surface of the flesh, where they are

(top) *Sarcostemma bilobum* showing close-up of inflorescence (right) [WJH].

(bottom left) *Stemmadenia donnell-smithii* [MB].

(bottom right) *Tabernaemontana alba* [GS].

extracted [A819, B2114, B2173, B2663, B3132]. To treat **sores**, **swelling**, or a **hernia**, the latex from the leaves is applied directly to the affected area [A42, A819]. To treat **internal swelling** in the lymph nodes of the groin said to be caused by an infected foot sore, the leaves are mashed and applied as a sticky poultice directly to the affected area [B1882].

To treat an **abscess**, **cyst**, or **carbuncle**, 9 leaves are boiled in 2 quarts of water for 10 minutes and the water used as hot as can be tolerated as a bath over the affected area [B2663]. Alternatively, 1 young, fresh leaf is warmed, mashed, and applied as a poultice to the affected area [B2663].

Tabernaemontana arborea Rose

horse balls [English]; *cojón de caballo, cojotón, huevo de caballo* [Spanish]; *ch'iich li ya, tz'epui* [Q'eqchi']

The latex is considered useful for many **ailments** [B2239]. To remove **beefworm**, **screwworm**, and **botfly** larvae, the latex from the leaves is mixed with tobacco (*Nicotiana tabacum*) and applied over the lesions, suffocating the larvae and driving them to the surface of the skin, where they are easily extracted [B2674]. To relieve **headache**, the latex is put on a cloth and pressed to the side of the face or head [B2239]. To treat a **cyst**, **carbuncle**, or an **abscess**, the leaves are prepared in water and used as a bath over the affected area [B2674]. Alternatively, 1 young leaf with the plant's milky latex is warmed and applied as a poultice directly to the affected area [B2674]. For **burns**, the leaves are mashed and applied with their latex twice daily, in the morning and in the evening [B3678]. However, for treating a **burn**, *gumbolimbo* (*Bursera simaruba*) is considered "100 percent better" [B3678].

Thevetia ahouai (L.) A. DC.

dog's tongue, grandfather's balls [English]; *cochetón, cogotone, cojón de mico, cojotón, taricambol* [Spanish]; *ch'iich li yak, tz'epui* [Q'eqchi']

The sweet, red fruit is eaten as a **food** [B1827, B1913, B3633].

To treat a **cyst**, **wart**, or **wound**, the latex from the stem and leaf is applied to the affected area by Maya healers of the Yucatan [Flores and Ricalde, 1996]. To halt **bleeding** or help a **boil** to erupt, the latex is applied directly to the affected area [B3633; Robinson and Furley, 1983]. To treat **swelling** or an **infection**, 9 leaves are boiled in 2 quarts of water for 10 minutes and used warm as a bath twice daily [A391]. To coax **herniated organs** back into place, the latex is applied to a piece of cloth already rolled into a ball, pressed into the area over the hernia, and wrapped in place. To apply pressure to the site, it is wrapped with a bandage or another cloth and left in place overnight [B1827, B1913].

For **beefworm**, the latex is applied directly to the affected area for 1–2 hours and the larvae squeezed out of the host's body [A391, A534, B2447, B3633].

It is said that "a fine **oil** which might be exploited commercially" is obtained from the seeds [Robinson and Furley, 1983].

Thevetia ahouai [IA].

Thevetia peruviana (Pers.) K. Schum.

good luck tree [English]
This tree is often called the "good luck tree" because it is believed that keeping the seeds in one's pocket will bring good **luck** to the bearer [B2157]. To treat a **wart**, **wound**, or **toothache**, the latex of the stem and leaf is used in an unspecified way by Maya healers of the Yucatan [Flores and Ricalde, 1996]. This species is highly **toxic** (Nelson, Shih, and Balick, 2007).

AQUIFOLIACEAE

Trees or shrubs, usually evergreen, rarely deciduous. Leaves alternate, sometimes opposite or subopposite, simple. Inflorescences axillary, in cymes, thyrses, or fascicles. Flowers often white or cream, sometimes green, yellow, pink, purple, red, or brown. Fruits a drupe. In Belize, consisting of 1 genus and 3 species. Uses for 1 genus and 1 species are reported here.

(top) *Thevetia peruviana* [IA].

(bottom) *Ilex guianensis* [GS].

Ilex guianensis (Aubl.) Kuntze
birdberry, broken ridge waterwood, dogwood [English]; *cassada, laurel del agua* [Spanish]; *kojl* [Q'eqchi']
To relieve **pain** and reduce **fever**, 1 large handful of leaves is boiled in 2 gallons of water for 20 minutes, steeped until warm, and used twice daily as a bath [B2682].

ARALIACEAE

Shrubs, trees, sometimes lianas or creeping herbs. Leaves alternate, usually palmate, sometimes simple. Inflorescences terminal, often compound and showy in umbels, heads, and sometimes in racemes or spikes. Fruits a drupe, sometimes a berry. In Belize, consisting of 4 genera and 7 species. Uses for 1 genus and 1 species are reported here.

Dendropanax arboreus (L.) Decne. & Planch.

lion's hand, mountain blossom berries, potatowood, white gumbolimbo [English]; *mano de léon* [Spanish]; *saq chacah* [Q'eqchi']; *tz'ub* [Mayan]

Bees love to come to the flower for **nectar**. The fruit provides **food** for birds; hunters know this and seek their quarry near this plant [A570, B2000].

The trees are used as **living fence posts**, but the wood is too soft to use in construction [A570, B2034, B3742]. This is an important secondary **hardwood** species [Munro, 1989]. Lamb (1946) recommended that "white gumbolimbo should be a useful general utility lumber. The straight grain and firm texture should make it suitable for rotary veneer for utility plywood ... The wood of an allied species is much used in Brazil for **carpentry** and for box boards."

Dendropanax arboreus [GS].

ARISTOLOCHIACEAE

Lianas or vines, sometimes perennial herbs or shrubs. Leaves alternate, distichous, simple. Inflorescences usually axillary or cauliflorous, in fascicles, racemes, or cymes. Flowers showy, reddish, purple, yellow, or pink, with a strong odor of carrion. Fruits a capsule. In Belize, consisting of 1 genus and 10 species. Uses for 1 genus and 4 species are reported here.

Aristolochia arborea Linden

tip te ák [Mayan]

To treat **gastritis (*ciro*)**, 1 piece of root along with one 6-inch piece of a dried sweet orange peel (*Citrus sinensis*), 1 garlic clove, and one 3-inch piece of cinnamon bark (*Cinnamomum verum*) are mixed, boiled in 4 cups of water for 10 minutes, strained, and 1 cup consumed 3 times daily until better [B2199].

Aristolochia grandiflora Sw.

bastard contrayerba, contribo vine [English]; *contribo, guaco* [Spanish]

For **fever, lack of appetite**, and **hangover**, 1 small handful of chopped vine is boiled in 3 cups of water for 10 minutes and ¼ cup consumed at room temperature twice daily, in the morning and in the evening [LR]. This plant is used to treat **malaria** and **depression** [B3669].

To treat **gastritis (*ciro*)**, one 3-inch piece of root and one 3-inch piece of stem along with one 3-inch piece of bay cedar bark (*Guazuma ulmifolia*), 1 head of garlic cloves, 2 tablespoons of grated ginger (*Zingiber officinale*), and one 6-inch piece of a dried sweet orange peel (*Citrus sinensis*) are combined, boiled in 2 quarts of water for 10 minutes, the person massaged with this decoction, and 1 full shot glass consumed at room temperature 3 times daily for 7 days [B3669]. For **gastritis** (ciro), non-insulin-dependent **diabetes** mellitus, **delayed menstruation**, **colds**, or **influenza**, refer to Billy Webb (*Acosmium panamense*).

To remove excess **mucus** in the prostate and uterus, 1 handful of chopped vine along with ½ handful of Billy Webb bark (*Acosmium panamense*), ½ handful of peeled raw peanuts (*Arachis hypogaea*),

Aristolochia grandiflora [MB].

1 handful of chopped balsam bark (*Myroxylon balsamum* var. *pereirae*), 1 handful of chopped wild yam tuber (*Dioscorea bartlettii*), and two 1-inch × 1-inch pieces of pine resin or pine wood (*Pinus caribaea* var. *hondurensis*) are combined in a container, covered with either *anisado* (anise liquor [*Pimpinella anisum*]) or Beefeater gin, soaked for 14 days, and ⅛ cup consumed twice daily, in the morning and at bedtime, for 1 week. After 1 week, the person will pass strings of white mucus in the morning urine. Treatment is then suspended for 1 week, followed by another week of treatment, and continued as such every month until this same dosage brings no more mucus. This is especially good for menopausal women and impotent men [LR].

For **snakebite**, one 3-inch piece of vine and one 3-inch piece of root along with an unspecified amount of *sarsa* root (*Smilax* sp.) are mashed in 1 quart of water, steeped, and the person encouraged to drink "a lot" to counteract the effects of the toxin, even though the infusion is extremely bitter [B3669].

Aristolochia maxima Jacq.

duck flower [English]; *contribo*, *guaco* [Spanish]

This woody vine, like others in the genus, is considered an important **medicine** in Belize. The many ailments for which it is used include **gastritis (*ciro*)**, **colds**, **influenza**, **fever**, **constipation**, **uterine congestion**, and **liver spots** [A511]. To treat these ailments, 1 handful of the chopped vine is boiled in 1 quart of water for 10 minutes, steeped for 10 minutes, strained, and ½ cup consumed warm or at room temperature 6 times daily [RA].

Aristolochia trilobata L.

duck flower [English]; *contribo, flor de pato* [Spanish]

Standley and Steyermark (1946) reported that the plant was used in Belize as a local remedy to treat **fever**. To treat **hangover, influenza, colds, constipation, stomach pain, indigestion, flatulence, gastritis (*ciro*), amoebas, colitis, high blood pressure, "heavy" heartbeat, loss of appetite, scanty or delayed menstruation**, and to **cleanse the urinary tract**, 1 small handful of chopped vine is boiled in 3 cups of water for 10 minutes, strained, and 1 cup consumed warm 3 times daily, before meals, as needed. If the above ailments are experienced by the weak or aged, the small handful of the chopped vine is soaked (not boiled) in 3 cups of water all day and 1 cup consumed 3 times daily before meals [Arvigo and Balick, 1998]. To induce a copious flow of mucus from **sinus congestion**, one 6-inch piece of vine is soaked in 1 quart of water all day in the sun, sipped over a 12-hour period until all is consumed, and repeated for a second day if necessary [Arvigo and Balick, 1998]. As a general **tonic** and to treat **colds**, refer to skunk root (*Chiococca alba*) [Mallory, 1991].

To treat **constipation** in infants, 1 handful of chopped vine is boiled in 3 cups of water and consumed by the nursing mother [Mallory, 1991]. To treat **constipation** in children, 2 cups of this same decoction are consumed spoonful by spoonful all day until finished [Mallory, 1991]. For **constipation** in adults, 1 cup of this decoction is consumed 3 times daily before meals [Mallory, 1991]. To treat **bloody dysentery**, the root is crushed, boiled in water, and consumed [Arnason et al., 1980].

As a **male aphrodisiac** reported to be used in the Toledo District, 1 entire plant is chopped, boiled in water, mixed with rum or skunk root, and consumed [Mallory, 1991]. As a **tonic** or **stomachic** (to stimulate the action of the stomach), 1 entire plant is boiled in water and consumed as needed [Mallory, 1991].

Caution is advised with this plant. Recent research and warnings about the carcinogenic activity of aristolochic acid found in species of this genus strongly suggest that this plant not be used in herbal medicine [Gold and Slone, 2003]. Aristolochic acid is considered a potent carcinogen and kidney toxin [Lewis and Alpert, 2000].

ASTERACEAE

Herbs, shrubs, sometimes trees, lianas, or epiphytes. Leaves alternate, opposite, or sometimes whorled, simple, rarely compound. Inflorescences with heads in panicles, racemes, cymes, or corymbs. Flowers variously colored. Fruits an achene. In Belize, consisting of 89 genera and 158 species. Uses for 42 genera and 49 species are reported here.

Acmella pilosa R.K. Jansen

orosus [Yucateca]

It is said that the flowers, when eaten, **numb** the mouth; children try to trick one another by encouraging each other to eat them [B1798].

For **headache**, 1 entire plant is steeped in 2 quarts of water for 1 hour and used cool to wash the head [B2127]. Alternatively, 1 entire plant along with 1 entire plant of wild ruda (*Diphysa* sp.) and 1 entire plant of *sinanche* (possibly *Zanthoxylum*

(top left) *Ageratum conyzoides* [MB].

(top right) *Bidens pilosa* [MB].

(bottom) *Borrichia arborescens* [RH].

juniperinum) are soaked in water and used to bathe the head [B2127]. For **cough**, the leaves are used to prepare a tea [Arnason et al., 1980].

Ageratum conyzoides L.

To relieve **itching** of scabies, other small sores, or itchy areas, 1 leaf is rubbed over the affected area [B3549].

Bidens pilosa L.

To treat **mouth blisters**, an exudate is used by the Maya of the Yucatan region of Mexico, but no specific details were provided [Flores and Ricalde, 1996].

Borrichia arborescens (L.) DC.

fisherman's tobacco [English]
A decoction made from the leaves is said to have unspecified **medicinal properties** [B1950].

Chromolaena odorata (L.)
R.M. King & H. Rob.

Jack in the bush [English]; *hatz, hotz, yaxhatz* [Mopan]

For **fever**, seven 3-inch pieces of branches with leaves are boiled in 1 quart of water for 10 minutes and consumed cold [RA]. To **promote sweat** and treat **nervousness**, this same decoction is consumed warm [RA]. To treat the "**culebrilla worm**" in babies, refer to *sikiya* (*Chrysophyllum mexicanum*) [B2115].

To treat **cough**, **colds**, and **depression**, 9 branchlets with leaves are boiled in 3 cups of water for 5 minutes and consumed as a tea [Arvigo and Balick, 1998]. For **anxiety**, the decoction described for cough is sipped hot throughout the day [Arvigo and Balick, 1998]. To treat **insomnia**, 1 cup of the decoction described for cough is consumed hot once daily, 1 hour before bedtime [Arvigo and Balick, 1998].

Chromolaena odorata [MB].

To treat many ailments, including **general pain**, **headache**, **nervousness**, and **insomnia**, 1 double handful of leaves is boiled in 1 gallon of water and used as a bath [B1769; Arvigo and Balick, 1998]. If the headache occurs during the daytime, the patient is bathed with a cooled decoction; if it occurs during the nighttime, a hot decoction is used [Arvigo and Balick, 1998]. To treat **lethargy**, **pain**, **muscle strain**, **headache**, **nervousness**, and **anxiety**, eighteen 1-inch pieces of the branches with leaves are boiled in 1 quart of water and consumed hot or cold once daily before bedtime. The person awakens feeling refreshed and well. For best results, it is said that the decoction should be made "strong," as it has no side effects and one cannot "overdose" while taking it [B2653]. In some cases, such as for **nervousness**, other plants may be mixed in with this species, but no details were provided [B2098]. To calm **nerves**, refer to wild sage (*Lantana trifolia*) [B2096].

This is considered to be the "female" counterpart of the *sim sim* plant (*Chamaecrista nictitans*) [A138].

(left) *Cirsium mexicanum* showing inflorescence [SH].

(right) *Clibadium arboreum* [MB].

Cirsium mexicanum DC.

wild San Carlos [English]; *Cardo Santo* [Spanish]

As a **libido suppressant**, the seeds are roasted, ground, and fed to "men who run around with women" [A334]. This plant is said to be very rare [RA].

Clibadium arboreum Donn. Sm.

corona, *pito sico* [Spanish]; *chal ché*, *zu pup* [Additional names recorded]

As a treatment for **fatigue** and muscle **pain** 1 handful of leaves and 3 stems are boiled in 2 gallons of water and used as a bath [B2622]. For **tired or sore feet**, the decoction described for fatigue is used warm to soak the affected area [B2622].

To treat a **wound** or **sore**, 9 leaves are boiled in 1 quart of water for 10 minutes and used hot as a bath over the affected area twice daily [A593]. Additionally, 1 handful of leaves is toasted over a heat source, powdered, the stems removed, castor oil (*Ricinus communis*) applied to the affected area, and the leaf powder sprinkled on top. This powder treatment is applied after each bath until better [A593]. To get rid of **internal infections** and **parasites**, the leaves are used to prepare a tea that is consumed twice daily [A593].

To treat *pasmo* **(congested blood; more than the usual amount in an area of the body)**, nine 6-inch branches are boiled in 2 gallons of water for 10 minutes and used tepid as a bath. The person also stays in the house for the rest of the day [BW]. To relieve **uterine congestion**, **uterine infection**, and to vaginally **pass blood clots or old blood**, nine 6-inch branches are boiled in 1 gallon of water and used as a vaginal steam bath for 20 minutes [BW]. For **menstrual cramps**, 3 leaves are boiled in 1½ cups of water for 5 minutes and consumed as hot as possible as needed. No cold food or drinks are consumed while using this tea for cramps [BW].

Cosmos caudatus Kunth

wild cosmos [English]

The leaves and flowers of this plant are used as **decorations** for tables and altars [A577].

Critonia morifolia (Mill.) R.M. King & H. Rob.

fish pot, green stick [English]; *palo verde* [Spanish]; *ya ax como che, yax che* [Mopan]

This is one of the major medicines in the Maya pharmacopoeia, and the leaves are considered powerful enough to be used alone as well as being an essential component of bath mixtures used for **serious illness** or **general malaise**, including **skin ailments, exhaustion, wound, insomnia, influenza, pain**, or babies with a **fever** [A809, B1806; Arvigo and Balick, 1998]. For **headache**, the leaves along with *madre de cacao* leaves (*Gliricidia sepium*) and *gumbolimbo* leaves (*Bursera simaruba*) are boiled in water and used to prepare a bath [B2202]. To treat **toothache** and **gum pain**, the root is chewed, releasing chemical compounds that deaden the pain [A809, B3687].

For **swelling, fluid retention, rheumatism, arthritis, paralysis**, and **muscle spasms**, the leaves are used to prepare a steam bath [Arvigo and Balick, 1998]. To treat a **skin growth, tumor, swelling, edema**, and all sorts of **pain**, including **wind (*viento*)**, the leaves are used to prepare poultices, baths, and steam baths [A108, B1806]. To reduce the size of a **skin growth** or

(top) *Cosmos caudatus* [GS].

(middle) *Critonia morifolia* [MB].

(bottom) *Cyanthillium cinereum* [MB].

tumor, the leaf is heated and applied as a poultice to the affected area [B1806]. To treat a **boil, tumor, cyst**, or **pus-filled sores**, the leaf is heated in oil and applied as a poultice to the affected area [Arvigo and Balick, 1998].

Dahlia pinnata [MB].

Egletes liebmannii var. *yucatana* [GS].

As a **fish poison**, the leaves and stems are mashed and then placed into streams and ponds, after which the stunned fish are easily harvested [A809].

Nash and Williams (1976) reported that the long, hollow stems of this species "are often used for temporary fencing, the sides of lowland huts, and other similar purposes" in Guatemala and elsewhere.

Cyanthillium cinereum (L.) H. Rob.
chaq e pim [Q'eqchi']
To treat **influenza**, 1 handful of fresh leaves is boiled in water for 5 minutes and consumed warm 3 times daily for 5 days [C53].

Dahlia pinnata Cav.
margarita [Spanish]
The red flower of this herbaceous plant makes it a beautiful garden **ornamental** [B2468].

Egletes liebmannii (Sch. Bip.) Klatt var. *yucatana* Shinners
swamp tomato [English]; *artemisia segunda* [Spanish]
To ease **menstrual cramps** and **spasms of painful menstruation**, 9 branches are boiled in 3 cups of water for 5 minutes, strained, and 3 cups consumed daily [A516]. To treat infantile **colic**, 1 cup of the decoction described for menstrual cramps is sipped daily [A516]. To treat **malaria**, 1 entire plant is boiled in 1 quart of water for 10 minutes and 3 cups consumed daily for 9 days [A338].

Egletes viscosa (L.) Less.

artemisia [Spanish]

To ease **stomach pain** and **menstrual cramps**, 1 entire plant is chopped, and 1 handful is boiled in 3 cups of water for 10 minutes and then consumed all at once [B1884]. For **colic** pain in babies, children, and adults, 1 handful of the entire plant along with 1 garlic clove, one 1-inch piece of ginger root (*Zingiber officinale*), and one 5-inch × 1-inch piece of dried sweet orange peel (*Citrus sinensis*) are boiled in 2 cups of water for 5 minutes and ¼ cup consumed warm every hour [LR].

Elephantopus mollis Kunth

To ease **headache**, the leaves are crushed and applied as a poultice to the head [B2678]. Alternatively, the leaves are crushed, steeped in water for 30 minutes, and used warm as a bath over the head [B2678].

Flaveria trinervia (Spreng.) C. Mohr

The entire plant is used in combination with *madre de cacao* (*Gliricidia sepium*), *gumbolimbo* (*Bursera simaruba*), and other plants to make a bath mixture for **general ailments** [B2149].

(top) *Egletes viscosa* [MB].

(bottom) *Fleischmannia pratensis* [GS].

Fleischmannia pratensis (Klatt) R.M. King & H. Rob.

purple malva [English]

As a treatment for **headache**, 1 entire plant is steeped in 2 quarts of water and used to bathe the head [A317]. Alternatively, the leaves and stems are pounded and the juice applied as a poultice to the head [A317]. For **hemorrhage**, 1 entire plant is boiled in 1 quart of water for 10 minutes, strained, and 1 cup consumed 3 times daily [A317].

Harleya oxylepis (Benth.) S.F. Blake

For **skin rash**, the leaves are crushed and rubbed on the affected area 3 times daily until better [B3638].

The wood is used in **construction** to make roofing rafters in houses [B3638].

Koanophyllon albicaule (Sch. Bip. ex Klatt) R.M. King & H. Rob.

black cordonsillo, old woman's walking stick, white cordonsillo [English]; *cordoncillo blanco, cordonsillo blanco* [Spanish]; *soscha, xolexnuc* [Additional names recorded]

The leaves are part of a Maya bath mixture used to treat just about **any ailment** [A814]. For **general pain** and to reduce **fever**, 1 quart of leaves is boiled in 2 gallons of water for 10 minutes and used as a steam bath [A904]. To reduce **swelling**, the roots, bark, and leaves are boiled in water and used as a bath over the affected area [A904]. For **postpartum swelling**, 1 double handful of leaves is boiled in 1 gallon of water for 10 minutes and used as a bath [A482]. For **nervousness** or **insomnia**, 1 handful of the leaves, flowers, and stems is boiled in 1 gallon of water for 20 minutes, cooled to a warm temperature, and used as a bath once daily at bedtime [A942].

As a first-aid response for **snakebite**, 1 piece of root equal in size to the affected person's forearm is dug from both this plant and red cockspur (*Acacia cornigera*), chewed, and the juice swallowed [A814].

The wood of this tree is used in **construction** to make house frames [A904].

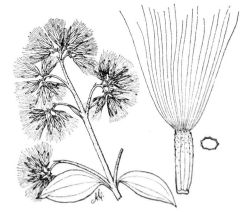

(top) *Koanophyllon albicaule* [NYBG].

(middle and bottom) *Lasianthaea fruticosa* [HW].

Koanophyllon galeottii (B.L. Rob.) R.M. King & H. Rob.

To break a "small **fever**" by causing a person to sweat, 3–6 leaves are boiled in 1½ cups of water for 5 minutes, strained, and 1 cup consumed as hot as possible once daily until better [B2558].

Lasianthaea fruticosa (L.) K.M. Becker

margarita, margarita del monte [Spanish]; *ish t, ish tá, ix ti pu, shti pè, shti pú, zta'ach* [Mayan]

The shrub is cultivated as an **ornamental** [B2440]. Bees favor the **nectar** of these flowers [A494].

To reduce **fever**, 1 entire plant is boiled in 2 gallons of water for 10 minutes and used very warm as a bath [A731]. For babies with "**green stools**," 1 handful of leaves is steeped in 1 gallon of water for 20 minutes and used as a bath [A731]. For children who have **difficulty sleeping**, the leaves along with *ixtipu* leaves (*Tagetes patula*) are used to prepare a bath [B2217]. To **ward off the evil eye** (mal ojo) in children, a large handful of leaves from both plants are boiled in a large pot, strained, and used as a warm bath at bedtime to make this charm [B2217].

Lepidaploa tortuosa (L.) H. Rob.
ich wâ, xorho nol [Mayan]

To treat the pain of **toothache**, the bark is put directly on the gum and in the "crack" of the tooth [B3524]. To ease **headache**, 1 quart of the green, non-woody stems, flowers, and leaves is mashed in 1 gallon of cool water; when the liquid is very green, it is strained and poured over the head and temples [A202]. For **nervousness**, 1 quart of stems, flowers, and leaves along with 18 large lime leaves (*Citrus aurantiifolia*) are crushed in 3 gallons of cold water, strained, and poured over the body as a bath [B2633]. Alternatively, this same formula may be boiled in water and used as a bath, depending on the healer and his or her tradition [B2633].

As a **fish poison**, the vine is mashed on a rock in a dammed-up waterway and then thrown into the water; the fish are temporarily stunned and float to the top of the water, where they are easily harvested [B3524].

Melampodium costaricense Steussy
yierba escaldadura [Spanish]

For **baby's rash**, 1 handful of leaves is boiled in 1 gallon of water for 20 minutes, cooled to room temperature, and used as a wash over the affected area. Usually, the rash goes away after 2 washes [B2286].

Melampodium divaricatum (Rich.) DC.
rabbit paw [English]; *dung ya* [Garífuna]; *k'inam* [Q'eqchi']

To treat **eyestrain** or **tired eyes**, 3 leaves are boiled in 1 cup of water for 5 minutes, strained, cooled, and 3 drops placed into the eyes 3 times daily [A142]. For **muscle pain**, 1 cup of stems, leaves, and flowers along with 1 tablespoon of sweet almond oil or olive oil and a pinch of salt are mashed, placed on a cloth, applied as a poultice to the affected area, wrapped up, left on for 8 hours, and then a second treatment given for a total of 2 applications before the evening, when the poultice is removed [W1841; RA].

Melampodium divaricatum [DA].

Melanthera nivea (L.) Small
Spanish needle [English]; *botón blanco* [Spanish]; *bega, tup lan xix* [Additional names recorded]
The entire plant is used as **fodder** for cattle and horses [A579]. The entire plant is mixed with green manure and laid along the rows in corn-fields as **mulch** for the soil, thereby improving corn yield [A579].

To relieve **headache** and **sinus ailments**, 1 entire plant is boiled in 1 gallon of water for 20 minutes and used tepid to bathe the head [A53]. To reduce **swelling**, 1 entire plant is mashed and applied as a poultice directly to the affected area [A892].

(top) *Melanthera nivea* [GS].

(right) *Mikania houstoniana* [NYBG].

Mikania aromatica Oerst.
To treat **hepatitis**, 1 large handful of fresh leaves is boiled in ½ liter of water for 15 minutes and consumed warm 3 times daily for 7 days [C77]. **Caution** is used with this treatment, as it may be toxic if too much of the plant is ingested [C77].

Mikania houstoniana (L.) B.L. Rob.
q'ehen [Q'eqchi']
To treat **seizures**, **fits**, or **fever**, 1 large double handful of fresh leaves is mashed in 1 gallon of water, set in the sun for 1 hour, strained, ½ cup consumed, and the remainder used as a bath [B3558].

Mikania sp.
bejuquillo, bejuco verde [Spanish]; *fish pot* [Creole]
To treat **toothache**, 1 small piece of the fresh root is applied directly to the affected area [C47]. To treat **chronic external sores**, **oval lesions**, **scaly spots**, **pale patches** (located mainly on the chest, shoulders, neck, and back), **skin "cancer" (open sore)**, and **fungal infections**, 9 leaves still attached to young stems are boiled in ½ quart of water and used cold as a wash over the affected area. Following this wash, 7 leaves are dried, powdered, and applied to the affected area. This treatment is repeated twice daily for 3 days [C47]. **Caution** is used with this plant, as it is toxic and should never be ingested [C47].

This plant is used as a **fish poison** [C47].

Montanoa speciosa with inset showing close-up of flower [VF].

Montanoa speciosa DC.

arnica [English]

This plant is considered to be highly effective in healing a **wound**, for which 9 fresh leaves are boiled over a low flame for 10 minutes in enough castor oil (*Ricinus communis*) to cover the leaves, steeped for 30 minutes, strained, and applied as a poultice or used as a wash over the affected area [B2091]. Additionally, the leaves are dried, powdered, and sprinkled over the cleansed wound [B2091]. For **itchy skin**, refer to wire wis (*Lygodium venustum*) [B2190]. To treat a **rash**, refer to *mañanita* (*Portulaca pilosa*) [B2100, B2099].

The leaves are said to be **smoked** for their intoxicating effect, which was not described in detail [B2091].

Neurolaena lobata (L.) Cass.

jackass bitters [English]; *mano de lagarto, tres puntas* [Spanish]; *kayabim* [Mopan]; *k'a'mank* [Q'eqchi']

To reduce **fever**, a cold-water infusion is used is an unspecified way [B3531]. Alternatively, refer to wild cherry (*Pseudolmedia spuria*) [B3733]. To relieve **pain** or **muscle soreness**, the leaves are warmed over a fire and applied as a poultice directly to the affected area [R52]. To treat **warts** and reduce **swelling**, fresh leaves are mashed and placed directly on the affected area [B3226].

For **wounds**, **sores**, and **skin ailments** such as **fungus**, the leaves are crushed and the fresh juice applied directly to the affected area [Arvigo and Balick, 1998]. For **chronic skin wounds**, such as **skin infection**, **ulceration**, and **fungus**, refer to cancer bush (*Acalypha arvensis*) [B1791; Arvigo and Balick, 1998] and *yerba del cáncer* (*Acalypha setosa*) [B1792]. To treat **leg wounds**, refer to *yerba del cáncer* [B1792]. For **skin ailments**, 1 handful of leaves is boiled in 1 gallon of water for 10 minutes, cooled, and used warm as a bath over the affected area [Arvigo and Balick, 1998]. For very serious **skin ailments**, **sores**, **infection**, and **fungal infection**, the leaves are toasted over a fire, powdered, and applied directly to the affected area twice daily after bathing with the leaf decoction [B1762, B1907, B2410; Arvigo and Balick, 1998].

Neurolaena lobata [MB].

To treat **parasites**, **worms** and **amoebas**, 3 leaves are boiled in 2 cups of water for 10 minutes and ½ cup consumed 3 times daily for 10 days [B1762, B1907; Arvigo and Balick, 1998]. For **skin sores**, **stubborn wounds**, **skin infection**, or other **skin ailments**, 1 slight handful of fresh leaves is steeped in 3 cups of boiling water for 20 minutes and poured over the affected area twice daily [B1762, B1907; Arvigo and Balick, 1998]. To treat **internal parasites**, including malaria, fungus, ringworm, amoebas, and intestinal parasites, and boring organisms, such as **beefworm** or **screwworm**, 1 fresh leaf is boiled in 1 cup of water for 10 minutes and 1 cup consumed 1–3 times daily [Arvigo and Balick, 1998]. To treat **ringworm**, **skin fungus**, **white fungal blotches**, **bee sting**, or other **stings**, 1 slight handful of fresh leaves is boiled in 3 cups of water for 10 minutes and used as a bath [B1762, B1907, B2410]. To treat **bee sting**, refer to *tah* (*Viguiera dentata*) [Arnason et al., 1980].

To treat **stomach pain** or **nausea**, 3 leaves are boiled in 1 cup of water and 1 cup consumed 3 times daily before meals [B2410, B2509]. Alternatively, 1 handful of leaves is mashed in 3 cups of cold water and 1 cup consumed 3 times daily before meals [B2410, B2509]. For **gastritis (*ciro*)**, non-insulin-dependent **diabetes** mellitus, **delayed menstruation**, **colds**, or **influenza**, refer to Billy Webb (*Acosmium panamense*) [P3]. As a **blood cleanser**, 1 slight handful of chopped roots is boiled in 3 cups of water for 10 minutes and 1 cup consumed 3 times daily, before meals, for 9 days [B1762, B1907; Arvigo and Balick, 1998]. To prevent non-insulin-dependent **diabetes** mellitus, the young leaves are chewed [P1].

To prevent **malaria**, 9 leaves of this species along with Billy Webb leaves (*Acosmium panamense*) are boiled in 1 quart of water for 10 minutes, strained, and 1 cup consumed once daily while exposure to mosquitoes is evident [B2632, P1; RA].

Alternatively, 1 slight handful of fresh leaves is boiled in 3 cups of water for 10 minutes and 1 cup consumed once daily during the malaria season [B1762, B1907]. To prevent **malaria** and eliminate **parasites**, 3 leaves are boiled in 1½ cups of water for 10 minutes and 1 cup consumed daily before breakfast for 9 days. On the 10th day of treatment, 1 tablespoon of castor oil (*Ricinus communis*) is usually consumed [R52]. To treat **malaria**, 1 slight handful of fresh leaves is boiled in 3 cups of water for 10 minutes and 1 cup consumed 3 times daily, before meals, for 9 consecutive days, after which the person should "feel better" [B1762, B1907]. Alternatively, 3 leaves are mixed in 1 pint of white rum and then soaked in the sun and dew for 3 days, after which 1 tablespoon is mixed with ½ cup of water and consumed once daily until the patient feels better [HR].

For **vaginal itching** and to treat **leukorrhea**, 9 leaves or stems are boiled in 2 cups of water for 10 minutes, strained, and used as a vaginal douche as needed [B2632; Arvigo and Balick, 1998].

For **head lice** (*Pediculus humanus*), 1 handful of leaves is boiled in 2 quarts of water for 10 minutes, and the scalp soaked in the decoction and then wrapped up with a towel, which left on for 30 minutes; this is repeated once daily for 3 days [RA; Arvigo and Balick, 1998]. To treat **snakebite**, a quantity of leaf extract sufficient to cover the area is placed directly on the affected region as antivenom [P1]. As a first-aid treatment for **snakebite**, the tips of the branches are chewed constantly until help is obtained. This slows the spread of venom through the body. When the bitterness is gone, these same leaves are applied as a poultice over the bite [LR].

As an **insecticide** or **fungicide**, the leaves are boiled in water, strained, and applied to infected plants in the home or garden [Arvigo and Balick, 1998].

Parthenium hysterophorus L.

artamesia, corriente, silantro [Spanish]; *kulant* [Q'eqchi']; *corriente ši'iu, silantro ši'iu* [Yucateca]

To treat **fever**, a tea is made from an unspecified plant part and consumed or used as a bath [Arnason et al., 1980]. Steggerda and Korsch (1943) reported that 1 entire plant was boiled in water and consumed once daily before breakfast to treat **bad breath**. To **build the blood**, the leaves are boiled in water and used as a bath [Arnason et al., 1980]. For **constipation** in infants, 3 leaves are boiled in 1 cup of water for 10 minutes and 1 teaspoonful consumed warm at a time when the baby is "straining to make stools" [B2174].

Parthenium hysterophorus [DA].

Pluchea carolinensis [MB].

Pluchea odorata [FA].

Perymenium gymnolomoides (Less.) DC.

For **fever**, the leaves and stem are boiled as part of a 9-leaf bath mixture [A370].

Piptocarpha poeppigiana (DC.) Baker

To treat **sleepwalking**, 1 large handful of the vine, leaves, and flowers (if available) is boiled in 2 gallons of water for 20 minutes and used warm as a bath once daily at bedtime [B2641].

Pluchea carolinensis (Jacq.) G. Don

pito sico [Spanish]; *chal ch, chal ché* [Mayan]

As a **postpartum therapy** and to **cleanse the genitals (vaginal area)**, **ensure that the internal organs return to their proper position**, staunch **prolonged hemorrhaging**, and **protect from infection**, 1 handful of leaves is boiled in 2 quarts of water for 10 minutes and used as a vaginal steam bath [B1846]. Additionally, ½ cup of this decoction is consumed once daily for 3 days following childbirth [A389, B1846, B2394]. To treat **swelling**, **burns**, or **rheumatism**, the decoction described above is used as a bath [A389, B1846]. To treat **stomach pain**, this decoction is consumed as a tea [A389, B1846].

The "male" counterpart to this plant is *chal che* (*Croton xalapensis*) [B1912].

Pluchea odorata (L.) Cass.

Santa María [Spanish]; *ix chal che* (female) [Mopan]

Steggerda (1943) reported that to treat **fever**, the leaves were covered in a tallow mixed with ground coffee (*Coffea arabica*) and tied to the bottoms of the feet. For **cough**, **colds**, and **influenza**, 3 leaves are boiled in 3 cups of water for 2 minutes, steeped for 15 minutes, and sipped throughout the day [Arvigo and Balick, 1998]. To treat **asthma**, 3 cups of the decoction described for cough are consumed over a 3-hour period [Arvigo and Balick, 1998].

To treat **swelling, tumors, inflammation,** and **bruises,** 1 large double handful of leaves is boiled in 1 gallon of water for 20 minutes and used as a bath over the affected area [Arvigo and Balick, 1998]. To treat **muscle soreness, rheumatism, neuritis,** and **arthritis,** a few leaves are warmed in oil, wrapped in a piece of cotton or flannel, and applied as a poultice over the affected area [Arvigo and Balick, 1998]. Steggerda (1943) reported that to treat **severe rheumatism,** the leaves were warmed, applied directly to the legs, and wrapped up tightly with a cloth. Steggerda (1943) reported that to treat **stomach pain, chest pain,** and to regulate **menstrual bleeding,** the leaves along with sour orange leaves (*Citrus aurantium*) and honey were boiled in water and 1 tablespoon consumed every hour. According to Steggerda (1943), pork, lard, or cold water were not consumed during this treatment. Nash and Williams (1976) reported that a decoction of the leaves was widely used to treat **stomach ailments** in Guatemala.

To treat **scanty or lacking menstruation,** 1 ounce of leaves and honey are boiled in water and divided into a 3-dosage treatment [Steggerda, 1943]. Steggerda and Korsch (1943) reported that to treat **ailments associated with menstruation,** the leaves, the "'hanging soot caused by the smoke in the kitchen,' some anise liquor (*Pimpinella anisum*) and seasoning," were mixed and consumed or that this decoction along with salt was boiled in water and used as a vaginal steam bath. Those authors also noted that for women in **labor,** an unspecified boiled leaf decoction was consumed. For **postpartum therapy**—for example, to ensure the uterus returns to its proper position, to prevent infection, and treat excessive hemorrhaging—1 large double handful of leaves is boiled in 1 gallon of water for 20 minutes and used as a vaginal steam bath for 30 minutes [Arvigo and Balick, 1998].

To treat **bad winds (*mal viento*),** the leaves are used to prepare a bath [Arnason et al., 1980]. Steggerda (1943) reported that to treat **twitching muscles,** "thought to be caused by evil winds of the woods," the leaves and branches were boiled in water and used as a bath.

Porophyllum punctatum (Mill.) S.F. Blake
squirrel's tail [English]; *piojillo*, *yierba del piojo* [Spanish]
To effectively eliminate **head lice** (*Pediculus humanus*) and their eggs without irritating the scalp, the leaves are mashed, soaked in cold water for about 1 hour, applied to the head, covered with a towel, left in place for 10 minutes, and repeated 3–4 times daily for 3 days [A953, W2062].

Pseudelephantopus spicatus (Juss. ex Aubl.) C.F. Baker
espinillo [Spanish]
To **slow hair loss** in men who are balding and strengthen the roots of hair, 1 entire plant is boiled in 2 quarts of water for 10 minutes and used as a hair rinse once daily [A332]. To treat **sores,** this same decoction is used as a wash over the affected area [A332].

Salmea scandens (L.) DC.

iklab [Mayan]

Thompson (1927) noted that the Mayan name referring to a hot chili pepper (*Capsicum annuum*) was most likely referring "to the effect caused by masticating it. It is said that if placed in the mouth, the whole of the mouth goes dead." This author reported that "this plant is used all through Central America as a **fish poison** under the Spanish name *Salta afuera*, as it is said to be so strong that it causes the fish to jump out of the water."

Standley and Record (1936) also reported that this plant was used as a **fish poison**.

Sinclairia deamii (B.L. Rob. & Bartl.) Rydb.

nieve [Spanish]; *tzik u lax* [Mayan]

Interestingly, the common names in both Spanish (*nieve*, meaning "snow") and Mayan (*tzik u lay*, meaning "white leaf") seem to reflect the plant's appearance (silvery-

(top left) *Porophyllum punctatum* [MB].

(top right) *Pseudelphantopus spicatus* [GS].

(bottom left) *Salmea scandens* [NYBG].

(bottom right) *Sinclairia deamii* [MB].

white undersides to the leaves) as well as its use (white fungal patches). To treat **skin fungus** characterized by white blotches, 1 handful of leaves is boiled in 2 quarts of water and used tepid as a wash over the affected area [B1772].

Smallanthus uvedalia (L.) Mackenzie
ixtipu [Mayan]

For **children who cannot sleep** and to help **ward off the evil eye (*mal ojo*)**, 1 quart of fresh leaves and flowers is boiled or mashed in 1 gallon of water for 10 minutes, strained, and used warm as a bath [B2216]. To prepare a stronger efficacy for **children who cannot sleep**, refer to the "male" counterpart, *margarita* (*Lasianthaea fruticosa*) [B2217; RA].

For **postpartum therapy**, **painful menstruation**, and **uterine congestion**, 1 quart of fresh leaves and flowers is boiled in 1 gallon of water for 10 minutes and used as a vaginal steam bath for 20 minutes either 3 days following childbirth or within 2 days before the onset of the menstruation [RA].

Sonchus oleraceus L.
chicalote [Spanish]

To **induce sleep**, 9 leaves and 1 flower are boiled in 3 cups of water for 5 minutes, strained, and ½ cup consumed once daily at bedtime. This treatment is for adults only; children should not consume it [A199].

Sphagneticola trilobata (L.) Pruski
rabbit's paw, wild sunflower [English]; *mano de conejo* [Spanish]; *dungya* [Garífuna]; *pasmo* [Mopan]; *kehil pím* [Q'eqchi']

(top) *Smallanthus uvedalia* [DA].

(middle) *Sonchus oleraceus* [KN].

(bottom) *Sphagneticola trilobata* [MB].

It is said that this plant will **pull "heat"** out of the body [Arvigo and Balick, 1998]. For **swelling, back pain, joint pain, muscle cramps, rheumatism, stubborn wounds,** or

sores, 1 large double handful of fresh stems and leaves is boiled in 2–3 gallons of water for 10 minutes and used as a bath 3 times weekly [A55, B1858; Arvigo and Balick, 1998]. For "**internal pus formations**," 1 cup of fresh leaves is boiled in 2 cups of water for 5 minutes and ½ cup consumed 4 times daily [A55]. To treat **foot fungus**, 1 quart of fresh leaves, stems, and flowers is boiled in 2 quarts of water, reduced to 1 quart, and used to soak the affected area twice daily [B3605]. Additionally, the leaves are mashed and applied as a poultice to the affected area [B3605].

For **arthritis**, the leaves and stems are mashed, applied as a poultice directly to the affected area, and secured with a warm covering [A898; Arvigo and Balick, 1998]. To treat **varicose veins** and *pasmo* (**congested blood; more than the usual amount in an area of the body**), 1 quart of fresh leaves and stems is boiled in 2 gallons of water for 10 minutes, steeped until tepid, and used as a bath [A55, B1858]. To treat **hepatitis, infection, burning while urinating, urine retention (stoppage of water)**, or **indigestion** due to sluggish liver, 1 cup of fresh stems, leaves, and flowers is boiled in 3 cups of water for 5 minutes and 1 cup consumed warm 3 times daily before meals [Arvigo and Balick, 1998].

Synedrella nodiflora (L.) Gaertn.
This is one of the components of a 9-leaf Maya mixture for baths used for various **ailments** [A321].

Tagetes erecta L.
marigold [English]; *flor de muerto* [Spanish]; *ix ta púl* [Mopan]; *tuhz* [Q'eqchi']; *štuph amarillo* [Yucateca] To relieve **fever, infantile colic, gastric pain, flatulence**, and **headache**, 3 flowers are steeped in 1 cup of hot, boiling water for 10 minutes and consumed [Arvigo and Balick, 1998]. To treat **fever** in babies, the leaf is rubbed on the skin [Arnason et al., 1980]. To **promote sweat** and as a **stimulant**, 3 flower heads are soaked in 3 cups of water, and ½ cup is consumed hot 6 times daily [Arvigo and Balick, 1998].

(top) *Synedrella nodiflora* [GS].

(bottom) *Tagetes erecta* [MB].

For **sores**, **abscess**, **cuts**, **wounds**, other **skin ailments**, and to treat infants and young children suffering from **malaise**, **colic**, **diarrhea**, **fever**, **colds**, and **influenza**, 1 entire plant is boiled in 2 gallons of water for 10 minutes and used as a bath [Arvigo and Balick, 1998]. For **sinus headache**, refer to warrie wood (*Diphysa americana*) [B2155]. To treat **back pain**, refer to *ix anal* (*Psychotria acuminata*) [Arvigo and Balick, 1998].

For **very painful varicose veins**, refer to *chacalpec* (*Salvia coccinea*) [Arvigo and Balick, 1998]. For **uterine ailments**, refer to wild basil (*Ocimum campechianum*) [A941]. To treat "**bad wind in the belly**" resulting from bewitching, refer to bay cedar (*Guazuma ulmifolia*) [Arvigo and Balick, 1998].

To cancel **evil spells (*obeah*)**, 9 fresh, young leaves along with 1 garlic clove are boiled in 1 cup water for 10 minutes and ½ cup consumed once daily for 9 days [Arvigo and Balick, 1998]. To treat **evil magic (*obeah*)**, refer to *begonia* (*Begonia sericoneura*) [BW]. For people who suspect they have been affected by **evil magic (*obeah*)**, refer to *dormilón* (*Mimosa pudica* var. *unijuga*) [HR]. To get rid of **evil spirits**, the flowers along with sweet orange peel (*Citrus sinensis*) are mashed by the Garifuna people in a bowl of water and then sprinkled around the body of a person exposed to spirits, such as at a funeral [Arvigo and Balick, 1998]. To treat someone possessed by **evil spirits**, refer to pine (*Casuarina equisetifolia*) [BW]. To treat the **evil eye (*mal ojo*)**, this plant is used in an unspecified way [Arnason et al., 1980].

While saying **healing prayers (*ensalmos*)**, a branch with leaves is held over the pulse of an infant [Arvigo and Balick, 1998]. To call the spirits during Maya **ceremonies**, the leaves and flowers are soaked in water and used by the priest as a hand and face wash [Arvigo and Balick, 1998]. To prepare a **hand wash** for funeral attendants, the flowers are soaked in water and used as a wash as needed [Arvigo and Balick, 1998]. To **keep out bats**, bunches of flowers are hung inside the house [Arvigo and Balick, 1998].

Tagetes patula L.

marigold [English]; *flor de muerto* [Spanish]; *ix ta pèl*, *ix ta púl*, *ixtipu* [Mopan]; *tuhz* [Q'eqchi']

This is cultivated as an **ornamental** [MB].

To treat **fever** in a child, 1 quart of fresh leaves, stems, and flowers (if available) is mashed in 2 quarts of water, strained, and used tepid as a wash [B2310]. Alternatively, 2 cups of fresh leaves, branches, and flowers (if available) are mashed in a bit of water and applied as a poultice to the soles of the feet, the back of the neck, and the forehead [RA].

For **varicose veins**, 9 flowers are soaked in 1 cup of sweet almond oil (*Prunus dulcis* var. *dulcis*) for 7 days and rubbed in an upward direction over the affected area [RA].

As a **postpartum treatment** done only if the mother's return to normalcy is uncomplicated by excessive bleeding, 1 large handful of leaves and flowers is boiled in 2 quarts of water for 5 minutes and used as a vaginal steam bath for 20 minutes 3 days after childbirth [RA].

(top left) *Tithonia diversifolia* [RH].

(top right) *Verbesina gigantea* [NYBG].

(bottom) *Vernonanthura patens* [GS].

Tithonia diversifolia (Hemsl.) A. Gray

The yellow-to-orange flower makes this a prized **ornamental** [B2307, B2320]. Because this plant **repels bees**, it is cultivated in dooryard gardens and along roadsides to discourage intruding bees, especially the "Africanized" ones, from establishing hives near areas inhabited by the local populace [R45].

Verbesina gigantea Jacq.

To treat **parasitic ailments**, such as **malaria**, **intestinal parasites**, and **skin fungus**, this plant is used by some people instead of *tres puntas* (*Neurolaena lobata*) [B2139].

Vernonanthura patens (Kunth) H. Rob.

corrimiento, inocenta [Spanish]

To treat **hepatitis** or **jaundice**, 1 handful of fresh leaves and young stems is boiled in water and consumed warm twice daily for 1 day. This treatment is considered to be very effective and is followed by some physical activity [C21].

To calm babies who **cry**, 1 entire plant is boiled in 1 gallon of water for 10 minutes and used warm as a bath [B2668].

Viguiera dentata [WLW].

Viguiera dentata (Cav.) Spreng.
tah [Yucateca]
To treat **nocturnal fever**, the leaves are soaked in water and used as a bath 3 times [Arnason et al., 1980]. To treat **bee sting**, this plant along with *tres puntas* (*Neurolaena lobata*) are used in an unspecified way [Arnason et al., 1980]. Steggerda (1943) noted that the stems of this plant were used by the Yucatan Maya "for sky rockets, and . . . to make corn bins by tying the stems together in the form of a mat."

Wedelia acapulcensis Kunth
orosus [Yucateca]
To treat **toothache**, the flower is chewed [Arnason et al., 1980].

BALSAMINACEAE

Herbs, with translucent, watery stems. Leaves alternate or opposite, simple. Inflorescences axillary, in racemes, cymes, or of solitary flowers. Flowers variously colored. Fruits a capsule. In Belize, consisting of 1 genus and 2 species. Uses for 1 genus and 2 species are reported here.

Impatiens balsamina L.
bálsamo, china, chino [Spanish]
This plant is cultivated for its colorful flowers as an **ornamental** [B2337].

Impatiens walleriana Hook. f.
  bálsamo, *china*, *chino* [Spanish]
  This plant is cultivated for its colorful
  flowers as an **ornamental** [B2326].

BASELLACEAE

Vines, usually fleshy and trailing along the
ground. Leaves alternate, simple. Inflorescences axillary or terminal, in spikes, dichasia, racemes, or panicles. Flowers small,
green, white, red, or yellow. Fruits nut-like.
In Belize, consisting of 2 genera and 2 species. Uses for 1 genus and 1 species are reported here.

Anredera vesicaria (Lam.) Gaertner
  suelda con suelda [Spanish]; *yiedra*
  [Additional name recorded]
  For a **fracture**, **sprain**, and **stubborn
  sores**, three 5-inch pieces of roots, 1
  handful of leaves, and four 6-inch pieces
  of vine are mashed, cooked in olive oil for
  2–3 minutes until soft, cooled, applied as
  a poultice to the affected area, wrapped
  with a cloth, and replaced every other day
  until better [B1905; LR].
      For **bleeding gums**, 1 small handful
  of the root is grated, mashed in 1 quart of
  water for 10 minutes, cooled to warm, and
  used as a mouthwash 3 times daily until
  better [BW].

(top) *Impatiens walleriana* [MB].

(bottom) *Begonia heracleifolia* [MB].

BEGONIACEAE

Herbs or small shrubs, frequently succulent, terrestrial or epiphytic. Leaves alternate,
simple, frequently lobed. Inflorescences axillary, in cymes, often showy. Flowers pink,
reddish, or white. Fruits a capsule. In Belize, consisting of 1 genus and 4 species. Uses
for 1 genus and 2 species are reported here.

Begonia heracleifolia Schltdl. & Cham.
  ch'ajum kahan [Mopan]
  For **measles**, 1 handful of fresh leaves and roots is chopped, mashed, boiled in 1 quart
  of water for 10 minutes, and consumed 3 times daily for 7 days [C65].

Begonia sericoneura Liebm.
begonia [Spanish]; *pah ulul* [Q'eqchi']

The flowers and stems are said to be high in vitamin C and are eaten as a **food** [HR].

To **promote sweat** to treat **blood acidosis**, 3 leaves with stems are boiled in 1 quart of water for 10 minutes and 1 cup consumed warm once daily, before a meal, for 3 days [HR]. To treat an **infected cut or wound**, 1 double handful of fresh leaves is mashed and ap-plied as a poultice directly to the

Begonia sericoneura [MB].

affected area once daily for 3 days [C41]. To treat **colds** and **influenza**, 1 double hand-ful of fresh leaves along with 1 large handful of fresh *Peperomia macrostachyos* leaves and 1 double handful of fresh *tep pim* leaves (*Peperomia costaricensis*) are mashed in ¼ liter of cold water and consumed once daily for 3 days [C41, C42, C43].

For any type of **swelling** or **pain**, the leaf is mashed and applied as a poultice to the affected area [LR]. To treat **swelling** in the feet, 3 leaves are mashed in ½ cup of cold water, mixed with an unspecified species of *Selaginella*, and applied 3 times daily until better [R24]. To treat **foot fungus**, 3 leaves are boiled in 2 quarts of water for 10 minutes and used warm to soak the affected area [BW]. Robinson and Furley (1983) reported that the leaves were used to make a hot poultice to treat "sprained ankles and feet." When **foot sores** caused either by **ringworm** or **foot fungus** are believed to result from **evil magic (*obeah*)**, 3 leaves are mashed in ½ cup of water and applied as a poultice directly to the affected area [B2554, B3366].

To treat **evil magic (*obeah*)**, 3 leaves along with one 9-inch branch of marigold (*Tagetes erecta*) and one 3-inch piece of *ruda* stem (*Ruta graveolens*) are boiled in 2 quarts of water for 10 minutes and used as a bath [BW].

BIGNONIACEAE

Usually lianas or trees, sometimes shrubs. Leaves usually opposite, sometimes alternate, with compound blades, rarely simple. Inflorescences terminal or axillary, in panicles, cymes, thyrses, or sometimes of solitary flowers. Flowers often showy, variously colored. Fruits usually a dry capsule with winged seeds, sometimes a hard pepo. In Belize, con-sisting of 29 genera and 49 species. Uses for 12 genera and 17 species are reported here.

Amphilophium paniculatum (L.) Kunth var. *paniculatum*
patito [Spanish]

The flowers are sucked, as the sweet nectar is a **sugar substitute** [B3207].

Arrabidaea floribunda (Kunth) Loes.

   bil in coc [Mayan]

   The Mayan name *bil in coc* is translated as "vine that goes a long way," an appropriate name because the plant's bark is peeled off and the stem is used as **cordage** to tie the roofing crossbars that hold thatch [B1910].

Arrabidaea podopogon (DC.) A.H. Gentry

   gerninda morada [Spanish]

   The flowers are used as a **decoration** in the home [A535].

Arrabidaea pubescens (L.) A.H. Gentry

   tie tie [English]

   The cross shape of the vine and the purple flowers are said to signify that this plant has medicinal uses, specifically to reduce **pain** and to **improve blood circulation**. One handful of the chopped vine, leaves, and stem is boiled in 3 cups of water for 10 minutes, cooled, and ½ cup consumed warm 6 times daily until better [A657].

   The vine is used as **cordage** to tie thatch and pole walls of local houses [Robinson and Furley, 1983]. When eaten by cows, this plant is said to give an unpleasant taste to their milk; however, cows often avoid it [Robinson and Furley, 1983].

(top) *Amphilophium paniculatum* var. *paniculatum* [TP] with inset showing close-up of leaf [RF].

(bottom) *Arrabidaea floribunda* [WJH].

Crescentia cujete L.

   calabash, calabash tree, savannah calabash, wild calabash [English]; *jícara* [Spanish]; *lek, luch* [Mopan]; *joom* [Q'eqchi']; *huaz* [Additional name recorded]

   The leaves are cooked and eaten as a **food**; the young shoots and leaves are cooked and eaten in stews, soups, and *atoles* or prepared with onions and egg and eaten with *tortillas* [Mallory, 1991].

   To treat **earache**, the flowers are heated over a flame and 1 drop of the juice placed into the ear every 20 minutes for 2–3 days [HR]. To relieve **hiccups**, the inside of 1 green fruit is cleansed, washed out, filled with water (preferably rainwater), and 9 sips consumed [Arvigo and Balick, 1998]. For **cough**, the pulp and seeds of 2 fruits are boiled in 2 quarts of water for 20 minutes, strained, boiled again for 5 minutes with

Crescentia cujete with inset showing fruit [MB].

an equal amount of brown sugar mixed well with 1 ounce of rum and some olive oil, and 1 tablespoon consumed 4 times daily for adults or 1 teaspoon consumed 3 times daily for children until better [A376]. To treat **colds**, the fruits are macerated in an unspecified way, and 1 small spoonful of the juice is "taken as a syrup" [Mallory, 1991].

To treat **cough**, **asthma**, **bronchitis**, and **lung congestion**, the inner stringy pith of 1 mature fruit along with 2 cups of sugar are boiled in 2 quarts of water for 30 minutes, strained, and 6 teaspoons consumed daily [Arvigo and Balick, 1998]. Alternatively, 1 handful of leaves or flowers along with some sugar are boiled in 3 cups of water for 30 minutes and 6 teaspoons consumed daily [Arvigo and Balick, 1998]. For **asthma** or **bronchitis**, the "heart" (or inside) of 1 large fruit is boiled in 3 cups of water for 20 minutes, strained, and consumed during an attack. Treatment is continued for 3 days, consuming 1 cup 3 times daily [A376]. To treat **asthma** when no fruits are available, 20 leaves are boiled in 3 cups of water for 10 minutes and consumed 6 times daily [A376].

To prevent **high blood pressure**, the inside of 1 green fruit is cleansed, washed out, the empty gourd taken to a river at noon, dipped in the water, 9 sips consumed, the remainder of the water thrown over the head, and the empty gourd tossed into the water, with the person not watching where it goes [Arvigo and Balick, 1998]. To "**build the blood**," the leaves are collected, boiled in saltwater or fried with onion and garlic in oil, and consumed [Mallory, 1991]. To treat **dysentery**, the leaves are soaked in water and consumed as needed, but no further details were provided [Mallory, 1991].

To treat an early stage **inguinal hernia**, 1 entire fruit is heated in ashes, wrapped in a cloth, and massaged over the affected area [HR]. Additionally, the center of 1 fruit is scooped out, filled with honey, roasted over coals until soft, and 1 tablespoon of the juice consumed once daily for 3 days [HR].

To expel a **retained placenta** following childbirth, 4 ounces of fresh, green bark or 9 leaves are boiled in 2 cups of water for 10 minutes and consumed [A376]. As a **female contraceptive**, refer to *Psychotria berteriana* [B2493]. This plant should not be taken during pregnancy, as it may have **abortifacient effects**.

The dried fruits make fine bowls and **containers** after the pith is removed [Arvigo and Balick, 1998]. Steggerda (1943) reported that the Yucatec Maya of Mexico used this plant to make **bowls** (known as *jícara*) and **dishes** and believed that

> the Luch [the cultivated variety of this species] must be planted on the 24th of June for St. John the Baptist; if planted on any other day, the fruits will fall off. If the tree does not bear, the Maya beat the tree (this holds true for other fruit trees as well) with a bejuco [vine], nine lashes, on the 24th of June; it must not be beaten any other time. The Maya also hang the heads of horses, the horns of cattle and pigs' heads in the Luch tree to make it ashamed of itself for not bearing fruit. (p. 204)

The gourds are used to hold *espacha*, "the local corn-wine, and impart a pleasant flavor to it" [Robinson and Furley, 1983]. Standley and Record (1936) noted that the wood, while not durable, was occasionally used in **carpentry** to make saddletrees and tool handles.

The fruit is used **ritualistically** in local folklore for health issues, social contacts, or to increase good fortune [Arvigo and Balick, 1998]. When the first egg laid by a young chicken (pullet) is collected inside a new fruit, it is believed that the chicken will lay a good amount of eggs [Arvigo and Balick, 1998]. In Maya tradition, to compel guests to return, they are served a sip of *atole* out of the same gourd used by the host [Arvigo and Balick, 1998]. Thompson (1927) reported that the tree was **sacred** to the Maya peoples and that its thorns were used to draw blood for sacrificial purposes. He also noted that as of the time he was observing the uses of this plant, "all memory of blood sacrifices is lost, [and] the reason for the sacredness of the tree is ascribed to the shape of the leaves, which bear some resemblance to a cross."

Cydista diversifolia (Kunth) Miers

wire herb, wire wis [English]; *alambre xiv* [Mayan]

For **eye pain**, one 12-inch piece of the vine is boiled in 1 pint of water for 15 minutes, strained well through muslin, and used lukewarm to bathe the eye with an eye cup 3 times daily [A133]. To treat **skin fungus**, the decoction described for eye pain is used as hot as possible to soak the affected area twice daily [A133]. To treat **leishmani-**

asis (**baysore**) or any **stubborn sores**, the leaves are toasted, powdered, and applied over castor oil (*Ricinus communis*) twice daily after bathing. The affected area is then wrapped with a clean cloth, and this treatment is repeated for 10 days [A133].

For **poor blood circulation**, when the veins have **painful knots** or the **hands and feet are cold or blue**, 1 large double handful of the leaves is boiled in 2 gallons of water for 10 minutes and used as a warm bath once daily, at bedtime, for 3 days [HR]. As a treatment for **gonorrhea** in men only, 1 large double handful of leaves is crushed well, soaked in over 1 quart of water, set to ferment for 1 day in the sun and 1 night in the dew, consumed in small doses throughout the entire next day, and repeated for 3 days [LR].

Macfadyena unguis-cati (L.)
A.H. Gentry
garlic vine [English]
Steggerda (1943) reported that the Yucatec Maya of Mexico used this vine as a **fiber** to make woven wicker doors and baskets and as **cordage** to tie beams and rafters in local houses. He noted that to treat **bronchitis** and **cough**, the young stems and leaves were boiled in water and the liquid consumed. He also noted that the leaves were crushed and applied to **wounds** as a styptic.

Mansoa hymenaea (DC.)
A.H. Gentry
garlic plant, garlic vine [English]
This plant is cultivated as an **ornamental** for its violet-colored flowers [B3204].

To treat **gastritis** (*ciro*), 5–6 inches of the thick vine are boiled in 2 quarts of water for 10 minutes and 1 cup consumed 3 times daily, before meals, for 3 days [A990].

(top) *Cydista diversifolia* [RH].

(middle) *Macfadyena unguis-cati* [JS].

(bottom) *Mansoa hymenaea* [MB].

Parmentiera aculeata with fruit and inset showing flower [MB].

For **general stomach ache**, 1 cup of the decoction described for gastritis (*ciro*) is consumed every 2 hours for 2 consecutive days [A990]. To "**sweat a fever**" and to treat all types of **evil magic (*obeah*)**, 1 large double handful of leaves is boiled in 2 gallons of water for 10 minutes and used as a bath once daily, at bedtime or noon, for 3 days. For evil magic, the bath is taken on a Friday only and repeated on the following Friday if needed [HR].

Paragonia pyramidata (Rich.) Bureau
tres gahos [Spanish]; *osh áy, osh y* [Mayan]
To keep **swollen cuts** from becoming infected, 1 handful of leaves and 9 ripe fruits are mashed in 1 gallon of cold water and used to bathe the affected area [B2218].

Parmentiera aculeata (Kunth) Seem
cow okra, snake okra, wild okra [English]; *pax te* [Yucateca]; *k'at* [Mayan]
The fruit is eaten as a **food** when very green or mature. It is eaten raw or, but if very hard or over-ripe, boiled first [A222]. The fruit is used as **fodder** for animals such as cows, horses, sheep, and goats [JB19].

To treat **kidney stones**, 1 large, mature fruit along with 1 large, mature pineapple (*Ananas comosus*) with the skins removed are boiled in 1 quart of water for 5 minutes, ⅓ cup of white rum added, the decoction boiled for an additional 5 minutes, cooled, and consumed with a pinch of bicarbonate soda once daily before breakfast. No food is eaten until noon, and the treatment repeated for 3 days. While it may cause internal pain, the full treatment must be continued. The passed stones may be observed on the fourth day by urinating into a pot covered with cheesecloth or by shining a light at night into a jar of urine to see them glitter [LR]. Steggerda and Korsch (1943) reported that to treat **urinary disease**, the Yucatec Maya utilized the roots along with the tender stems of *roble* (*Ehretia tinifolia*) and the roots of another unidentified species, which were ground, heated, and applied as a poultice over the loin and pubic area.

For **magic**, the mature seeds of 9 fruits along with the mature seeds from 9 bird pepper fruits (*Capsicum annuum*) are crushed and sprinkled around the house of a neighbor you want to get rid of. Special, secret prayers are recited while sprinkling the seeds on a Friday [RA].

Pseudocatalpa caudiculata (Standl.) A.H. Gentry

For **sore joints**, 1 double handful of the stem of the vine and 9 leaves are boiled in water for 10 minutes and used warm as a bath over the affected area [B2631]. Alternatively, one 12-inch piece of chopped vine and 9 leaves are steeped in a 1-quart bottle of brandy for 9 days and then applied with cotton to the affected area 3 times daily until better [B2631].

Tabebuia chrysantha (Jacq.) G. Nicholson

Cortéz tree, yellow mayflower [English]; *cortez che'* [Q'eqchi']; *hahauche* [Yucateca]
This is an important secondary **hardwood** species in Belize [Munro, 1989]. Standley and Record (1936) noted that the olive-brown wood was very hard, heavy, tough, durable, and said to be used locally in **construction** to make truck parts.

Tabebuia guayacan (Seem.) Hemsl.

black mango [English]
The wood of this tree is used as lumber in **construction**, although it is said to be soft [B3134; LR].

Tabebuia heterophylla (DC.) Britton

roble [Spanish]
Lamb (1946) reported that the wood was used in **construction** to make ox-yokes and in **carpentry** to make furniture.

Tabebuia rosea (Bertol.) A. DC.

May bush, mayflower [English]; *xna' cortez* [Q'eqchi']
To treat **biliousness** caused by gallbladder and liver congestion, 1 handful of mashed bark is boiled in 1 quart of water for 10 minutes and consumed throughout the day until better [B1888]. To make a mild and effective **laxative**, one 6-inch × 1-inch strip of bark is boiled in 2 cups of water for 10 minutes and 1 cup consumed warm twice daily, in the morning and in the evening [A359].

The wood is used in **construction** to make house posts [B3583]. Standley and

(top) *Tabebuia heterophylla* [IA].

(bottom) *Tanaecium tetragonolobum* [MB].

Record (1936) reported that the strong and "fairly durable" wood of this tree was used locally in **construction** to make cattle yokes and should be investigated as a potential raw material for furniture and interior trim.

Tanaecium tetragonolobum (Jacq.) L.G. Lohmann

duppy beans [Creole]; *kenq' i che'* [Q'eqchi']; *š ki'iš* [Yucateca]; *an lol* [Mayan]
To treat **biliousness**, 1 handful of chopped stem is boiled in 3 cups of water for 10 minutes and 1 cup consumed warm 3 times daily between meals [B1918]. To stop **bleeding** of a cut, the leaf is chewed and applied as a poultice to the affected area [Arnason et al., 1980].

Tynanthus guatemalensis Donn. Sm.

allspice titai [English]; *gernanda, monja blanca* [Spanish]

Water can be obtained from cutting this forest vine and letting it drip into one's mouth or a water jug [RA].

To treat **gastritis (*ciro*)**, one 1-inch × 18-inch piece of vine is boiled in 1 gallon of water for 10 minutes and consumed cool as drinking water [RA]. This same decoction is consumed hot as a delicious **beverage** [RA].

The flowers and vine of this plant were used as **decoration** on altars for Maya ceremonies; this is rarely a practice today [A502].

Tynanthus guatemalensis [NYBG].

BIXACEAE

Trees, shrubs or subshrubs, often with red or orange sap. Leaves alternate, simple. Inflorescences terminal, in thyrses or racemes. Flowers white to pink or yellow. Fruits a capsule. In Belize, consisting of 2 genera and 2 species. Uses for 2 genera and 2 species are reported here.

Bixa orellana L

achiote, achote, annatto, recado [Spanish]; *annatto, atta, kuxub, xayaw* [Q'eqchi']; *cu shèb, ku xub* [Mopan]; *ku šu* [Yucateca]

The tree is considered decorative and is cultivated as an **ornamental** [B2401].

The seed is ground and used as a **spice** for foods such as tamales [A237]. To prepare a cooking sauce, raw seeds are mashed into a pulp and added to black pepper (*Piper nigrum*), salt, garlic, and cumin (*Cuminum cyminum*) and then used with, for example, beef and chicken [A212].

The red pulp that surrounds the seed is used as a **food coloring** to "paint the food" a yellow-to-orange color while not affecting the food's flavoring [B2342, B3573]. The fruit pulp is used to give boiled rice (*Oryza sativa*) a red color [Mallory, 1991]. Thompson (1927) recorded that this plant was used to color chewing gum (*Manilkara zapota*) and, in Honduras, to color chocolate. According to Lundell (1938), "the ancient Maya undoubtedly utilized vegetable dyes, not only in coloring foods and dying, but also in painting and decorating, and it is probable that this tree was one of the important sources of red dye."

To treat **fever** in a baby, the leaves are placed in the bed, as they are considered to be "cooling" [Arnason et al., 1980]. Alternatively, the cooled leaves are placed on the forehead or in the bed [Mallory, 1991]. To draw the heat out of a **fever**, the leaves and fruits are made into a bed on which the person rests [A237]. Alternatively, 1 ounce of the root is boiled in 1 cup of water for 10 minutes and 1 cup consumed 3 times daily until better [A237]. For **fever** and as a **laxative**, 9 leaves and 3 fruits are boiled

(left) Seeds of *Bixa orellana* covered by orange-red aril used for coloring [MB].

(right) *Bixa orellana* [NYBG].

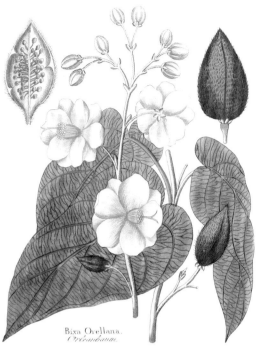

Bixa Orellana.
Orleanbaum.

in ⅔ cup of water for 10 minutes, 1 cup of plain water added, and ½ cup consumed warm 3 times daily [A237].

For **swelling**, the leaves are used to prepare a bath [Arvigo and Balick, 1998]. For **eye swelling** and **pain**, 9 leaves are mashed in 1 quart of cool water and used to bathe the head and eyes once daily [B2390]. To treat **sore throat** and **swollen glands**, the red seed coating is mixed with urine, an unspecified amount consumed, and the remainder rubbed on the affected area [Arnason et al., 1980]. To treat **sores**, **rash**, or **infected insect bites**, 1 handful of leaves is crushed in water, steeped in the sun all day, strained, and used cool as a wash over the affected area [Arvigo and Balick, 1998]. To treat **rash** in a baby or **chigger bites**, the leaf is boiled in water and used as a wash over the affected area [Arnason et al., 1980]. Standley and Williams (1961) noted that this plant was used in Guatemala to **prevent scars** from forming on sores and burns.

For ***uich*** sickness (a condition that presents symptoms looking like measles and produces fever and rash), 6 leaves and ⅓ cup of human urine are mixed in ½ cup of water and 1 ounce consumed 3 times daily [A212]. Steggerda (1943) noted that the Yucatan Maya of Mexico washed the seeds in warm water and consumed this during the early stages of **measles**; this was thought to cause the measles to develop and then resolve more quickly.

To treat **liver biliousness**, 1 handful of leaves along with 1 handful of *piss-a-bed* leaves and flowers (*Senna uniflora*), 7 pumpkin flowers (*Cucurbita pepo*), and 7 *bu kút* leaves (*Cassia grandis*) are boiled in 2 quarts of water for 10 minutes, cooled, and 1 cup consumed once daily. Treatment is stopped when loose stools are present [LR].

For **infant jaundice**, 9 drops of the decoction described above for liver biliousness are mixed with ⅛ cup of water and consumed once daily for 2–3 days until the yellow is gone. The stool will first turn yellow and then return to normal color when treatment is effective [LR]. For **adult jaundice**, 9 leaves are boiled in 2 cups of water for 10 minutes and consumed warm twice daily, before breakfast and dinner, for 2 days [HR]. To **build the blood**, the leaves and salt are boiled in water and eaten frequently [Mallory, 1991]. When **vomiting blood**, 3 older leaves are boiled in 3 cups of water for 10 minutes and consumed [Arvigo and Balick, 1998].

For **diarrhea** or **dysentery**, 3 young leaves are mashed in 1 cup of water, strained, and consumed as needed [Arvigo and Balick, 1998]. For **urine retention (stoppage of water)**, 9 seedpods are boiled in 3 cups of water for 10 minutes and 1 cup consumed 3 times daily before meals [Arvigo and Balick, 1998]. To **induce labor**, 1 handful of young leaves along with 3 Spanish thyme leaves (*Lippia graveolens*), and another unspecified plant are mashed in ⅓ cup of water and consumed frequently until the onset of delivery [B3573].

For **grief** (*pesar*) in infants, 3 leaves are boiled in 1 gallon of water for 10 minutes, cooled to lukewarm, and used as a bath once daily, before noon, outside on sunny days for 3 consecutive days [Arvigo and Balick, 1998]. To treat **insanity**, refer to tamarind (*Tamarindus indica*) [Steggerda and Korsch, 1943].

Cochlospermum vitifolium (Willd.) Spreng.

cotton flower, wild cotton [English]; *amapa, chum, comasuche, pahote* [Additional names recorded]

For **urinary ailments**, such as **burning in the urine** and **urine retention (stoppage of water)**, 3 inches of root are boiled in 1 quart of water for 10 minutes and 1 cup consumed warm 3 times daily, before meals, until better [A368]. The stamens of the flowers are used as a **saffron substitute** [Marsh, Matola, and Pickard, 1995]. The bark fiber is used as **cordage** to make rope [Marsh, Matola, and Pickard, 1995].

Cochlospermum vitifolium flower (top) and leaves (bottom) [LK].

BORAGINACEAE (INCLUDING HYDROPHYLLACEAE)

Herbs to shrubs, sometimes trees or lianas. Leaves usually alternate, rarely opposite. Inflorescences terminal or axillary, in helicoid or scorpioid cymes, sometimes reduced to 1 flower. Flowers usually white, yellow, blue, or pink. Fruits a capsule, drupe, or nutlet. In Belize, consisting of 8 genera and 35 species. Uses for 6 genera and 16 species are reported here.

Bourreria huanita (Lex.) Hemsl.
> *sombra de ternero* [Spanish]; *ter ech mach* [Mayan]
> The fruit is eaten as a **food** by the keel-billed toucan (*Ramphastos sulfuratus*) in the wild [B2406].

Bourreria oxyphylla Standl.
> laurel, night bloom, wild craboo [English]; *lima del monte* (male), *roble* [Spanish]; *chi' che'* [Q'eqchi']; *ter ech m sh*, *ter ech másh* [Mayan]; *sac bay eck* [Additional name recorded]
> The fruit is eaten as a **food** by birds, especially wild pheasants and wild pigeons [A603, A659, B2453].
>
> The leaves are a component of a Maya bath used for treatment of a **variety of ailments**, such as **skin ailments**, **fever**, **swelling**, and **nerves** [A163]. For **headache**, the leaves are crushed and applied as a poultice over the temples [A670]. To treat **burns** and **wounds**, the orange-brown juice is extracted from the young stems and applied to the affected area [A670]. Alternatively, the leaves are mashed and applied to the affected area [A670].
>
> The trunk is used in **construction** for house posts and as a source of **fuel** [A603, JB80].

Cordia alliodora (Ruiz & Pav.) Oken
> laurel, samwood, salmwood [English]; *laurel blanco*, *poma rosa* [Spanish]; *suuchaj* [Q'eqchi']; *bohun* [Additional name recorded]
> The leaves are a component of one of the Maya bath mixtures used to treat a **variety of ailments** [A746].

Bourreria oxyphylla [GS].

To treat **fever**, **rash**, and **insomnia**, 1 large handful of leaves is boiled in 1 gallon of water for 20 minutes and used warm over the affected area twice daily [A746]. To bring **boils** to a head, the resin is applied as a poultice directly to the affected area [A746].

The trunks are used in **construction** for house posts and lumber [A106, R69]. Standley and Record (1936) reported that the local uses in Belize at that time included "logging truck parts, piling and railway ties; lining of furniture and chests (as protection against insects)." As the Latin name indicates, it was reported that the crushed leaves had an odor similar to garlic [Standley and Record, 1936].

Cordia curassavica (Jacq.) Roem. & Schult.
ak sheb [Q'eqchi']
To alleviate **toothache**, 3 small branches with leaves are boiled in 1 cup of water for 10 minutes and used cool as a mouthwash [A666].

Cordia dentata Poir.
According to Standley and Record (1936), the sticky fruit pulp of this species is very sweet and eaten by birds and "sometimes" by people as a **food**. To treat **stomach ulcers**, the juice from the ripe fruit is used, but no further details were provided [Flores and Ricalde, 1996].

Cordia diversifolia Pav. ex DC.
macajuita [Spanish]; *latzb'hl* [Q'eqchi']; *chischis* [Additional name recorded]
To relieve **body, muscle, and joint pain**, 1 quart of leaves is boiled in 1 gallon of water for 20 minutes and used hot as a bath [B2715].

The sap and juice are removed from the berries and used as an **adhesive** (glue) for paper [A572, A962]. The trunk is used in **construction** for fence posts [A572].

Cordia dodecandra DC.
sericote, zericote [Spanish]
Lundell (1938) reported that the fruit was eaten as a **food**, either raw or made into *dulces* (sweets).

This tree provides dark, strong heartwood, which is widely used by **wood carvers** in Belize [Marsh, Matola, and Pickard, 1995]. The wood is used in **construction** to make cabinets and furniture [Marsh, Matola, and Pickard, 1995]. Standley and Record (1936) noted that the wood of this tree was good in the **construction** of fine furniture and articles produced on a lathe.

Cordia diversifolia [GS].

Cordia sebestena L.

ciricote, ziricote [Spanish]

The ripe fruit is used to make **jam** [RA].

As a **purgative**, the ripe fruit is eaten raw [B1966]. To expel a **retained placenta** following childbirth, 9 leaves are boiled in 1 cup of water for 10 minutes, strained, and 1 ounce consumed warm every 10 minutes until the placenta comes out [RA].

To relieve **cough**, 2 ounces of bark and 9 leaves along with 3 ounces of sugar are boiled in 1 pint of water for 10 minutes, strained, and 1 tablespoon consumed 3 times daily until better [A427]. Alternatively, for **cough**, the bark and leaves are chewed raw [A427].

The wood, which is dark in color, is used to make beautiful **handicrafts** and **carvings** [B1966].

(top) *Cordia sebestena* [MB].

(bottom) *Cordia spinescens* [GS].

Cordia spinescens L.

bejuco negro [Spanish]; *box ak', 'ek shee b,'ek sheeb, q'eqxe'eb' k'aham* [Q'eqchi'];
arroz xiv [Additional name recorded]

The plant is used as a component of a bath mixture for **general ailments** associated with the **skin**, **fever**, and **other conditions** [A207].

To treat **burns**, the leaves are dried, toasted, crushed into a powder, and applied directly to the affected area [B3658]. To treat **fear** and **fright** (*susto*), when the "worms get stirred up," 9 leaves are boiled in 1 cup of water for 10 minutes and 1 cup consumed warm once daily for 3 days. This is said to "sedate the worms" [A580].

Ehretia tinifolia L.

roble [Spanish]; *kawil che'* [Q'eqchi']; *bec* [Yucateca]

Standley and Record (1936) reported that the fruit was eaten as a **food**.

To treat **sores** and **wounds**, Steggerda (1943) reported that a bath was made from the leaves. For **urinary disease**, refer to wild okra (*Parmentiera aculeata*) [Steggerda and Korsch, 1943]. Additionally, Steggerda (1943) noted that for the treatment of **pyorrhea** (a periodontal disease), the leaves of this tree were "cooked and the liquid used as a mouthwash" applied 2–3 times daily until better.

Steggerda (1943) noted that the wood from this tree was used in **construction** to make furniture, such as benches and tortilla tables. He also reported the Yucatan Maya of Mexico saying that to use the wood, it "must be cut by the full moon to prevent its decaying."

Heliotropium angiospermum Murray

scorpion tail, monkey tail [English]; *cola de alegran, rabamíco, rabomico* [Spanish]; *xtye i max* [Q'eqchi']

To treat **diarrhea**, the roots of 3 smaller plants or 2 larger plants and three 6-inch branches with leaves along with one 3-inch piece of a dried sweet orange peel (*Citrus sinensis*) are boiled in 1 cup of water for 10 minutes and ⅓ cup consumed hot twice daily for 1 day [B2294].

For **stomach pain** due to food poisoning, three 6-inch branches with leaves along with three 6-inch pieces of a dried sweet orange peel (*Citrus sinensis*) are boiled in 1½ cups of water from a fresh, immature coconut (*Cocos nucifera*) and ¼ cup consumed every 2 hours until finished [B1939].

To relieve spasms in babies with **colic**, two 3-inch branches along with one 1-inch piece of a sweet orange peel (*Citrus sinensis*) and 1 garlic stem are boiled in 8 ounces of water for 10 minutes, strained, 1 teaspoon of honey added, 1 ounce consumed as needed, and the baby burped [B1901]. To reduce **fever**, this plant is soaked in water, but no further details were provided [Robinson and Furley, 1983]. To treat **skin eruptions** and **erysipelas**, this plant is macerated in cold water and used as a bath [Williams, 1986].

Heliotropium indicum L.

scorpion tail [English]; *cola de alegran, rabomico* [Spanish]

For **infant diarrhea**, **malaise**, or **vomiting**, 1 entire plant is boiled in 1 gallon of water for 5 minutes and used as a warm bath at bedtime [Arvigo and Balick, 1998]. For infants suffering from **colic diarrhea**, 1 ounce of leaves is boiled in 2 cups of water for 5 minutes and sipped [A95]. To **wash the eye**, the same decoction as described for colic diarrhea is strained through a cloth and used as a wash [A95; Arvigo and Balick, 1998].

For **skin ailments**, three 6-inch pieces of stem with leaves are boiled in 3 cups of water for 5 minutes and used as a bath over the affected area [Arvigo and Balick, 1998]. For **scanty or painful menstruation**, the decoction described above for skin ailments is consumed warm and unsweetened [Arvigo and Balick, 1998]. This plant is said to be **toxic** if consumed regularly in large doses [Arvigo and Balick, 1998].

Heliotropium indicum [RH].

(left) *Nama jamaicensis* [NYBG].

(right) *Tournefortia hirsutissima* [GS].

Nama jamaicensis L.

coriemiento xiv, *xanab mucuy* [Mayan]

For **pain**, 1 part (i.e., a handful) of fresh plant material is mashed and soaked in 2 parts of a mixture of equal parts water, alcohol, and urine. This infusion is steeped for 30 minutes, placed on a cloth, applied as a poultice, and massaged onto the affected areas once daily. It is said that the urine must be obtained from an infant who is the same gender as the person being treated [B1968].

Tournefortia glabra L.

tke n m, *tke nám* [Mayan]

For **skin inflammation** that develops into red patches, the juice of 1 entire plant is squeezed into 1 quart of cold water and applied to the affected area [B2428].

Tournefortia hirsutissima L.

crocus, poison fish tie tie [English]; *rabo de mico* [Spanish]; *chik* [Q'eqchi']

The fruit is eaten as a **food** [A622, A677].

To relieve **cough**, 1 handful of leaves and 1½ cups of sugar are boiled in 1 quart of water for 10 minutes and consumed until better [A677].

The presence of this plant is an **indicator** of good, rich soil [A622].

As a **fish poison**, the leaves are mashed, placed into a fish trap, and used to bring fish to the surface of the water for harvesting [A483].

Tournefortia maculata Jacq.

mai pan [Q'eqchi']

To treat an **allergic reaction** from an insect sting, the leaves are mashed and applied directly to the site of the sting [A620].

BURSERACEAE

Trees, sometimes shrubs or epiphytes, often containing aromatic resin. Leaves alternate, compound, sometimes unifoliolate. Inflorescences axillary to subterminal, in panicles or racemes. Flowers small, pale green to greenish-yellow. Fruits a compound drupe or pseudocapsule. In Belize, consisting of 3 genera and 7 species. Uses for 2 genera and 5 species are reported here.

Bursera simaruba (L.) Sarg.

birch, red gumbolimbo, tourist tree [English]; *chino, gumbolimbo, gumbolimbo rojo, Indio desnudo, Indio peludo, sirvella simarona, xaka palo* [Spanish]; *kaqajl, ká ka gumbolimbo* [Q'eqchi']; *čacah, xa ka* [Yucateca]; *hukup, jiote* [Additional names recorded]

The young, tender leaves are eaten as a **food** [Arvigo and Balick, 1998]. To make a non-medicinal **beverage**, 9 young branches with leaves or one 12-inch × 3-inch strip of bark are boiled in 1 gallon of water for 10 minutes and consumed as a tea [B3546].

The leaves are 1 of 9 different leaves in a bath mixture used to treat a **variety of ailments** [B2188]. Alternatively, refer to *Flaveria trinervia* [B2149].

To cool an **overheated** person before he or she becomes sick, the leaves are used in an unspecified way and said to be "cooling" [Arnason et al., 1980]. For **fever**, 9 young branches with leaves or one 12-inch × 3-inch strip of bark are boiled in 1 gallon of water for 10 minutes, cooled, and used as a sitz bath [A745c]. To treat

Bark of *Bursera simaruba* [MB].

fever, **internal infections**, **urinary tract ailments**, **sunstroke**, **colds**, **influenza**, or to **purify the blood**, the decoction described above for fever is consumed as a tea [Arvigo and Balick, 1998]. For **fever, pain**, or **colds**, this same decoction is used warm as a bath 3 times daily as needed [A745c, B1815]. For **fever**, refer to *annona* (*Annona reticulata*) [B2106] and my lady (*Aspidosperma cruentum*) [A489; LR]. For **fever, pain**, or **anemia**, refer to buttonwood (*Piper amalago*) [B3673, B3674, B3677]. To treat **cough** and **colds**, refer to *maravilla* (*Mirabilis jalapa*) [B2283].

For **headache**, 9 leaves are mashed in ½ cup of water and applied directly to the forehead [B1815]. Alternatively, 9 leaves are wrapped on the head [B1815; Arvigo and Balick, 1998]. In a third reported treatment, 9 leaves or 1 large handful of bark are boiled in water and used cool to bathe the head [B1815]. For a fourth treatment, refer to *palo verde* (*Critonia morifolia*) [B2202]. For **sinus headache**, refer to warrie wood (*Diphysa americana*) [B2155].

Steggerda and Korsch (1943) reported that a Yucatan Maya healer recommended squeezing 3 drops of the white sap into the ear to treat **earache**. For **gastrointestinal pain**, **malaise**, and to reduce **fever**, **itching**, and discomfort from **measles**, one 12-inch × 3-inch piece of bark is boiled in 2 quarts of water for 10 minutes and ½ cup consumed throughout the day [B1815]. To treat **nosebleed**, the decoction described above for gastrointestinal pain is cooled and used externally [B1815]. To treat **rash** and reduce **itching**, the white sap from the bark is applied directly to the affected area [RA]. To eliminate **bumps on the hand** due to swelling from an injury, the hand is rubbed on a smooth section of the trunk [B3546]. To treat **skin rash**, refer to Pine Ridge bay cedar (*Trema micrantha* var. *floridana*) [A481].

This species usually grows near the poisonwood tree (*Metopium brownei*), serving as a handy first-aid item [RA]. To treat **burns**, **sunburns**, **measles**, **insect bites**, **skin infections**, **rash**, **skin sores**, or as an **antidote** to plants such as poisonwood sap, which causes **rash**, **blisters**, **swelling**, **itching**, and **severe discomfort**, leading to **severe dermatitis**, 9 young branches with leaves or one 12-inch × 3-inch strip of bark are boiled in 1 gallon of water for 10 minutes, cooled, and used as a bath over the affected area 3 times daily. This treatment is said to "keep burns cool and healing" [A745c, A909, A927, B1815, B2412, B2352; Arvigo and Balick, 1998]. To treat **contact dermatitis** resulting from contact with poisonwood sap, the inner bark is scraped to collect clear resin, which is applied directly to the affected area [RA]. To treat **contact dermatitis** resulting from contact with the *manchineel* tree, refer to John Charles (*Hyptis verticillata*) [Arvigo and Balick, 1998].

To treat non-insulin-dependent **diabetes** mellitus, 1 handful of fresh bark is mashed in cold water, steeped for 12 hours, and consumed as a beverage 7 times daily for 30 days [C51]. To help sweat out the toxins from **typhoid**, a steaming hot pot of the decoction described for diabetes is placed under a chair with slits, and the person sits over the steam for 20 minutes while being covered with a blanket [A909; Arvigo and Balick, 1998]. Additionally, for **typhoid**, the person is placed on a bed of the leaves [Arvigo and Balick, 1998].

To treat **digestive disorders** such as **internal gas** and **indigestion**, it is said that 1 handful of fresh, paper-like bark is eaten [A927]. For an **internal infection**, 1 double handful of seeds is mashed in a clean cloth, dropped into 1 quart of boiling water for 5 minutes, and ½ cup consumed twice daily for 3 days. One quart should be finished in 3 days, followed by an intestinal purge [LR]. As a **diuretic**, for **stomach pain**, or to treat **kidney infection**, 9 leaves are boiled in 1 cup of water for 10 minutes, strained, and consumed warm before meals [A745c]. As a **diuretic** or for **urine retention (stoppage of water)** caused by a kidney infection, nine 3-inch pieces of young branches with leaves and 2 ounces of crushed bark are boiled in 1 quart of water for 10 minutes, strained, and sipped all day until the urine is clear or there is no settlement of solid material in urine [A909, B1815]. To treat **kidney ailments, pain**, and **anemia** and to **cleanse the kidneys**, to remove any **infection** that may be present, and as a **tonic** to aid the weak and elderly, one 6-inch × 12-inch piece of bark is boiled in 3 quarts of water for 10 minutes and consumed all day in place of water [Arvigo and Balick, 1998]. For another reported **kidney** treatment, refer to *coqueta* (*Rivina humilis*) [B2123].

As a **blood cleanser, blood builder**, or for **back pain**, 9 branches with leaves and one 6-inch × 2-inch piece of bark are boiled in 2 quarts of water for 10 minutes, strained, and 1 cup consumed twice daily [B1815]. To treat **postpartum vaginal burning**, the decoction described as a blood cleanser is used very warm as a douche [B1815]. To treat **hemorrhage** from a wound, menstruation, or following childbirth, 4 ounces of bark are boiled in 2 cups of water for 10 minutes, cooled, and 1 cup consumed every 20 minutes [HR].

To make **white lime**, refer to *botán* (*Sabal mauritiiformis*) [BW]. To **cool down horses**, the leaves are placed around their necks [B1815]. To facilitate the de-shelling of coffee beans (*Coffea arabica*), fresh leaves are placed over the ripe coffee beans that have been collected in a bucket or pan, and this is covered with a cloth and allowed to sit for 2 days, after which the outer coating of the coffee seeds is easily removed [HR].

As a **hand wash** for those who have recently attended to or touched the dead, such as undertakers, to ensure that crops planted by the attendant will continue to yield good harvests and not be tainted, the fresh leaves are mashed in a basin of water and used to wash the hands [Arvigo and Balick, 1998]. To **ward off evil spirits**, bouquets of this plant and *zapato de la virgen* (*Pedilanthus tithymaloides*) are placed in homes and the leaves from both plants spread all around the house and doorstep [HR]. Alternatively, to **ward off evil spirits**, refer to *zapato de la virgen* (*Pedilanthus tithymaloides*) [RA].

Standley and Steyermark (1946) reported that the wood of this tree was used in Guatemala to make the soles of **sandals (*saitas*)** worn by many Indians at that time. They also reported on the harvest of **resin** from this tree by notching the trunk and collecting the sap in gourds placed beneath the cup. The sap was boiled in water, and the resin rose to the surface, where it was skimmed and added to cold water to harden. "It is shaped into oblong blocks, very hard and brittle, which are wrapped in corn husks, tied at the ends with strips of corn husk and in this form taken to market, to be used as incense in the churches." Those authors also reported that a poultice from the leaves of this plant was used to prevent the spread of **gangrene** resulting from snakebite. It was also said that the **resin** could be used to mend broken china and glass and that the Caribs used it to **paint** their canoes to prevent them from being destroyed by worms. Finally, they noted that the name used in Belize, *gumbolimbo*, "is believed to be a corruption of the Spanish *goma elemi*, sometimes given by the Spaniards to the resin or copal."

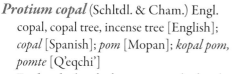

(left) *Protium copal* bark and resin [MB].

(right) *Protium copal* [NYBG].

Protium copal (Schltdl. & Cham.) Engl.
    copal, copal tree, incense tree [English];
    copal [Spanish]; *pom* [Mopan]; *kopal pom,
    pomte* [Q'eqchi']

For **headache**, the leaves are mashed and ap-
plied as a poultice to the forehead [A822]. To treat **colds**, 2 ounces of bark along with
3 ounces of balsam bark (*Myroxylon balsamum* var. *pereirae*) are boiled in 1 quart
of water for 10 minutes and ¼ cup consumed 3 times daily until better [A507]. To
treat a **wound**, **sore**, or **infection**, the fresh bark is scraped, powdered, wrapped with
gauze, applied directly to the affected area, and the dressing changed daily [B1859;
Arvigo and Balick, 1998].

For **tooth removal**, 1 piece of fresh resin is applied directly into the tooth cavity
until the resin swells and breaks the tooth apart, usually in a few days [A507, B1859;
Arvigo and Balick, 1998]. To aid in **thorn removal**, fresh resin liquid is applied to an
embedded thorn and wrapped in gauze for 4 hours [A822].

For **stomach ailments** and **intestinal worms**, one 2-inch × 2-inch piece of bark
is boiled in 3 cups of water for 10 minutes and 1 cup consumed 3 times daily before
meals [B1859; Arvigo and Balick, 1998]. To treat **illness**, the resin is burned as incense under the bed of the sick person [MB]. Alternatively, refer to corn (*Zea mays*)
[Thompson, 1927].

In ancient times, to **ward off witchcraft**, **envy** (*envídia*), the **evil eye** (*mal ojo*),
and **evil** in general, cloth rags were dipped in the resin and burned as incense while
saying prayers at a *primicia* (a sacred ceremony) [B1859; Arvigo and Balick, 1998].
To **ward off evil spirits**, such as those caused by the evil eye (*mal ojo*), the resin is
burned as incense [A822, B1859]. The resin is extracted by making spiral cuts in the
tree on nights of the full moon [A822; Arvigo and Balick, 1998]. To treat the **spiritual
sickness** that follows shock, refer to Bethlehem star (*Hippobroma longiflora*) [BW].

Thompson (1927) noted that to purify a patch of forest before it was converted
to a crop field, the resin of copal was burned and prayers offered as conciliation to

the spirits of the forest to be destroyed. A free translation of one of these prayers is as follows:

> O God, my mother, my father, Huitz-Hok, Lord of the Hills and Valleys, Che, Spirit of the forests, be patient with me, for I am about to do as my fathers have ever done. Now I make my offering to you that you may know that I am about to trouble your very soul, but suffer it I pray you. I am about to dirty you—to destroy your beauty—I am going to work you that I may obtain my daily bread. I pray you suffer no animal to dog my footsteps, nor snake to bite me. Permit not the scorpion or wasp to sting me. Bid the trees that they fall not upon me, and suffer not the axe or knife to cut me, for with all my heart I am about to work you. (p. 45)

The resin is used as a **varnish** and for **nail polish** [A822]. To wrap and protect the resin, refer to *Chamaedorea ernesti-augustii* [B3496; MB] and monkey tail (*Chamaedorea pinnatifrons*) [B2474].

For a **fish poison** tradition, refer to *barbasco* (*Serjania lundellii*) [Thompson, 1930].

Protium costaricense (Rose) Engl.
copal [Spanish]
The resin is burned as **incense** [B2711].

For **back pain**, **back spasms**, or to **remove spines or thorns** from the skin, the soft, fresh, liquid resin is applied with a piece of cotton directly to the affected area as a back plaster or a poultice [B2711].

To treat someone with "**fits**," having a seizure like epilepsy with sudden fainting and shaking, 1 teaspoon of dry, grated resin is steeped in 1 cup of hot water for 10 minutes and 1 cup consumed warm once, repeated twice only if necessary [B2711].

Protium multiramiflorum Lundell
copal, copal colorado [Spanish]
The resin is extracted and burned for **ceremonies**, specifically as a part of healing ceremonies [B2036]. It is also burned by a person wanting to **attract a lover** [RA].

Protium schippii Lundell
mountain copal [English]; *copal, copal macho, macho limoncillo* [Spanish]
To treat **blood clots**, 9 leaves are boiled in 1 quart of water for 10 minutes and applied warm with a piece of cotton to the clotted area until better [B2615]. To clear **fungal infections** of the skin, 9 leaves are boiled in 1 quart of water and 1 cup consumed when a person bathes, once daily, until better, usually for 9 days [B2615].

CACTACEAE

Trees, shrubs, or vines, usually with spines and succulent stems. Leaves small, reduced or ephemeral. Inflorescences usually of solitary flowers, sometimes in panicles or cymes. Flowers often brightly and variously colored. Fruit a berry, often spiny or bristly. In Belize, consisting of 7 genera and 12 species. Uses for 4 genera and 5 species are reported here.

Cereus sp.

joining bone [English]; *chiqb'al b'aq* [Q'eqchi']

For **bone pain**, the entire plant is boiled in water and consumed as a tea [R32]. This treatment is prepared only if the treatment described for *rah li ch'och* (*Tectaria panamensis*) has been prepared and is unsuccessful [R32]. Steggerda (1943) reported that the Yucatan Maya crushed the stem of *Cereus undatus* Haw. (a species not found in Belize) in cold water for use as a **shampoo**.

Epiphyllum phyllanthus (L.) Haw.

chiqb' al b'aq [Q'eqchi']

To treat **colds** or **influenza**, 1 large handful of the entire plant is boiled in 1 liter of water for 15 minutes and consumed warm 4 times daily for 5 days [C44].

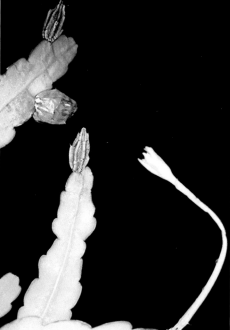

Epiphyllum phyllanthus with close-up (right) [RF].

Opuntia cochenillifera (L.) Mill.

prickly pear [English]; *nopal, tuna* [Spanish]; *cagineel, scogineel* [Creole]; *pa'kam* [Mopan]

The young cactus pads are peeled, chopped, and eaten as a **food** either raw, pickled, or sautéed with onion, garlic, and tomatoes [BW]. The fruit is highly esteemed and also eaten as a **food**. The spiny, outer portion is peeled off, and the red or yellowish seedy center is consumed, having a crunchy, sweet taste [Arvigo and Balick, 1998].

For **pain, headache, fever, bruises,** or **swelling,** the older cactus pads are split in half and the inside surface applied directly to the affected area. The pads are left on for 1–2 hours daily. A headache generally receives a 1-hour application. A bruise or swelling generally receives a 2-hour application 3 times daily until better [B3363; Arvigo and Balick, 1998].

To alleviate **arthritis**, the pads are peeled, steamed, chilled, and eaten in salads during a physic nut (*Jatropha curcas*) treatment [Arvigo and Balick, 1998]; for this treatment, refer to physic nut. To treat **high blood pressure, fever,** or **malaise,** 1 pad is boiled in 3 cups of water for 5 minutes and 1 cup consumed 3 times daily before meals [Arvigo and Balick, 1998]. For **high blood pressure,** 1 older pad is roasted, cut into small pieces, soaked in 1 quart of water with the juice of a lime, and ½ cup consumed 3 times daily as needed [BW].

For **burns,** 1 pad is cut in half and the inner gel scraped out and applied to the affected area 3 times daily [BW]. To treat **skin ulcers,** 1 pad is sliced in half and directly applied over the affected area as needed [Arvigo and Balick, 1998]. For **eye infections, conjunctivitis (pinkeye),** or to **cleanse bruised blood** out of eye injuries, 2 drops of the gel from the pad are dropped into the eye 3 times during the day and once at bedtime [LR].

For **kidney ailments** and to treat "**heat in the body,**" 1 medium-sized pad is sliced, steeped in 1 quart of tepid water for 20 minutes, and 1 cup consumed 3 times daily, be-

fore meals, for 1–3 days or until better. Heat in the body is an indication that the kidneys are not working properly [B2105]. To treat a **swollen spleen** or "**pain in the internal organs,**" 1 pad the size of the person's foot is cut out in the outline of a foot and hung over the fire hearth, and while the foot mold dries, the spleen becomes normal or the pain goes away [BW; Arvigo and Balick, 1998].

Opuntia cochenillifera [MB].

To **prevent hair loss**, 5 fresh pads are crushed, soaked in 1 gallon of water, and used as a hair rinse [Arvigo and Balick, 1998]. To **sheen hair** and to **stimulate hair growth**, 1 older pad is squeezed into 1 quart of water and used as a hair wash [BW]. As a **hair conditioner**, producing soft, lustrous results, 1 pad is peeled, mashed, spread on the hair, covered in plastic wrap for 1 hour, and then rinsed thoroughly out of the hair [Arvigo and Balick, 1998].

To prepare for an easy **childbirth**, ¼ cup of juice is consumed or 1 small pad eaten daily for 7–10 days before delivery [BW; Arvigo and Balick, 1998]. To ease **delivery** at the onset of childbirth, 1 cup of juice from a fresh pad is consumed [Arvigo and Balick, 1998]. To treat **bladder ailments**, 5 fresh pads are crushed, soaked in 1 gallon of water, and consumed as a tea [Arvigo and Balick, 1998].

To **starch clothes**, 9 pads are mashed in one 5-gallon bucket of water, strained, and used as a rinse to make clothes shiny and starched [HR].

The small, hair-like spines, called "glochids," found on the outside of the fruit are not eaten, as they can be potentially very **irritating** to the body [Arvigo and Balick, 1998]. **Caution** is advised when handling the leaf pads, as these almost-invisible spines are highly irritating to the skin and difficult to remove [RA].

Selenicereus testudo (Karw. ex Zucc.) Buxbaum

For **sprains** and **swelling**, the entire plant is mashed and applied as a poultice twice daily to the affected area [B3246].

Selenicereus testudo with inset showing fruit [MB].

Selenicereus sp.

devil's guts [English]; *pitahaya* [Spanish]

The mature, ripe fruit is eaten as a **food** [B1927].

To prevent the blistering, scarring, and infection of **burns**, the inner gel is scraped out, mashed, and applied as a poultice to the affected area twice daily [LR].

For **hair loss**, 3 pieces of the pad, each equal to the length of the person's hair, are chopped, soaked in 1 quart of water, and used as a rinse for the hair [BW]. To treat **hair loss** due to malnutrition or nervousness, one 12-inch piece of the pad is mashed in 1 quart of warm water, steeped for ½ day, strained, and used as a hair rinse. It is rubbed deeply into the scalp after normal shampoo and repeated 2–3 times weekly, until the hair stops falling out [RA].

CAMPANULACEAE

Herbs, shrubs, trees, or lianas with milky latex present. Leaves alternate, sometimes opposite or whorled, usually simple. Inflorescences terminal or axillary, of solitary flowers or in racemes. Fruits a capsule or berry. In Belize, consisting of 2 genera and 4 species. Uses for 1 genus and 1 species are reported here.

Hippobroma longiflora (L.) G. Don

Bethlehem star [English]

For **swelling**, 1 entire plant is boiled in 1 gallon of water for 20 minutes and used warm as a bath once daily [A335]. For **pain** in the body, 15 flowers are soaked in 1 bottle of bay rum for 14 days and rubbed all over the body twice daily, in the morning and in the evening, until better [HR].

For **toothache**, the leaves are chewed or mashed in a mortar and applied over the gums [A335]. For **toothache**, the leaf juice should be spit out after treatment and not swallowed, as the plant's sap is **toxic** [HR].

To treat **stomach and body "chills,"** 9 leaves are reported to be boiled in 3 cups of water for 10 minutes and 1 cup consumed tepid to warm 3 times daily for 1 day [B2564]. However, this plant is very **toxic** [Nelson, Shih, and Balick, 2007] and poses danger if used internally. Thus, alternatively, for **stomach and body "chills,"** 2 handfuls of leaves are steeped in 2 quarts of hot water for 1 hour and poured over the body as a bath once daily for 3 days [B2564].

To treat the **spiritual sickness** that follows shock, characterized by "lost and confused people who don't sit still or con-

Hippobroma longiflora [MB].

stantly want to be on the move," 1 handful of flowers along with three 6-inch branches of *epasote* (*Chenopodium ambrosioides*) are boiled for 10 minutes, or steeped in the sun for 1 hour, in 1 gallon of water and then during a 20-minute period is poured over the head down to the feet once daily for 3 days [BW]. Additionally, incense, such as copal (*Protium copal*) is burned while bathing [BW].

Children use the flower buds as **toy "arrows"** [B3200].

Caution is taken when using this plant, as it is known to be toxic [A795]. Robinson and Furley (1983) described it as "very poisonous" and cited various works noting the presence of a compound, isotomine, that can stop the heartbeat.

CARICACEAE

Small, sparsely branched trees with milky latex. Leaves alternate, simple, trifoliolate or palmately compound. Inflorescences axillary, sparsely to much-branched cymes. Flowers white, yellow, or green. Fruits a berry. In Belize, consisting of 2 genera and 2 species. Uses for 2 genera and 2 species are reported here.

Carica papaya L.

papaya, wild papaya [English]; *papaya* [Spanish]; *papaw* [Creole]; *put* [Mopan]; *papay, putul* [Q'eqchi']; *čič pu'ut* [Yucateca]

The ripe fruit is eaten as a **food** [A890, B1967, B2332; Arvigo and Balick, 1998] and used to make a **beverage** [Arvigo and Balick, 1998]. Green fruits are boiled and eaten

Carica papaya [MB] with inset showing flowers and young fruit [IA].

like zucchini squash (*Cucurbita pepo*) [Arvigo and Balick, 1998]. The root is roasted and eaten as a **survival food** [BW].

To treat **constipation**, **sluggish liver**, **indigestion**, **high blood pressure**, and for use as a **diuretic** or **purge**, the ripe fruits are peeled and eaten [Arvigo and Balick, 1998]. To treat **intestinal parasites**, the green fruit is eaten as a regular part of the diet [Arvigo and Balick, 1998], while Standley and Williams (1961) reported that the same therapeutic effect was obtained by using dry and powdered seeds or milky sap. To treat **eye cataracts** and **fever**, the stem, leaf, and fruit latex are diluted in water and used, but no further details were provided [Flores and Ricalde, 1996]. As a **purgative** for people, the ripe fruits and roots are used in an unspecified way [Robinson and Furley, 1983]. As a **purgative** for ill horses, the leaves are used in an unspecified way [Robinson and Furley, 1983]. Standley and Williams (1961) noted that the fruit was used to produce syrup employed in the region as a **cough** remedy.

For **back pain**, 1 leaf is cut, placed on the floor or a bed, and the person lies down on the leaf for 1 hour [B2203]. To help "draw out" the pain of **back pain**, 1 leaf is chopped, mixed in a bit of oil, and applied directly to the affected area [AR]. To treat **wounds**, **cuts**, and **infections**, sliced fruit and crushed seeds are mixed and applied directly to the affected area [Arvigo and Balick, 1998]. To dissolve **foot corns**, juice from the sliced stems is applied directly to the affected area [Arvigo and Balick, 1998]. To treat **irritated or red skin** caused by contact with fire coral (*Millepora* spp.), very ripe or even rotting fruit is rubbed directly onto the affected area [Arvigo and Balick, 1998].

To treat **rash** and **unopened sores**, the "milk" (the white latex that runs from green, unripe fruits when scored or cut) is directly applied to affected area 3 times daily [B1967]. To treat **warts**, **leishmaniasis (baysore)**, and **itching**, sap from the leaf base is dripped directly onto the affected area 3 times daily, left on the skin between applications, and repeated for 9 days [A836, A890, B3234]. For **warts**, the milky sap from sliced green fruits is applied to the affected area [Arvigo and Balick, 1998]. Alternatively, equal parts of latex from the fruit and baking soda are mixed to form a paste that is applied daily until the warts drop off [BW].

To treat **snakebite**, one 12-inch piece of root is ground, toasted, fried in oil, placed as a poultice directly over the bite twice daily, and left on between applications [A135]. To remove **venom** from a snake, a piece of stem is cut, the bark scraped off, and the stem given to the snake to bite, into which the venom will be released [LR]. To kill a **snake**, the end of a stem is sharpened and thrown like a lance at it [B3234]. Standley and Williams (1961) reported that the sap of this species was applied to the affected area to poison tropical **chiggers** (*Trombicula* spp.).

For venereal **diseases**, the root is boiled in water and used in an unspecified treatment [Arvigo and Balick, 1998]. Alternatively, refer to *Sida antillensis* [B2658]. For non-insulin-dependent **diabetes** mellitus and **venereal diseases** such as **gonorrhea**, refer to "male" chicken weed (*Chamaesyce hypericifolia*) [A315] and chicken weed (*Euphorbia armourii*) [A322]. As a **female contraceptive**, 3 ounces of seeds are roasted, ground, and 1 teaspoon of the powder mixed in ½ cup of warm water, which is consumed once daily for 3 days before menstruation. For permanent **sterility**, this treatment is consumed consecutively for 2½ years [Arvigo and Balick, 1998]. For antidotes to **sterility**, refer to skunk root (*Chiococca alba*) [Arvigo and Balick, 1998] and *piss-a-bed* (*Senna alata*) [Arvigo and Balick, 1998].

As a **meat tenderizer**, the leaves are mashed and placed on the meat for 20 minutes [A890]. Alternatively, the leaves are wrapped around the meat for 1–2 hours [Arvigo and Balick, 1998; Lundell, 1938].

A **belief** of the Maya is that the man of the household will become lazy if this tree is planted too close to the house or bedroom [Arvigo and Balick, 1998].

Jacaratia dolichaula (Donn. Sm.) Woodson

papaya cimmaron, wild pawpaw [English]

This is the so-called wild papaya, which Standley and Record (1936) noted could be found throughout much of Central America. It is said to grow to 36 feet high with an 8-inch-diameter trunk, which was used in Costa Rica for making "rough **casks** in which grain is stored."

Standley and Williams (1961) reported that the raw fruits of this species (here under the name *Jacaratia mexicana*) were sometimes eaten as a **food** in salads in Mexico and that "when maize is scarce, the Mayas of Yucatan are said to pulverize the pith of the plant and mix it with maize in making tortillas, although this seems unlikely." We have never observed this reported use during our studies in Belize [MB].

To treat **eye cataracts** and **fever**, the stem, leaf, and fruit resin of "*Jacaratia mexicana*" (synonym for *J. dolichaula*) is diluted in water and used in an unspecified way [Flores and Ricalde, 1996].

Jacaratia dolichaula leaf (left), fruit (top right), and halved fruit (bottom right) [FM].

CASUARINACEAE

Trees. Leaves whorled, reduced to scale-like teeth. Inflorescences in spikes or heads. Flowers much reduced. Fruits a samara, appearing capsule-like. In Belize, consisting of 1 genus and 2 species. Uses for 1 genus and 1 species are reported here.

Casuarina equisetifolia L.

pine [English]

To treat **gonorrhea**, 1 small handful of bark along with 1 small handful of balsam bark (*Myroxylon balsamum* var. *pereirae*) are boiled in 1 quart of water for 20 minutes, steeped for 1 hour, strained, and sipped warm throughout the day for 9 days [B1958].

Casuarina equisetifolia with inset showing close-up of fruit [MB].

For male **impotence**, 1 ounce of the resin is steeped in ½ quart of gin for 10 days and ½ ounce consumed twice daily, in the morning and in the evening, until finished [HR].

To **ward off evil**, the resin is burned as incense [BW]. To allow **evil spirits** to depart from a possessed person, 9 pine "needles" along with three 6-inch branches of *zapato de la virgin* (*Pedilanthus tithymaloides*), 7 red roses (*Rosa chinensis*), and three 12-inch branches of marigold (*Tagetes erecta*) are boiled in 2 gallon of water for 10 minutes or steeped in the sun for 1 hour, ½ cup consumed, and the remainder used as an outdoor bath, repeated twice if necessary [BW].

CELASTRACEAE (INCLUDING HIPPOCRATEACEAE)

Small trees, shrubs, vines, and lianas. Leaves alternate, subopposite, or opposite, sometimes very reduced. Inflorescences terminal or axillary, in panicles, corymbs, cymes, fascicles, racemes, umbels, or rarely of solitary flowers. Flowers usually small, greenish, or white. Fruits a capsule, drupe, berry, woody capsule, schizocarp, or samara. In Belize, consisting of 11 genera and 16 species. Uses for 2 genera and 3 species are reported here.

Crossopetalum eucymosum (Loes. & Pittier) Lundell

xi lik [Q'eqchi']

To treat **stomach pain**, 9 leaves are boiled in 3 cups of water for 10 minutes, steeped until warm, and 1 cup consumed at a time as needed [B2484].

To treat **itchy and bumpy skin**, caused by passing through a place where the devil has been, 1 large handful of leaves is boiled in 2 gallons of water for 20 minutes, steeped until warm, 1 cup consumed cool, and the remainder used as a bath [B2484].

(left) *Crossopetalum eucymosum* [NYBG].

(right) *Hemiangium excelsum* [RH].

Crossopetalum parviflorum (Hemsl.) Lundell

ye bok che [Mopan]

To treat **influenza**, 1 handful of fresh leaves is boiled in water for 15 minutes and consumed warm twice daily for 5 days [C66]. Additionally, this decoction is used as a bath [C66].

Hemiangium excelsum (Kunth) A.C. Sm.

roble [Spanish]; *tie tie* [Creole]

To **promote sweat** to treat **fever**, 1 handful of leaves and young stems is boiled in 1 gallon of water for 10 minutes, 1 cup consumed hot, and the remainder used as a bath once daily, at noon or bedtime [A381].

CHRYSOBALANACEAE

Trees, sometimes shrubs. Leaves alternate, simple. Inflorescences axillary, terminal, occasionally cauliflorous, in racemes, panicles, spikes, or cymes. Fruits a drupe. In Belize, consisting of 4 genera and 9 species. Uses for 4 genera and 5 species are reported here.

Chrysobalanus icaco L.

caye caulker plum, coco plum, coco-plum, cocoplum, hicaco plum, jicaco plum [English]; *icaco, jicaca* [Spanish]; *rhum ka ka tà, kocho* [Mayan]

The fruit is eaten raw as a **food** by people and birds [A976, B1954, B2088]. The ripe fruit may be stewed with sugar to make **jam** [A414].

Chrysobalanus icaco [MB].

To treat thick, white **vaginal discharge** in women, 1 ounce of ground-up seeds is blended with 2 cups of water, strained, and 1 cup consumed twice daily after menstruation [A348].

The fruit is used as **fish bait**. "When the fruit falls you catch [the] fish as they congregate near this" [B3495].

Standley and Record (1936) noted that the bark and leaves were **astringent** and that the seeds had high **oil** content.

Couepia polyandra (Kunth) Rose
baboon cap, baboon cup, monkey cap [English]
The fruit is said to be yellow and eaten as a **food** when ripe [N46898].

Hirtella americana L.
blossom berry, pigeon plum, wild coco-plum [English]; *limoncillo* [Spanish]
The fruit is eaten as a **food** by birds and other animals but not by people [A548, B3243].

The branches are tied together and used as **brooms** [BW].

Standley and Record (1936) noted that "the bark of some species of *Hirtella* is said to have been used for **tanning**," but no further information was provided.

Licania hypoleuca Benth. var. *hypoleuca*
pigeon plum [English]; *sir ch* [Mayan]
The wood is used in **construction** to make house rafters [B3639].

Licania platypus (Hemsl.) Fritsch
monkey apple [English]; *holob'oob'* [Q'eqchi']; *succotz* [Yucateca]; *sunco, urraco* [Additional names recorded]

Hirtella americana [RF].

Licania hypoleuca var. *hypoleuca* [RF].

Licania platypus [RF].

The ripe fruit was reported to be eaten as a **food** by people [B2017]. According to Lamb (1946), this was not a popular food because the flesh was too fibrous and not tasty. The fruits provide **food** for monkeys; hunters know this and seek their quarry by these trees [B2017]. It is curious to note that Standley and Record (1936) reported that the fruit "is **edible** but little esteemed, especially because of a belief that it causes fevers and other ailments."

To treat **diarrhea** in babies, ¼ seed is chopped, mixed with the skins of 1 garlic clove (not the clove itself), boiled in 1 pint of water for 10 minutes, and 1 teaspoon consumed every 10 minutes until better [BW]. To treat **diarrhea** in children, ½ seed is chopped, mixed with 1 garlic clove and skins, boiled in 1 quart for 10 minutes, and 2 tablespoons consumed every 20 minutes until better [BW]. To treat **diarrhea** in adults, 1 entire seed is chopped, mixed with 1 garlic clove and skins, boiled in 1 quart of water, and ½ cup consumed every 20 minutes until better. This is a strong **astringent** and usually works after only 1 dosage [BW].

To prepare for **snakebite**, the outer shell of the large inner seed is removed, dried, powered, and carried into the bush. One fruit shell will yield about 6 teaspoons of powder [LR]. To treat a **snakebite**, 1 teaspoon of this seed powder is swallowed with saliva every 30 minutes until it is possible to find hot water. Then, 2 teaspoons are mixed in ½ cup of hot water and consumed once every hour until better (when the area is no longer painful). A poultice is still required to remove the snake's "tooth scale" [LR].

Lamb (1946) reported that the easy-splitting wood was used for **firewood**.

CLUSIACEAE

Trees, shrubs, or herbs containing yellow, orange, white, or clear latex. Leaves simple, usually opposite, sometimes whorled or alternate, often thick and leathery. Inflorescences terminal or axillary, in cymes but sometimes racemes, fascicles, or of solitary flowers. Flowers often yellow or white, sometimes pink or red. Fruits a capsule or berry. In Belize, consisting of 9 genera and 22 species. Uses for 6 genera and 8 species are reported here.

Calophyllum brasiliense Cambess. var. *rekoi* (Standl.) Standl.
   Santa María [Spanish]
   To treat **leishmaniasis (baysore)**, the milky bark sap is applied directly to the affected area 3 times daily until better [B2204]. To treat **snakebite**, one 5-inch × 8-inch piece of bark is boiled in 1½ quarts of water for 5 minutes and 1 cup consumed warm every hour until the bleeding stops from the pores, gums, and fingernails. This treatment is only to stop the **bleeding**; another

Calophyllum brasiliense var. *rekoi* [RF].

formula must be used to stop the venom. An unspecified poultice is additionally applied to remove the snake's "tooth scales" [LR]. Lamb (1946) noted that "the seeds yield an **oil** used in parts of Central America for illumination. It is not utilized in British Honduras."

Lamb (1946) also reported that the wood was used in **construction** for making house frames, boat masts, and boat spars. Additionally, he observed that the wood was occasionally used in **construction** to make dugout canoes, although its tendency to be heavy and warp easily were limiting factors for this. During mahogany (*Swietenia macrophylla*) logging, Lamb (1946) noted that its strength made it useful in **construction** for bridge timbers and railroad ties. Today, it is used in **construction** for house posts and roof beams [B2204]. This plant was once used in **construction** for ship building in the royal dockyards of British Honduras, but again, as it had a tendency to warp, it was not extremely useful for this purpose [Marsh, Matola, and Pickard, 1995].

Clusia flava Jacq.

matapalo [Spanish]; *chunup* [Yucateca]

Standley and Record (1936) noted that "the latex, which is yellow at first, as in related species, is sometimes used to **adulterate chicle**."

Steggerda (1943) reported that the Yucatan Maya heated the leaves of this species and applied them to children with a **protruding navel**.

Clusia lundellii Standl.

matapalo [Spanish]; *hub'ub'* [Q'eqchi']; *chunup* [Yucateca]

As a **fish poison**, the flowers are mashed and placed in the river, after which the stunned fish float to the surface, where they can be easily collected [R54].

The fruit juice produces a **yellow-red dye** often used to decorate cloth with a print block of jaguar paws [A574].

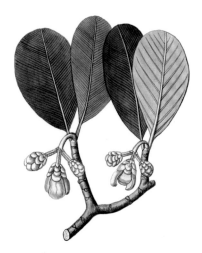

Clusia flava [NYBG].

Clusia lundellii [GS].

Garcinia intermedia [RF].

(top) *Hypericum terrae-firmae* [MB].

(bottom) *Symphonia globulifera* [MB].

Garcinia intermedia (Pittier) Hammel

waika plum [English]; *caimito, jocomico* [Spanish]; *k'a re che'* [Q'eqchi']

The fruit is eaten as a **food** when ripe [B3570], but Standley and Record (1936) noted that "it is of inferior quality and flavor, and is little esteemed." The wood (referred to in the following reference as *Rheedia edulis*) is a strong wood used in **construction** for making tool handles and wooden tools [Horwich and Lyon, 1990].

Hypericum terrae-firmae Sprague & L. Riley

Pine Ridge flower [English]; *la ámor, pericon* [Spanish]

To treat **headache** or **fever** or as a **face wash** at bedtime, 1 handful of leaves and stems is steeped in 2 quarts of cold water for 20 minutes and used to wash the head as needed [B2260].

To **keep away the devil**, refer to morning glory (*Ipomoea carnea* subsp. *fistulosa*) [RA].

Symphonia globulifera L.f.

waika chewstick [English]; *leche amarillo macho* [Spanish]; *q'an i leech* [Q'eqchi'];
corbán, wycot [Additional names recorded]

The fruit is eaten as a **food** by wild and domesticated pigs [B3487].

To treat **back pain**, the yellow latex is used as a plaster directly on the back [B3487].

The yellow latex is used for **caulking** and for **torches** [B3487].

Lamb (1946) reported that the wood was used in **construction** of house beams, railway sleepers, and boat keels. Today, the wood is used in **construction** to make house beams [G14]. Standley and Record (1936) reported that the wood of this species was exported in small quantities for use as **veneer**.

Vismia camparaguey [MB].   Vismia macrophylla [MB].

Vismia camparaguey Sprague & L. Riley

can't-be-helped, old William, ringworm tree, wild annato [English]; *llora sangre, achotio* [Spanish]; *q'an parakway* [Q'eqchi']; *ka'n k'arhy, sak' am pa ra guáy* [Mayan]

To control the **bleeding** of wounds and cuts, the sap is applied directly to the affected area [B3547, B3621]. To treat **ringworm**, the leaves are used to prepare a bath for the affected area [B2628]. Additionally, the red latex is applied directly to the round lesions [B2628]. Alternatively, 9 fresh leaves and a pinch of salt are mashed with a bit of water, applied as a poultice directly over the affected area, wrapped up, changed twice daily, and repeated for 3–7 days [LR].

The hard, long-lasting wood is used in **construction** for house posts and roof poles [A571, A963]. Standley and Record (1936) reported that one of the local names, as explained by J.B. Kinloch, referred to the fact that "the wood is poor for building huts, but if nothing else is available this is used as 'it can't be helped.'"

Vismia macrophylla Kunth

ringworm plant (female) [English]; *ambara guay, kaq' q'an pakway* [Q'eqchi']

To stop the **bleeding** of wounds and cuts, the orange sap from the inner bark and stems is applied directly to the affected area [B3620]. To treat **ringworm**, 1 handful of leaves is boiled in 1 quart of water for 20 minutes and poured as hot as possible over the affected area [B2623]. Additionally, the sap is applied directly to the white patches on the skin twice daily [B2623]. For **skin fungus**, 3 leaves are mashed in water and applied as a poultice to the affected area for hours [B2508].

COMBRETACEAE

Trees, shrubs, subshrubs, or lianas. Leaves opposite or alternate, simple, pubescent. Inflorescences terminal and/or axillary, in racemes, spikes, or panicles. Flowers small, greenish-white, pink, red, or orange. Fruits an indehiscent false drupe, false nut, or false samara. In Belize, consisting of 6 genera and 10 species. Uses for 5 genera and 6 species are reported here.

(left) *Bucida buceras* [FA].

(right) *Combretum fruticosum* with close-up of inflorescence (bottom) [RF].

Bucida buceras L.

bullet tree, bully tree [English];
q'an xan [Q'eqchi']; *puk te* [Mayan]
This plant makes excellent **charcoal** [BW]. The wood is said to be the "hardest lumber" and, in fact, will sink in water. Lamb (1946) noted that because it maintained its durability even when wet, it was considered an excellent choice for outdoor **construction** and was reportedly used for a variety of such purposes, including pilings, house posts, porches, bridge timbers, railroad ties, boat timbers, and water vats. It is still used in many of the same ways today [A215]. Horwich and Lyon (1990) reported that the bark was used in **tanning**. Standley and Record (1936) stated that one of the uses for this plant was as a **fuelwood**, along with **charcoal**.

Combretum fruticosum (Loefl.) Stuntz

chew stick, curassow comb, ixtobai comb, monkey brush, monkey brush tie tie, tie tie (male), yellow brush [English]; *chupa mile, sepillo, sepillo amarillo* [Spanish]
The vine is used to make **wine**, known as "chew stick wine" [LR].

For **general illness**, 1 small handful of the root is boiled in 3 cups of water for 10 minutes and 1 cup consumed 3 times daily for 3 days [B2240]. As a **sedative**, 1 cup of chopped vine and root is boiled in 3 cups of water for 10 minutes and 2 cups consumed daily for 9 days [A745].

For non-insulin-dependent **diabetes** mellitus, 1 ounce of leaves is boiled in 3 cups of water for 5 minutes and 1 cup consumed 3 times daily, before meals, for 3 days. Following this, the blood sugar is checked, and if it is not better, the treatment is continued by consuming 1 cup twice daily, in the morning and in the evening, until better [A745].

To treat **high blood pressure**, ½ cup of chopped roots and ½ cup of chopped vine are boiled in 2 quarts of water for 10 minutes and ½ cup consumed 6 times daily until the blood pressure decreases [W1852]. To treat **high blood pressure** when it "feels as if [your] head wants to burst," two 3-inch pieces of bark are boiled in 3 cups of water for 10 minutes and ½ cup consumed, hot or cold, all day long for a total of 6 doses each day, until better [B3294].

To **stimulate faster and better hair growth**, the hair is brushed with a flower of this species on Good Friday in 9 downward strokes and repeated for 3 consecutive days [BW].

Conocarpus erecta L.

button wood, button-bush, buttonwood, white mangrove [English]; *botoncillo* [Spanish]; *kanche* [Additional name recorded]

To treat **sores**, one 12-inch strip of bark is roasted until very dry and crushed through a sieve to make powder. Sores are covered with castor oil (*Ricinus communis*), sprinkled with this bark powder, covered with a cloth, and the poultice changed once daily until better [B1949].

The dry wood is used as a **fuel** [A418], and Standley and Record (1936) referred to it as "the favorite fuelwood of the Colony." This tree is used to make **charcoal** [Horwich and Lyon, 1990]. Fresh wood, especially the black heart from mature trees, is used in **construction** for making stems for boats and dock posts [A418].

Conocarpus erecta [MB].

Laguncularia racemosa (L.) C.F. Gaertn.

white mangrove [English]; *mangle blanco* [Spanish]; *saq' b'ut' nut'* [Q'eqchi'];
zacalcom [Additional name recorded]
The dry wood is used as **fuel** for firewood [A611]. This tree is said to be one of the
"prized woods for **corner posts** and **framing**" if a person lives near the coast, where
it is found [Horwich and Lyon, 1990].

Terminalia amazonia (J.F. Gmel.) Exell

bullywood [English]; *almendro, nargusta* [Spanish]
Lamb (1946) reported that the wood was used in **construction** to make railway
sleepers, bridge decks, house walls, and furniture. Standley and Record (1936) noted
that this species was used in **construction** for railway ties and paneling. Lamb (1946)
noted that the bark was used for **tanning** leather, to which it imparted a yellow-brown
color. This tree is used for making **charcoal** [Horwich and Lyon, 1990].

Terminalia catappa L.

almond, almond tree [English]; *almendra* [Spanish]; *hammon* [Creole]
The fruit and seeds are eaten as a **food** by people and animals because of the good flavor. The "seed" is almond-like in appear-
ance and taste [B1964, B2214; Arvigo and
Balick, 1998].

To treat **headache** and to draw the heat
out of a **fever**, 1 leaf is mixed with a bit of
olive oil or sweet almond oil and bound
to the palms of the hands and the soles
of the feet [Arvigo and Balick, 1998]. For
swelling, 1 leaf is heated, "Vicks" (Vicks
VapoRub® topical ointment) is added to
the leaf, and the leaf is applied as a poultice to the affected area for 2 hours and
repeated if necessary [LR].

To treat **high blood pressure** and
heart trouble, 3 leaves are boiled in 3 cups
of water for 10 minutes and 1 cup consumed before breakfast every other day. It
is said that the dosage should not exceed 3
cups weekly, as the sex drive is thought to
be weakened by consuming more [Arvigo
and Balick, 1998].

Standley and Williams (1962) noted
that the bark and fruit had a high content
of **tannins** and that the seeds were used to
dye textiles in El Salvador.

(top) *Laguncularia racemosa* [DA].

(bottom) *Terminalia catappa* [MB].

CONNARACEAE

Small trees, shrubs, or lianas. Leaves alternate, compound, rarely unifoliolate, impari-pinnate. Inflorescences axillary, pseudoterminal, or terminal, in panicles, racemes, or spikes. Flowers white, light yellow, or light pink. Fruits a follicle, often red. In Belize, consisting of 3 genera and 4 species. Uses for 1 genus and 1 species are reported here.

Connarus lambertii (DC.) Sagot
can shúl, can shèl [Q'eqchi']
As a **poison** for dogs, the leaves and fruits are put in the food, and they "get crazy and die" [B3607]. In ancient times, this plant was said to be used to **poison** people [B3607].

Connarus lambertii [GS].

CONVOLVULACEAE

Herbs, shrubs, lianas, vines, or rarely trees, milky latex present. Leaves alternate, mostly simple. Inflorescences terminal or axillary, of solitary flowers or in clusters. Flowers variously colored, often large and showy. Fruits usually a capsule, sometimes a berry. In Belize, consisting of 10 genera and 50 species. Uses for 4 genera and 14 species are reported here.

Evolvulus alsinoides L.
To treat **venereal diseases**, especially "runny V.D.," defined as "when pus is present," 3 entire plants are boiled in 3 cups of water for 15 minutes and 3 cups consumed daily, before meals, for 9 days [A373].

Ipomoea alba L.
amapola [Spanish]
As a **beverage**, the bark is mashed in water and the liquid consumed as a person would drink water, whenever thirsty [B3699].

Ipomoea batatas (L.) Lam.
sweet potato [English]; *batatas, camote* [Spanish]; *k'aham is, ki'il is* [Q'eqchi']
The cooked tuber is eaten as a **food** [B2325].

To grow **bigger female breasts**, the leaves are mashed with a bit of water on a full moon and applied as a plaster under the bra, left on all day, and repeated for 3 days of each full moon for 3 consecutive months [BW].

To **develop breast milk** in a woman who adopts a baby and wants to nurse, 9 leaves are boiled in water for 10 minutes and consumed warm once daily [HR]. Additionally, the tuber is cooked, grated, and eaten as porridge daily. Milk should begin to flow within 3 days, and the baby is put to breast feed as often as possible during this time [HR].

(top) *Evolvulus alsinoides* [GK].

(bottom left) Flower of *Ipomoea alba* [RH].

(bottom right) *Ipomoea batatas* [MB].

Steggerda (1943) reported that the Yucatan Maya were said to use the leaves of this plant to treat **snakebite**.

Ipomoea carnea Jacq. subsp. *fistulosa* (Mart. ex Choisy) D.F. Austin
blue bell, mauve bell, morning glory [English]
The flowers and stems are used as an **ornamental** indoors and as a trailing **decorative** outdoors [B2459].

For **sprains**, 1 quart of leaves is boiled in 1 gallon of water for 10 minutes, cooled to tepid, and used to bathe the affected area [A358]. Additionally, for sprains, 1 handful of leaves and a pinch of salt are mashed with a bit of water, placed into a cloth, and used as a poultice over the affected area once daily until better [A358].

If you are growing this plant close to your house, it is said to cause the **devil** to come to your house. Sometimes, Pine Ridge flowers (*Hypericum terrae-firmae*) or *ruda* (*Ruta graveolens*) are planted near this plant to keep away the devil [RA].

Steggerda and Korsch (1943) reported that the Yucatan Maya used a bath of warm water with the leaves and vines of this plant to treat **insanity**. However, they specified that before the bath, "the patient must be pricked on the forehead with a rattlesnake's fang."

Ipomoea clavata (G. Don) Ooststr. ex J.F. Macbr.

morning-glory [English]; *gloria de mañana* [Spanish]

For **general ailments**, 1 quart of leaves is boiled in 1 gallon of water for 10 minutes, cooled to tepid, and used as a daily bath at bedtime for 1–3 days [B3227].

It is said that touching the flowers too much will bring on a **fever** [BW].

To treat **swelling** and **sprains**, 9 flowers are crushed by rolling a bottle over them, applied as a poultice to the affected area, and covered by a leaf, with the backside of the leaf pressed against the skin. This treatment is changed once daily until better [HR]. Additionally, the affected area is massaged daily with any type of oil [HR].

Ipomoea heterodoxa Standl. & Steyerm.

chokonob' [Q'eqchi']; *chin na ha* [Mayan]

To treat **headache** or **pain** in the body, 1 large handful of leaves is mashed in 1 gallon of cool water, 1 cup consumed, and the remainder poured over the body or head as a bath once daily, at bedtime or at noon, until better [A164, B2219, B2490].

Ipomoea nil (L.) Roth

According to Steggerda (1943), this vine was gathered for **horse fodder**.

Ipomoea heterodoxa [MB].

(top) *Ipomoea pes-caprae* [RH].

(bottom) *Ipomoea quamoclit* [NYBG].

Ipomoea nil [NYBG].

Ipomoea pes-caprae (L.) R. Br.

cowslip [English]

To **fix dyes** in fabric, 1 quart of chopped stem is boiled in 1 gallon of water for 20 minutes and used to soak cloth [B2661].

For **general ailments**, the flowers, stem, and leaves of the plant are boiled in water for 10 minutes and used as a bath once daily, at bedtime or noon [B2661].

Ipomoea quamoclit L.

pacaya vine [English]; *waiy pim* [Mayan]

As a treatment for unspecified **tumors**, 1 handful of the vine is dried, boiled in 3 cups of water, and 1 cup consumed 3 times daily before meals [A587]. Alternatively, the stem is boiled in 1 gallon of water for 10 minutes and ⅓ cup consumed 3 times daily until better [A587].

Ipomoea sepacuitensis Donn. Sm.

mountain rose, wood rose [English]; *chun* [Yucateca]

This plant is used as **cattle fodder** [B3322].

Ipomoea triloba [PA].   *Merremia tuberosa* [PA].

Ipomoea triloba L.
The entire plant is used for **animal fodder** [B1975].

Ipomoea umbraticola House
camotillo [Spanish]
This plant is used as **cattle fodder** [B3323].

Itzaea sericea (Standl.) Standl. & Steyerm.
green wis [English]
The stem is used as **cordage** to tie packages [B2198].

Merremia tuberosa (L.) Rendle
seven-fingers [English]
This plant is part of a bath mixture for **general ailments** [B3231].

CRASSULACEAE

Herbs, subshrubs, or shrubs. Leaves alternate, opposite, whorled, or in basal rosettes, usually simple, often succulent. Inflorescences terminal or axillary, usually cymes, sometimes panicles or of solitary flowers. Flowers variously colored. Fruits in separate follicles, sometimes a capsule. In Belize, consisting of 1 genus and 3 species. Uses for 1 genus and 2 species are reported here.

Kalanchoe integra (Medik.) Kuntze
This plant is cultivated as an **ornamental** [B2470].

Kalanchoe pinnata (Lam.) Pers.

life everlasting, tree of life [English]; *hoja de aire, hoja de la bruja, quiebra vientos, siempre viva* [Spanish]

The young leaves are eaten as a **food** in salads and eaten with salt **medicinally** [BW; Arvigo and Balick, 1998]. It is said that "to **maintain good health**, one fresh leaf is eaten every day" [Mallory, 1991].

This plant is considered a **panacea**, said to be good for "whatever ails you" [BW; Arvigo and Balick, 1998]. As a curative bath, 9 leaves along with 1 quart of unspecified leaves are boiled in 1 gallon of water for 10 minutes and poured hot over the body once daily at bedtime [A170].

To treat **malaise**, 1 handful of leaves is boiled in 3 cups of water for 10 minutes and 1 cup consumed 3 times daily, before meals, for 3 days [B1825]. For **cough**, **colds**, **sore throat**, **influenza**, and **weakness**, 9 large leaves are mashed in 1 quart of warm water, strained, and ½ cup consumed 6 times daily until better [B2322; Arvigo and Balick, 1998]. To treat **colds** and **congestion**, the leaves are crushed and applied as a poultice to the chest and left on overnight [Mallory, 1991]. To alleviate **wheezing** associated with **asthma**, the leaves are heated, the juice extracted, honey added, and consumed by the spoonful as needed until better [Arvigo and Balick, 1998]. For **headache**, 3 large leaves are mashed with a bit of salt and olive oil, applied as a poultice to the forehead for 1 hour, and repeated if the leaf dries before the headache dissipates [BW; Arvigo and Balick, 1998]. Alternatively, refer to broomweed (*Sida acuta*) [Mallory, 1991]. Steggerda and Korsch (1943) noted that a Yucatan Maya woman treated **nosebleed** by moistening the macerated leaves with vinegar and applying it as a poultice to the forehead of the person.

To reduce and soothe **swelling**, **bruises**, **pain**, **cuts**, and to prevent infection of **skin wounds**, 1 handful of leaves and a pinch of salt are mashed with a bit of water, placed on a cloth, applied as a poultice directly over the affected area, and changed twice daily [B1825, B2322; Arvigo and Balick, 1998]. To cool and heal **burns** or draw the pus out of **boils** or **infected sores**, 9 leaves are mashed with a bit of water, applied as a poultice directly to the affected area, and changed once daily [LR; Arvigo and Balick, 1998]. As a **skin moisturizer**, the leaf juice is applied directly to the skin [LR].

Kalanchoe pinnata [IA].

For **eye infection** or **conjunctivitis (pinkeye)**, 1 leaf is heated, the juice extracted, and 1 drop placed in the eye once daily at bedtime [LR]. For non-insulin-dependent **diabetes** mellitus, 1 leaf is eaten after every meal for 7 days, after which the blood sugar level checked and the treatment repeated until insulin is no longer needed [HR]. To treat **mastitis**, the leaves are mashed with castor oil (*Ricinus communis*) and applied to the breasts [Arvigo and Balick, 1998]. To treat *bazo* (a condition of the spleen) 9 large leaves are washed, chopped, juiced in a blender with water, strained, and 1 cup consumed once daily, in the morning or at bedtime. After 7 days of treatment, a laxative is taken for purging [LR].

To treat **bad winds (*mal viento*)** or **bad spirits**, 1 large handful of leaves and stems is boiled in 2 gallons of water for 10 minutes and used warm as a bath once daily at bedtime [HR]. To **ward off evil and bad energy**, this is planted around your house. If one of these plants gets sick, it is a warning that evil is coming, and the plant is ripped up and thrown in the river [BW]. Steggerda (1943) reported that the Yucatan Maya believed "that if a baby plays with the flowers of the Siempreviva he will have ill luck with raising chicken when he grows older. The flowers, they believe, leave an invisible stain on the hands which later effects the eggs."

CUCURBITACEAE

Vines with coiled tendrils. Leaves alternate, simple, or palmately compound. Inflorescences axillary, of solitary flowers or in cymes, racemes, panicles, or fascicles. Flowers showy, yellow, greenish, orange-yellow, or white. Fruits a berry or capsule. In Belize, consisting of 17 genera and 29 species. Uses for 7 genera and 7 species are reported here.

Cionosicyos excisus (Griseb.) C. Jeffrey

mountain mellon, rat pumpkin, wild squash [English]; *calabasillo* [Spanish]

To treat **arthritis** and **rheumatism**, three 12-inch stems with leaves along with three 12-inch stems with leaves of cow foot (*Bauhinia divaricata*) are boiled in 2 gallons of water for 20 minutes. This is poured into a tub of water and the whole body or affected area soaked for 20 minutes, 3 times weekly, until better [B3242].

For **head lice** (*Pediculus humanus*), the inner flesh from the fruit is removed, massaged into the scalp, the hair tied up with a scarf for 30 minutes, the scalp shampooed, and the lice combed out. Only 1 treatment is applied [HR]. Additionally, after shampooing, 1 handful of chopped banana stem (*Musa acuminata*) is pounded, mashed well in 2 quarts of water, allowed to sit for 1 hour, strained, and used to rinse out the hair. Following this 2-part treatment, it is said that "no more lice will stay in your head ever again" [HR].

For **dog mange**, 1 quart of fresh leaves is mashed in 1 quart of water until the water is very green, and the soaked leaves are rubbed on the dog twice daily for 7 days [B2169].

Caution is used with this plant, as it is known to be toxic and should not be eaten [LR].

Cionosicyos excisus (top left) leaf
and fruits; (top right) leaf and
flower; (lower right) proximal leaf
(left) and distal leaf (right) [WJH].

(bottom) *Cucurbita moschata*
showing fruit [MB].

Cucurbita moschata Duchesne

pumpkin flower, squash [English]; *calabaza* [Spanish]; *k'um* [Q'eqchi']

The fruit is eaten as a **food** known locally as pumpkin or squash [Mallory, 1991].

To treat **jaundice**, the flowers or fruit are boiled in water and an unspecified amount consumed as needed [Mallory, 1991]. To treat **skin blisters**, the leaf or fruit resin is applied to the affected area [Flores and Ricalde, 1996].

Lagenaria siceraria (Molina) Standl.

Thompson (1927) suggested that this plant was one of very few species known to occur in both the Old and New Worlds before the Spanish colonization. He discussed the traditional use of two varieties: *lek*, a large, round cultivar that was "chiefly

Lagenaria siceraria with inset showing fruit [AC].

employed for storing tortillas," and *tšu*, which was very long, had the shape of the number 8, and was used to carry water. Steggerda (1943) reported that the leaves were placed on the stomach of babies who had **diarrhea**.

Luffa cylindrica M. Roem.

luffa, Spanish towel, sponge gourd, sponges [English]

The young fruit is eaten as a **food** [R46].

The coarse, sturdy inside of the fruit is used as a **sponge** after the fruit has ripened, dried, and the skin removed [R46].

Momordica charantia L.

balsam pear [English]; *cundiamor, sorosi* [Spanish]; *sorosee* [Creole]

The young, green, bitter fruits are sautéed and eaten as a **vegetable** with curry [HR; Arvigo and Balick, 1998]. For an additional **food** recipe, refer to turmeric (*Curcuma longa*) [B2113]. The entire plant is fermented to produce an **alcoholic beverage** [Robinson and Furley, 1983].

For **sore throats** and **mouth sores**, the raw leaves are chewed [B1760; Arvigo and Balick, 1998]. To prevent **colds**, **worms**, or **general sickness** in children, 1 green fruit is mashed in ½ cup of water and the juice consumed once daily, 1 day a week [B2124]. As a general household **tonic** given by mothers to their children, one 12-inch piece

(top left) *Luffa cylindrica* [FA].

(top right) *Luffa cylindrica* showing fruit (left) and fibrous mesocarp of fruit used as a sponge (right) [IA].

(bottom) *Momordica charantia* leaf and flower (left) and fruit (right) [MB].

of vine and leaves along with ½ of a grapefruit skin (*Citrus × paradisi*) are boiled in 3 cups of water for 20 minute, and 1 cup consumed 3 times daily as needed [A63, B1760].

For non-insulin-dependent **diabetes** mellitus, **high blood pressure**, early stages of **"cancer" (internal growths), constipation, tiredness, colds, skin ailments, intestinal parasites, amoebas, worms, anemia, menstrual cramps, painful menstruation, delayed menstruation**, as a **tonic** for internal organs, or as a **blood cleanser**, 1 handful of the vine is boiled in 3 cups of water for 10 minutes and 1 cup consumed 3 times daily, before meals, as needed according to the specific malady, for a maximum of 10 days [A63, B1760; Arvigo and Balick, 1998]. To **build the blood** and to treat non-insulin-dependent **diabetes** mellitus, the leaves are boiled in water and an unspecified amount consumed [Arnason et al., 1980].

To treat **skin ailments, skin infection, sores, wounds**, and infestations of **ticks** and **chiggers**, 1 large double handful is boiled in 1 gallon of water for 15 minutes, cooled to tepid, and used as a bath over the affected area [Arvigo and Balick, 1998]. Alternatively, for **tick bites**, one 12-inch piece of vine and leaves is mashed with ¼ teaspoon of salt and applied directly to the skin 1–3 times daily [B1760]. To **prevent mosquito bites**, 1 handful of leaves is boiled in 3 cups of water and consumed [Mallory, 1991]. Dieterle (1976) reported that "Standley wrote: '... In [the] Yucatán the fruit is applied as a poultice to cure **itch, sores**, and **burns.'**"

As a treatment for **malaria**, six 12-inch pieces of stem, vines, and leaves are steeped in 1 quart of rum for 3 days and 1 tablespoon consumed in ½ cup of water 3 times daily as needed until better [HR]. To prevent **malaria**, this same decoction and dosage as described above for malaria is consumed once daily [HR]. Alternatively, one 12-inch piece of vine and leaves along with ½ of a grapefruit skin are boiled in 3 cups of water for 20 minutes and 1 cup consumed 3 times daily until better [B2124]. As a **blood tonic**, 1 cup of the second decoction described for malaria is consumed warm once daily each morning [A63, B2287].

As an **emmenagogue**, three 12-inch pieces of the vine, leaves, and fruits are boiled in 1½ quarts of water and 1 cup consumed every 2 hours, all day long, for 3 days [A63]. For **fever**, 1 cup of the decoction described for emmenagogue is consumed twice daily [LR].

The fruits are used as a **soap substitute**, but it is not clear whether this was a use known in Belize or from elsewhere [Mallory, 1991].

Caution is used with this plant. It is not to be consumed by pregnant women, as it may have abortifacient effects [BW]. The TRAMIL 4 Workshop on Caribbean Medicinal Plants noted the toxicity of the fruit and recommended "to discourage its use" internally [Robineau, 1991].

Psiguria warscewiczii (Hook. f.) Wunderlin

For **tremors** and **shaking**, the leaves and stems are mashed in cold water and used as a beverage and bath until cured [B2483].

Psiguria warscewiczii [RF].

Sechium edule [MB].

Sechium edule (Jacq.) Sw.

cho cho [Creole]; *chayote* [Spanish]; *chi'ma* [Q'eqchi']; *chu um, huisquil* [Mayan]
The fruit is eaten as a **food**, either boiled or grated raw in salads [B2379]. The tuber is potato-like and is boiled and eaten as a **food** [B2379]. Lundell (1938) reported that "the fruits, tender shoots, flowers, and tuberous roots are all **edible**, being boiled and eaten as a vegetable."

To treat **cough** and **colds**, refer to *maravilla* (*Mirabilis jalapa*) [B2283]. For **painful and difficult urination**, 1 fruit with the skin is chopped, boiled in 1½ quarts of water for 20 minutes, strained, and sipped all day until urination normalizes [B2379].

To reduce **high blood pressure**, 1 entire fruit is eaten daily for 9 days (or more, if needed) of each month [B2379]. Alternatively, the fruit is grated, the juice squeezed out, and 1 teaspoon consumed daily for 3 days [LR]. Consuming too much of this fruit is said to cause the blood pressure to drop too much [LR].

DICHAPETALACEAE

Trees, shrubs, or lianas. Leaves alternate, simple. Inflorescences axillary, in fascicles, corymbs, or cymes. Flowers small, whitish. Fruits a drupe. In Belize, consisting of 1 genus and 3 species. Uses for 1 genus and 1 species are reported here.

Dichapetalum donnell-smithii Engl. var. *donnell-smithii*

> The leaves are a component of one of the Maya bath mixtures for **general ailments** [B2636].

DILLENIACEAE

Lianas, creeping shrubs, or sometimes trees. Leaves alternate, simple, often rough to the touch. Inflorescences terminal or axillary, in panicles, fascicles, or thyrses. Flowers white or yellow. Fruits a berry. In Belize, consisting of 5 genera and 10 species. Uses for 3 genera and 5 species are reported here.

Curatella americana L.

> sandpaper tree [English]; *chaparro* [Spanish]; *saha*, *yaha* [Mayan]
>
> The seeds are roasted and eaten as a **food** and were once used to **flavor chocolate** [Marsh, Matola, and Pickard, 1995].
>
> For **skin ailments**, 1 quart of fresh leaves is boiled in 1 gallon of water for 20 minutes and used to bathe the affected area [A341]. For **diarrhea** in babies, 3 cups of leaves are boiled in 1 cup of water for 10 minutes and sipped throughout the day for 1 day; this may be repeated for a second day if necessary [A341].
>
> The wood is made into **charcoal** [Marsh, Matola, and Pinkard, 1995]. Lundell (1938) reported that the Petén Maya used the rough leaves, which contain silica, to **smooth wood** and other materials. Standley and Record (1936) reported that the leaves of this species were used as "fine sandpaper" and that the bark was used for **tanning**.

Davilla kunthii A. St.-Hil.

> cow's tongue [English]; *chaparo, chaparro, lava plato, lengua de vaca, oregano de monte* [Spanish]; *si hob* [Mayan]; *levisa* [Additional name recorded]

(top) *Dichapetalum donnell-smithii* [GS].

(bottom) *Curatella americana* [NYBG].

To expel a **retained placenta** following childbirth, 6 leaves are boiled in 1 cup of water for 10 minutes and 1 cup consumed; only 1 cup is needed [A344]. For **stomach pain**, 3 leaves are boiled in 1 cup of water for 10 minutes and 1 cup consumed hot 3 times daily as needed [B2259].

The rough leaves are used as a **scouring pad** to wash dishes and pots [A230, A289].

Davilla nitida (Vahl) Kubitzki
chaparo, chaparro, lava platos [Spanish]
For **coughs**, 9 leaves are boiled in 3 cups of water for 10 minutes and 1 cup consumed cold with honey twice daily, in the morning and in the evening, until better [B2620].

To expel a **retained placenta** following childbirth, 1 handful of leaves is boiled in 2 quarts of water for 10 minutes and used as a vaginal steam bath immediately after childbirth [B2620; RA].

The leaves are used as a **scouring pad** to wash dishes [RA].

Tetracera volubilis L. subsp. *mollis* (Standl.) Kubitzki
lava platos, lebica, lengua de danto [Spanish]; *sa hab* [Q'eqchi']
The mature, red fruits are placed in kerosene lanterns as a **decoration** [A602].

The rough leaves are used as a **scouring pad** to wash dishes and pots [A602].

Tetracera volubilis L. subsp. *volubilis*
Sarah coat [English]; *lava platos, lavaplato* [Spanish]
Standley and Record (1936) stated that "the Tetraceras are well known in Central America as water vines, the stems yielding **potable sap** when cut."

(top) *Davilla kunthii* [GS].

(middle) *Davilla nitida* [MB].

(bottom) *Tetracera volubilis* subsp. *volubilis* fruit [DA] with inset showing flowers [GS].

To **cleanse wounds**, an unspecified treatment is prepared from this plant [Robinson and Furley, 1983].

The leaves are used as a **scouring pad** to wash dishes [B1830, W1365]. This species is used as **sandpaper** [Robinson and Furley, 1983].

EBENACEAE

Trees and shrubs, often with black heartwood. Leaves alternate, simple. Inflorescences axillary, ramiflorous or cauliflorous, in cymes, fascicles, or of solitary flowers. Flowers white or cream, sometimes greenish or light pink. Fruits a berry. In Belize, consisting of 1 genus and 5 species. Uses for 1 genus and 2 species are reported here.

Diospyros digyna Jacq.
zapote negro [Spanish]; *tauch* [Yucateca]
According to Lundell (1938), this tree grew on ruins in northern Belize, as relics from cultivation. Standley and Record (1936) reported that the mature fruit had "an **edible** pulp of poor flavor, that is soft, black, and of most disgusting appearance." Furthermore, Standley and Williams (1967) stated:

> The immature fruits are intensely bitter and pucker the mouth. The ripe fruits are sweet and **edible**, but the black mushy flesh is repulsive in appearance, reminding one of dirty axle grease. In spite of this, the ripe fruit is much eaten in Mexico and some other countries, and is sometimes made into preserves, or fermented to produce a kind of brandy. The wood is reported to have the qualities of typical ebony. (p. 248)

Diospyros salicifolia Humb. & Bonpl. ex Willd.
guayabillo, *silion* [Spanish]
Steggerda (1943) noted that the fruit of this tree was a particular favorite **food** for parrots.

The wood is used for **firewood** [B2189]. Steggerda (1943) reported that the Yucatan Maya burned this wood as **firewood** in their cooking ovens, used it to make **lime**, and said that it was "rarely used for house **construction**, as it disintegrates very rapidly."

Diospyros digyna [IA].

(left) *Erythroxylum guatemalense* [MB].

(right) *Acalypha arvensis* [MB].

ERYTHROXYLACEAE

Shrubs or small trees, evergreen or deciduous. Leaves alternate, usually distichous, simple. Inflorescences axillary, with few to many fascicles or of solitary flowers. Flowers greenish, white, or yellowish-white. Fruits a drupe. In Belize, consisting of 1 genus and 4 species. Uses for 1 genus and 1 species are reported here.

Erythroxylum guatemalense Lundell

redwood, ridge redwood, swamp redwood [English]

According to Standley and Record (1936), the **wood** of this shrub or small tree was "hard, heavy, fine-textured, irregularly grained, highly durable."

EUPHORBIACEAE

Herbs, shrubs, trees, vines, and lianas. Leaves usually alternate, rarely opposite, simple, entire to palmately lobed when compound, then palmate. Inflorescences terminal or axillary, sometimes cauliflorous or ramiflorous, in cymes. Flowers not usually very showy. Fruits usually a schizocarp, sometimes a berry or drupe. In Belize, consisting of 34 genera and 113 species. Uses for 20 genera and 42 species are reported here.

Acalypha arvensis Poepp.

cancer bush, cancer herb, cat tail [English]; *cordiemiento, hierba del cáncer, yerba del cáncer* (female) [Spanish]; *mis xiv* [Mopan]

The common names *hierba del cáncer* and cancer herb reflect the local use of the word "**cancer**," meaning a type of **open, oozing sore** that does not heal, as in an ulcer [Arvigo and Balick, 1998]. This term is also used to describe **internal growths** or **tumors.**

For **toothache**, 1 handful of the leaves is boiled in 1 liter of water for 10 minutes and used as a warm gargle 3 times daily as needed [B1792, B2301]. To reduce **swelling**, 3 entire plants are boiled in 1 gallon of water for 10 minutes and used as a warm bath once daily in the evening [Arvigo and Balick, 1998]. To reduce stubborn **swelling**, refer to dog's tongue (*Psychotria tenuifolia*) [Arvigo and Balick, 1998].

To treat **chronic skin ailments**, such as **rash, blisters, peeling skin, deep sores, ulcers, fungus, ringworm, inflammation**, or **labial itching and burning**, 1 entire plant is boiled in 1 quart of water for 10 minutes, strained, and used very hot as a wash over the affected area 3 times daily [Arvigo and Balick, 1998]. For **chronic skin wounds**, such as **skin infection, ulceration, sores, boils**, and **fungus**, ½ quart of fresh leaves and ½ quart of fresh *tres puntas* leaves and flowers (*Neurolaena lobata*) are mixed, toasted, sieved through a screen into a powder, and applied to the wound twice daily as needed until better [B1792; Arvigo and Balick, 1998]. To treat **leg wounds**, 1 quart of fresh leaves is boiled in 1 gallon of water for 10 minutes and poured as hot as can be tolerated over the affected area [B1792]. Additionally, castor oil (*Ricinus communis*) is applied to the sore, and the dried, sieved powder described above for skin wounds sprinkled over the sore twice daily until better [B1792]. Standley and Steyermark (1949) reported that in Guatemala, it was "used commonly in treating **sores, cutaneous and venereal diseases**, and the **bites** of poisonous animals."

For **stomach pain**, 1 entire plant is boiled in 2 quarts of water for 10 minutes and 1 cup consumed warm 4 times daily as needed [B2301]. To treat **stomach ailments** or **urinary infection**, 1 medium-sized plant is chopped, boiled in 3 cups of water for 10 minutes, and 1 cup consumed slowly and as hot as possible 3 times daily before meals [B1972; Arvigo and Balick, 1998]. For **intestinal gas** causing stomach pain, honey or brown sugar is added to the decoction described above for stomach ailments or urinary infection and 1 cup consumed hot 3 times daily before meals [Arvigo and Balick, 1998].

The "male" counterpart to this plant is *yerba del cáncer* (*Acalypha setosa*) [B1972].

Acalypha diversifolia Jacq.

costilla del danto [Spanish]; *kok' uk'ub'* [Q'eqchi']

For **toothache**, 1 leaf and a pinch of salt are mashed in a few drops of water and applied as a poultice to the affected tooth until better [B2656]. To stop the **bleeding** of a wound within minutes, 9 leaves are mashed in ¼ cup of water and applied on a cloth as a poultice directly over the affected area [B3520].

Acalypha mortoniana Lundell

mahava, moho [Spanish]; *mahaua* [Mayan]

To treat a "**fever blister or cold sore** on the gum," one 2-inch piece of the root is mashed in ¼ cup of water and applied directly to the affected area, usually for 3 days [A820].

The bark strips are used as **cordage** in the forest to tie roofing, baskets, and packages and for **weaving** [A755]. The young stems are used in **carpentry** and to make kitchen utensils [A820].

(left) *Acalypha diversifolia* [MB].

(right) *Amanoa guianensis* [NYBG].

Acalypha setosa A. Rich.

yerba del cáncer (male) [Spanish]

For **toothache** or **headache**, 1 handful of leaves is boiled in 1 liter of water for 10 minutes and used as a warm gargle 3 times daily as needed [B1791, B2302].

For **internal ulcers** or **stomach pain**, 1 entire plant is boiled in 2 quarts of water for 10 minutes and 1 cup consumed hot 4 times daily as needed [B2302]. Alternatively, for **stomach ailments**, 1 medium-sized plant is chopped, boiled in 3 cups of water for 10 minutes, and 1 cup consumed slowly and as hot as possible 3 times daily before meals [B1791].

For **skin wounds**, such as **chronic skin infections**, **ulceration**, or **fungus**, ½ quart of fresh leaves and flowers and ½ quart of fresh *tres puntas* leaves and flowers (*Neurolaena lobata*) are mixed, toasted, sieved to a powder, and applied directly to the affected area twice daily as needed until better [B1791]. To treat **leg wounds**, 1 quart of fresh leaves is boiled in 1 gallon of water for 10 minutes and poured as hot as can be tolerated over the affected area [B1791]. Additionally, castor oil (*Ricinus communis*) is applied to the sore and the dried, sieved powder described above for skin wounds sprinkled over the sore twice daily until better [B1791].

The "female" counterpart to this plant is cancer herb (*Acalypha arvensis*) [B1791].

Amanoa guianensis Aubl.

swamp icaco [English]; *lemonario* [Spanish]

Standley and Record (1936) reported that the wood of this tree was hard, heavy, and strong, and while there were no local uses reported, it was "probably suitable for **tool handles**."

Chamaesyce blodgettii [MB].

Bernardia interrupta (Schltdl.) Müll. Arg.

Waika ribbon [English]

Standley and Record (1936) reported that the **wood** of this species, though not resistant to decay, was moderately hard, strong, and easy to work and "probably suitable for tool handles."

Chamaesyce blodgettii (Engelm. ex Hitchc.) Small

chicken weed [English]

For **malaria**, 1 entire plant is boiled in 3 cups of water for 30 minutes and 1 cup of the strong decoction consumed twice daily until better [B1957].

To **repel lice and fleas**, the entire plant is placed in the nests of domestic fowl. However, there is an insect that inhabits this plant that is toxic to fowl, so it is necessary to keep that insect from eating the plant [B1957].

Chamaesyce hirta (L.) Millsp.

chicken weed [English]

Steggerda (1943) reported that the milky sap from this plant was used "consistently from Colonial times to the present to alleviate **sore eyes**." A number of healers he consulted with stated that the sap from this plant was applied directly into the eyes to reduce **inflammation**. Other uses from the Yucatan Maya that he reported included consuming the boiled leaves to treat **dysentery** and collecting 3–4 plants to be made into a tea as a diuretic for **urinary and kidney ailments**, including **kidney stones**.

To treat **insect bites**, the stem latex is applied to the affected area by the Yucatan Maya of Mexico [Flores and Ricalde, 1996].

Steggerda and Korsch (1943) reported that the pulverized leaves of this plant were applied as a poultice to the forehead of a child to treat **spasms**.

Chamaesyce hypericifolia (L.) Millsp.
chicken weed (male) [English]; *golondrina, hierba de zapo* [Spanish]

To treat **cough** and **loosen the mucus** from the chest, 1 handful of stems and leaves is boiled in 2 cups of water for 5 minutes and consumed warm twice daily, in the morning and in the evening, for 2–3 days until better [LR].

For non-insulin-dependent **diabetes** mellitus and **venereal diseases**, such as **gonorrhea**, 2 entire plants along with one 6-inch piece of papaya root (*Carica papaya*) and one 1-inch × 6-inch piece of pine wood (*Pinus caribaea* var. *hondurensis*) are boiled in 2 quarts of water for 10 minutes, with 3 cups consumed daily for 7 days, after which a doctor is consulted [A315].

For **kidney stones** or **urine retention (stoppage of water)**, one 12-inch branch with leaves, an unspecified amount of "female" chicken weed leaves (*Chamaesyce mendezii*), and an unspecified amount of *tan chi* leaves (*Capraria biflora*) are boiled in 3 cups of water for 10 minutes and 1 cup consumed 3 times daily, before meals, until the stones pass [B2120]. Alternatively, for **urinary ailments**, refer to little bamboo (*Commelina erecta*) [B2118]. To treat **constipation**, **urine retention (stoppage of water)**, or **kidney stones**, refer to *tan chi* (*Capraria biflora*) [B2116].

(top) *Chamaesyce hypericifolia* [MB].

(bottom) *Chamaesyce lasiocarpa* [GS].

Chamaesyce hyssopifolia (L.) Small
chicken weed (female) [English]; *wild piss-a-bed* [Creole]; *yòok mukuy* [Yucateca]

To treat all types of **sores**, freshly squeezed sap from the leaves and stems is applied twice daily, left on the sore all day, and repeated for as many days as needed until better [B3604]. To treat **eye soreness**, the sap is applied directly into the eye [Arnason et al., 1980].

Chamaesyce lasiocarpa (Klotzsch) Arthur
red herb [English]

To dissolve **warts**, the stem latex is applied directly to the skin until better [A561].

Chamaesyce mendezii (Boiss.) Millsp.

chicken weed (female) [English]

For **kidney stones** or **urine retention (stoppage of water)**, refer to "male" chicken weed (*Chamaesyce hypericifolia*) [B2119, B2120; RA]. For **urinary ailments**, refer to little bamboo (*Commelina erecta*) [B2118].

Chamaesyce mesembryanthemifolia (Jacq.) Dugand

chicken weed [English]

For non-insulin-dependent **diabetes** mellitus, 1 entire plant is boiled in 3 cups of water for 5 minutes and 1 cup consumed at room temperature 3 times daily, before meals, every other day as needed [A420].

Cnidoscolus aconitifolius (Mill.) I.M. Johnst.

Standley and Record (1936) reported that the young leaves of this plant were cooked and eaten as a **food**, like spinach (*Spinacia oleracea*). Lundell (1938) noted that the leaves were boiled for **food**, being served as a vegetable, a use that continues in the present times. Steggerda (1943) reported that the Yucatan Maya used the sap of this species as **mucilage** and in the treatment of **urinary diseases**. He also noted that a branch of this particular tree was used to beat some of the **evil spirits** that take the form of cows, goats, pigs, and sheep.

To treat **mouth blisters** and **fever**, the leaf latex is used by the Yucatan Maya, but no further details were provided [Flores and Ricalde, 1996]. According to Thompson (1927), one of the **legends** of the San Antonio (Toledo District) Maya involving the creation of the sun, moon, and Venus, explained how this plant became **edible**. According to the legend, Lord Xulab, who was later to become the planet Venus, was out hunting, and after capturing animals, his hands were bloody. To cleanse them, he wiped his hands on this plant, which then became edible. Lord Xulab did so "that the people might have more to eat to replace the tame animals that were no more."

Cnidoscolus chayamansa McVaugh

chaya [Spanish]

The leaves are eaten as a **food** just as one would eat spinach—for example, raw, cooked, or steamed [B2399].

To **build the blood**, the leaves are boiled in water with salt or cooked in oil with onion or garlic and then eaten [Mallory, 1991]. This plant is high in iron and is often eaten to treat **anemia** [B2399].

For **increasing breast milk**, 12 leaves are added to 1 quart of soup and 1 bowl consumed twice daily [LR].

To treat non-insulin-dependent **diabetes** mellitus, 9 large leaves are boiled in 1 quart of water for 10 minutes, with 1 cup consumed after each meal for 7 days, after which the doctor is consulted. This decoction is consumed regularly to control non-insulin-dependent diabetes mellitus [HR].

To treat **rheumatism** and **warts**, the thorns and glandular juice is used by the

Yucatan Maya, possibly by brushing the person with plant parts containing thorns, but the exact method of administration was not described [Flores and Ricalde, 1996].

Cnidoscolus multilobus (Pax) I.M. Johnst.

tea [English]; *tza* [Additional name recorded]

To stop **hemorrhaging** or **heavy menstruation**, the root is used in an unspecified treatment [B3689, W1839]. To treat **arthritis** and **headache**, the leaves are mashed and the juice massaged onto the affected area [B3689, W1839].

Croton aff. *arboreus* Millsp.

sak pa [Mayan]

The fruit is eaten as a **food** by birds [A499].

To **stimulate the appetite** and **aid in digestion**, 1 quart of chopped, sour fruit and ¼ teaspoon of black pepper (*Piper nigrum*) are covered by equal proportions of vinegar and water, soaked for 10 days, and eaten [A499].

(top) *Cnidoscolus chayamansa* [MB].

(bottom) *Cnidoscolus multilobus* [MB].

Croton argenteus L.

higuerella [Spanish]

For **vaginal discharge**, 1 entire plant is chopped, boiled in 1 gallon of water for 10 minutes, and used as a vaginal steam bath for 20 minutes, repeated every 2 weeks as needed [A517].

Croton niveus Jacq.

uvitas [Spanish]

For **burns**, 1 quart of leaves is mashed in ½ cup of water until soft, and the green juice is applied as a poultice to the affected area twice daily as needed [A387]. Steggerda and Korsch (1943) reported that to treat **stomach pain**, the Yucatan Maya mixed the scraped bark and an unspecified part of *Zanthoxylum fagara* in hot, salty water and consumed an unspecified amount twice daily.

Croton schiedeanus Schltdl.

wild cinnamon [English]; *chi' che', zununkil che'* [Q'eqchi']; *iri skutě* [Yucateca]

For **adult diarrhea**, 1 small handful of bark is boiled in 1 liter of water and ½ cup of the cooled decoction consumed 3 times daily, before meals, until better [B2525]. To **build the blood**, the leaves are boiled in water and used as a bath [Arnason et al., 1980].

Croton xalapensis Kunth

pito sico, Santa María [Spanish]; *chal che* (male) [Mopan]

To treat **joint pain**, fresh leaves are heated in oil and applied as a poultice directly to the affected area [Arvigo and Balick, 1998]. For **fever** and **pain**, 3 branches with leaves are mashed, soaked in 2 gallons of water, set outside where there is partial sun and partial shade (e.g., in a lightly branched area or under a slitted chair), the mixture warmed slowly, ½ cup consumed, and the remainder used as a bath once daily for 3 consecutive days [Arvigo and Balick, 1998]. For **burns**, 1 quart of mashed leaves is steeped in 1 gallon of water for 30 minutes and used as a bath at room temperature over the affected area twice daily [B1912].

To assist in **healing** after childbirth, to **ease menstruation**, and to treat **uterine ailments**, 1 small handful of leaves is boiled in 3 cups of water for 5 minutes and 1 cup consumed 3 times daily [Arvigo and Balick, 1998]. To **cleanse the genitals**, **prevent postpartum infection**, **ensure that internal organs return to their proper positions**, or **staunch prolonged bleeding** following childbirth, 1 handful of leaves is boiled in 2 quarts of water for 10 minutes and used as a vaginal steam bath for 30 minutes [B1912; Arvigo and Balick, 1998]. Often following childbirth and before the vaginal steam bath, a midwife will give the new mother a traditional uterine massage [Arvigo and Balick, 1998].

It is believed that the most effective use of *chal che* occurs when the "male" and "female" plants are used in combination. The "female" plant is *pito sico* (*Pluchea carolinensis*) [B1912].

Croton xalapensis
[GS].

Dalechampia schippii Standl.

la salud [Spanish]

The leaves of this plant are a component of a Maya bath mixture for **general ailments**.

As a **general analgesic** for relief of pain, 9 leaves are steeped in 3 cups of hot water for 20 minutes and 1 cup consumed tepid 3 times daily until better [B2251].

Drypetes brownii Standl.

bullhoof, bullhoof macho, wild monkey apple [English]; *succoutz* [Mayan]; *luín* [Additional name recorded]

The ripe fruit is eaten as a **food** [A474, A824].

To treat **foot fungus**, the leaves are toasted, powdered, and 1 quart sprinkled twice daily to cover the feet until better [A474].

Standley and Record (1936) noted that the stem of this tree was used in **construction** to make railroad ties and house frames and that it was also suitable for tool handles and similar implements. Lamb (1946) reported that the wood was used in local house **construction**, such as for posts, rafters, beams, and windowsills, and to make shafts for agricultural tools. The wood is still used today in **construction** and is prized as "extra strong lumber that lasts a long time" [B3212].

(top) *Dalechampia schippii* [MB].

(bottom) *Drypetes brownii* [RF].

Euphorbia armourii Millsp.

chicken weed [English]

The plant is a component of one of the Maya bath mixtures [A322].

This plant is used in the same manner and to treat the same ailments as "male" chicken weed (*Chamaesyce hypericifolia*) [A322]. For **cough**, **kidney stones**, **constipation**, **urinary ailments**, and **urine retention (stoppage of water)**, refer to "male" chicken weed (*Chamaesyce hypericifolia*) [A322].

For non-insulin-dependent **diabetes** mellitus and **venereal diseases**, such as **gonorrhea**, 2 entire plants along with one 6-inch piece of papaya root (*Carica papaya*) and one 1-inch × 6-inch strip of pine wood (*Pinus caribaea* var. *hondurensis*) are boiled in 2 quarts of water for 10 minutes and 3 cups consumed daily [A322].

Euphorbia heterophylla L.

redhead, wild poinsetta [English]; *flor de navidad* [Spanish]

For **sinus infections**, 1 small handful of fresh leaves and stems is boiled in water for 5 minutes and applied in a piece of cloth as a warm poultice over the face twice daily for 5 days [C19].

To treat **warts** and **insect bites**, the stem latex is applied to the affected area [Flores and Ricalde, 1996].

To **induce lactation** in nursing women whose milk has stopped flowing, 3 entire plants are steeped in 2 quarts of water and used to bathe the breasts for 9 evenings [A54]. Additionally, 9 entire plants are braided together and worn as a wreath around the new mother's neck [A54]. However, this plant is **toxic** and must not be consumed in any manner. This treatment has no toxic effects on the baby because the breasts are first washed before trying to nurse [A54].

Euphorbia lancifolia Schltdl.

ixbut [Mayan]

Rosengarten (1978) reported that this plant was a traditional **galactagogue** (a substance that stimulates the flow of milk in nursing women) in Belize, Guatemala, and Honduras. Regarding the folklore and traditional information on this species as well as chemical data, he noted:

Euphorbia lancifolia [MB].

> In this connection, one still hears countless tales about the wondrous powers of ixbut as a galactagogue, some of which border on the sensational; for example, there are numerous claims that aged grandmothers, or even great-grandmothers, after taking ixbut, have been able to suckle newly born infants through their withered breasts when the young mothers died in childbirth. (pp. 277–8)

Standley and Steyermark (1949) mentioned this plant being effective in **stimulating milk production** in cattle: "[I]t is said to double the quantity of milk given by cows that eat it." As Rosengarten (1978) mentioned in his paper, a Guatemalan product known as "GALAC-LATEX," which had this plant as the primary ingredient, was sold as a supplement for cattle feed during the early 1900s. He listed numerous

Euphorbia pulcherrima [MB].

Euphorbia tirucalli [MB].

other cases of using this plant to promote milk production in animals and, as above, in humans.

Euphorbia pulcherrima Willd. ex Klotzsch

wild poinsetta [English]; *flor de pascua del monte* [Spanish]; *pas cua* [Mayan]

The plant is cultivated as an **ornamental** [B2319].

To **increase lactation** and to relieve **swelling** of the breasts, 9 entire plants are braided in a necklace that is worn for 3 days [Arvigo and Balick, 1998]. Additionally, 9 entire plants are boiled in 1 gallon of water for 10 minutes and used as a warm bath around the breasts twice daily [Arvigo and Balick, 1998].

Standley and Steyermark (1949) reported that to relieve **pain**, the leaves were applied as a poultice directly to the affected area. To remove **body hair**, the milk of the stem is applied to the skin [Arvigo and Balick, 1998].

Caution is used with this plant, as it is toxic and must not be consumed [Arvigo and Balick, 1998].

Euphorbia tirucalli L.

horse tail [English]; *cola de caballo*, *esqueleto* [Spanish]

This plant is cultivated as an **ornamental** [A351].

To reduce **swelling**, 1 quart of leaves is boiled in 1 gallon of water for 20 minutes and used as a bath over the affected areas [A351]. For **hernia**, the leaves are mashed and applied with a cloth as a poultice to the affected area all day long, usually for 9 days [A351].

Gymnanthes lucida showing fruits (left) and inflorescences (right) [PA].

Gymnanthes lucida Sw.
false lignum-vitae [English]
Standley and Record (1936) noted that the wood was hard, heavy, strong, and used in the **construction** of walking sticks, handles, and articles of turnery.

Hippomane mancinella L.
manchineel [English]
Standley and Record (1936) reported that the milky latex caused **dermatitis**, resulting in blistering and swelling of the skin. To treat **contact dermatitis** caused by this plant, refer to John Charles (*Hyptis verticillata*) [Arvigo and Balick, 1998].

Standley and Record (1936) also noted that the wood was excellent in the **construction** of furniture and cabinetwork and that the fruits of this tree are **toxic**.

Jatropha curcas L.
hazel nut, physic nut [English]; *piñon* [Spanish]; *sakilte'* [Q'eqchi']; *xkakalche* [Mayan]
To relieve **headache**, the leaf is rubbed directly on the head [B2111]. For **baby's thrush**, the clear sap from the stem and leaf ends is mixed with 1 cup of sterile water and used as a mouthwash [B2111]. To treat **mouth sores** and **baby's thrush**, the clear sap from the stem and leaf ends is rubbed directly on the mouth membranes [Arvigo and Balick, 1998]. For **mouth sores** of the gum and throat, 1 leaf is boiled in 1 cup of

(top) *Hippomane mancinella* with inset showing fruit [RH].

(bottom) *Jatropha curcas* [MB].

water for 5 minutes and used as a mouthwash as needed [Arvigo and Balick, 1998]. According to Standley and Steyermark (1949), this sap was applied to **wounds** or **sores** to promote healing and placed in the teeth to relieve the pain of **toothache**.

For "**heat**" in the body that causes **rash**, **pimples**, or **boils**, 9 leaves are boiled in 1 gallon of water for 10 minutes, cooled, and 2 cups consumed daily until 1 gallon is finished. The decoction is refrigerated between doses, and it is a good sign when the urine becomes as clear as water [LR]. To treat **back pain**, **ovarian inflammation**, **urine retention (stoppage of water)**, **constipation**, and **hot flashes** caused by menopause in women or anger in men, 1 leaf is boiled in 1 cup of water for 5 minutes and 1 cup consumed as needed until better [Arvigo and Balick, 1998]. To treat **stomach pain**, the leaves or one 1-inch piece of peeled stem is boiled in water, cooled, and consumed [Mallory, 1991].

To treat **arthritis**, the roots of 9 young trees are chopped, soaked in 2 quarts of water all day, and sipped throughout the day, that day and then every other day, until better [Arvigo and Balick, 1998]. Additionally, refer to prickly pear (*Opuntia cochenillifera*). For **spleen ailments**, 3 leaves and 9 chopped young limes (*Citrus aurantiifolia*) are boiled in ½ gallon of water for 10 minutes and consumed daily in place of water [Arvigo and Balick, 1998]. For **vaginitis**, one 3-inch × 3-inch piece of bark and 6 leaves are boiled in 2 cups of water for 5 minutes and used as a vaginal douche [Arvigo and Balick, 1998].

To **cleanse the intestines**, 1 seed is dried, ground, boiled in 1½ cups of water for 5 minutes, and ½ cup consumed at a time as needed. The dosage is as such to see how the person reacts, as it can be a very powerful **laxative** [B2111; LR; Arvigo and Balick, 1998]. Standley and Steyermark (1949) noted that around Cobán, Guatemala, nursing mothers sometimes placed the heated leaves on their breasts with the belief that this would **increase the flow of milk**.

To treat **insect bites** and **skin sores**, the latex is applied to the affected area [Robinson and Furley, 1983].

Caution is used, as this plant is a strong purgative and treatment may be toxic to children, the aged, or the weak [B2111; LR; Arvigo and Balick, 1998].

Standley and Steyermark (1949) reported that in Guatemala, the seeds were pressed to yield clear oil used as **fuel** for lanterns and for **soap making**. They noted that the oil had very strong **purgative** properties and the seeds were **dangerous** or even **fatal** when eaten, "at least in the case of small children." They also described a very curious use of this species in Mexico

> as a host plant for certain lac [a resinous secretion]-producing scale insects known by the name Axi or Axin . . . The lac thus produced is highly esteemed as varnish for finishing guitars and other articles of wood. In Guatemala, an infusion of the leaves is used commonly by some of the Indians, for setting the dyes of cotton and perhaps other textiles. (p. 127)

Jatropha integerrima Jacq.
The plant is cultivated as an **ornamental** for its red flowers [B2473].

Manihot esculenta Crantz
cassava, white cassava [English]; *yuca, mandioca* [Spanish]; *tz'in* [Q'eqchi']
The tuber is eaten as a staple **food** of Belize and is grown extensively in traditional cultivation systems that include corn, beans, and squash [R95].

Margaritaria nobilis L.f.
clawberry [English]; *bastard madre cacao, mato palo, mora, ramón macho* [Spanish]; *ininche* [Mayan]
The wood is used in **construction** and **carpentry**, such as for carvings and posts [A829].

(top) *Manihot esculenta* with inset showing tubers [MB].

(bottom) *Margaritaria nobilis* [PA].

Pedilanthus tithymaloides [MB].

Phyllanthus liebmannianus [MB].

Pedilanthus tithymaloides (L.) Poit.

zapato de la virgen [Spanish]; *but il* [Mayan]

To reduce **swelling**, **infections**, and **inflammation of the breasts (mastitis)**, 1 handful of the leaves is mashed in 2 cups of water and applied to the breasts as a poultice twice daily until better [B2472]. To treat an erupted **swelling** between the toes, 1 handful of leaves is mashed in 2 cups of cold water and used at room temperature as a wash over the affected area twice daily, once in the morning and once in the afternoon [B1911].

To **ward off evil spirits**, the leaves along with the leaves of *gumbolimbo* (*Bursera simaruba*) are mashed in water, steeped in the sun for 1–3 hours, a sip is consumed, and the remainder used as a bath [RA]; alternatively, refer to *gumbolimbo* [HR]. To treat someone possessed by **evil spirits**, refer to pine (*Casuarina equisetifolia*) [BW].

Phyllanthus acidus (L.) Skeels

wild plum [English]

Standley and Record (1936) reported that the fruit of this tree, though very sour, was eaten as a **food**, particularly as preserves.

Phyllanthus grandifolius L.

monkey rattle [English]; *pix thon* [Yucateca]

Steggerda (1943) reported that the Yucatan Maya used this plant for bathing babies who had **diarrhea** and that it was said to be particularly effective on Mondays and

Fridays. It was also noted as a treatment for **evil eye** (*mal ojo*). Children made a **toy** from the fruit. In addition, Steggerda also reported that the leaves were used as a bath to treat **pellagra** (caused by a vitamin deficiency, usually a lack of niacin). A **ceremonial** bath with the boiled leaves was given to a child when they were weaned.

Phyllanthus liebmannianus Müll. Arg.

ojo [Spanish]; *chin chin pol ojo* [Mayan]

To treat **stomatitis**, **internal infections**, **kidney stones**, or **urine retention (stoppage of water)**, 1 entire plant is boiled in 3 cups of water for 10 minutes, strained, and 1 cup consumed 3 times daily before meals [B1761; Arvigo and Balick, 1998]. For **abdominal pain**, 9 stems are boiled in 3 cups of water for 10 minutes and 1 cup consumed 3 times daily before meals [B2461].

To treat green-colored **diarrhea** in infants (**meconium**), several handfuls of leaves are placed in a pot of water, steeped in the sun all day until warm, and poured all over the body once daily until better [B2285]. To treat **red lips**, **thirst**, and **green feces** in babies caused by the evil eye (*mal ojo*), which is said to be evident in babies because one eye is smaller than the other, one 12-inch long branch is boiled in 3 cups of water for 10 minutes and 1 ounce consumed at room temperature 4 times daily until the stool is no longer green [A948]. This plant is used as a treatment because the position of the flower resembles an eye [A948].

To treat the **evil eye** (*mal ojo*), **constantly crying babies**, or **sick babies**, 1 entire plant is boiled in 3 cups of water for 10 minutes, strained, and used as a warm bath once daily, at bedtime, for 3 consecutive nights [Arvigo and Balick, 1998]. Additionally, for the **evil eye**, 7 branches are gathered and placed in a cross formation over the baby's wrist and prayers said simultaneously. This is repeated for 7 consecutive days, and if the situation does not improve, the treatment is extended for a total of 9 days [Arvigo and Balick, 1998].

Ricinus communis L.

castor bean, castor plant [English]; *iguerra* [Spanish]; *š kotč* [Yucateca]

There are two forms of this plant found in Belize. The one having "whitish" branches is said to be preferable for medicinal use in the household [Arvigo and Balick, 1998]. This plant is the source of **castor oil** as well as the deadly **poison** known as ricin [MB].

For **pain** in the body, **headache**, **head colds**, or **fever**, 1 leaf and 1 tablespoon of castor oil are mixed, warmed over a flame, and applied as a poultice to the affected area of the body and left overnight [A837; Arvigo and Balick, 1998]. For **headache** and **sinus congestion**, 1 leaf, a small amount of "Vicks" (Vicks VapoRub® topical ointment), and a cloth are applied to the forehead for 1 hour or, if the treatment occurs at bedtime, overnight [Arvigo and Balick, 1998]. To aid in **tooth extraction**, the leaves are used in an unspecified way [JB117].

To treat **rheumatism**, 1 leaf with salt and sweet almond oil is heated, applied as a poultice to the affected area, wrapped up, and left on overnight. This is repeated for 7 consecutive nights, followed by a 7-day rest period, and then another 7 nights of treat-

Ricinus communis [IA].

ment. During this entire treatment, cold-water baths are avoided [Arvigo and Balick, 1998].

To eradicate **tapeworm**, refer to coconut (*Cocos nucifera*) [Arvigo and Balick, 1998]. To treat **ringworm**, refer to trumpet tree (*Cecropia peltata*) [A938]. To remove **intestinal worms**, refer to *epazote* (*Chenopodium ambrosioides*) [A153]. Similarly, to treat **intestinal parasites**, refer to *epazote* [Arvigo and Balick, 1998]. To prevent **malaria** and to eliminate **parasites**, refer to *tres puntas* (*Neurolaena lobata*) [R52].

To treat a **wound**, 1 leaf is placed face down on the affected area as a bandage [B2176]. To reduce **swelling**, 1 leaf is placed against the affected area [B2176]. Alternatively, refer to *Coccocypselum herbaceum* [B2613]. To treat a **cut**, **sore**, or **swelling**, fresh leaves are crushed, mashed into a paste, and applied directly to the affected area [Arvigo and Balick, 1998]. To treat a **wound** or **sore**, refer to *pito sico* (*Clibadium arboreum*) [A593], arnica (*Montanoa speciosa*) [B2091], and nightshade (*Solanum torvum*) [B2144]. To treat a **leg wound**, refer to cancer bush (*Acalypha arvensis*) [B1791] and *yerba del cáncer* (*Acalypha setosa*) [B1792]. For **sores**, refer to button wood (*Conocarpus erecta*) [B1949].

To alleviate **itching** and **prevent scarring** on children with **measles**, 5 large leaves are boiled in 2 gallons of water for 10 minutes and used as a bath [Arvigo and Balick, 1998]. To treat **skin ailments**, refer to *pega ropa* (*Teucrium vesicarium*) [A496]. For serious **skin ailments**, refer to jackass bitters (*Neurolaena lobata*) [B1762]. To treat **leishmaniasis (baysore)** or any **stubborn sores**, refer to wire wis (*Cydista diversifolia*) [A133].

To treat **infant bleeding**, **hernia**, or **infections**, refer to *coronilla* (*Cissus verticillata*) [A404]. For **stiff knee joints**, refer to *coronilla* [A404]. To treat **"cancer" (internal growth)** in the abdomen, 1 fresh leaf is heated over a flame, castor oil rubbed on it, and the leaf applied warm to the affected area and left on all day and all night [HR].

For **constipation**, 1 young shoot is peeled, cut into 3-inch pieces, dipped in sweet almond oil, and inserted about 1 inch into the anus; a bowel movement should follow in approximately 10 minutes [Arvigo and Balick, 1998]. As a **purgative**, ½ leaf and 5 dried seeds are boiled in 2 cups of water for 5 minutes and consumed warm [A837; Arvigo and Balick, 1998]. Similarly, as a **laxative**, ½ leaf and 2 dried, roasted seeds are boiled in 2 cups of water for 5 minutes and consumed warm [Arvigo and Balick, 1998].

For **women's ailments**, 9 leaves along with 9 sweet orange leaves (*Citrus sinensis*), and 9 *ob'el* leaves (*Piper auritum*) are boiled in 1 gallon of water for 10 minutes and used as a vaginal steam bath for 20 minutes; usually, only one treatment is needed

[B2176]. To reduce **breast cysts**, to treat **breast "cancer" (internal growth)**, or to **dry up the milk** when a mother cannot or will not nurse, 1 fresh leaf is heated over a flame, castor oil rubbed on it, and placed warm against the breast and left on all day and all night [HR, LR]. Steggerda (1943) reported that the Yucatan Maya woman who wished to **stop the flow of milk** and wean her children "selects 13 small twigs of the X-koch [*škotč—Ricinus communis*] and ties it around her neck." To treat **mastitis**, refer to life everlasting (*Kalanchoe pinnata*) [Arvigo and Balick, 1998].

For **snakebite**, refer to *cramantee* (*Guarea glabra*) [LR].

Caution is used with this plant, as the seeds are extremely toxic and must be properly prepared before ingesting any remedy prepared with them [Arvigo and Balick, 1998].

Standley and Steyermark (1949) reported that the oil "is highly esteemed as a **lubricant**, in **soap manufacture**, in **dyeing and printing** cotton goods, and for **dressing** tanned hides."

Sapium macrocarpum Müll. Arg.

To treat **warts**, the stem latex is used by the Yucatan Maya of Mexico but no further details were provided [Flores and Ricalde, 1996].

Sebastiania tuerckheimiana (Pax & K. Hoffm.) Lundell

poison-wood, ridge white poison-wood, white poison [English]; *chemchem blanco*, *reventadillo* [Spanish]

Based on the various local names, it is likely that this species could cause **dermatitis** if the latex comes in contact with the skin [MB].

Standley and Record (1936) noted that the timber of this tree was not utilized but was probably useful in the **construction** of tool handles.

Sebastiania tuerckheimiana with inset showing white latex exuding from cut stem [MB].

Tragia yucatanensis Millsp.

p'oop'os [Yucateca]

To treat **burns**, the leaf is crushed and placed on the affected area [Arnason et al., 1980]. For **rheumatism**, the leaf is wrapped around the affected area with its stinging hairs penetrating the skin [Arnason et al., 1980].

Steggerda (1943) noted that the Yucatan Maya "formerly used this plant to **whip** their children when they were naughty, especially if they were inclined to run

Tragia yucatanensis [WJH].

away from home. Rubbing the leaf on an aching part of the body causes an **irritation** which feels good." He also noted that the roots were boiled and consumed to treat **gonorrhea** and that the water from the boiled roots consumed to treat **stomach pain**.

FABACEAE: SUBFAMILY CAESALPINIOIDEAE

Trees, shrubs, sometimes climbing, or rarely herbs. Leaves usually pinnate, sometimes bipinnate. Inflorescences axillary or terminal. Flowers organized into a butterfly shape. Fruits a legume. In Belize, consisting of 15 genera and 63 species. Uses for 11 genera and 28 species are reported here.

Bauhinia divaricata L.

cow hoof, cow foot, goat hoof [English]

To prepare a **tonic**, the leaves and stems are boiled with "Wizard oil," a locally produced medicinal oil (more commonly used in the past), and the decoction consumed [B2181]. Steggerda (1943) reported that a tea made from the flowers of this species and mixed with sugar was effective in treating **dry cough** and that **liver and kidney ailments** were treated with a boiled leaf decoction.

As a first-aid treatment for **snakebite**, 1 large handful of a 9-leaf bath mixture (including this plant) is boiled in 2 quarts of water for 5 minutes and used as hot as possible as a bath 3 times daily

Bauhinia divaricata [MB].

until a person can find specialized attention for the bite. This treatment is a first-aid remedy only [B2101].

For **arthritis**, 1 handful of a 9-leaf bath mixture (including this plant) is boiled in 2 gallons of water for 10 minutes, ½ cup consumed, and the remainder poured warm over the affected area once daily for 3 days of each week as needed [B3240]. To treat **arthritis** and **rheumatism**, refer to wild squash (*Cionosicyos excisus*) [B3242].

The fibrous bark is harvested and used as **cordage** for tying beams, rafters, and posts of houses together [Horwich and Lyon, 1990].

Bauhinia erythrocalyx Wunderlin
pata de vaca [Spanish]
To prevent **postpartum infection**, 1 double handful of leaves is boiled in 1 gallon of water for 10 minutes and used once as a vaginal steam bath for 20 minutes on the third day after childbirth [A480].

Bauhinia herrerae (Britton & Rose) Standl. & Steyerm.
cow foot, cow foot vine [English]; *pata de vaca* [Spanish]; *kibix* [Mopan]
The leaves are a component of traditional bath mixtures used to treat many **general ailments** [Arvigo and Balick, 1998].

For **rheumatism** and **wind (*viento*)**, 1 large handful of a 9-leaf mixture that includes this plant is boiled in 2 gallons of water for 10 minutes and used warm as a bath or hot as a steam bath once daily, at bedtime, as needed, usually 3 times a week. Following this, the person is wrapped up with towels or blankets [B2463, BW].

To treat **headache**, 1 handful of leaves is mashed in 1 quart of water, steeped in the sun for 1 hour, and used to wash the head [Arvigo and Balick, 1998].

To treat **dysentery**, to stop **bloody diarrhea**, and to reduce **internal hemorrhaging** caused by ulcers, injuries, or heavy menstruation, 1 handful of chopped vine is boiled in 3 cups of water for 10 minutes and ½ cup consumed 6 times daily until the bleeding stops [B1816, B2178; Arvigo and Balick, 1998]. For **chronic diarrhea**, refer to guava (*Psidium guajava*) [B2138]. For **vaginal hemorrhage** or **internal bleeding** due to falls or wounds, 2 cups of the decoction described above for dysentery are consumed in 30 minutes [Arvigo and Balick, 1998]. For **bleeding or infected wounds**, this same decoction is used as a wash over the affected areas [Arvigo and Balick, 1998].

To treat **uterine tumors**, 1 handful of chopped vine along with one 6-inch piece of *negrito*

Bauhinia herrerae [MB].

bark (*Simarouba glauca*) are boiled in 3 cups of water for 10 minutes and ½ cup consumed as hot as possible 3 times in the morning. In the afternoon, 1 handful of chopped vine along with one 6-inch piece of the skunk root (root) (*Chiococca alba*) are boiled in 3 cups of water for 10 minutes and ½ cup consumed as hot as possible 3 times in the afternoon. This 6-doses-per-day treatment is continued until tumors pass vaginally, not to exceed 30 days [B1898, B2220].

As a **female contraceptive**, 1 handful of chopped vine is boiled in 3 cups of water for 10 minutes. The key to its efficacy as a contraceptive comes from the dosage consumed. For 6 months of infertility, 3 cups are consumed warm daily during the first 3 days of the menstruation. For 12 months of infertility, this dosage is repeated for a second menstruation. For 18 months of infertility, the dosage is repeated for a third consecutive menstruation. Though a 3-month treatment is the recommended limit, a woman may safely resume this process after abstaining from the treatment for 6 months. Consumption of this tea for 9 consecutive menstrual cycles will produce irreversible **sterility** for the woman [B1898, B2220; Arvigo and Balick, 1998]. It is said that for this treatment to be an effective contraceptive, the woman must not consume fresh fruits during her menses, which is the traditional custom of many villages [RA].

As a **male contraceptive** or to "**dry up the testicles**," 3 cups of the decoction described above for the female contraceptive are consumed over 3 consecutive days every month for a total of 9 cups per month [B1848, B2220]. For a purge that can reverse and "wash out" this contraceptive in men, refer to Billy Webb (*Acosmium panamense*) [B1848, B2220].

Standley and Steyermark (1946) reported that the Yucatan Maya employed the bark to make **cordage** for tying the crossbars and roof timbers of houses and have since ancient times. The bark was peeled from the stem, doubled over on its inner surface, and rolled up for storage. For cordage, the rolls were moistened with water to become pliable for use.

Bauhinia ungulata L.
cow foot tree [English]
To prepare a bath for children to treat a **variety of ailments**, 1 handful of leaves is boiled in 2 gallons of water for 10 minutes and used warm at bedtime as needed [W1853]. Steggerda (1943) reported that the Yucatan Maya made a tea from the leaves to treat **urinary disorders** and **diarrhea**.

Steggerda (1943) also reported that the Yucatan Maya used the young stems as **cordage** to tie together the poles forming the walls of their houses.

Caesalpinia gaumeri Greenm.
axe master [English]; *quiebra hacha, ruda del monte* [Spanish]
For **anemia**, 1 handful of bark is boiled in 3 cups of water for 10 minutes and 1 cup consumed 3 times daily for 9 days [A477]. For "**old age**" and **forgetfulness**, the same decoction as described for anemia is consumed as needed [A477].

The trunk is used in **construction** for house posts [A523].

Caesalpinia gaumeri [GS].

Caesalpinia pulcherrima [MB].

Caesalpinia pulcherrima (L.) Sw.

bird of paradise flower, flambeau flower, virgin's flower [English]; *flor de la virgen, irritación, sinquin colorado* [Spanish]; *can zink in, chink in, zink in* [Mopan]

The green seeds are eaten as a **food**. A good-quality **honey** is yielded from the flowers [Arvigo and Balick, 1998].

To treat **headache**, the leaves are crushed, mixed with rubbing alcohol, and placed directly on the forehead [Arvigo and Balick, 1998]. To treat **arthritis**, 1 double handful of leaves is boiled in 2 gallons of water for 5 minutes and used warm as a bath over the affected area once daily as needed [A319].

To treat *irritación* in infants (characterized by **fever**, **swollen belly**, **cold hands and feet**, **sweat**, and **diarrhea**), 1 large double handful of leaves is squeezed into 1 gallon of hot water, steeped in the sun all day, used warm as a bath for 3 nights, and ¼ cup consumed by the infant following each bath [Arvigo and Balick, 1998]. To treat **sadness** (*tristeza*) and **grief** (*pesar*) in adults and children, the same infusion as described for *irritación* is used as a bath [Arvigo and Balick, 1998]. For **calming people close to death**, the leaves and flowers are used to prepare a bath [Arvigo and Balick, 1998]. Alternatively, for *irritación*, **sadness** (*tristeza*), **grief** (*pesar*), and for people close to **death**, 7 branches are held in a cross formation on the wrist of the sick person and prayers said [Arvigo and Balick, 1998].

For **evil eye** (*mal ojo*), the leaves along with *ruda* leaves (*Ruta graveolens*) are boiled in water and used as a face wash [Arnason et al., 1980].

To treat **fever** in infants, the leaves are rubbed on the head and the chest, as this plant is considered "cooling" [Arnason et al., 1980].

Children with **bad tempers** are "lashed" with a young branch to stop them from misbehaving. Following the lashing, the switch is thrown over the roof of the house at noon [Arvigo and Balick, 1998].

Steggerda (1943) reported that the Yucatan Maya of Mexico used different parts of this tree to treat **dysentery**, **ulcers**, **scanty or lacking menstruation**, and **venereal diseases**. One healer he spoke to discussed drinking a tea from the roots as a **purgative**, a second described a tea made from the flowers for treating **bronchitis** and **lung trouble**, while a third recommended mixing the leaves with water and giving it to a person with **diarrhea**.

Standley and Steyermark (1946) noted that the flowers are fragrant and yield good-quality **honey**, the fruit was used for **tanning**, and in Guatemala, people mashed the leaves for use as a **fish poison**, throwing them into a stream to stupefy the fish.

This tree is commonly planted in **cemeteries** [Arvigo and Balick, 1998]. The flowers are used as a **decoration** on home altars to **honor the dead** [Arvigo and Balick, 1998]. On All Soul's Day, the flowers are gathered and used to decorate the home to honor the deceased [Arvigo and Balick, 1998].

Caesalpinia yucatanensis Greenm. var. *yucatanensis*
chink in [Mopan]; *can chin* [Mayan]

For a baby with **measles** or **fever**, the leaves and flowers of nine 12-inch branches are boiled in 2 gallons of water for 10 minutes and used warm as a bath once daily, at bedtime, until better [A734].

Cassia grandis L.f.
beef-feed, stinking toe [English]; *carqué* [Spanish]; *bu kút* [Mopan]; *b'ukut* [Q'eqchi']; *carao* [Additional name recorded]

The ripe fruit is eaten as a **food** [B3357, B2050]. The flowers smell like chocolate [B3357]. Horwich and Lyon (1990) noted that the fruiting pods have an unpleasant odor, hence the name "stinking toe."

For **measles** or a **light fever**, 1 handful of leaves is boiled in 2 gallons of water and used as a bath at bedtime until better [A630]. To treat **ringworm**, **fungus**, or other **skin ailments**, the juice is squeezed from the young leaves and applied directly to the affected area 3 times daily for 2 days [A630; Arvigo and Balick, 1998]. To treat **vitiligo** (a **skin pigmentation disorder**) and **skin ailments**, the leaf is rubbed on the affected area [Arnason et al., 1980].

For non-insulin-dependent **diabetes** mellitus, the leaves are soaked in water and consumed regularly [Arvigo and Balick, 1998]. Alternatively, 2 handfuls of fresh leaves and young stems are mashed in 2 quarts of cold water and steeped for 1 night, with 2 quarts entirely consumed once daily, in the morning before eating, and repeated for 20 days. If the first dose causes severe diarrhea, the dosage is reduced to 1 quart [C89].

To treat **anemia**, **tiredness**, and **malaise**, the seeds are removed from the pods, dark juice is strained from the seed pulp and mixed with water or milk (1:1 ratio; ½ pulp juice, ½ other liquid), and 1 cup consumed daily [Arvigo and Balick, 1998]. For an "iron rich" **blood tonic**, 3 pods from the fruit are placed in a damp spot and allowed to soften, the dark brown sticky fluid extracted from the pods, 2 quarts of water mixed with it, and 3 cups consumed daily until the blood improves [B2050]. To **build blood** and as a **vitamin source**, the seeds are dried, roasted, powdered, 1 teaspoon added to 1 cup of water, steeped, and consumed as a pleasant **beverage**, like coffee, for 9 days [B3203]. Robinson and Furley (1983) reported that the fruits were made into a medicine to treat **low blood pressure** and that a poultice made from the leaves was used to treat **fever**. As a tonic to **build the blood**, the fruit is soaked in water to loosen the dark resin (the aril) around the seeds, which is then consumed [Arnason et al., 1980].

As a **blood cleanser** and a **uterine cleanser**, 3 handfuls of leaves are boiled in 1 gallon of water for 10 minutes and used as a bath at bedtime, once weekly, for 3 months [BW]. To treat **kidney ailments**, **water retention**, **back pain**, or **biliousness**, 3 small branches with leaves are boiled in 3 cups of water for 10 minutes and sipped all day instead of water as needed [Arvigo and Balick, 1998]. For **liver biliousness**, refer to *achiote* (*Bixa orellana*) [LR]. To eliminate **toxins** from the body tissue and as a **diuretic**, ½ cup of fresh leaves is soaked in 3 cups of water and consumed as needed [Arvigo and Balick, 1998].

(top) Cassia grandis [MB].

(bottom) Branch and seedpod of *Cassia grandis* [MB].

As a mild **laxative** or **blood tonic**, ½ cup of fresh leaves is boiled in 1 cup of water for 2 minutes and consumed as a tea [Arvigo and Balick, 1998]. To prepare a **laxative**, 10 leaves are boiled in 1 cup of water for 10 minutes, strained, and consumed; this is not to exceed 2 cups per week [B2050]. For a **laxative** treatment and to **loosen stools** resulting from constipation, 1 handful of leaves is boiled in 1 quart of water for 5 minutes and 1 cup consumed once daily, although this is not a purge [LR]. Mallory (1991) reported that a tonic made from the sweet pulp and consumed "cleans one out" when a person feels **malaise** or **constipation**. For **postmenopausal women**, 1 handful of leaves is boiled in 1 quart of water for 10 minutes and sipped all day for 2 days each month for 3 months, starting when the woman first notices cessation of her period [BW].

The leaves are used as a **soil fertilizer** [BW]. Leaves are laid on the ground to rot or mixed with wood ashes until rotten and then used as **compost** [BW]. The ash from the wood of this tree is used in making **soap** [Horwich and Lyon, 1990].

Chamaecrista nictitans (L.) Moench var. *jaliscensis* (Greenm.) H.S. Irwin & Barneby
sim sim [Mayan]

To treat **headache**, 1 small handful of leaves is boiled in 2 quarts of water for 10 minutes and used tepid to bathe the head at bedtime or as needed [A138].

This is considered to be the "male" counterpart of the *hotz* plant (*Chromolaena odorata*) [A138].

Delonix regia (Bojer ex. Hook.) Raf.
flamboyante, guacamay [Spanish]

The large tree is used as a common **ornamental** [A607, B2266].

Tasty **honey** is collected from the hives of bees that feed on the pollen [A607, B2266].

Chamaecrista nictitans var. jaliscensis [NYBG].

Delonix regia [MB].

Dialium guianense (Aubl.) Sandwith

ironwood [English]; *paleta* [Spanish]

Lamb (1946) reported that the fruits were said to be eaten as a **food** with a flavor similar to that of tamarind. Standley and Record (1936) reported that the fruits are said to be an important **food** for animal species.

This tree was one of the most valued timbers in **construction** for making corner posts and house frames [Horwich and Lyon, 1990]. Standley and Record (1936) reported that the timber was used in **construction** for railroad ties, fence posts, and parts for logging trucks. They also reported that the wood was hard, heavy and strong, highly resistant to decay and insects, and valued for heavy and durable **construction**, including repairs to logging-cart wheels.

Haematoxylum campechianum L.

logwood [English]; *palo tinta* [Spanish]; *ek* [Additional name recorded]

Logwood was one of the very important early economic plant species of Belize, being valued as a **dye** for the clothing industry. It was logged by the Spanish and English buccaneers in the 16th and 17th centuries, known as "haematoxylon dye," and exported in great quantity at a substantial price. With the development of synthetic dyes in the 1700s, prices for this tree dropped, and other species, such as the mahogany (*Swietenia macrophylla*), became important export products [Marsh, Matola, and Pickard, 1995].

Hymenaea courbaril L.

broken ridge locust, locust, South American copal [English]; *guapinol* [Spanish]

The taste of the pulp that surrounds the seed is said to resemble that of a banana [Horwich and Lyon, 1990].

Dialium guianense [NYBG].

Haematoxylum campechianum [MB].

Hymenaea courbaril [MB].

Standley and Record (1936) reported that a pale yellow or reddish gum was produced from the trunk, known in commerce as South American copal. Occasionally, this was found buried in the soil and dug up as "fossil" gum. The resin was used to make **varnish** and **incense**. As with many members of this family, the sweet pulp around the seeds was eaten as a **food**. The wood was strong and used for heavy and durable **construction** and wheel-wright work.

Schizolobium parahyba (Vell.) S.F. Blake

quam, quamwood [English]; *plumajillo, tambor, zorra* [Spanish]

The wood of this species was used as a "light timber" for house **construction**, although it was not very durable [Robinson and Furley, 1983]. Standley and Record (1936) reported that during Colonial times, this wood was studied for its potential as a source for pulp in making **paper**. The conclusion was that the pulp was of good strength and quality, with a good yield, and could be used to produce paper of satisfactory quality. This tree was considered an important secondary **hardwood** species and was once exported as lumber to the United Kingdom [Munro, 1989].

Senna alata (L.) Roxb.

Christmas candle, shrimp flower [English]; *barraja, hoja de barajas* [Spanish]; *piss-a-bed* [Creole]

To treat **liver congestion, liver spots, kidney ailments,** and to **purge the lymph**

Schizolobium parahyba (top) trunk with buttresses (left), leaves (upper right), and tree in flower (lower right) [RF].

(bottom) *Senna alata* with inset showing close-up of open fruit [MB].

system, 3 flowers and 9 leaves are boiled in 2 quarts of water for 10 minutes and ½ cup consumed tepid for 3 days [Arvigo and Balick, 1998].

To treat **urinary tract ailments**, 3 bunches of flowers are boiled in 3 cups of water for 2 minutes, steeped for 15 minutes, and a total of 3 cups sipped all day [Arvigo and Balick, 1998]. To treat **constipation**, 3 dried flower heads (or 9 leaves when no flowers are available) along with 1 ounce of anise seed (*Pimpinella anisum*) and 1 ounce of chamomile flowers (*Chamaemelum nobile*) are boiled in 1 quart of water, reduced to ½ quart, and ½ cup consumed twice daily as needed. Beans, pork, cold foods, and cold baths are avoided during the treatment [Arvigo and Balick, 1998].

As a **uterine cleanser** to counteract the papaya seed (*Carica papaya*) sterility treatment, 2 tablespoons of the root along with 1 garlic clove are boiled in 1 pint of water for 10 minutes, and 1 cup consumed twice weekly, at bedtime, until conception, which should occur within 3 months. This remedy is said to be effective only if the papaya treatment has been used for less than 2 years [Arvigo and Balick, 1998]. For female **infertility**, 1 cup of mashed roots are soaked in 1 quart of *anisado* (anise liquor [*Pimpinella anisum*]) for 5 days and 1 ounce consumed daily for 10 days before the onset of menstruation [Arvigo and Balick, 1998].

Standley and Record (1936) reported that this species was used to make an ointment to treat **ringworm**.

Senna atomaria (L.) H.S. Irwin & Barneby
barba de jolote [Spanish]; *Xtuab* [Yucateca]
Steggerda (1943) noted that local people used the leaves to treat **nosebleeds** by shredding the leaves and inhaling the aroma, which was believed to stop the hemorrhage.

Steggerda (1943) also reported that the wood of this tree was used in **construction** to build local houses in the Yucatan region.

Senna hayesiana (Britton & Rose) H.S. Irwin & Barneby
John Crow bead [English]; *dulcesillo, frijolillo, frijolillo del monte* [Spanish]; *karab'ans che'* [Q'eqchi']; *la mush* [Mayan]
The leaves are a component in Maya bath mixtures [A141].

For **general pain**, the leaves are steeped in hot water and used as a healing bath [B2135]. To treat **headache**, **bone pain**, and **toothache**, 1 pound of fresh leaves is boiled in 1 cup of water, then consumed warm and/or used warm as a bath 3 times daily for 3 days [C24].

For mild **anemia**, locally known as a person who is "half pale," 1 large handful of leaves is mashed, boiled in 2 gallons of water for 5 minutes, 5 tablespoons consumed hot, and the remainder used cool as a bath [B2497]. For babies who have **difficulty falling asleep**, the decoction described for anemia is used as a bath and the flowers placed under the baby's pillow [A333].

Senna obtusifolia (L.) H.S. Irwin & Barneby
To treat **hepatitis**, 2 large handfuls of fresh leaves are mashed in water and used as a bath [C76]. Additionally, 1 handful of fresh leaves is boiled in an unspecified amount of water for 15 minutes and ½ cup consumed 2–3 times daily for 5–7 days [C76].

Senna occidentalis (L.) Link
rat bean, yama bush [English]; *frijolillo* [Spanish]; *llana bush* [Creole]; *buul che, ya me bu* [Mopan]
To treat a child or adult who is **tired**, **weak**, or **exhausted**, especially due to illness or old age, the leaves are toasted, ground, mixed with egg white and brandy, and applied

as a poultice to the wrist pulse [Arvigo and Balick, 1998]. For **heart weakness**, the leaves are mashed, mixed with egg and rum, wrapped in a cotton cloth, and laid as a poultice against the skin, directly over the heart, for 1 hour daily [Arvigo and Balick, 1998].

As a **heart tonic, blood thinner**, and for **depression**, 1 entire plant is chopped, boiled in 2 quarts of water for 5 minutes, and ½ cup consumed 3 times daily for 3 days followed by 3 days of no treatment. This 6-day cycle is repeated until

Senna occidentalis [MB].

better [A801, B1899]. As a **heart tonic** and for many other **ailments**, 1 entire plant is boiled in 3 cups of water for 10 minutes and 2 cups consumed daily [Arvigo and Balick, 1998]. For **heart ailments** and as a **coffee substitute**, the fruits are collected, dried, roasted, ground, mixed with water, and consumed several times during the day as a beverage for 9 days [BW; Arvigo and Balick, 1998].

For **fever** and **influenza**, 1 root is boiled in 2 cups of water for 10 minutes and consumed warm twice daily [Arvigo and Balick, 1998]. To treat **chills**, 1 leaf is warmed with a small amount of oil and rubbed over the body; this leaf is considered very "warm" [Arnason et al., 1980].

To treat **hepatitis**, 1 handful of fresh leaves and 3 pieces of fresh root along with 3 pieces of fresh yellow-wood bark (*Morinda panamensis*) the size of one's hand from the trunk and wood are boiled in water for 10 minutes and consumed warm 4 times daily for 4 days [C50].

For **painful menstruation**, the roots of 1 plant are boiled in 3 cups of water for 10 minutes and 3 cups consumed hot before meals [Arvigo and Balick, 1998]. As a **pregnancy test**, women urinate on the leaves; midwives say that a baby is on the way if the leaves look "scorched" [Arvigo and Balick, 1998].

To treat a child with a **bad temper**, the branches are used for spanking and are then thrown away [Arvigo and Balick, 1998].

Senna pendula (Humb. & Bonpl. ex. Willd.) H.S. Irwin & Barneby
musty currents [English]
To treat **skin ailments** and "**itchy babies**," 1 large handful of leaves is boiled in 2 gallons of water for 10 minutes and the entire decoction used as a bath once daily until better [A353].

Senna peralteana (Kunth.) H.S. Irwin & Barneby
The seeds are eaten as a **food** [A505].

Senna racemosa (Mill.) H.S. Irwin & Barneby var. *racemosa*
yellow poy [English]; *cante* [Spanish]; *cassia* [Additional name recorded]
The wood is used for **fuel** and in **construction** to make house posts [A66].

Senna reticulata (Willd.) H.S. Irwin & Barneby
To treat an **internal infection** of any organ or system, 9 leaves are boiled in 1 quart of water for 5 minutes and ½ cup consumed 6 times daily until better [R92].

Senna racemosa var. *racemosa* [DA].

Senna reticulata [MB].

Senna sophera (L.) Roxb.

As a **heart tonic** and to make a **coffee substitute**, the seeds are collected, roasted, ground, and 1 teaspoon mixed with 1 cup of water, which is then consumed [B3096].

Senna spectabilis (DC.) H.S. Irwin & Barneby var. *spectabilis*

piss-a-bed [Creole]

Mallory (1991) reported that to treat **colds**, a tea was made from 2 dry leaflets and that these same leaflets were boiled and eaten as needed for a **purge**. The same author also reported that for male **virility**, the root was eaten raw.

Senna undulata (Benth.) H.S. Irwin & Barneby

John Crow bead, John Crow bean [English]; *karab'ans che'* [Q'eqchi']; *dulia* [Additional name recorded]

The fruit is eaten as a **food** by birds [B3484].

This is one of the many plants in Belize used in bath mixtures to treat a **variety of ailments** [B2683].

To treat **jaundice**, 9 leaves and 10 flower clusters are steeped in 3 cups of water for 20 minutes and 1 cup consumed 3 times daily until the jaundice has passed, usually in 3 days [B2683].

Senna uniflora (Mill.) H.S. Irwin & Barneby

Christmas candle, shrimp flower [English]; *barraja* [Spanish]; *piss-a-bed* [Creole]

As a **general analgesic**, 1 large handful of leaves is steeped in 2 gallons of water for 20 minutes and poured over the head as a bath at bedtime [B2195].

Standley and Steyermark (1946) reported that to treat **ringworm**, **scabies**, or other **skin diseases**, leaf juice was applied directly to the affected area. To treat **liver congestion**, **liver spots**, **kidney ailments**, and to **purge the lymph system**, 3 flowers and 9 leaves are boiled in 2 quarts of water for 10 minutes and ½ cup consumed tepid for 3 days [BW]. For **liver biliousness**, refer to *achiote* (*Bixa orellana*) [LR].

For children who **wet their beds**, 3 flowers or one 6-inch piece of root (if there are no flowers) are boiled in 1 quart of water for 10 minutes and ½ cup consumed after school for 3 consecutive days [BW]. Additionally, 1 piece of green stem is placed on the fire, and as it develops foam while it burns, this foam is rubbed on the belly button every 10 minutes until the foaming from the wood stops. This treatment is repeated twice. During this 2-part treatment, the child should not eat anything all day long and is kept warm with a blanket or clothing [BW].

As a **purge**, 4 flowers and 3 leaves are boiled in 1 quart of water for 10 minutes and ½ cup consumed cool as needed. There is usually a stool after the first dose [BW].

To **cleanse the uterus** after childbirth, 3 flowers along with 1 teaspoon of anise seed (*Pimpinella anisum*) and 1 teaspoon of dried rosemary leaves (*Rosmarinus officinalis*) are boiled in 1 quart of water for 10 minutes and sipped all day on the third day following childbirth. It is said that the clots and mucus will pass after 1 quart is consumed [BW].

Tamarindus indica [MB].

Tamarindus indica L.

tamarind [English]; *tamarindo* [Spanish]

To relieve **headache**, the leaves are crushed and applied as a poultice to the head [Arvigo and Balick, 1998]. For **sore throat**, the leaves are soaked in water and used as a gargle [Arvigo and Balick, 1998]. Alternatively, the bark is used to prepare a gargle [Arvigo and Balick, 1998]. For a **wound**, **boil**, or **rash**, a leaf decoction is made using 1 small handful of leaves boiled in 1 quart of water for 10 minutes, allowed to steep until it gets to room temperature, and then used as a wash over the affected area twice daily [Arvigo and Balick, 1998]. For **scabies**, the leaves are mashed, made into a paste, and applied as a poultice over the affected area [Arvigo and Balick, 1998]. To treat a **boil** or **skin ulcer**, the leaves are dried, powdered, and sprinkled over the affected area [Arvigo and Balick, 1998]. To treat an **ulcer**, the bark is dried, ground into a powder, and dusted onto the affected area [Arvigo and Balick, 1998].

To help **regulate and cleanse the system**, the pulp of the ripe fruit is eaten, as it contains the nutrients calcium, phosphorous, and iron [Arvigo and Balick, 1998]. To relieve **biliary colic** and as a **laxative**, 1 cup of the pulp of the ripe fruit is soaked in 1 quart of water and consumed as a beverage all day until it is finished [Arvigo and Balick, 1998]. For **morning sickness**, a piece of pulp is chewed with a bit of salt and black pepper [Arvigo and Balick, 1998].

For **scorpion sting**, the bark is mashed into a paste and used as a poultice over the affected area [Arvigo and Balick, 1998].

It is said that sleeping under this tree will cause **ill health** [Arvigo and Balick, 1998]. Additionally, it is reported that few garden plants can grow under or near this tree, as the roots are **toxic** to other plants [Arvigo and Balick, 1998].

Steggerda and Korsch (1943) reported that the Yucatan Maya treated **insanity** by giving a person a beverage made from the fruit of this plant or *Cassia fistula* for 9 days along with a dram (⅛ ounce) of *achiote* (*Bixa orellana*) mixed with the person's urine before breakfast in the morning.

FABACEAE: SUBFAMILY MIMOSOIDEAE

Trees, shrubs, lianas, or rarely herbs. Leaves usually bipinnate, sometimes pinnate, the leaflets often numerous and small. Inflorescences axillary or terminal, often in spikes or capitate, often showy. Flowers small. Fruits a legume. In Belize, consisting of 20 genera and 79 species. Uses for 12 genera and 28 species are reported here.

Acacia collinsii Saff.

black cockspur [English]; *zubín* [Spanish]; *sub'in* [Q'eqchi']

For **cough**, 9 thorns along with nine 2-inch pieces of bay cedar bark (*Guazuma ulmifolia*) are boiled in 1 cup of water for 5 minutes and consumed orally [A342]. Alternatively, 9 black stinging ants that live in the thorns are squeezed into 1 cup of water and consumed orally [A342].

To treat **nervousness**, 1 handful of bark is boiled in 2 cups of water for 5 minutes and ⅓ cup consumed 3 times daily until better [LR].

For **bloody urine**, particularly in men, 9 thorns and 9 leaves along with one 6-inch × 3-inch piece of the waha leaf (*Calathea lutea*) and 9 tea box leaves (*Myrica cerifera*) are boiled in 1 quart of water for 10 minutes and 1 cup consumed 3 times daily before meals [A342; RA].

To treat **snakebite** while in the bush, 1 piece of bark the length of the person's forearm is peeled, chewed, placed on the bite as a poultice, and secured with a bandage until help is available [B2187, B3320].

Acacia cornigera (L.) Willd.

cockspur [English]; *cuerno de vaca* [Spanish]; *zubi* [Mopan]; *sub'in (k'ix)* [Q'eqchi']

The seeds inside the ripe pod are eaten as a **food**, especially by children [Arvigo and Balick, 1998].

For **skin ailments** such as **acne**, one 2-inch × 2-inch piece of bark is boiled in 3 cups of

Acacia cornigera [MB].

water for 10 minutes and used as a wash over the affected area [B1860]. To treat **skin rash** and **skin infection**, the fresh bark is used to prepare a salve that is applied to the affected area [A951]. To treat **fungus, leishmaniasis (baysore)**, and **hemorrhage**, the bark is roasted, powdered, and applied directly to the affected area [A951]. For **hemorrhage**, 4 ounces of bark are boiled in 3 cups of water for 10 minutes, cooled, and sipped all day long [A951].

To treat **headache, poisoning, cough, lung congestion**, and the onset of **asthma** attacks, 9 thorns containing ants are boiled in 3 cups of water for 10 minutes and 1 cup consumed 3 times daily before meals [Arvigo and Balick, 1998]. For **headache**, the thorns are gently pricked along the forehead until there is a slight burning sensation [A951]. Additionally, the bark is roasted, powdered, and sprinkled on the forehead following the thorn treatment [A951]. To prevent an **asthma** attack, 9 thorns are boiled in 1 cup of water for 10 minutes and sipped constantly until relieved [B1860].

To treat **infantile catarrh**, 9 ants collected from the thorns are squeezed into ½ cup of boiled water, strained, and 1 teaspoon consumed every 30 minutes for 1 day until finished [B1860; Arvigo and Balick, 1998]. For **infections**, one 9-inch piece of bark, 9 branches with leaves about 6-inch long, 9 thorns, and the ants that live in them are boiled in 2 quarts of water for 10 minutes and used cool as a bath [A951].

To **prevent swelling of the prostate gland**, one 9-inch piece of bark is boiled in 2 cups of water for 10 minutes and ½ cup consumed 4 times daily, before each meal and at bedtime, 2–3 times weekly for 2 weeks of every month [A951]. For an unspecified **prostate gland treatment**, nine 6-inch branches with leaves are boiled in 2 cups of water for 10 minutes and ½ cup consumed 4 times daily, before each meal and at bedtime, 2–3 times weekly for 2 weeks of every month. Because of their spines, the branches are more difficult to use than the bark [A951]. For male **impotence**, one 2-inch × 2-inch piece of bark is boiled in 3 cups of water for 10 minutes and 1 cup consumed 3 times daily, before meals, for 7 days. If results are slow, the strength of the tea can be doubled and the treatment repeated for 3 more days [B1860; Arvigo and Balick, 1998].

To treat **snakebite**, 1 piece of bark equal in size to the length of the person's forearm is cut, chewed, the bark juices swallowed, and the leftover bark fibers applied as a poultice to the bite [A951, B1860; Arvigo and Balick, 1998]. Additionally, while walking to get help, the root is chewed and the root juices swallowed. This 2-part procedure is said to slow the toxic effects of the bite by 3–8 hours, allowing extra time to get help [A951, B1860; Arvigo and Balick, 1998]. Alternatively, refer to *cordoncillo blanco (Koanophyllon albicaule)* [A814]. For **snakebite** or **bee sting**, the sap is applied to the bite or sting and a piece of the inner bark used as a bandage to cover the affected area [A951].

To counteract **bad luck** or **envy (envidia)**, 1 large piece of bark is boiled in 1 gallon of water for 10 minutes, 1 cup consumed, and the remainder used as a bath on a Friday [Arvigo and Balick, 1998].

Acacia dolichostachya S.F. Blake

junco wood, wild tamarind [English]; *tamarindo del monte* [Spanish]; *guin, jesmo* [Additional names recorded]

For **tanning** leather, the bark is mashed in water and used with white lime in the preparation and preservation of the hide [A492, A840]. Wood from the larger trees is used in **construction** to make marimba keys [A492, A840]. The wood is also used in **construction** for making houses [A828].

Acacia farnesiana (L.) Willd.

cashaw, cuntich [Additional names recorded]

To treat **toothache**, the stem resin was used by the Yucatan Maya of Mexico [Flores and Ricalde, 1996]. Steggerda and Korsch (1943) reported that the "juice" was used to relieve **toothache**.

Standley and Record (1936) stated that this plant was cultivated in Southern Europe, where its flowers (called "cassie flowers" in commerce) were harvested to make **perfume**. They also noted that "in some regions **ink** is made from the pods for local use," but it is not clear whether this referred to Belize.

Acacia gentlei Standl.

red cockspur [English]; *zubín* [Spanish]

The ripe fruit is eaten as a **food** [B3600].

As a first-aid treatment for **snakebite**, 1 piece of bark the length of the person's forearm is peeled, chewed, and applied as a poultice to the bite until help is obtained [B2197].

The leaves are said to be smoked as a **marijuana substitute** [B2348].

The wood is used for **fuel** [B2460].

(top) *Acacia dolichostachya* [GS].

(bottom) *Acacia gentlei* [MB].

Acacia globulifera Saff.

cockspur [English]; *zubín* [Spanish]

As a first-aid treatment for **snakebite**, 1 piece of bark the length of the person's forearm is peeled, chewed, and applied as a poultice to the bite until help is obtained [B3350].

Acacia polyphylla DC.

bastard prickly yellow, Jim Crow, prickly yellow, white tamarind, wild tamarind [English]; *salam* [Additional name recorded]

The trunk is used in **construction** to make fence posts [B1818]. Standley and Record (1936) reported that the nearly white wood was suitable for interior **construction** and **veneer** but was not resistant to insects or decay.

For **tanning** leather, the bark is macerated, added to a dugout canoe containing 1½ gallons of water, and layered in between the pieces of hide. The hide is sprinkled with white lime, and another layer of crushed bark is added. A second hide is placed on top of the lime and crushed bark. These layers are repeated until they reach the top of the canoe, which is set undisturbed in the shade for about 7 days. When tanning is complete, the hair is removed from the hide, which has become flexible [B1818; LR].

Balizia leucocalyx (Britton & Rose) Barneby & J.W. Grimes

tamarind, wild tamarind [English]; *tamarindo silvestre* [Spanish]

The wood is used in **construction** for making houses, basins, and dugout canoes [A600, A964].

Cojoba arborea (L.) Britton & Rose var. *arborea*

red tamarind [English]; *barba de jolote, barba jalote* [Spanish]; *barba jelóte* [Mayan]

The plant is cultivated as an **ornamental** because of the colorful fruits [R66].

The seeds are eaten as a **food** by wild animals, including gibnut (*Agouti paca*) and peccary (*Pecari tajacu* and *Tayassu pecari ringens*) [B3218].

Lamb (1946) reported that the bark was used for **tanning** leather. The bark is still used for **tanning** leather today [B3218]. Lamb (1946) also noted that the wood was used in **construction** to make dugout canoes, railroad ties, bridge timbers, flooring, ceiling posts, and for general **carpentry**.

Cojoba graciliflora (S.F. Blake) Britton & Rose

John Crow bead [English]; *frijolillo, hierba del frijol* [Spanish]; *bul xíu, wak út* [Mayan]; *ch c lac t, chá lac té, poma, sha'subin* [Additional names recorded]

To treat **any ailment**, the leaves are used in Maya bath mixtures [A393, B2244].

To treat a skin condition known as *empacho de agua*, referring to an intestinal obstruction in which swellings form small pustules on the body, the leaves are rubbed on the skin, so that the sap drips directly onto the pustules, twice daily until better [B1889]. To treat **skin sores**, the bark is prepared and used as a bath over the affected area [A393].

(left) Leaves and mature fruit of *Cojoba graciliflora* [MB].

(right) Leaves and fruit of *Desmanthus virgatus* [PA].

To protect babies from the **evil eye (*mal ojo*)**, the seeds are strung and placed around the baby's neck [A397]. To treat babies with the **evil eye (*mal ojo*)**, 1 handful of leaves is boiled in 2 gallons of water for 5 minutes and used warm as a bath once daily, at noon or bedtime [A397].

Desmanthus virgatus (L.) Willd.
ironweed [English]
The entire plant is used as **fodder** for horses [B2180].

Enterolobium cyclocarpum (Jacq.) Griseb.
guanacaste, *tubroos* [Spanish]
The dried seeds are removed from their pod with a knife, roasted on the fire in a pot, and eaten as a **food**. The seed shells pop off when they are ready [HR]. Alternatively, the seeds are ground up and combined with chili [HR]. Green seedpods are boiled in water until soft, when they are opened and the green seeds eaten like beans [HR]. Lamb (1946) reported that the pods were used as **fodder** for cattle.

The seeds are used to make **jewelry** for sale in local markets [Marsh, Matola, and Pickard, 1995].

Traditional dory being carved from *Enterolobium cyclocarpum* wood [MB].

Havardia albicans [GS].

Inga pavoniana [MB].

Lamb (1946) noted that the bark was used for **tanning**. He also reported that due to its durability and resistance to insect invasion, the wood was used in **construction** to make large dugout canoes, cattle troughs, and mortars "for hulling rice and macerating oil-seeds." The wood is very durable, even though it is light, and a good canoe (also known as a dory) made from it can last for 10 years [Horwich and Lyon, 1990].

Havardia albicans (Kunth) Britton & Rose
Standley and Record (1936) reported that in the Yucatan, the bark was said to be used for **tanning** and the wood for **construction**.

Inga affinis DC.
guamo [Spanish]; *bribri* [Additional name recorded]
The fleshy pulp around the seeds is eaten as a **food**, as are the stamens in the large flowers [Howrich and Lyon, 1990].

Inga pavoniana G. Don
guama [Spanish]; *tzotz ni* [Mayan]
The fleshy pulp around the seeds is eaten as a **food** [B2037].
To treat **swelling**, 1 large handful of leaves is boiled in 2 gallons of water for 5 minutes and used cool to bathe the affected area twice daily [B1917, B2037].

Inga punctata Willd.
choochok [Q'eqchi']
The fleshy pulp around the seeds is eaten raw as a **food** [B3521].

Gregory Shropshire collecting a specimen of *Inga vera* along the Macal River [MB].

Inga thibaudiana DC.
broken ridge [English]; *bri bri, bribri* [Creole]; *b'itz'* [Q'eqchi']
The fleshy pulp around the seeds is eaten as a **food** [B2669].

Inga vera Willd.
wama [English]; *bitis* [Spanish]; *bri bri, bic gri* [Creole]; *b'itz'* [Q'eqchi']; *chalum* [Mayan]
The fleshy, white pulp around the seeds is very sweet and eaten as a **food** [A69, A797, JB66]. The entire plant is used as **fodder** [A69].
  For **foot sores** and other **skin ailments** caused by an overexposure to water, three 12-inch strips of bark are boiled in 2 gallons of water for 20 minutes and used cool as a bath over the affected area twice daily, in the morning and in the evening, until better [B1921].

Leucaena leucocephala (Lam.) de Wit subsp. *leucocephala*
wild tamarind [English]; *anil* [Mayan]
The leaves are used as **fodder** for cows and horses and are eaten as a **food** by wild animals [B1986, B2163]. Interestingly, Steggerda (1943) noted the Yucatan Maya believed that "when horses eat the leaves of the tree, the hairs of their tails fall out." This was said to be a universal belief in the region at that time. He also noted that

Leucaena leucocephala
subsp. *leucocephala* [MB].

because Maya women used wood ashes to soften water, they would not burn this wood in their cooking fires because its ashes, when mixed with water, were said to be extremely irritating to the skin.

Lysiloma latisiliquum (L.) Benth.
rain tree [English]; *chalan* [Mayan]; *salaam, salem, salom* [Additional names recorded]
To **cure leather** and **bleach** it white, the hide is soaked with the bark [B1938, B2148].

Mimosa bahamensis Benth.
logwood brush, raintree [English]
The presence of this plant is considered an **indicator** of "good soil" [B2166].

Mimosa hondurana Britton
guana teeth, iguana leaf, tear coat [English]; *diente de iguana* [Spanish]
Locally, this plant is regarded as a **hazard** because it "cuts you when you are working in the field," hence its common name of "tear coat" [B3685]. The vine is rather like a saw with curved teeth [RA].

Lysiloma latisiliquum showing fruit (left) and flowers (right) [RF].

Mimosa pudica var. *unijuga* with inset showing flower [MB].

Mimosa pudica L. var. *unijuga* (Duchass. & Walp.) Griseb.
twelve o'clock prickle, prickle, sensitive plant, sleepy head [English]; *dormillon, dormilón, duermidillo* [Spanish]
mútz, xmutz [Mopan]; *waral k'ix* [Q'eqchi']; *šmuȼ'ic* [Yucateca]
For **fever**, 1 handful of leaves and branches is boiled in 1 gallon of water for 10 minutes, ¼ cup consumed tepid, and the remainder used as a bath [B3575]. For **toothache**, the root is mashed, boiled in water, and applied as a poultice to the affected area [Arvigo and Balick, 1998]. To **cleanse a "film" on the eye** that might develop, the sap from the stem is dropped into the eye [Mallory, 1991]. To relieve **back pain**, **muscle spasms**, and **nervous irritability**, the leaves are dried, powdered lightly, rolled, and smoked like a cigarette [B1822; Arvigo and Balick, 1998]. This same leaf powder described above for back pain is sprinkled on food as a **sedative** [Arvigo and Balick, 1998]. As an **antispasmodic**, **relaxant**, **pain reliever**, **diuretic**, and **sleep inducer**, 9 branches with leaves are boiled in 3 cups of water for 5 minutes and ½ cup consumed 3–6 times daily as needed [Arvigo and Balick, 1998]. To treat **depression**, the leaves are boiled in water and consumed until better [Mallory, 1991].

To treat **stomach pain**, 1 entire plant is boiled in water, cooled, and consumed by the spoonful until better [Mallory, 1991]. To treat **stomach pain** in infants, ½ to 1 spoonful of this same decoction is consumed [Mallory, 1991]. **Caution** is advised, as overdosing from this plant (when consumed as a liquid) is said to result in lethargy [Mallory, 1991].

For **insomnia**, bunches of leaves are placed under the pillow in a cross formation: 4 crosses for children and 9 crosses for adults [B1822; Arvigo and Balick, 1998]. To **calm** babies and help them **sleep**, 1 entire plant is wrapped in cloth and put under a pillow [B2429, B3575]. To **induce sleep** in adults, nine 6-inch branches are boiled in 1½ cups of water for 10 minutes and 1 cup consumed once daily before bedtime [B1822].

To **improve blood circulation**, 1 bundle of ten to twelve 12-inch branches with leaves along with 1 small handful of balsam bark (*Myroxylon balsamum* var. *pereirae*), 1 small handful of wild yam tuber (*Dioscorea bartlettii*), and 1 large handful of the entire strong back plant with root (*Desmodium adscendens*) are boiled in 1 gallon and 1 cup of water for 10 minutes, steeped under a low flame for 10 minutes, allowed to cool, strained, and ½ cup consumed twice daily, in the morning and in the evening, until finished [LR]. Alternatively, refer to *golondrina* (*Alternanthera flavogrisea*) [LR]. For children with **urinary tract ailments**, 1 entire plant is boiled in 1 quart of water, reduced to ½ quart, and sipped all day until better [W1844]. For **venereal diseases**, refer to *Sida antillensis* [B2658].

As part of a **snakebite** formula, the root is mashed and used as a poultice [LR]. Alternatively, refer to wild mahogany (*Mosquitoxylum jamaicense*) [LR] and snakeroot (*Pentalinon andrieuxii*) [P2].

For people who suspect they have been affected by **evil magic (*obeah*)**, 1–2 roots along with four 12-inch stems and leaves of basil (*Ocimum basilicum*), four 12-inch stems and leaves of marigold (*Tagetes erecta*), nine 3-inch sprigs of *ruda* (*Ruta graveolens*), and 1 bottle of bay rum are mashed in 1½ gallons of water, steeped in the sun for 2 hours, and ½ gallon used warm as a bath in the evening, heated if necessary, 3 times weekly, on Monday, Wednesday, and Friday, until better [HR].

Mimosa tarda Barneby
dandelion, sensible weed [English]

To calm **nervousness** or relieve **anxiety**, 1 entire plant is boiled in 1 gallon of water for 10 minutes, 1 cup consumed cool, and the remainder used as a bath once daily at bedtime [B2639].

Pithecellobium keyense Britton
x coy, *xa coy* [Mayan]

The fleshy part around the seed is eaten as a **food** that children in particular enjoy [A416, B1951].

To ease a **sore throat**, the fruit is eaten [B1951]. To treat **diarrhea**, the fruit is added to a tea made from the leaves [B1951]. To relieve **rheumatism**, 1 small handful of leaves is boiled in 3 cups of water for 5 minutes and 1 cup consumed 3 times daily, before meals, for 3 days of each week as needed [B1951].

Pithecellobium lanceolatum (Humb. & Bonpl. ex Willd.) Benth.

For a **bloody stool**, one 6-inch piece of bark is boiled in 3 cups of water for 10

(top) *Pithecellobium keyense* [MB].

(bottom) *Samanea saman* [IA].

minutes, strained, cooled, ¼ cup consumed 12 times daily, and repeated until the blood clears up, usually in 3 days [A686].

Pithecellobium macrandrium Donn. Sm.

prickle wood, *puma* [Additional names recorded]

To lower **fever** in children who have measles, 1 large handful of leaves and young branches is boiled in 2 gallons of water for 10 minutes and used cool as a bath once daily at noon until better [A650].

Samanea saman (Jacq.) Merr.

The plant is eaten as a **food** by farm animals [Standley and Steyermark, 1946].

Standley and Record (1936) reported that cross sections of the large trunks of this tree were used throughout Central America in the **construction** of cart wheels and the wood used in **carpentry** for interior trim, furniture, and musical instruments.

FABACEAE: SUBFAMILY PAPILIONOIDEAE

Usually herbs, occasionally vines, sometimes trees, shrubs, or lianas. Leaves often pinnate, sometimes palmate or trifoliolate, occasionally simple. Inflorescences axillary or terminal. Flowers with 3 free petals and 2 fused petals that form a keel. Fruits a legume. In Belize, consisting of 52 genera and 178 species. Uses for 26 genera and 47 species are reported here.

Abrus precatorius L.

lukum pim [Q'eqchi']

Standley and Record (1936) reported that the characteristic red and black seeds were used to make **necklaces** and **bracelets** as well as other **ornamental articles**. It should be noted that these seeds are extremely **toxic**; a single ingested seed, if it is broken open, is sufficient to kill a child. However, as the seed coat is so hard, the intact seed will usually pass through the body [MB; Nelson, Shih, and Balick, 2007]. Steggerda (1943) noted that the Yucatan Maya used the seeds of this species to make the eyes for **dolls**.

To treat **hepatitis**, 1 handful of fresh root is boiled in ½ liter of water for 20 minutes and consumed warm 3 times daily for 5 days [C39]. Steggerda (1943) stated that a bath of the leaves of this plant was used to treat **diarrhea** in babies caused by evil eye (*mal ojo*) or bad winds (*mal viento*). Additionally, the leaves were roasted, ground, and made into a salve placed on the head of the sick baby.

Acosmium panamense

(Benth.) Yakovlev

Billy Webb [English]

For **cough**, refer to John Charles (*Hyptis verticillata*) [A355, A479] and Spanish cedar (*Cedrela odorata*)

Abrus precatorius open fruit showing mature seeds [RL].

Acosmium panamense [MB].

[B2179]. For **fever**, refer to wild cherry (*Pseudolmedia spuria*) [B3733]. To remove **excess mucus** in the prostate and uterus, refer to *guaco* (*Aristolochia grandiflora*) [LR]. To heal **wounds**, the bark is applied as a poultice to the affected area [Robinson and Furley, 1983]. To **stimulate the appetite**, the bark is made into a tea [Robinson and Furley, 1983].

For **gastritis (*ciro*)**, non-insulin-dependent **diabetes** mellitus, **delayed menstruation**, **colds**, or **influenza**, the following decoction is prepared with this bark either alone or in combination with unspecified parts of *tres puntas* (*Neurolaena lobata*), white rum, *chicoloro* (*Strychnos panamensis* var. *panamensis*), Spanish cedar (*Cedrela odorata*), *guaco* (*Aristolochia grandiflora*), or 1 of at least 5 other *Aristolochia* species. In total, 1 small handful of any or a combination of all of the ingredients above mentioned is boiled in 1 quart of water for 10 minutes, steeped for 20 minutes, and 1 cup consumed warm 3 times daily, before meals, until better [P3]. For **gastritis** (ciro), 1 small handful of equal parts of all the ingredients mentioned above for gastritis (*ciro*) is boiled in 1 gallon of water for 10 minutes and ½ cup consumed warm twice daily, before the morning and noon meals, and continued until better or the tea is finished [LR].

To treat non-insulin-dependent **diabetes** mellitus, **malaria**, **sluggish digestive system**, **postmenopausal ailments**, and **bilious-tasting mouth**, 1 small handful of fresh or dried leaves is boiled in 3 cups of water for 10 minutes and either 1 cup consumed 3 times daily before meals, ½ cup consumed 6 times daily, or sipped all day until 3 cups are consumed [B1909]. For non-insulin-dependent **diabetes** mellitus, 1 small handful of bark is boiled in 3 cups of water for 10 minutes and 1 cup consumed 3 times daily before meals [RA]. Alternatively, refer to snake root (*Pentalinon andrieuxii*) [B2128]. This plant should not be used during pregnancy, as it may have **abortifacient effects**.

For **skin fungus**, the bark is boiled in water and used to treat the affected area. For example, for **foot fungus**, the foot is soaked in the pot, while for **skin fungus**, the bark and water are wrapped in a cloth and applied hot, as a poultice, to the affected area [B1909]. To prevent **malaria**, refer to *tres puntas* (*Neurolaena lobata*) [B2632, P1]. To **purge, reverse, or "wash out" a male contraceptive**, an unspecified treatment is made with this plant along with skunk root (*Chiococca alba*), *copalchi* (*Ocotea veraguensis*), and *chicoloro* (*Strychnos brachistantha*) [B1848, B2220].

For **snakebite**, one 6-inch piece of bark is boiled in 3 cups of water for 5 minutes and the decoction consumed immediately, followed by ½ cup per day for the next 3–4 days. This eliminates or slows the venom and is only a first-aid remedy, until the person is able to get help [P3]. Alternatively, refer to *cramantee* (*Guarea glabra*) [LR] and blossom berry (*Eugenia capuli*) [A911].

Standley and Record (1936) reported that the wood of this tree was used for **construction** to make truck parts, wheels, and cart shafts. This is one of the important secondary **hardwood** species of Belize [Munro, 1989]. It is considered to be excellent for making **floors** [RA].

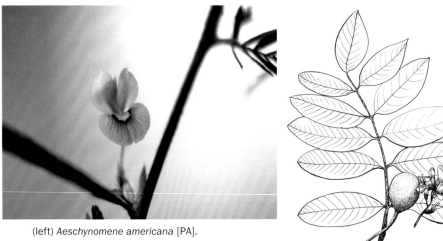

(left) *Aeschynomene americana* [PA].

(right) *Andira inermis* [FA].

Aeschynomene americana L.

zarab' 'zotz' [Q'eqchi']

For **heavy menstruation**, 1 double handful of leaves is mashed, steeped in cool water for 10–20 minutes, and 1 cup consumed twice daily, once in the morning and once in the evening, until the condition improves [B3626].

Aeschynomene deamii B.L. Rob. & Bartlett

cork wood [English]

The wood is used in the **construction** of bottle stoppers [A366].

As a **laxative** to promote bowel movement, 3 sprigs containing both leaves and flowers are boiled in 1 cup of water for 5 minutes and consumed hot once daily in the morning [A366].

Aeschynomene rudis Benth.

waral k'ix [Q'eqchi']

To numb the pain of a **toothache**, one 6-inch piece of root is boiled in 2 cups of water and swished around the mouth as a mouthwash at a cool temperature for 1 minute [B3628].

Andira inermis (W. Wright) Kunth ex. DC.

cabbage-bark [English]; *iximche* [Additional name recorded]

Standley and Record (1936) noted that "the bark has a nauseous odor, and is used sometimes as a **vermifuge**, **purgative**, and **narcotic**, but in large doses it is reported to be a dangerous **poison**."

The wood is used in **construction** to make house posts [B3133]. The wood is also used in **construction** for making logging truck parts, skids for hauling lumber, bridges, and house frames [Mars, Matola, and Pickard, 1995].

Centrosema pubescens Benth.

white bell [English]; *kenq tyuk* [Q'eqchi']

For **difficult menstruation**, one 12-inch piece of vine is mashed in 2 cups of cold water, strained, and 1 cup consumed warm twice daily, in the morning and in the evening, for 1–2 days [B3648].

Crotalaria cajanifolia Kunth

tzib che (male) [Mayan]

This plant is considered a major medicine for **spiritual diseases**, such as **fright (*susto*)**, grief (*pesar*), **envy (*envidia*)**, or the **evil eye (*mal ojo*)**, and is used as a bath or by waving part of the plant over the body [A169].

For protection against v*iento caliente de los Mayas* ("**hot wind of the Mayas**"), 9 branches with leaves are boiled in 1 quart of water for 5 minutes and used as a bath [A169]. For **protection against harmful spirits** during ceremonies, a total of 9 crosses, made from the branches, are waved over the front and back of a person's body while prayers are said [A169].

Crotalaria verrucosa L.

cascabel [Spanish]

The pod is used as a **toy** to play "shop," to "sell" to the "customer" [A406].

Dalbergia glabra (Mill.) Standl.

logwood brush [English]; *cibix, muk* [Additional names recorded]

Steggerda (1943) noted that the roots were said to be useful in treating **dysentery**. The fresh stem is used medicinally for an unspecified purpose [A924].

Crotalaria verrucosa [MB].   *Dalbergia glabra* [GS].

Steggerda (1943) also reported that the Yucatan Maya produced a **rope** from the bark of this vine to tie the beams of their houses together and to lift water containers from their wells.

Dalbergia stevensonii Standl.

rosewood [English]

The wood of this tree is heavy and hard, and the heartwood is very durable [Marsh, Matola, and Pickard, 1995]. The wood has long been used in **construction** to make the bars for marimbas and xylophones [Marsh, Matola, and Pickard, 1995]. Standley and Record (1936) reported that the wood was used in **construction** for house posts and lintels.

Dalbergia sp.

logwood bush [English]

For **diarrhea**, one 6-inch piece of root or 3 young branches along with 2 garlic cloves with skins are boiled in 3 cups of water for 10 minutes and 1 cup consumed cool 3 times daily, before meals, until better [A664].

Open flowers are a **sign** of rain for that same day [A664].

Desmodium adscendens (Sw.) DC.

strong back [English]; *chim pim, pash p'am* [Q'eqchi']

To treat **back pain**, **muscle pain**, **kidney ailments**, and **impotence**, 1 entire plant is boiled in 3 cups of water for 10 minutes and 1 cup consumed 3 times daily, before meals, for 3–5 days [Arvigo and Balick, 1998]. For **joint and muscle pain** and to treat **headache**, this same decoction as described above for back pain is used as a

bath [Arvigo and Balick, 1998]. To relieve **muscle spasms**, 1 entire plant about 12 inches long is boiled in 3 cups of water for 10 minutes and sipped all day until 3 cups have been consumed [Arvigo and Balick, 1998]. For **back pain**, 1 entire plant is soaked in rum for 24 hours and 2 tablespoons consumed 3 times daily for 7–10 days [Arvigo and Balick, 1998]. To **improve blood circulation**, refer to *dormilón* (*Mimosa pudica* var. *unijuga*) [LR]. For a **bladder infection**, refer to yellow malva (*Malvastrum corchorifolium*) [LR].

As a **female contraceptive**, 9 leaves along with 9 *să rá* leaves (*Desmodium* sp.), 9 *pega ropa* leaves (*Priva lappulacea*), and 9 leaves of an unspecified plant are mashed in 3 cups of cold water and 1 cup consumed 3 times daily "until the period disappears" [B3643, B3644, B3645].

Desmodium adscendens [MB].

Desmodium axillare (Sw.) DC. var. *acutifolium* (Kuntze) Urb.

strong back [English]

To treat **kidney ailments** and relieve **back pain**, 1 entire plant is boiled in 1 quart of water for 5 minutes, steeped for 20 minutes, cooled, and sipped all day for 3 days [A785].

Desmodium incanum (Sw.) DC.

strong back [English]; *chim pim, să rá* [Q'eqchi']

As a **stimulant** to treat **fatigue**, a few branches are boiled in 3 cups of water for 5 minutes and consumed hot or warm 2–3 times daily as needed [B3225]. This plant is 1 of 4 plants that can be used as a tea to treat **heavy menstruation** [B3574]. This plant is considered a "male" plant [B3574].

For **back pain** and **muscle spasms**, 3 entire plants are boiled in 3½ cups of water for 10 minutes and 3 cups consumed daily for 3 days [A728].

(top) *Desmodium incanum* [MB].

(bottom) *Dioclea wilsonii* [MB].

To treat **snakebite**, the leaves were chewed and the juice swallowed. This treatment is believed to delay the action of the poison so that the person can reach a doctor [Mallory, 1991]. Alternatively, the leaves are prepared into a tea [Mallory, 1991].

Desmodium tortuosum (Sw.) DC.

petch mam [Additional name recorded]

The leaves and stems of this plant are used as **fodder** for horses [B2164].

Desmodium sp.

chim pim, să rá [Q'eqchi']

As a **female contraceptive**, refer to strong back (*Desmodium adscendens*) [B3643, B3644, B3645].

Dioclea wilsonii Standl.

horse's eye [English]; *t'oro' koch, se ru caway* [Q'eqchi']

For **epileptic seizures**, the seed is grated and 1 teaspoon of the powder mixed in ½ cup of water and consumed warm [B2626].

The seeds are used as **beads** for necklaces and to **smooth** clay pots when they are "newly made and still wet" [A596].

Diphysa americana (Mill.) M. Sousa

warrie wood, wild rue, wild ruda [English]; *ruda del monte* [Spanish]; *k'an té, q'ante'* [Q'eqchi']; *susuk, tsutsuc* [Additional names recorded]

For **sinus headache**, 1 large handful of combined leaves of equal parts of this plant along with skunk root (*Petiveria alliacea*), gumbolimbo (*Bursera simaruba*), *madre de cacao* (*Gliricidia sepium*), marigold (*Tagetes erecta*), and little bamboo (*Commelina erecta*) is mashed in 2 gallons of water, steeped in the sun for 1 hour, and used to bathe the head 3 times daily, in the morning, at noon, and at night [B2155]. For **fever**, 1 handful of leaves is mashed in 2 gallons of warm water and used as a bath once daily, at noon or bedtime, as needed [B3649]. To treat **fever** in babies, the leaves are used to prepare a bath [B1851].

For **colds, coughs, influenza, stomach pain,** and **nervousness**, particularly in babies, 1 small handful of leaves or nine 6-inch branches are mashed in 3 cups of water for 5 minutes and 1 cup consumed 3 times daily for adults or ½ cup consumed 3 times daily for children [A388]. To soothe **stomach pain** or **induce menstrual flow**, 1 handful of leaves and small branches is mashed, steeped in 3 cups of water, squeezed, and 1 cup consumed 3 times daily before meals [B1851].

To **dispel bad magic** or **witchcraft**, 1 handful of leaves and small branches is mashed, steeped in 3 cups of water, strained, and used as a bath or consumed as a tea [B1851]. Alternatively, 1 handful of small branches with leaves is worn as an amulet by or used for flogging the person affected [B1851].

The wood, which is said to last 100 years, is used in **construction** to make house posts [A388, B1851].

Diphysa carthaginensis Jacq.

warrie wood, wild rue, wild ruda [English]; *¢u ¢uk, susuk, tsutsuc* [Yucateca]

For **diaper rash** and **any illness in babies**, the juice from nine 12-inch branches containing leaves is squeezed into cool water and used cool as a bath once or twice daily as needed [A647].

For **bad winds** (*mal viento*), the juice from 9 small branches with leaves along with 9 sprigs of *ruda* (*Ruta graveolens*) are squeezed into 3 cups of water and consumed warm 3 times daily, in the morning, noon, and night, for 3 days [A647]. Additionally, the same mixture is prepared and used as a bath once daily, in the evening, for 3 days [A647]. To treat **evil eye** (*mal ojo*) in babies, the leaf is boiled in water and used as a face wash; this is considered a mild cure [Arnason et al., 1980]. For a total cure of **evil eye** (*mal ojo*) in babies, the leaf juice is placed in the eyes for 9 consecutive Fridays [Arnason et al., 1980]. To treat **fever** in an infant, the leaf is boiled in water and used as a bath; the leaf is considered "cooling" [Arnason et al., 1980]. It is said that this plant is "one of the most celebrated of Maya herbal medicines" [Arnason et al., 1980].

According to Steggerda (1943), this plant

has been used consistently throughout Yucatán history for **sores** and **wounds**, and independent statements from five Indian herb doctors, stress the value of

Diphysa carthaginensis [GS].

the sap of this tree in such cases today. One adds, "Nine drops of the raw juice from the leaves of this plant when taken in a small amount of water are taken for red **dysentery**." Another says that five or six applications of the sap of this tree will cure the **chiclero ulcer**. (pp. 204–5)

The wood is used in **construction** for house posts and is a prized lumber said to last 150 years. It is also used as a **living fence post** in Belize [A647].

Erythrina folkersii Krukoff & Moldenke
coral tree, John Crow bead, tiger tree [English]; *pito* [Spanish]
For non-insulin-dependent **diabetes** mellitus, 4 ounces of the leaves are boiled in 1 quart of water, reduced to 1 pint, cooled, and ⅓ of the decoction consumed daily for 3 days [B2616]. To draw a **fever** "out from the body" and to keep it from going to the head, the bark is placed as a poultice on the feet for 3 hours [B1834].

As **bleach** for clothes, 1 large handful of leaves is boiled in 1 gallon of water for 10 minutes and then applied to the clothes [B1834]. To keep **fuel** clean, the seeds are placed in the fuel tank of kerosene lamps [B1834].

Erythrina standleyana Krukoff
pito [Spanish]
The red flowers are eaten as a **food** [A251].

To treat **fever**, a small amount of oil is placed on the leaf and then applied to the forehead [Arnason et al., 1980].

To **keep fuel clean**, the pods and seeds are placed inside kerosene lamps to absorb dirt that accumulates in fuel [A251].

Standley and Record (1936) reported that this tree was used as a **living fence post**.

Erythrina standleyana (left) with close-ups of seed pod (top right) and spines (bottom right) [WJH].

Gliricidia maculata (Kunth) Kunth ex. Walp.
madre de cacao [Spanish]

The flowers are eaten as a **food** and good for **weakness** [A377].

For a general bath mixture, refer to *Flaveria trinervia* [B2149]. To treat **cataracts** or **tired, wounded eyes**, one 6-inch piece of bark is soaked in 2 cups of water and used as an eyewash [A377]. For **skin infection**, 1 large handful of leaves is boiled in 1 gallon of water for 10 minutes and used as a cool wash twice daily over the affected area [A377].

For **head lice** (*Pediculus humanus*), 1 double handful of leaves is crushed in 1 quart of water, soaked for 2 hours, and used to wet the head. Then, these leaves are applied directly to the head and the head wrapped up all day. In the evening at bedtime, the head is unwrapped, the hair combed to remove the dead lice. This treatment repeated twice [LR].

Gliricidia sepium (Jacq.) Kunth ex. Walp.
madre cacao, madre de cacao [Spanish]; *k'ante'* [Q'eqchi']; *sayab* [Mopan]; *hutz* [Additional name recorded]

The fresh flowers are cooked with eggs and eaten as a **food** [Arvigo and Balick, 1998]. Standley and Steyermark (1946) reported that the leaves were eaten as a **fodder** by cattle, although they were **toxic** if eaten by rodents and dogs.

For **headache**, 1 large handful of leaves is mashed in 1 gallon of water, steeped for 1 hour, cooled, and used to bathe the head for about 10 minutes [B2140, B2236]. Alternatively, refer to *palo verde* (*Critonia morifolia*) [B2202]. For **sinus headache**, refer to warrie wood (*Diphysa americana*) [B2155].

To treat **wounds**, **skin ulcers**, **boils**, or **diaper rash**, fresh leaves are mashed and applied as a poultice to the affected area [Arvigo and Balick, 1998]. Standley and Steyermark (1946) reported that to treat **ulcers**, **tumors**, or **sores** that had begun to become **gangrenous**, the fresh leaves were macerated and applied as a poultice to the affected area.

To treat **eye ailments**, such as **cataracts**, **tiredness**, **injuries**, **irritation**, **pus**, or **mucus**, one 1-inch × 1-inch piece of bark is steeped in 1 cup of

Gliricidia sepium [MB].

water for 1 hour and used as a cold wash once daily until better [B1835]. For **tired, burning, or irritated eyes**, one 3-inch × 1-inch piece of bark is boiled in 1 cup of water for 10 minutes, cooled, strained twice through a cloth, and used as an eyewash [Arvigo and Balick, 1998].

Standley and Steyermark (1946) noted that the fresh leaves were placed in hens' nests to treat **parasites**, which "take refuge from the leaves, which are then removed and destroyed." To **poison** rats, the seeds and bark are pulverized, mixed with ground corn, and left out for rats to eat [Arvigo and Balick, 1998].

In Mexico, to make *cacahuananche*, an excellent brown **soap** suitable for washing hair, skin, and laundry, the bark is used in an unspecified way [Arvigo and Balick, 1998].

Standley and Record (1936) reported that this tree was used in **construction** for house posts and living fences. Steggerda (1943) noted that the wood apparently was so hard that it was able to break the axe of the person who cut it. The wood was said to be highly valued in **construction** for making corner posts and framing of houses [Horwich and Lyon, 1990].

Indigofera suffruticosa Mill.

Steggerda (1943) noted that this plant was formerly the source of **indigo**, which was used as a **blue dye**. He also reported that the Yucatan Maya occasionally used a leaf extract of this plant to produce blue markings on the foreheads of children who suffered

Indigofera suffruticosa with close-up of inflorescences [ST].

from **gastritis (*ciro*)** as the result of **evil eye (*mal ojo*)**. Steggerda and Korsch (1943) reported that the Yucatan Maya mixed *ruda* (*Ruta chalepensis*) leaves with this plant, soaked it in wine, and applied the liquid to the nose, hands, feet, and coccyx of a child suffering **spasms**. The application was made on Tuesdays and Fridays. Following the treatment, a warm bath of 9 chopped plants of *chacalpec* (*Salvia coccinea*) was given.

Lecointea amazonica Duke
tango [Spanish]
Standley and Record (1936) reported that the wood of this species was hard, heavy, very strong, and used in the **construction** of tool handles.

Lonchocarpus castilloi Standl.
black cabbage bark, cabbage bark [English]; *man chich* [Mayan]
The plant is used as medicine for many unspecified **ailments** [B2044].

Lamb (1946) reported that the wood was heavy, durable, long-lasting when in contact with the ground, and resistant to attack by insects and fungi. He also noted that in the 1940s, the wood was said to be commonly used in **construction** for making bolsters, drawbars, wheels, house posts, and wagons used to haul mahogany (*Swietenia macrophylla*).

Lecointea amazonica [RF].

Lonchocarpus castilloi [MB].

Lonchocarpus hondurensis Benth.

ink wood, logwood, swamp dogwood, waterside turtle-bone [English]; *jabin del agua*, *palo tinto* [Spanish]

For **rheumatism**, **colds**, and **influenza**, 1 large handful of stems and leaves is boiled in 2 gallons of water for 10 minutes and used warm as a bath once daily at bedtime [A380].

To treat **loose stools**, one 3-inch piece of stem or 1 thin branch about ¼ inch in diameter is boiled in 1 cup of water for 10 minutes and 1 teaspoon consumed cool every 30 minutes for children 1–6 years old and 1 ounce consumed 3 times daily for adults. This treatment is repeated for 2 days [HR].

In previous times, the wood was used as a source of **blue-purple dye** [A74]. As a **dye** for cloths, 1 quart of fresh bark is boiled in 2½ quarts of water for 5 minutes and the cloth soaked in the water for 30 minutes [LR]. Alternatively, one 12-inch × 5-inch branch of fresh wood, not dried, is chopped into 4 pieces, soaked in water overnight, and used as a strong fabric **dye** that does not run [LR].

Lonchocarpus luteomaculatus Pittier

b'itz' [Q'eqchi']

The wood is used for **fuel** [A616].

Lonchocarpus minimiflorus Donn. Sm.

white cabbage bark [English]; *palo de guzano* [Spanish]; *man chich* [Additional name recorded]

For **gum ailments**, such as **gingivitis**, the resin from the exposed bark is gathered on a finger and applied to gums twice daily until better [A831].

The wood is used in **construction** for house posts, frames, beams, and flooring [A831].

Lonchocarpus rugosus Benth.
black cabbage-bark, twisted mouth [English]; *ca'an xin, canacin, canasín* [Q'eqchi']; *selek chi* [Mayan]; *chaperon* [Additional name recorded]
The wood is used as **firewood** [A604].

Lonchocarpus xuul Lundell
cabbage bark (male), dogwood, turtle-bone [English]
The wood is used in **construction** to make fence posts [B3253]. According to the Maya, a tree cut during the full moon contains bitter, toxic sap in the trunk, and this deters insects, creating a higher-quality wood that lasts much longer [B3253; Arvigo and Balick, 1998].

Lonchocarpus xuul [MB].

Mucuna argyrophylla Standl.
pa chu [Mayan]
The fruit is thought to bring **bad luck**, and elders say "when you touch this fruit your calabash will break," referring to encountering bad luck [B1778].

Mucuna pruriens (L.) DC.
cow-itch, cowitch, nettle [English]; *pica pica, picapica* [Spanish]; *hul im* [Mayan]; *chican, chiican* [Additional names recorded]
To treat **breast swelling** and **pain** in nursing women, 1 small handful of leaves is boiled in 3 cups of water for 10 minutes, 1 cup consumed daily, and the remainder used as a bath over the breasts twice daily until better. Because the leaves are boiled, this bath does not cause itching on the woman's breasts [A140].

To **cause violent itching**, "**bother people you don't like**," **chase people away from public places**, or **cause a nuisance**, the leaves are powdered and applied to the skin or sprinkled on the floor of a public place [A302].

Muellera frutescens (Aubl.) Standl.
mother of cacao, swamp dogwood [English]; *madre de cacao* [Spanish]
The white, pink, and purple flowers are eaten as a **food** [R8].

To treat **itching**, 1 double handful of leaves is mashed in a 5-gallon bucket of water, steeped in the sun for 2 hours, and poured over the affected area. Oil or lotion is ap-

Mucuna pruriens flowers (left) [RH] and fruit (right) [PA].

plied after the bath so that the affected area does not cause the skin to crack [HR]. To treat **foot fungus**, the feet are soaked in the infusion described above for itching for 30 minutes once daily and grease applied after the bath [HR].

This tree is commonly inter-cropped with *cacao* (*Theobroma cacao*) and coffee (*Coffea arabica*) because of its ability, like most legumes, to **fertilize** the soil via nitrogen fixation [R8].

Myroxylon balsamum (L.) Harms var. *pereirae* (Royle) Harms

balsam, balsam seed, balsam tree [English]; *bálsamo* [Spanish]; *na bá* [Mopan]

To treat **colds**, refer to copal (*Protium copal*) [A507]. To **improve blood circulation**, refer to *dormilón* (*Mimosa pudica* var. *unijuga*) [LR]. For **kidney ailments**, one 1- to 3-inch piece of bark is boiled in 3 cups of water for 10 minutes and sipped all day long until better [B2077]. To treat **gonorrhea**, refer to pine (*Casuarina equisetifolia*) [B1958].

Myroxylon balsamum [MB].

To treat **stomach pain**, **burns**, and **wounds**, the resin from the bark is used by the Yucatan Maya [Flores and Ricalde, 1996].

To treat **urinary tract infections**, other **urinary ailments**, **kidney stones**, **burning in the bladder**, and **liver congestion**, 1 teaspoon of the chopped bark is boiled in 3 cups of water for 10 minutes and 1 cup consumed 3 times daily before meals. The bark does not need to be fresh and can be reused for 5 consecutive days [B1789]. As a **prostate treatment**, to "**clean the lines**" and **remove mucus**, 1 cup of the decoction described above for urinary tract infections is consumed twice daily, in the morning and at bedtime, until better [LR]. To **clear mucus** from the male urinary tract quickly and to treat **prostate ailments**, refer to *flor azul* (*Miconia albicans*) [LR]. To **remove excess mucus** in the prostate and uterus, refer to *guaco* (*Aristolochia grandiflora*) [LR].

To **dispel witchcraft and evil spirits**, the resin or dried bark is burned as incense [B1789]. The bark is used as a **charm** for protection, but the protection only lasts 6 months and then the charm must be remade [B1789]. The bark is used as **incense** accompanied by incantations, especially at *primicia* festivals, which are sacred Maya ceremonies [B1804, B2077].

Animals are said to rub their horns and skin against this tree for **perfume**, "to get a nice smell" [B1804].

According to Standley and Record (1936):

> By tapping the tree there is obtained the Balsam of Peru, a fragrant aromatic liquid variously employed in industry, and an official drug of the United States Pharmacopoeia. Almost all of this product comes from the so-called Balsam Coast of the Republic of [El] Salvador. The wood, though of excellent quality, is of no commercial importance in British Honduras because of its scarcity. (p. 189)

The tree is considered an important secondary **hardwood** species and was once exported to the United Kingdom [Munro, 1989].

Pachyrhizus erosus (L.) Urb.
yam bean [English]
The tubers are boiled and eaten as a **food** [B2210].

Pachyrhizus erosus [NYBG].

Phaseolus vulgaris [MB].

Pachyrhizus ferrugineus (Piper) M. Sørensen
wild jicama [English]; *la disgracia* [Spanish]

To relieve **grief** (*pesar*), 9 leaves and ½ cup of ground roots are steeped in 3 cups of water for 20 minutes and 1 cup consumed warm 3 times daily as needed [B2250, B3252]. As a **calmative**, a hot bath is prepared from the leaves, with the option of adding other plants to the decoction [B3252].

Phaseolus vulgaris L.
bean [English]; *frijol* [Spanish]

This plant is one of the very important cultivated **food** crops in Belize, and many varieties and colors are found [MB]. To prepare a recipe, refer to *epasote* (*Chenopodium ambrosioides*) [BW; Arvigo and Balick, 1998].

To treat babies with **grief** (*pesar*), refer to cross vine (*Paullinia tomentosa*) [Arvigo and Balick, 1998].

Piscidia piscipula (L.) Sarg.
dogwood, iguano blossom, wormwood [English]; *habim, jabin, jabine, palo de gusano* [Spanish]; *kaqla' che', tiaxib* [Q'eqchi']

As an **astringent for general ailments** and to treat **diarrhea**, **dysentery**, or **heavy menstruation**, one 2-inch × 2-inch piece of bark is boiled in 3 cups of water for 10 minutes and 1 cup consumed 3 times daily, before meals, until better, generally

Piscidia piscipula [MB].

for 1–2 days [B1850, B1869; Arvigo and Balick, 1998]. For **diarrhea** and **hemorrhage**, 1 small handful of chopped bark is boiled in 3 cups of water for 10 minutes and 1 cup consumed cool 3 times daily [A97, A111].

Steggerda (1943) noted that to treat **coughs** and **colds**, 9 young leaves were boiled in water and sugar and consumed. To reduce **fever** in an infant, 1 small handful of leaves is boiled in 2 gallons of water for 5 minutes and used warm as a bath once daily at bedtime [A747]. For **sores**, 3 leaves are mashed with a bit of water, applied directly to the affected area as a poultice, and changed once daily [LR]. To treat **wounds**, **rash**, **skin sores**, **leishmaniasis (baysore)**, or other **skin ailments**, 1 handful of bark is boiled in 3 cups of water for 5 minutes and used cool as a bath twice daily for 3 days [A111; Arvigo and Balick, 1998]. For **bleeding gums**, this decoction described above for wounds is used as a mouthwash [Arvigo and Balick, 1998]. To treat **wounds** and **toothache**, the juice from the bark is used by the Yucatan Maya [Flores and Ricalde, 1996].

According to Standley and Steyermark (1946):

The dry bark, especially that of the root, is reported to have a strong and disagreeable odor resembling that of opium, and it produces a burning sensation in the mouth. It contains various narcotic substances, one of which has been named piscidin. An extract of the bark, applied locally, has been used in tropical America to relieve **toothache**, and in Jamaica to cure **mange** in dogs. The most remarkable use of the tree, however, is as a **fish poison**, the bark and leaves being crushed and thrown into the water, where they soon stupefy the fish, causing them to float upon the surface. *Piscidia* bark is said to be used in this manner in Baja Verapaz and probably in other parts of Guatemala. The specific name assigned by Linnaeus refers to the poisonous properties of the genus. (pp. 336–7)

The heartwood is used in **construction** for house posts [A89, A97, A111]. It is said that the wood is very durable and, if cut on the full moon, can last up to 50 years [Arvigo and Balick, 1998]. Steggerda noted that the wood was once used for the **construction** of railway sleeper cars. Standley and Record (1936) reported that the wood was once used in **construction** for wheel-wright work.

Pterocarpus officinalis [RH].

Pterocarpus officinalis Jacq.

swamp, swamp kaway [English]; *sangre* [Spanish]; *kaway* [Creole]

The strong wood from this species is used in **construction** to make mortar sticks for pounding grain, tool handles, and wooden tools [Horwich and Lyon, 1990].

The red sap dries into a red resin was once important in **medicine** and sold under the name "dragon's blood" [Horwich and Lyon, 1990].

Rhynchosia longeracemosa M. Martens & Galeotti

vaca [Spanish]; *corrimiento xiv, ibis* [Additional names recorded]

For **toothache**, the leaves are chewed, giving off a numbing quality [A208]. To **improve blood circulation**, 1 entire plant is boiled in 2 gallons of water for 10 minutes, ½ cup consumed, and the remainder used as a bath once daily for 3 days [A396].

Swartzia cubensis (Britton & P. Wilson) Standl. var. *cubensis*

bitter wood, northern rosewood [English]; *llora sangre, sangre de toro* [Spanish]

For **snakebite** and **malaria**, 1 handful of chopped bark is boiled in 3 cups of water for 10 minutes and 1 cup consumed warm 3 times daily, before meals, for 3 days [A544].

Swartzia cubensis var. *cubensis* [GS].

Vigna unguiculata [PA].

Vigna vexillata var. vexillata [MB].

Swartzia simplex (Sw.) Spreng.

Standley and Record (1936) reported that the wood of this species was highly colored, hard, heavy, and useful in heavy and durable **construction**.

Tephrosia multifolia Rose

ch'al pim [Q'eqchi']

In the Toledo District of Belize, this is 1 of 3 plants used as a **fish poison** [B3572]. One burlap bag is filled with leaves and stems, placed in a stream, and the leaves are mashed, causing the fish to be paralyzed and rise to the surface [B3572]. Alternatively, the leaves are dried, powdered, and mixed in the stream while stirring the water vigorously [B3572].

Vigna unguiculata (L.) Walp.

pelon [Spanish]; *sh pelú, sh pelè* [Mayan]

The pods are eaten as a **food** [B2430].

Vigna vexillata (L.) A. Rich. var. *vexillata*

corrimiento, pato [Spanish]

To treat a **headache** that comes during the daytime, 9 leaves are mashed in 2 quarts of water, boiled for 5 minutes, and used warm to bathe the head [B2650]. To treat a **headache** that comes in the nighttime, 9 leaves are mashed with a pinch of salt and wrapped around the head overnight as a poultice [B2650].

FAGACEAE

Trees and shrubs. Leaves alternate, simple. Inflorescences axillary or rarely terminal, in catkins, spikes, or heads. Flowers small. Fruits a nut or large achene, subtended or enclosed by a scaly or spiny cupule. In Belize, consisting of 1 genus and 8 species. Uses for 1 genus and 1 species are reported here.

Quercus oleoides Schltdl. & Cham.

live oak, oak [English]

The fruit is eaten as a **food** by many animals, including squirrels, gibnut (*Agouti paca*), and peccary (*Pecari tajacu* and *Tayassu pecari ringens*) [Marsh, Matola, and Pickard, 1995]. The fruit is roasted and eaten as a **food** by humans [Marsh, Matola, and Pickard, 1995].

The wood of this tree is excellent **firewood**, as it is slow burning and can also be made into **charcoal** [Marsh, Matola, and Pickard, 1995].

GENTIANACEAE

Small herbs, shrubs, or trees. Leaves usually opposite, sometimes alternate, simple, often somewhat leathery. Inflorescences usually terminal or axillary, in dichotomously branched cymes, spikes, sometimes racemes, or of solitary flowers. Flowers often showy and brightly colored. Fruits a capsule or berry. In Belize, consisting of 7 genera and 15 species. Uses for 3 genera and 3 species are reported here.

Coutoubea spicata Aubl.

This plant is believed to have **magical powers**. To **induce someone to do a favor**, 1 entire plant is mashed in 2 gallons of water with "a bit of rosewater and holy water from church," steeped for 1 hour in the sun, and used as a bath, either by immersion or by pouring over the head, depending on if you have a tub, shower, or an outside stall [B2686].

(top) *Quercus oleoides* [GS].

(bottom) *Coutoubea spicata* [NYBG].

Lisianthus axillaris (Hemsl.) Kuntze

Spanish girl earring [English]; *maravilla del monte* (male) [Spanish]; *red chilar* [Additional name recorded]

For **grave illnesses**, the leaves are 1 component of a Maya bath mixture boiled in water, cooled, and used as a bath at room temperature once daily in the evening [A159].

(top) *Lisianthus axillaris* [MB].

(bottom) *Codonanthe crassifolia* [RF].

To treat *pasmo* (**congested blood; more than the usual amount in an area of the body**), 1 quart of leaves is boiled in 1 gallon of water and used as a bath. Depending on the site of bruising, the bath could be by immersion or by pouring over the affected area [A345].

Caution is used with this plant. It is not to be taken internally, as it is toxic and causes nausea and vomiting [LR].

Schultesia lisianthoides (Griseb.) Benth. & Hook. f. ex Hemsl.
k'es ru pim [Q'eqchi']
To treat **colds** and **influenza**, 9 leaves are mashed in ½ cup of cool water, soaked until soft and the water is very green, placed in a cloth, and applied directly to the forehead for 1 hour daily as needed [B3519].

GESNERIACEAE

Herbs or subshrubs, sometimes lianas, rarely shrubs or small trees, terrestrial or epiphytic. Leaves usually opposite, equally to strongly unequal, sometimes alternate, rarely whorled, simple. Inflorescences axillary or terminal, in cymes, racemes, or of solitary flowers. Flowers inconspicuous to large and showy, yellow, red, orange, or greenish to white. Fruits a capsule, sometimes a berry. In Belize, consisting of 8 genera and 14 species. Uses for 2 genera and 2 species are reported here.

Codonanthe crassifolia (H. Focke) C.V. Morton
tep pim, kaq ipa [Q'eqchi']
For **pain** and **swelling** from **bites** or **skin irritation**, 1 leaf is torn in half and rubbed on the affected area 3 times daily. It is the juice from the leaf that stays on the skin as a treatment [B3562].

Drymonia oinochrophylla (Donn. Sm.) D.N. Gibson
  sak eh chók [Q'eqchi']
  To treat a **cough** or a **sore throat**, 1 entire plant is boiled in 1 quart of water for 10 minutes, steeped for 10 minutes, strained, and ½ cup consumed every 2 hours until better [B3556; RA]. Additionally, this same decoction is used as a bath for 20 minutes once daily with the oral treatment [B3556; RA].

HAMAMELIDACEAE

Shrubs or trees. Leaves alternate, distichous, or rarely spiral, simple. Inflorescences axillary, in spikes, heads, thyrses, or panicles. Flowers usually small, yellow, reddish, white, or green. Fruits a capsule. In Belize, consisting of 1 genus and 1 species. Uses for 1 genus and 1 species are reported here.

Liquidambar styraciflua L.
  liquidambar [English]
  Standley and Record (1936) noted that this tree was the source of a **fragrant balsam**, employed in local and traditional **medicine** in Honduras and elsewhere in Central America, that is obtained by making a gash in the trunk.

Liquidambar styraciflua [NYBG].

LAMIACEAE

Herbs, sometimes shrubs, rarely trees or vines, often aromatic. Leaves opposite or in whorls of 3 or more per node, usually simple or rarely compound. Inflorescences axillary or terminal, of solitary flowers or in dense clusters. Flowers showy. Fruits a mericarp or nutlet. In Belize, consisting of 10 genera and 29 species. Uses for 5 genera and 9 species are reported here.

Hyptis capitata Jacq.
  charlston, charleston [English]; *xk'ot kaway* [Q'eqchi']
  As a **panacea** to treat all sicknesses, nine 12-inch branches are boiled in 1½ gallons of water for 10 minutes, steeped away from the heat, and used comfortably warm as a bath for the entire body once daily, at noon or bedtime, 3 times weekly [A356].
  For **pain**, 1 leaf is rubbed on the skin or 1 entire plant used in a bath [B2612, B3488].
  For **dry cough**, one 12-inch branch with leaves is boiled in 3 cups of water for 5 minutes, steeped away from the heat for 20 minutes, and 1 cup consumed hot 3 times daily, before meals, until better [B2612].

Hyptis pectinata (L.) Poit.

John Charles [English]

To improve the action of other herbs in alleviating ailments, the leaves and branches are added to herbal formulas to work as a **catalyst for healing** [A190]. This plant is considered one of the **"thinking herbs"** used to determine just what a person requires from a formula [A190].

To treat **coughs** and as a **panacea** to treat all ailments, nine 12-inch branches are boiled in 1½ gallons of water for 10 minutes, steeped away from the heat, and used comfortably warm as a bath for the entire body once daily, at noon or bedtime, 3 times weekly [A190].

Hyptis suaveolens (L.) Poit.

Jamaican spikenard, oregano [English]

For **kidney infection**, three 12-inch branches with leaves and one 6-inch piece of root are boiled in 3 cups of water for 10 minutes and 1 cup consumed once daily, in the early morning before breakfast, for 9 days [B2200]. Alternatively, 1 large handful of leaves and branches is boiled in 2 gallons of water for 10 minutes and used warm as a bath once daily, at bedtime, for 9 days [B2200].

Hyptis verticillata Jacq.

John Charles, John Charles weed [English]; *hierba Martín, hoja de Martin, Juan Carlos* [Spanish]; *xkis kaway, xk'ot kaway* [Q'eqchi']

People add this leaf to **herbal medicines** to strengthen the effect and to improve the taste. Specific quantities depend on the formula; the aromatic leaves lessen the bitter taste of other medicinal plants, making the decoction easier to consume [B1902; Arvigo and Balick, 1998].

For **malaise** of children and infants or to treat **general sickness**, 1 entire plant is boiled in 1 gallon of water for 20 minutes and used as a bath [Arvigo and Balick, 1998]. To treat **colds**, **spinal cord ailments**, and **excess mucus** in the bowels or urine,

three 9-inch branches are boiled in 1 quart of water for 5 minutes and 1 cup consumed 3 times daily before meals [A479, A798; HR].

To treat **colds**, 1 full handful of leaves, young stems, and flowers is mashed in cold water and consumed twice daily for 5–6 days until better [C29]. To treat **high fever**, 1 small handful of leaves and branchlets along with an unspecified amount of

Hyptis verticillata [MB].

lemon grass leaves (*Cymbopogon citratus*) are steeped in 3 cups of water for 20 minutes and ⅓ cup consumed warm throughout the day as needed [Arvigo and Balick, 1998]. To treat **fever** and **malaria**, an unspecified part of the plant is soaked in water, but no further details were provided [Robinson and Furley, 1983].

For **cough**, **mucous ailments**, onset of **asthma**, **fever**, **tonsillitis**, **uterine fibroids**, or **bronchitis**, 1 small handful of leaves and branchlets is steeped in 3 cups of water for 20 minutes and ⅓ cup consumed warm throughout the day as needed [Arvigo and Balick, 1998]. For **cough**, nine 6-inch sprigs, optionally prepared with 2 ounces of Billy Webb leaves (*Acosmium panamense*), are boiled in 3 cups of water for 10 minutes and the decoction sipped all day until 3 cups are consumed [A355, A479].

For **cradle cap sores** in infants, the leaves are used to bathe the affected areas [Arvigo and Balick, 1998]. Additionally, the leaves are toasted, powdered, and sprinkled over the sores until better [Arvigo and Balick, 1998]. For **contact dermatitis** from the *manchineel* tree (*Hippomane mancinella*), 1 entire plant along with an unspecified amount of *gumbolimbo* (*Bursera simaruba*) bark are boiled in water and used as a bath [Arvigo and Balick, 1998].

For **stomach pain**, 1 handful of leaves and inflorescence is boiled in 3 cups of water for 5 minutes and sipped hot all day until better [B1902]. To treat **kidney and ovary ailments**, nine 6-inch sprigs are boiled in 3 cups of water for 10 minutes and ½ cup consumed warm 6 times daily until better [A355, A479]. For **severe gastric distress** or to **"cool" kidneys** that are hot, referring to a **burning in the urine**, **kidney infection**, or **kidney inflammation**, the roots are boiled in water and consumed as a tea [A479; Arvigo and Balick, 1998]. For **uterine ailments**, refer to wild basil (*Ocimum campechianum*) [A941]. To relieve **pain** following childbirth, 1 small handful of either the roots, leaves, or both is boiled in 3 cups of water for 10 minutes and 1 cup consumed 3 times daily until better [Arvigo and Balick, 1998].

To rid animals of **worms**, the root is chopped up and mixed into the feed [Arnason et al., 1980]. Standley and Record (1936), referring to information ascribed to noted botanist W.A. Schipp, reported that the leaves were mashed and put in chicken's nests to "drive away vermin."

As an **insecticide**, refer to wild annona (*Annona primigenia*) [LR].

Ocimum basilicum L.

basil [English]; *albahaca* [Spanish]; *balsley*, *barsley* [Creole]; *ca cal tun* [Mopan]
The fresh or dried leaves are used as a **spice** or **flavoring**—for example, in soups, meat sauces, tomato dishes, and salads [R70; Arvigo and Balick, 1998]. The leaves are also used to prepare a **beverage** [R70].

To treat **headache**, 3 branches are boiled in 1 quart of water for 10 minutes and used to bathe the head at bedtime [Arvigo and Balick, 1998]. For **earache**, juice extracted from the leaf is placed in the ear [Arvigo and Balick, 1998]. To discourage **cataracts**, clear **eye mucus**, or treat **phlegm in the eyes**, 1 seed is dried, placed in the corner of each eye, and left there overnight. In the morning, it is removed or falls out in the stream of mucus from the eye [R70; Arvigo and Balick, 1998].

Ocimum basilicum [MB].

To **promote sweat** during a **fever**, to treat **intestinal parasites**, or for **stomach pain**, 1 small handful of fresh leaves is boiled in 2 cups of water for 2 minutes, steeped for 20 minutes, and consumed [Arvigo and Balick, 1998]. For **sores**, especially caused by worms or larvae, the leaves are dried, powdered, and applied directly to the affected area [Arvigo and Balick, 1998].

To **promote delayed menstruation**, ease **painful or difficult menstruation**, or facilitate **childbirth**, 1 entire plant is boiled in 3½ cups of water for 5 minutes, steeped for 10 minutes, and 1 cup consumed hot every hour as needed [RA; Arvigo and Balick, 1998]. Following childbirth or on the onset of menstruation, three 12-inch branches with leaves are boiled in 1 gallon of water for 5 minutes and used as a vaginal steam bath for 20 minutes once after childbirth or once monthly on the first day of menstruation, serving as a **uterine lavage** [RA; Arvigo and Balick, 1998].

To treat **spiritual ailments**, such as **fright** (*susto*), **envy** (*envidia*), **grief** (*pesar*), and **evil magic** (*obeah*), fresh leaves and stems are used to prepare a bath [Arvigo and Balick, 1998]. Alternatively, refer to bay cedar (*Guazuma ulmifolia*) [RA]. To prevent these **spiritual ailments** listed above, a sprig is carried around [Arvigo and Balick, 1998]. For people who suspect they have been affected by **evil magic** (*obeah*), refer to *dormilón* (*Mimosa pudica* var. *unijuga*) [HR].

Ocimum campechianum Mill.

basil, wild basil [English]; *albahaca, albahaca silvestre, albahaca cimarrona, simarona* [Spanish]; *barsley, basu* [Creole]; *b'enk, ish bénk, orrek, ratz um canán, t'eb, zupub' li zaab'* [Q'eqchi']; *ke kel tèn, ke kel tún, cal tun, ish bnk, orr réh* [Mayan]

The leaves are used as a **condiment**, **spice**, and **marinade** to season food, including soups and meat dishes [A225, A560, B3625, G24].

To treat a **fever** or **low back pain**, 1 handful of fresh leaves is boiled in water and consumed warm 3 times daily for 14 days [C18]. To **promote sweat**, the plant is boiled in water and consumed [Robinson and Furley, 1983]. For **headache**, the leaves are steeped in hot water and poured warm over the head as a bath at bedtime [B2112, W1840]. Alternatively, the leaves are mashed in 1 gallon of water and the head bathed with a cool infusion in the morning or a warm infusion in the evening [B2432]. For **earache**, 3 drops of juice extracted from the warmed leaves are placed in the ear 2–3 times daily until better [B2112].

For **blurred vision, cataracts, "bad eyes," weak eyes**, or a **burning sensation**,

Ocimum campechianum [GS] with close-up of inflorescence (right) [MB].

dust, or dirt in the eyes, 1 small seed is placed in each eye, the eyes are closed, and the combination of the herb and the tearing releases the infection. Seeds are left in the eye overnight, and they will turn from brown to white. They are said to draw out mucus from the eyes and the cataracts. This is repeated daily until better [A941, B2112, B3693, C18, W1840]. To **cleanse the eyes** or **delay cataracts**, 9 leaves are mashed in 1 cup of water, strained well, and used as eyewash [G24].

For **skin sores**, juice from the leaves is applied directly to the affected area [B3625]. To treat **skin ailments**, especially **hand sores** or a **baby's sun rash**, fresh leaves are heated, mashed, and applied directly to the affected area [G24]. For **kidney infections**, **burning urine**, or as a **blood purifier**, 1 handful of fresh leaves is boiled in water and consumed warm 3 times daily for 14 days [C18].

To treat **stomach gas** or **stomach pain**, three 6-inch branches with leaves are boiled in 1½ cups of water for 5 minutes and ½ cup consumed warm every 30 minutes until better [A225, A941]. For **stomach gas**, 1 small handful of leaves is boiled in 2½ cups of water for 5 minutes and 1 cup consumed cool twice daily until better [B2526]. For **baby colic**, 1 entire plant is boiled in 1 gallon of water for 10 minutes, 1 teaspoon consumed, and the remainder used warm as a bath once daily, at naptime or bedtime, for 9 days [B3693]. For **diarrhea**, six 12-inch branches are boiled in 4 cups of water for 5 minutes and sipped all day long until better [W1840]. Steggerda (1943) reported that the Yucatan Maya boiled the entire plant in water and used this as a beverage to treat **dysentery**.

For **uterine ailments**, nine 12-inch branches with leaves, used alone or in combination with John Charles (*Hyptis verticillata*), marigold (*Tagetes erecta*), or Spanish thyme (*Lippia graveolens*), are steeped for 5 minutes in 2 gallons of previously boiled water and used as a sitz bath once monthly, 2 days before the onset of menstruation [A941]. To **prevent vaginal infections**, the same combination of leaves as described above for uterine ailments is boiled in 2 gallons of water for 5 minutes and used as a vaginal steam bath for 20 minutes once monthly, 5 days before the onset of menstruation [B2432].

To treat **snakebite**, 1 cup of fresh leaves is mashed in ¼ cup of water, applied as a poultice, and changed every hour until festering decreases or the "tooth scales" come out [W1840].

Steggerda (1943) reported that the Yucatan Maya used the leaves to rub on their horses "to prevent the **bites of horseflies**."

To **dispel witchcraft**, 1 entire plant is steeped in water and sprinkled around the house while saying prayers [A225]. Additionally, the plant is hung in the doorway [A225].

Plectranthus amboinicus (Lour.) Spreng.

English thyme, Spanish oregano, Spanish thyme [English]

The plant is used as a **condiment** for foods [A214].

To **loosen phlegm** and ease a **cough**, 3 leaves are chewed raw or steeped in 1 cup of water for 10 minutes and 1 cup consumed 3 times daily until better; usually, a 1-day treatment is sufficient [A214]. To ease **earache**, 3 drops of juice from the warmed leaves are squeezed into the ear, the ear wrapped loosely with a cotton cloth, and repeated 3 times daily until better [A214, B2095]. For **swelling**, 1 handful of leaves is mashed and applied directly to the affected area [B2095].

For **stomach gas**, 9 leaves and one 3-inch tip of the stem are mashed, soaked in 1 cup of cool water for 30 minutes, strained, and ½ cup consumed as needed until better [B2315]. For **stomach pain**, 1 teaspoon of extracted juice from the heated leaves is consumed as needed [B2315].

Plectranthus amboinicus [MB].

Salvia coccinea Buc'hoz ex. Etl.

clavel, fósforo, pasmo [Spanish]; *chac te pec, chac tepec, chacalpec* [Mopan]

To treat **varicosities**, a type of ***pasmo* (congested blood; more than the usual amount in an area of the body)** that is found any place on the body having **bruises, blood clots, congested blood**, or **gout**, 1 handful of the entire fresh plant is boiled in 1 quart of water for 10 minutes, ½ cup consumed warm, and the remainder used to bathe the affected area; this treatment is repeated once daily until better [B1767]. For **very painful varicose veins**, ¼ pint of flowers along with ¼ pint marigold flowers (*Tagetes erecta*) are soaked in 1 pint of olive oil for 7 days and rubbed over the affected area [Arvigo and Balick, 1998]. For **spasms**, refer to *Indigofera suffruticosa* [Steggerda and Korsch, 1943].

Teucrium vesicarium Mill.

pega ropa [Spanish]; *mosote* (male) [Mayan]

For **general ailments**, 1 large double handful of leaves is used in a Maya bath mixture. Plants used in the mixture might include John Charles (*Hyptis verticillata*), skunk root (*Chiococca alba*), marigold (*Tagetes erecta*), and/or Spanish thyme (*Lippia graveolens*). Other plants are chosen from hundreds of medicinal leaves and are used in a formula of 4 or 9 different species. There is no specific formula—it is up to the traditional healer to decide which mixtures of leaves to use [A205].

(top) *Salvia coccinea* [MB].

(bottom) *Teucrium vesicarium* [GS].

For **skin sores** and **rashes**, ½ cup of leaves is mashed and applied as a poultice directly to the affected area until better [A231]. To treat **skin ailments**, the leaves are toasted, powdered, and applied on top of castor oil (*Ricinus communis*) [A496]. For **stomach pain**, three 12-inch branches with leaves are boiled in 3 cups of water for 10 minutes and 1 cup consumed warm 3 times daily before meals [A231]. For **diarrhea**, 9 stems with leaves are boiled in 1 cup of water for 5 minutes and consumed 3 times daily, before meals, for 3 days [A496].

Both the leaves and seeds stick to clothing, as the name *pega ropa* suggests [RA].

LAURACEAE

Trees, sometimes shrubs, or rarely vines. Leaves usually alternate or in spirals, sometimes pseudo-opposite or opposite, simple, often aromatic. Inflorescences axillary, usually compound, in racemes, umbles, or thyrses. Flowers small. Fruits a drupe. In Belize, consisting of 8 genera and 31 species. Uses for 5 genera and 7 species are reported here.

Cassytha filiformis L.
jaundice tie tie [English]
The mature, white fruit is eaten as a **food** [A375].

For **hepatitis**, 9 fruits along with 9 fruits of "may plum" (an unidentified species) are boiled in 2 quarts of water for 20 minutes, strained, and ½ cup consumed 6 times daily for 3 days. This treatment is repeated until better [A375].

Licaria peckii (I.M. Johnst.) Kosterm.
bastard oak, laurel, timber sweet [English]; *tzo otz né* [Mayan]
The trunk is used in **construction** to make house posts [B2457].

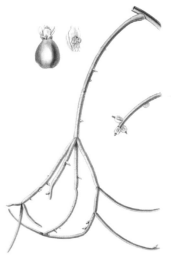

Nectandra salicifolia (Kunth) Nees
black water, black wattle, laurel, timber sweet [English]; *aguacatillo, el amor (segundo)* [Spanish]
This is one of the most respected herbs, used commonly by Don Eligio Panti [B1896] to treat a **variety of ailments**; 1 small handful of leaves is a component of a Maya bath mixture [B1896, B2245].

To treat **nervousness**, 1 large handful of leaves with equal portions of this plant and *ginda* (*Calyptranthes chytraculia* var. *americana*) is boiled in 2 gallons of water for 20 minutes and poured warm over the person as a bath at bedtime as needed [A399]. For **ringworm**, 1 handful of

Cassytha filiformis [NYBG].

leaves is crushed with a bit of salt and applied as a poultice to the affected area twice daily for 9 days [A957].

The wood is used for **fuel** [A469].

This plant is considered the "female" counterpart of *eremuil* (*Malmea depressa*) [B1896].

Nectandra salicifolia [MB].

Nectandra sp.

timber sweet [English]

The trunk is used in **construction** for house posts [A81].

Ocotea veraguensis (Meisn.) Mez

copalchi [Spanish]; *copal chi* [Mayan]

For **bladder infection**, ½ cup of bark is boiled in 1 quart of water for 10 minutes, steeped for 15 minutes, and 1 cup consumed 3 times daily, before meals, until better [A841]. To treat non-insulin-dependent **diabetes** mellitus, ½ cup of the decoction described above for bladder infection is consumed 4 times daily for 5 days of each week and the blood sugar level continually checked [A841]. For **dysmenorrhea**, **fertility**, and to **cleanse the uterus**, 1 cup of the decoction described above for bladder infection is consumed warm by women 3 times daily, before meals, 5 days before menstruation for 3 cycles [A841]. To **purge, reverse, or "wash out" a male contraceptive**, refer to Billy Webb (*Acosmium panamense*) [B1848, B2220].

For **cuts**, the decoction described above for bladder infection is used cool to bathe the affected area 3 times daily as needed [A841]. As a first-aid remedy to treat **snakebite**, 1 cup of the decoction described above for bladder infection is consumed cool 3 times daily and immediate help is sought [A841]. For **snakebite**, refer to *cramantee* (*Guarea glabra*) [LR]. To treat non-insulin-dependent **diabetes** mellitus and **snakebite**, one 8-inch piece of bark is boiled in 1 gallon of water for 10 minutes, cooled, and consumed in small doses twice daily for 15 days [C90].

Persea americana Mill.

avocado, avocado pear, butter-pear, pear [English]; *aguacate*, *aguacote* [Spanish]; *on* [Mopan]; *o* [Q'eqchi']

The fruit is eaten as a **food** [B2343, B2411].

For **cough**, refer to *guanábana* (*Annona muricata*) [B2107, B2108, B2400, B2403, B2411, B2458]. To treat **cough** or a **bruise**, 1 handful of leaves along with 1 handful of mango leaves (*Mangifera indica*) are boiled in 2 cups of water and 1 cup consumed cool twice daily for 3 days [B3324]. To treat **cough** and **colds**, refer to *maravilla* (*Mirabilis jalapa*) [B2283].

Persea americana [MB].

For **sprains**, **rheumatism**, or **headache**, the leaves are mashed and applied as a poultice to the affected area [Arvigo and Balick, 1998]. Alternatively, the leaves are coated with cooking oil, placed on the temples, and secured with a cloth or bandana [B2397]. When suffering from **new or infected wounds or sores**, it is advised *not* to eat the fruit, as it is believed to encourage pus formation, thus delaying recovery [Arvigo and Balick, 1998].

To treat **coughs**, **colds**, **fever**, **diarrhea**, **high blood pressure**, or **painful menstruation**, 3 leaves are boiled in 3 cups of water for 10 minutes and 1 cup consumed 3 times daily before meals [Arvigo and Balick, 1998]. Steggerda (1943) reported that the Yucatan Maya applied fresh leaves directly to the feet of a person suffering from **fever**. For **diarrhea**, 1 handful of bark is boiled in 2 cups of water for 10 minutes and 2 cups consumed cool daily until better [B3324]. Alternatively, the seed is cut up, roasted, boiled in 3 cups of water, and 1 cup consumed 3 times daily, before meals, until better [Mallory, 1991].

For **intestinal obstructions** (*empacho*), the seeds are ground or mashed, boiled in 2 cups of water for 10 minutes, and 1 cup consumed hot twice daily [Arvigo and Balick, 1998]. To ease onset of **asthma** and the **wheezing** caused by it, honey is added to the decoction described for intestinal obstructions (*empacho*) and consumed 3 times daily [Arvigo and Balick, 1998]. To treat the "**culebrilla worm**" in babies, refer to *sikiya* (*Chrysophyllum mexicanum*) [B2115].

To treat non-insulin-dependent **diabetes** mellitus ("strong or sweet urine"), the seeds are prepared and consumed as a tea as needed [Mallory, 1991].

As a **female contraceptive**, the seeds are dried, ground into a powder, 1 teaspoon mixed in ½ cup of warm water, and ½ cup consumed 3 days after the menstruation. This treatment is repeated every 2 weeks but gives protection only while dosing [LR]. Alternatively, 3 seeds are dried, chopped, boiled in 3 cups of water for 10 minutes, and consumed orally. This treatment is repeated monthly [Arvigo and Balick, 1998]. It is said that this treatment can lead to **sterility** after 1 year of constant use [Arvigo and Balick, 1998].

The seed yields oil good for the hair as a **shampoo** [LR]. As a **hair conditioner**, the fruit is mashed and applied directly to the hair and skin [Arvigo and Balick, 1998].

Persea schiedeana Nees
wild pear [English]
Standley and Record (1936) reported that while the skin of the species was thick, the flesh was of excellent flavor with an oily texture and used as a **food**.

LOGANIACEAE

Herbs, shrubs, lianas, or small to large trees. Leaves opposite, simple. Inflorescences axillary or terminal, usually in cymes. Flowers showy. Fruits a capsule or berry. In Belize, consisting of 4 genera and 9 species. Uses for 2 genera and 5 species are reported here.

Spigelia humboldtiana Cham. & Schltdl.

lombricera [Spanish]

Standley and Record (1936) reported that this plant was widely used in Central America and elsewhere "to expel **tapeworms** and other **intestinal parasites**."

Strychnos brachistantha Standl.

chicoloro [Spanish]

For **cough**, one 9-inch piece of the stem is soaked in 1 quart of water and 1 cup consumed 3 times daily, before meals, until better [B2038].

To treat non-insulin-dependent **diabetes** mellitus, one 8-inch piece of the vine and one 5-inch piece of the root are boiled in ½ gallon of water for 10 minutes, cooled, and consumed as a beverage twice daily for 10 days [C91]. **Caution** is used with this treatment, as overdosage can be toxic [C91]. To **purge, reverse, or "wash out" a male contraceptive**, refer to Billy Webb (*Acosmium panamense*) [B1848, B2220].

As a first-aid remedy to treat **snakebite**, one 9-inch piece of root is boiled in 1 quart of water for 20 minutes and the entire decoction consumed as hot as possible all at once [B2038].

Strychnos panamensis Seem.

var. *panamensis*

chicoloro [Spanish]

The fruit is eaten as a **food** [B2172].

For a Maya bath formula used as a **panacea** and especially to treat **stomach ailments**, 1 double handful of a leaf mixture (including this plant and other species) is boiled in 2 gallons of water, ¼ cup consumed, and the remainder poured in cupfuls over the body as a bath once weekly; sometimes 3 baths are given over a period of 7–10 days until better [B1788].

For **female ailments, stomach disorders**, and to **cleanse the spleen**, 1 small handful of the chopped vine is boiled in 3 cups of water for 10 minutes and ½ cup consumed once or twice daily before meals [B1788]. Dosage should not exceed more than 1 cup daily, otherwise the treatment is **toxic** [LR].

As a first-aid remedy to treat **snakebite**, one 9-inch piece of root is boiled in 1 quart of water for 20 minutes and consumed as hot as possible all at once. Used alone to

(top) *Spigelia humboldtiana* [DA].

(bottom) *Strychnos panamensis* var. *panamensis* [MB].

treat snakebite, it is considered a "weak medicine" [B2172]. Alternatively, refer to blossom berry (*Eugenia capuli*) [A911] and Spanish cedar (*Cedrela odorata*) [Arnason et al., 1980].

For **gastritis** (ciro), non-insulin-dependent **diabetes** mellitus, **delayed menstruation**, **colds**, or **influenza**, refer to Billy Webb (*Acosmium panamense*) [P3]. To treat non-insulin-dependent **diabetes** mellitus, refer to periwinkle (*Catharanthus roseus*) [B2358].

Strychnos cf. *panamensis* Seem. var. *hirtiflora* Standl.
o max [Q'eqchi']
To treat **influenza**, 1 small handful of fresh root and 1 handful of fresh stem are boiled in 1 quart of water for 10 minutes and consumed warm 6 times daily for 8 days [C38].
This plant is also used to **dispel witchcraft** [C38].

Strychnos peckii B.L. Rob.
holob'oob' k'aham [Mayan]
For **dizziness** or **light-headedness**, 1 double handful of leaves is boiled in water and 1 cup consumed cool 3 times daily for 3 days [B3613].

LORANTHACEAE

Shrubs, sometimes small trees or vines, hemiparasitic on woody dicots, evergreen. Leaves frequently opposite, sometimes alternate or whorled, simple, leathery or fleshy. Inflorescences axillary and/or terminal, in spikes, racemes, or umbels. Flowers sometimes large and brightly colored. Fruits a berry or drupe. In Belize, consisting of 4 genera and 9 species. Uses for 2 genera and 3 species are reported here.

Psittacanthus mayanus Standl. & Steyerm.
matapalo (female) [Spanish]
For **headache**, **fever**, or feeling **overheated**, 1 handful of leaves along with 1 handful of wild ruda leaves (*Diphysa* sp.) are mixed, squeezed in 2 gallons of water, steeped in the sun for 1 hour, and poured over the head as a cooling bath [B2160]. To treat **headache**, refer to God Almighty bush (*Struthanthus orbicularis*) [B2161; RA].
This is considered the "female" counterpart to the God Almighty bush [B2161; RA].

Struthanthus cassythoides Millsp. ex Standl.
God Almighty bush, scorn the earth [English]; *matapalo* [Spanish]
As a treatment for **colds**, **influenza**, or **asthma**, one 12-inch piece of vine is boiled in 3 cups of water for 5 minutes, steeped for 10 minutes, and sipped as hot as possible for 1 day. The asthma attack may stop after consumption of the first cup [A680].
To reduce **swelling**, nine 12-inch pieces of the entire vine are boiled in 2 gallons of water for 10 minutes and used cool as a wash over the affected area 3 times daily [B1823]. To prevent **infection** or **swelling** from a dog bite, 1 handful of leaves is

Psittacanthus mayanus [DWS].   *Struthanthus cassythoides* [MB].

mashed with 2 garlic cloves, applied as a poultice directly to the affected area, wrapped with a cloth, and changed once daily for 4–5 days or until completely healed [LR].

For **skin ulceration**, especially of the leg, one 12-inch piece of the entire vine is either fried in ½ cup of oil for 5 minutes or soaked in rum for 24 hours and applied as a liniment directly to the affected area [B1823].

To treat **internal tumors** (growths or perhaps "cancer"), 1 large handful of vines and leaves along with one 12-inch piece of *callawalla* (*Phlebodium decumanum*) are boiled in 1 gallon of water for 10 minutes and 1 cup consumed twice daily for 30 days. Following the 30-day treatment, a laxative is used [LR]. "This is supposed to be the cure," and it is said to dissolve the growths or tumors [LR].

Struthanthus orbicularis (Kunth) Blume

God Almighty bush, [English]; *matapalo* (male) [Spanish]; *scorn de earth* [Creole] For **headache**, ½ quart of fresh leaves and ½ quart of fresh wild ruda leaves (*Diphysa* sp.) are mashed in 1 gallon of water, steeped for 30 minutes, and used tepid to bathe the head at bedtime [B2161]. Additionally, *matapalo* (*Psittacanthus mayanus*) may be added to the infusion [B2161; RA]. To **lower blood pressure**, the plant is boiled in water and an unspecified amount consumed [Robinson and Furley, 1983].

This plant is considered the "male" counterpart to *matapalo* [B2161; RA].

LYTHRACEAE (INCLUDING PUNICACEAE)

Herbs, shrubs, or trees. Leaves usually opposite or subopposite, simple. Inflorescences axillary or terminal, in thyrses, cymes, umbels, or of solitary flowers. Flowers red, purple, white, or yellow. Fruits a capsule or berry. In Belize, consisting of 6 genera and 13 species. Uses for 3 genera and 3 species are reported here.

Cuphea calophylla Cham. & Schltdl.
hog bush, hog weed, Mexican heather [English]; *corriente, corrimiento, mañanita* [Spanish]; *schma mo cal, x'mamocal* [Mopan]

To treat **nervousness** in babies, 1 small plant about 6 inches high is boiled in 2 quarts of water for 5 minutes, 1 ounce consumed, and the remainder used warm as a bath once daily at bedtime [B2611].

To treat **diarrhea** resulting from dysentery, exhaustion, or chronic tiredness, 1 entire plant is boiled in 2 quarts of water for 10 minutes and sipped cool all day until better [Arvigo and Balick, 1998] or used as a bath [B2328; Arvigo and Balick, 1998].

For **exhaustion**, 1 double handful of leaves is boiled in 3 gallons of water for 10 minutes and used as a bath once daily, at noon or bedtime [A642, B1829].

Cuphea calophylla [MB].

To aid the **fertility** of weak or undernourished women, 1 plant is boiled in 1 quart of water for 5 minutes and ½ cup consumed 6 times daily for 9 days of every month before ovulation. This treatment is repeated as needed [RA]. For **infertility** due to general weakness, 9 branches are boiled in 3 cups of water for 5 minutes and 1 cup consumed 3 times daily, before meals, for 9 days preceding menstruation [Arvigo and Balick, 1998].

For a **snakebite** treatment, refer to wild mahogany (*Mosquitoxylum jamaicense*) [LR].

Lagerstroemia speciosa (L.) Pers.
crepe myrtle [English]

As a **diuretic**, 2 leaves are boiled in 3 cups of water for 10 minutes and sipped all day, not to exceed 6 cups weekly [Arvigo and Balick, 1998]. To treat **wounds** and **infections**, one 3-inch × 1-inch piece of bark is boiled in 2 quarts of water for 10 minutes and used as a bath over the affected area [Arvigo and Balick, 1998].

Punica granatum L.
pomegranate [English]; *granada, granadillo* [Spanish]

To treat **diarrhea**, ½ cup of the outer skin of the fruit is boiled in 1 quart of water for 5 minutes and ½ cup consumed every 3 hours until better [B1973].

(top) *Lagerstroemia speciosa* with inset showing inflorescence [MB].

(bottom) *Punica granatum* [FA].

To treat **anxiety**, **insanity**, and for use as a **tranquilizer**, 1 small handful of either fresh or dried bark is boiled in 1 quart of water for 10 minutes and ½ cup consumed 6–8 times daily until better [B2434].

For **itchy skin**, refer to wire wis (*Lygodium venustum*) [B2190]. To treat a **rash**, refer to *mañanita* (*Portulaca pilosa*) [B2099, B2100].

To treat unspecified **women's ailments**, the leaf is boiled in water and the liquid used in an unspecified way [Arnason et al., 1980].

MALPIGHIACEAE

Trees, shrubs, or herbs. Leaves usually opposite, sometimes whorled, occasionally subopposite or alternate. Inflorescences terminal or axillary, in racemes or panicles. Flowers yellow, pink, white, or rarely blue. Fruits usually a drupe or berry, sometimes a schizocarp, mericarp, or nutlet, sometimes winged. In Belize, consisting of 10 genera and 36 species. Uses for 6 genera and 8 species are reported here.

Bunchosia lindeniana A. Juss.

café silvestre [Spanish]; *wild coffee* [Creole]

To treat **back pain** and **muscle spasms**, 9 fresh leaves are mashed and applied to the affected areas as a plaster twice daily for 2–3 days [C17]. For **insomnia**, **hysteria**, **nightmares**, **back pain**, and **muscle spasms**, 18 fresh leaves are boiled in water and used as a bath twice daily for 2–3 days [C17].

    Caution is advised, as the leaves may irritate the skin [C17].

Bunchosia lindeniana [GS].

Byrsonima bucidaefolia Standl.

craboo [English]

Standley and Record (1936) reported that the fruit was eaten as a **food** and occasionally sold in local markets. Steggerda (1943) reported that the fruit had a sour taste but that the Yucatan Maya ate it with salt, vinegar, and anise liquor (*Pimpinella anisum*) or cooked it with sugar as a dessert.

Byrsonima crassifolia (L.) Kunth

craboo, sour craboo, wild craboo [English]; *crabu, grabo, nance* [Spanish]; *chi'* [Q'eqchi']; *chà, nanchi, nanci, zacpa* [Additional names recorded]

The yellow fruit is eaten as a **food** [A241, A939, B1821, B2269, B2402, JB95]. Standley and Record (1936) reported that the fruit was pickled and used as a **food** and that the wood was used for general **construction**.

    To treat **sores** or **diaper rash**, two 3-inch pieces of bark are boiled in 2 cups of water for 5 minutes, cooled, and 1 tablespoon consumed without any sugar 3 times daily or used as a bath over the affected area. Special care must be taken not to boil the bark for too long, as this may concentrate the tannins to a point exceeding the therapeutic dosage [A939, B2269]. To treat **infected, stubborn sores** or **skin rash**, two 3-inch pieces of bark and 9 leaves are boiled in 2 gallons and 2 cups of water for 10 minutes, used as a wash or bath twice daily, and the sores left to air dry [B2402]. To **cleanse wounds**, refer to guava (*Psidium guajava*) [B2273].

Byrsonima crassifolia [MB].

Heteropterys brachiata showing winged seeds (left) and flowers (right) [WJH].

For **stomach pain** and **vomiting**, one 3-inch piece of bark is boiled in 3 cups of water for 10 minutes and sipped throughout the day [A241]. To treat **dysentery**, 1 ounce of seeds is pounded, boiled in 3 ounces of water for 5 minutes, cooled, and 1 tablespoon consumed every 2 hours until better [B1821]. For **loose stools** and **diarrhea**, ½ pound of bark strips is boiled in 2 quarts of water and 1 cup consumed warm 3 times daily until better [A939, B2269, B3603]. For **diarrhea**, refer to *Sagittaria lancifolia* subsp. *media* [B2171] and guava (*Psidium guajava*) [B2273]. To treat **bloody diarrhea**, refer to *Sagittaria lancifolia* subsp. *media* [B2171]. To treat **diarrhea ("bad stomach")**, refer to sweet orange (*Citrus sinensis*) [B2279].

For a **postpartum treatment** for the vulva, the leaves are used in an unspecified manner as a bath [JB95].

The inner bark can **stain** the skin permanently [A939].

This tree is one of many used to make **charcoal** [Horwich and Lyon, 1990]. The skin of the fruit is used to make a **dye** for inks and dying cotton [Horwich and Lyon, 1990]. The bark is used in **tanning** [Horwich and Lyon, 1990].

Callaeum malpighioides (Turcz.) D.M. Johnson

This is 1 of 9 plants in a bath mixture used as a **panacea** [B1852].

To treat **gastritis (*ciro*)**, **cirrhosis of the liver**, **snakebite**, **diarrhea**, and several other unspecified **ailments**, 1 slight handful of the vine is boiled in 3 cups of water for 10 minutes and 1 cup consumed warm 3 times daily, between meals, until better [B1852].

Heteropterys brachiata (L.) DC.

mi corazón [Spanish]; *in vida ech* [Additional name recorded]

To treat **heart pain**, the leaf is used to prepare a bath [B2234]. Additionally, an unspecified amount of the root is boiled in water and consumed warm [B2234].

(left) *Hiraea reclinata* [GS].

(right) *Malpighia glabra* with inset showing flower [IA].

Heteropterys lindeniana A. Juss

The presence of the plant is an environmental **indicator** signifying water is close by [A658].

The leaves are used to **preserve** fresh meat and fish [A658].

Hiraea reclinata Jacq.

To treat **urine retention (stoppage of water)**, 1 handful of chopped roots is boiled in 3 cups of water for 10 minutes and 1 cup consumed warm 3 times daily, before meals, as needed until better [A748].

Malpighia glabra L.

wild craboo [English]; *nance* [Spanish]; *simche, uzte* [Yucateca]

Standley and Record (1936) reported that the acidic fruit was eaten as a **food** and that the bark was said to have been used for **tanning** leather. They also noted that the fruits bore a few needle-like hairs that could cause intense **irritation** when they penetrated the skin.

MALVACEAE (INCLUDING BOMBACACEAE AND STERCULIACEAE)

Shrubs, trees, or sometimes herbs. Leaves alternate, simple, or palmately compound. Inflorescences axillary, terminal, ramiflorous or cauliflorous, in panicles, cymes, racemes, or of solitary flowers. Flowers variously colored, some large and showy. Fruits a capsule, samara, schizocarp, aggregate of follicles, or sometimes a berry. In Belize, consisting of 35 genera and 75 species. Uses for 27 genera and 38 species are reported here.

Abelmoschus moschatus [FA].

Abutilon hirtum [MB].

Abelmoschus moschatus Medik.

wild okra [English]; *ruh shúl*, *ruxul* [Mayan]

For **pain** all over the body, 9 leaves are boiled in 3½ cups of water for 5 minutes, cooled, and 1 cup consumed warm 3 times daily, before meals, until better [B2528]. To renew **damaged cartilage in joints**, 9 young fruits are boiled in 2 cups of water until the water turns pink, the fruits cooked and eaten, and the water from the pot sipped lukewarm throughout the day until finished [LR].

Abutilon hirtum (Lam.) Sweet

wild cotton [English]; *amor-ano* [Spanish]

For **hemorrhoids**, **anal itching**, and **fistulas**, 9 leaves are heated in ½ cup of hot olive oil for 3 minutes and applied to inside of the anus once daily, before bedtime, for 3 days [B1856].

Abutilon permolle (Willd.) Sweet

arepa [Additional name recorded]

The fruit is used to **sweeten** and **garnish** tortillas and the flower as an **imprint** on tortillas [B2377].

Allosidastrum pyramidatum (Cav.) Krapov., Fryxell, & D.M. Bates

tulipan del monte [Spanish]

This plant is used in baths for a **variety of ailments**, especially **nervousness** and **skin ailments** [A206].

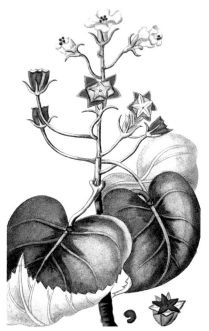

Allosidastrum pyramidatum [NYBG].

Bernoullia flammea
[DA].

Bernoullia flammea Oliv.

red mapola [English]; *amapola, mapola* [Spanish]

Standley and Steyermark (1949) reported that the seeds were "stated to be **edible**." Lamb (1946) noted that the wood was light and perishable and used in **carpentry** for making boxes and veneer.

Byttneria aculeata (Jacq.) Jacq.

tear coat [English]

A bath made from this plant is said to be a treatment for **any ailment** [A363].

Ceiba aesculifolia (Kunth) Britten & Baker f.

Steggerda (1943) reported that the young fruits were boiled and the seeds roasted and eaten as a **food**. He also noted that the young shoots of the tree were said to be an ancient **famine food** of the Maya.

Steggerda and Korsch (1943) reported that the Yucatan Maya collected the stem thorns, boiled them in water, and used the decoction as a wash to help **boils** open.

(left) *Byttneria aculeata* [MB].

(right) *Ceiba pentandra* [FA].

Steggerda (1943) reported that the cottony **fiber** was harvested from the fruits and used to make pillows by the Yucatan Maya of Mexico. That author noted that apparently, this plant was considered somewhat dangerous, as "the Maya believe that as the cotton is blown about by wind at noon, it is set on fire by the hot sun, and that if this blazing cotton falls on a thatch house, the house will burn."

Ceiba pentandra (L.) Gaertn.

ceiba, cotton tree [English]; *inup* [Q'eqchi']; *yax che* [Mayan]

To treat **burning in the urine** or **urine retention (stoppage of water)**, 4 ounces of the root are boiled in 2 cups of water for 10 minutes and 1 cup consumed hot twice daily, before meals, for 3 days [HR].

Standley and Record (1936) noted that oil from the seeds was once used as a **fuel** in lanterns and to make **soap**. Lamb (1946) reported that the wood was used in **construction** to build large dugout canoes, expected to last not more than 2–3 years because the quality was not very hard or durable. He also noted that the cottony **fiber** surrounding the seeds was the source of kapok floss. Thompson (1927) reported that the trunk was used in **construction** to make canoes and that a coarse cotton **fiber** was obtained from the fruit.

Thompson (1927) described in great detail a **ceremony** in Succotz (on March 19th) known as "bringing-in of the ceiba," during which a young tree was harvested, returned to the village, placed in the plaza, and became the central point of the festival. He also observed that this plant was a **sacred** tree of the ancient Maya. This was said to be the habitat of *Ixtobai*, a mythical female creature of the Maya underworld who stands 10 feet tall, has 3-toed feet that face backward, and comes out on nights of the full moon to seduce unfaithful husbands and drag them into the underworld, where they suffer many trials [RA].

Corchorus siliquosus L.

chi chi bé [Mayan]

The white variety of this plant is used as an **ornamental** [B2436].

Steggerda (1943) reported that chickens, turkeys, and pigs ate the leaves of this plant as a **food**. It was also said to be used in an unspecified way to relieve the **bite of the pic bug** (also known as **chagas bug** [*Triatoma dimidiata*]).

Gossypium hirsutum L.

cotton, cotton tree, wild cotton [English]; *algodón* [Spanish]; *noq* [Q'eqchi']; *ta man* [Yucateca]

To treat **pain** or **muscle spasm** with *ventosa* **(cupping)**, cotton from the dried pods is attached to a ball of wax and placed inside a thick drinking glass. This is then lighted with a match and applied to the affected area as in Chinese cupping techniques, by pulling the flesh up inside the glass [RA]. For **cough**, refer to *guanábana* (*Annona muricata*) [B2107, B2108, B2400, B2403, B2411, B2458].

Guazuma ulmifolia Lam.

bastard cedar, bay cedar, box cedar, wild bay cedar [English]; *caulote, guacimo, pechote, tapa culo* [Spanish]; *pixoy, simaron del tzibche* [Mayan]

(top) *Corchorus siliquosus* [PA].

(bottom) *Gossypium hirsutum* [NYBG].

The fruits and leaves are used as **fodder** for grazing animals as well as a **food** for birds, squirrels, and other forest animals [A44, A713, B1764; Arvigo and Balick, 1998]. The sweet fruit is eaten as a **food** that children in particular enjoy [A713, A815; Arvigo and Balick, 1998]. Eating the fruit is said to cause **constipation** due to the tannic acid content [A815; Arvigo and Balick, 1998]. The common name *tapa culo* refers to this effect of constipation [Robinson and Furley, 1983].

Standley and Record (1936) reported that the mucilaginous sap from the branches was used to **clarify sugar syrup**.

The leaves are part of a bath used for **general ailments** [B1870]. For **itchy skin**, 1 pound of leaves is boiled in 2 gallons of water for 10 minutes and used cool to soak the entire body for 20 minutes [A933]. For **cough**, refer to black cockspur (*Acacia collinsii*) [A342].

To treat non-insulin-dependent **diabetes** mellitus, one 12-inch piece of fresh "inner" bark is mashed, soaked in 1 gallon of cold water, steeped for 12 hours, boiled, and consumed 3 times daily for 15 days [C92]. For this treatment, the outer surface of the bark is scraped away until a red-colored layer appears, and this inner bark is cut for use [C92]. Alternatively, refer to wild damiana (*Turnera scabra*) [A674].

For **gastritis (ciro)**, refer to *guaco* (*Aristolochia grandiflora*) [B3669]. To treat **constipation**, 1 handful of the ripe fruits is soaked in 1 quart of water for 1 hour without crushing the fruits and ½ cup consumed 3 times daily [HR]. Alternatively, refer to blue moho (*Hibiscus tiliaceus*) [Steggerda and Korsch, 1943]. Ripe fruits are sucked on to treat **diarrhea** [B1764].

To treat **dysentery**, **diarrhea**, **vomiting**, or **upset stomach**, one 2-inch × 2-inch piece of bark is boiled in 3 cups of water for 10 minutes and ¼ cup consumed every hour for 1 day [A44, A793, A920, A933, B1764; Arvigo and Balick, 1998]. Alternatively, 1 cup of this decoction is consumed 3 times daily before meals [Arvigo and Balick, 1998]. To treat **prostate ailments**, honey is added to this decoction and 1 cup consumed 3 times daily before meals [Arvigo and Balick, 1998]. As an aid to **labor**, this decoction is sipped slowly all day long [Arvigo and Balick, 1998]. To help lubricate a dry **labor** during childbirth, one 3-inch × 3-inch piece of inner bark is soaked in 1 quart of water for 1 hour and 1 cup consumed at a time [A933].

For **urine retention (stoppage of water)** when there is a blocked urinary flow, one 3-inch × 3-inch piece of bark along with 1 entire *tan chi* plant (*Capraria biflora*) and 1 entire little bamboo plant (*Commelina erecta*) are boiled in 2 quarts of water for 10 minutes and sipped all day until better [B2132]. As a **purgative**, the bark is soaked in water and consumed [A933].

To treat **skin sores**, **infections**, and **rash**, 1 large handful of chopped bark is boiled in 1 gallon of water for 10 minutes, used cool as a bath 3 times daily over the affected area, and left to air dry [Arvigo and Balick, 1998].

To stop **bleeding**, the bark is chewed and applied as a poultice directly over the affected area [A815]. As an antidote to **snakebite**, bark from the "east" and "west" sides of the tree is powdered and used to cover the bite [A910]. To slow the circulation of poison following a **snakebite**, 1 piece of the inner bark is tied directly around the bite [HR].

Worms that feed on this plant are 3 inches long, flat as a finger, black and white like a caterpillar, and hang in bunches on the tree. They are a favorite food of the Yucatan Maya, and are eaten fresh, dried, or fried in oil with salt and spices [A933].

Guazuma ulmifolia [MB].

To treat **spiritual ailments**, such as **fright** (*susto*), **envy** (*envidia*), **grief** (*pesar*), and **evil magic** (*obeah*), 1 double handful of the leaves along with 1 double handful each of *ruda* leaves (*Ruta graveolens*), basil leaves (*Ocimum basilicum*), and *dama de noche* leaves (*Cestrum nocturnum*) are crushed, soaked in water, and poured as a bath over the body once daily for 3 days [RA].

To treat "**bad wind in the belly**" from **bewitching**, one 3-inch piece of bark, one 6-inch branch of flowers and leaves of marigold (*Tagetes erecta*), and 9 *epasote* leaves (*Chenopodium ambrosioides*) are mashed and soaked in 1 quart of water for 1 hour and taken in sips all day until consumed. This treatment may be repeated for 3 more days if needed [Arvigo and Balick, 1998].

Another example of how this plant is used in **healing** regards the worms that live on it. If a person is really sick, the worms are asked if the person will die. It the worm raises its head and hits the tree, the person will die. If the worm bobs its head from side to side, the person will live [A933].

The bark is cut into long strips and used as **cordage** to tie beams and rafters of houses together, and the tree is used to produce **charcoal** [Horwich and Lyon, 1990].

Hampea trilobata Standl.

saltwater moho [English]; *majana, majua, majava, mohara blanca, moho* [Spanish]; *bak el man* [Mayan]; *kajana* [Additional name recorded]

To treat **nervousness** or **insomnia**, 1 handful of leaves is boiled in 1 gallon of water for 10 minutes and used as a warm bath once daily at bedtime [A651]. To treat **gum ailments**, one 6-inch piece of root and 9 young leaves are mashed in a little bit of water (enough to moisten the mixture and create a semi-liquid state) and 2 tablespoons used 3 times daily as a mouthwash [A902].

The bark yields a yellow-orange **dye** and is also used as a **fiber** [A651]. The bark is soaked in water to "rot" the softened fibers, which are peeled away, and the **cordage** woven to produce ropes, cords, beds, and hammocks [A902, B2026].

The lumber is used in **construction** for making houses because it is said to be long-lasting [B2177].

Hampea trilobata [DA].

Helicteres guazumifolia Kunth

moho [Spanish]; *mahaua* [Mayan]

To treat **diarrhea** and **dysentery**, one 16-inch × 2-inch piece of bark and two 6-inch pieces of root are boiled in 1 quart of water for 10 minutes and ½ cup consumed cool

Helicteres guazumifolia with inset showing fruit [MB].

Heliocarpus americanus [NYBG].

4 times daily [A739, A803; HR]. To staunch excess bleeding, ½ cup of this decoction is consumed every 30 minutes until better [A739, A803; HR].

The stripped, fresh bark is used as a **cordage** for rope and for weaving. This fiber must be stripped on nights of the full moon and is said to last a very long time [HR].

Heliocarpus americanus L.

broadleaf moho, high ridge moho, mountain moho, white moho [English]; *majao, moho* [Spanish]; *chai* [Additional name recorded]

The **latex** is used for making the rubber head of the marimba mallet [JB20].

The bark is used as **cordage** to make rope [B20]. To prepare the rope, the bark is peeled from the straight branches, beaten with a stick, washed in water, dried, and the rough, uneven pieces of fiber combed out. These are then braided with 3 sections to make a "very strong" rope [HR]. Fresh bark is used to make a splint for **broken bones**. Strips of the bark are used to tie around the splint and the broken bones [HR].

Heliocarpus mexicanus (Turcz.) Sprague

broadleaf moho [English]; *cajete* [Spanish]

This plant is used as an **ornamental** [B2449].

When nothing else is available, this wood is used in **construction** to make house posts, although it is not recommended because of its tendency to rot quickly [A98].

Hibiscus costatus A. Rich.

snake okra, wild okra [English]; *woyo'* [Q'eqchi']

To **improve blood circulation** after a snakebite, 10 leaves and three 3-inch pieces of root are boiled in 1 gallon of water for 10 minutes and poured over the entire body [B2675]. For **snake-bite**, 1 ounce of crushed seeds along with one 3-inch piece of a "soft candle" (a homemade candle made of animal fat) and 3 garlic cloves are mashed together and applied as a poultice twice daily following the bath described above for blood circulation. Treatment duration depends on the condition of the person [B2675].

Hibiscus rosa-sinensis L. var. *rosa-sinensis*

Chinese hibiscus, red bell [English]; *tulipán* [Spanish]; *kaq chi atz'um* [Q'eqchi']

Flowers are eaten as a **food**, added to salads for taste and color, and are rich in iron [RA; Arvigo and Balick, 1998].

For **headache** and **fever**, the leaves are mashed and applied to the head as a poultice [Arvigo and Balick, 1998]. For **loose stools with blood**, 1 double handful of leaves is boiled in 1 quart of water for 10 minutes, cooled to lukewarm, and 1 cup consumed 3 times daily until better [B3585].

As a **blood tonic** to treat **anemia**, 3 flowers are eaten daily until better [LR]. To treat **painful menstruation**, the flowers are eaten daily during menstruation [Arvigo and Balick, 1998]. To staunch **heavy menstruation** and to **prevent miscarriage**, 9 leaves, 5 open flowers, and 4 closed flowers are boiled in 3 cups of water for 10 minutes and 1 cup consumed cool 3 times daily [Arvigo and Balick, 1998]. For **postpartum hemorrhages**, the same decoction as described above for heavy menstruation is sipped constantly until the bleeding stops [Arvigo and Balick, 1998].

(top) *Hibiscus costatus* with inset showing root [MB].

(bottom) *Hibiscus rosa-sinensis* var. *rosa-sinensis* [MB].

To stop **internal hemorrhaging**, such as after childbirth or during menstruation, refer to red China root (*Smilax ornata*) [Arvigo and Balick, 1998].

To treat various **skin ailments**, 1 large handful of leaves and flowers are mashed in a bucket of water and poured as a cool wash or used to soak the affected area [Arvigo and Balick, 1998]. For **dry or cracked skin**, **hair loss**, **dandruff**, or to **stimulate hair growth**, 1 quart of leaves and flowers is mashed, mixed in 1 gallon of water, and used as a hair rinse or skin lotion [LR].

To treat **uncertainty** and **indecisiveness**, refer to *alamanda* (*Allamanda cathartica*) [B2651]. To entice a **strayed husband** to return home to his wife, 9 leaves and 9 open flowers are boiled in 3 quarts of water for 10 minutes and 1 cup given by the wife to the husband daily, with the remainder used by her to bathe him. This treatment is repeated for 9 days [Arvigo and Balick, 1998].

Hibiscus rosa-sinensis L. var. *schizopetalus* Dyer

For **loose, bloody stools**, 1 double or triple handful of leaves is boiled in a pot of water, cooled to lukewarm, and 1 cup consumed 3 times daily until better [B3586].

Hibiscus sabdariffa L.

sorrel [English]; *roselle* [Additional name recorded]

This plant is cultivated for its fleshy calyx, which is eaten as a **food** or boiled in water and used as a **beverage** [N46876].

Hibiscus sabdariffa with inset showing dried petals [MB].

(left) *Hibiscus tiliaceus* [MB].

(right) *Luehea speciosa* [GS].

Hibiscus tiliaceus L.

blue moho [English]; *xholol* [Yucateca]

Steggerda and Korsch (1943) reported that a **beverage** made from the boiled bark along with an unspecified part of bay cedar (*Guazuma ulmifolia*) was used to treat **constipation**.

Standley and Record (1936) reported that the bark fiber was used extensively to make **rope** in some regions, presumably along the coast, where it was found in swampy areas. The bark of this tree is stripped and used as **cordage** to tie beams, rafters, and posts together [Horwich and Lyon, 1990].

Luehea speciosa Willd.

broadleaf bay cedar, bay cedar (male), mountain moho [English]; *caulote, mahaua* [Spanish]; *c'am pác* [Mopan]; *b'alam max* [Q'eqchi']; *pixoy* (male) [Mayan]

The fruit is eaten as a **food** by birds [B3126].

The bark from large branches is stripped and pounded into **rope** for carrying loads [A392, B3194]. The trunk is cut into timber used for **roof poles** because it is said that it "will not last on the ground," meaning that it will rot when put in contact with the ground [A392, B3267]. The wood is also used as **firewood** [A625]. Ash from the wood is used to remove **stains** from clothes [A392].

Malachra fasciata Jacq.

blood of Christ [English]

For **anal** or **rectal bleeding**, 9 leaves are heated in ¼ cup of olive oil for 30 minutes and applied internally into the affected area 3 times daily [B1828].

Malvastrum corchorifolium (Desr.) Britton ex Small

broomweed, yellow malva [English]; *malva amarilla* [Spanish]; *chi chi be* [Mopan]

For **headache**, the leaves and flowers and a pinch of salt are mashed with a bit of water and applied as a poultice to the forehead for 30 minutes [HR].

To treat **fever** and for children with **measles**, 1 entire plant is boiled in 1 quart of water for 10–15 minutes and ½ cup consumed at room temperature 6 times daily until better [B2354].

As a treatment for **skin rash**, 2 entire plants are boiled in 2 gallons of water for 10 minutes and 1 gallon used cool twice daily as a bath over the affected area as needed until better [HR].

To treat **nervousness** and to cool a person suffering from **sun overexposure**, 1 entire plant along with 1 small vervain plant (*Stachytarpheta jamaicensis*) are boiled in 2 quarts of water for 5 minutes, cooled, and 1 cup consumed 3 times daily as needed [LR].

For a **bladder infection**, 1 large or 2 small plants along with 1 handful of chopped wild yam roots (*Dioscorea bartlettii*) and one 12-inch piece of strong back vine (*Desmodium adscendens*) are boiled in 2 quarts of water for 10 minutes and 1 cup consumed cool 6 times daily until better [LR].

To **stimulate hair growth**, 6 entire plants are mashed, boiled for 10 minutes in 1 gallon of water, steeped for 1 hour, and used as a hair rinse once daily for 9 days [LR].

Malvastrum coromandelianum (L.) Garcke
yellow malva [English]; *malva* [Spanish]

For **jaundice**, 1 entire plant is boiled in 1 gallon of water for 15 minutes and 1 cup consumed 3 times daily, before meals, until finished [A318]. For **headache**, the decoction described for jaundice is poured cool over the head while the head is tipped back once daily before bedtime [A318]. To treat **hemorrhage**, 1 handful of young leaves is mashed in 1 cup of water, strained, and 1 teaspoon consumed every 5 minutes for hemorrhage until the bleeding stops [A317].

(left) *Malachra fasciata* [NYBG].

(right) *Malvastrum coromandelianum* [DA].

Malvaviscus arboreus var. mexicanus [MB].   Muntingia calabura [RH].

Malvaviscus arboreus Cav. var. *mexicanus* Schltdl.

bootblack flowers, old man's apple, white moho, wild hibiscus [English]; *tulipán*, *tulipan del monte*, *tulipancillo* [Spanish]; *catusa* [Additional name recorded]

This plant, with attractive red flowers, is cultivated as an **ornamental** [B2471].

The fruit is eaten as a **food**, and children suck the **nectar** out of fresh flowers [Arvigo and Balick, 1998].

To stop **hemorrhaging**, 9 leaves along with 9 leaves of *maguey sylvestre* (*Tradescantia spathacea*) are boiled in 2 cups of water for 10 minutes and ½ cup consumed every 10 minutes until the bleeding stops [B2357]. Alternatively, 9 leaves and 1 flower along with one 3-inch piece of cinnamon stick (*Cinnamomum verum*) are boiled in 2 cups of water for 10 minutes, cooled, and sipped all day until finished [HR]. For a third reported treatment, refer to *maguey sylvestre* (*Tradescantia spathacea*) [B2327].

To **stimulate hair growth**, 9 flowers are mashed in 1 quart of water, rubbed onto the head, and left on the balding area for 5 minutes 3 times daily [B2137].

To **promote sweat**, 1 double handful of leaves is boiled in 1 gallon of water for 10 minutes and used hot as a bath. Following the bath, the person returns to bed, is covered, and will sweat there [B2137].

Muntingia calabura L.

capuleen [English]; *capulin* [Spanish]; *e'ek eeb, may pim* [Q'eqchi']; *mahoa* [Mayan]

The fruit is eaten as a **food** [A908, G21]. Interestingly, Standley and Record (1936) noted that "the fruit is **edible** but so intensely sweet as to be rather unpleasant."

To treat **stomach pain**, 2 ounces of ashes from burned branches are soaked in 1 cup of water and 2 tablespoons of the strained water consumed after meals [A679].

Hammocks, bags, and other items are made from the bark **fibers** [A679]. Standley and Record (1936) reported that the bark fiber was very strong and locally used as **cordage** for making rope.

Ochroma pyramidale (Cav. ex. Lam.) Urb.

balsa [Spanish]; *polka* [Additional name recorded]

Standley and Record (1936) reported that this plant was a source of kapok for **stuffing** pillows and other objects. They also noted the wood was said to be used to **sharpen razors** and make **insulation**, and the hardest portion of the stem was suitable for making **paper**.

Ochroma pyramidale [MB].

Pachira aquatica Aubl.

provision bark [English]; *bobo, Santo Domingo, sapote bobo, zapote bobo* [Spanish]; *cuyche, zapoton* [Mopan]; *uacut* [Additional name recorded]

As a **food substitute** when food stores were low during the era of *chicle*, rubber, and mahogany (*Swietenia macrophylla*) camps, the bark was used to make a **tea** for the workers and thus originated the common name "provision bark" [Arvigo and Balick, 1998]. The large seeds are eaten as a **food**, either boiled or roasted, and said to taste like green corn [R1; Arvigo and Balick, 1998].

Pachira aquatica with inset showing flower [MB].

To treat **kidney ailments** or a **pain in body organs**, such as **peritonitis**, 1 fruit or 1 cup of chopped bark is boiled in 2 quarts of water for 10 minutes and consumed warm [B1897]. For **kidney pain**, 1 seed is quartered and boiled in 1 cup of water for 5 minutes and consumed once daily, before breakfast, for 3 consecutive days [Arvigo and Balick, 1998].

This is probably the most famous **blood tonic** of Belize [RA]. To **build the blood** and to treat **low blood pressure**, 9 leaves are steeped in 3 cups of boiled hot water for 20 minutes and 1 cup consumed 3 times daily for 9 days [B1908]. To **build the blood** in old age and to treat **anemia**, **exhaustion**, and **low blood pressure**, one 1-inch × 4-inch piece of bark is boiled in 3 cups of water for 10 minutes and consumed 6 times daily [Arvigo and Balick, 1998]. For a **blood tonic** treatment, refer to red China root (*Smilax ornata*) [W1864].

For **tuberculosis**, the center of 1 young, unripe fruit and 4 ounces of bark are boiled in 2 quarts of water, reduced to 1 quart of liquid, strained, 4 ounces of honey and 1 quart of *anisado* (anise liquor [*Pimpinella anisum*]) added, and 1 tablespoon consumed 3 times daily, before meals, until better [HR].

For **uterine "cancer"** (**sores, abscesses, or ulcers on or in the uterus**), refer to *cocano boy* (*Bactris major* var. *major*) [A330]. As part of a **male tonic** for **energy**, **vitality**, and **sexual potency**, refer to red China root (*Smilax ornata*) [W1864; LR].

Pseudobombax ellipticum (Kunth) Dugand
poma rosa [Spanish]
To bring **boils** to a head, the leaves are mashed with a pinch of salt and applied as a poultice directly to the affected area [B3704].

Quararibea funebris (La Llave) Vischer
Lundell (1938) noted that the Maya of the Petén region used the small branches of this tree as "**swizzle sticks**" to produce a foam in cacao beverages (*Theobroma cacao*).

Sida acuta Burm. f.
broomweed, wire weed [English]; *escoba*, *escobilla*, *malva* [Spanish]; *chi chi be* [Mopan]; *mesh'eel* [Q'eqchi']
The leaves are sometimes used as **fodder** for horses and cows to **accelerate labor** [B2288].

For **general ailments**, refer to *malva* (*Sida glabra*) [B2346]. To treat **burns** and **dandruff**, 1 entire plant is boiled in 1 gallon of water for 20 minutes and used cool as a bath twice daily [A895]. For **urinary ailments**, 1 quart of the decoction described for burns is sipped warm

Inflorescence of *Pseudobombax ellipticum* [KN].

Quararibea funebris [RF].   *Sida acuta* [PA].

all day long or consumed in place of water until better [A895]. For **stomach pain** and other **pain**, 1 cup of the decoction described for burns is consumed hot 3 times daily before meals [B2288]. To treat **headache**, the leaves along with life everlasting leaves (*Kalanchoe pinnata*) are mashed in cold water and applied as a plaster to the head overnight [Mallory, 1991]. To treat **headache** and other **ailments**, the leaves are applied to the head in the Yucatan [Mallory, 1991].

For **boils** or **sores**, 1 handful of the green leaves and a pinch of salt are mashed in enough water to dampen the leaves, wrapped in a clean cloth, applied as a poultice directly to the affected area, changed twice daily, and repeated until better [B2102, B2182, B3581].

The leaves are dried and smoked as a **tobacco substitute** [A895, B2102].

The stem is used in **carpentry** and for making **broom bristles**, hence the common name [B2469]. This plant is said to produce fine bast fibers from its stem for **cordage** or **cloths** [Horwich and Lyon, 1990]. The flowers are said to act as **nature's clock** because they open at about noon [B3581].

Sida antillensis Urb.

For **venereal diseases**, ½ cup of chopped roots and 1 cup of chopped young stems along with 1 entire *dormilón* plant (*Mimosa pudica* var. *unijuga*) and one 6-inch piece of papaya root (*Carica papaya*) are boiled in 1 quart of water for 15 minutes and 1 cup consumed 4 times daily, before each meal and at bedtime, until tests and symptoms are better. If symptoms do not subside after 3 days of treatment, it is advised to see a physician [B2658].

Sida glabra Mill.

malva [Spanish]; *chi chi be* [Mayan]

For **general ailments**, 1 entire small plant (about 12 inches high) along with 1 entire small broomweed plant (*Sida acuta*) are boiled in 1 gallon of water for 15 minutes and 1 quart sipped all day or consumed in place of water for thirst as needed until better [B2346]. To treat **sores**, **rash**, **burns**, or **boils**, the decoction described for general ailments is gradually poured cool over the affected area [B2346]. For **injuries**, **sprains**, and **bruises**, the decoction described for general ailments is used warm to soak the affected area [B2346]. To **promote urination** that may be linked to a swollen prostate in men, the decoction described for general ailments is consumed throughout the day [B3682].

Sida rhombifolia L.

broomweed [English]; *escoba*, *malva* [Spanish]; *chichibe* [Mopan]; *mesb'eel* [Q'eqchi']

To reduce **fever**, 9 roots are boiled in 1 quart of water for 10 minutes and used as an enema [Arvigo and Balick, 1998]. For **sprains**, the root is mashed and applied as a poultice directly to the affected area [Arvigo and Balick, 1998].

To treat **burning in the urine**, **urine retention (stoppage of water)**, **gonorrhea**, and to **loosen dry coughs**, 1 cup of fresh leaves is boiled in 3 cups of water for 5 minutes and 1 cup consumed 3 times daily before meals [Arvigo and Balick, 1998]. To prepare "**old man's tonic**," 9 roots along with three 1-inch pieces of pine wood (*Pinus caribaea* var. *hondurensis*) are soaked in 1 quart of rum and one-shot doses consumed twice daily until the phlegm has passed in the urine [Arvigo and Balick, 1998].

For **scorpion sting** or **snakebite**, 1 entire plant without roots is boiled in 1 gallon of water for 15 minutes, 1 cup consumed twice daily, and the remainder refrigerated and used gradually as needed [B2529]. Note that this is considered an adjunct treatment to aid the kidneys in eliminating toxins; it is not a cure from the poison of snakebite or scorpion sting [B2529].

Fresh stems and leaves are tied to a pole and used as a **broom** for the house and yard [Arvigo and Balick, 1998]. Standley and Steyermark (1949) reported that the leaves of this plant were macerated in water to produce foam used as **soap** for washing clothes.

Sida rhombifolia [MB].

Theobroma cacao fruit (left) and flowers (right) [IA].

Theobroma cacao L.

cacao, chocolate [English]; *cacao* [Spanish]; *kakaw* [Q'eqchi']; *kuku* [Mayan]

This is an important traditional crop in Belize, and the seeds in its fruits, mixed with red chili pepper (*Capsicum annuum*), corn (*Zea mays*), and vanilla (*Vanilla planifolia*), are made into a **beverage**. During fieldwork for this project, we were frequently served a warm cacao beverage, sometimes containing sugar and/or black pepper (*Piper nigrum*), in a number or households we visited. On numerous occasions, we observed cacao beans (*Theobroma cacao*) drying on the Maya *comal* (a traditional baking tray suspended over the cooking fire) for future sale or household use [MB]. Generally, the dried seeds are roasted for about 30 minutes, the outer shells peeled off, and the remaining seed is ground into a powder and mixed with sugar, vanilla, chili, corn, or other food items, depending on the intended use. Some make **ceremonial offerings** to the Maya spirits from the roasted, powdered seeds boiled in water with sugar. Cacao plays a role to this day in some extant Maya **ceremonies**, such as the *primicia* and to celebrate the first steps of an infant [RA]. To flavor cacao, refer to vanilla (*Vanilla planifolia*) [B3555; Lundell, 1938], *guanábano* (*Cymbopetalum mayanum*) [Standley and Record, 1936], *achiote* (*Bixa orellana*) [Thompson, 1927], and *chaparro* (*Curatella americana*) [Marsh, Matola, and Pickard, 1995].

According to Thompson (1927), this was

a plant that has played a role of the utmost importance in the religious, commercial, and gastronomical life of the Maya. Cacao was the general **currency** among the Maya before the conquest, and, in fact, still is in some of the remoter parts of Guatemala. Even in Yucatan it has not long ceased to be used for this purpose. A young Maya of Valladolid informed me that his father had told him that when he was a young boy, cacao-beans still served as the general currency of that district. As cacao-beans were the general medium of exchange, it is not surprising to find that the patron deity of Maya merchants was the spirit of cacao, Ekchuah. (pp. 185–6)

For a tree that is commonly inter-cropped with cacao, refer to mother of cacao (*Muellera frutescens*) [R8]. For a **fish poison** tradition, refer to *barbasco* (*Serjania lundellii*) [Thompson, 1930].

Trichospermum grewiifolium (A. Rich.) Kosterm.

balsa wood [English]; *lagroso, macapal, moho* [Spanish]; *chahib, makapal* [Q'eqchi']; *mahaua, mahaua mahaua* [Mayan]

The stripped bark and stem fibers are used as **cordage** for making rope to weave hammocks and tie house posts [A493, A628, B3271, B3292, B3594]. As a cordage substitute for this plant, refer to belly full tie tie (*Philodendron radiatum* var. *radiatum*) [B3539]. The wood is used in **construction** for making split boards, house posts, walls, fences, and rustic furniture [A493, A628, B3271, B3292, R60].

Triumfetta lappula L.

bur [English]

Standley and Record (1936) noted that the mucilaginous sap was used to **clarify sugar syrup**.

Urena lobata L.

wild cotton [English]; *ka khe cwo yo'* [Mayan]

To **bleach sugar**, the root, commonly found in pastures, is added to the boiling sugar water mixture [R91].

(left) *Trichospermum grewiifolium* [GS].

(right) *Triumfetta lappula* [JV].

(left) *Urena lobata* [NYBG].

(right) *Urena lobata* with inset showing close-up of flower and fruit [PA].

MARCGRAVIACEAE

Lianas or shrubs, terrestrial or epiphytic. Leaves alternate, sometimes distichous or spiral, simple. Inflorescences terminal, sometimes cauliflorous, in racemes, pseudospikes, or pseudoumbels. Flowers often with brightly colored bracts. Fruits a capsule. In Belize, consisting of 3 genera and 7 species. Uses for 1 genus and 1 species are reported here.

Souroubea sympetala Gilg

white strangler fig [English]; *tolox* [Q'eqchi']

To treat **ringworm** or **leishmaniasis (baysore)**, 1 leaf is squeezed and the juice applied directly to the affected area twice daily, in the morning and in the evening [A598].

MELASTOMATACEAE

Herbs, shrubs, small trees, or rarely large trees. Leaves opposite, simple. Inflorescences terminal and/or axillary, in cymes, umbels, or rarely of solitary flowers. Flowers white, pink to reddish, or bluish-purple, sometimes yellow or orange. Fruits a capsule or berry. In Belize, consisting of 20 genera and 97 species. Uses for 11 genera and 24 species are reported here.

Aciotis rostellata (Naudin) Triana

For **fever**, 1 entire plant is mashed, boiled in 1 gallon of water for 10 minutes, 1 quart sipped cold all day long, and the remainder used as a bath [B3557]. For **morning sickness**, 3 branches and 3 leaves are chopped, boiled in 3 cups of water for 15 minutes, and 1½ cups consumed cold once daily in the morning before rising from bed [B2607].

Acisanthera quadrata Pers.

To treat **chest colds** of young children, 2–3 ounces of the entire plant are boiled in 4 pints of water for 5 minutes and 1 teaspoon consumed 3 times daily for a single treatment [B2621]. This remedy is considered potentially **dangerous** and could cause harm if too strong a mixture is used or the child is too young.

Arthrostemma ciliatum Pav. ex D. Don

top yuk [Mayan]; *pin win* [Additional name recorded]

The leaves and stem are eaten raw as a **food** and are "sour like lime; children love it!" If eaten in excess, they can be **toxic**, causing diarrhea [A291, A588; RA]. The leaves and flowers provide enjoyable **fodder** for red brocket deer (*Mazama americana*), locally known as "antelopes" [A291].

To treat a **baby's diaper rash**, 1 quart of fresh leaves is crushed in 2 quarts of warm water and used as a bath twice daily until better [B2660].

Arthrostemma parvifolium Cogn.

The immature fruit is eaten as a **food** and has a somewhat sour taste. If eaten in excess, it can be **toxic** and cause diarrhea [B2043; RA].

Bellucia grossularioides (L.) Triana

black moir [English]; *maya, sir in* [Additional names recorded]

The fruit is eaten as a **food** [G11].

To treat **swelling** caused by bruises and injuries, 1 quart of fresh leaves is boiled in 1 gallon of water for 10 minutes and used cool as a bath 3 times daily [B2684].

Bellucia pentamera Naudin

asi'r, plo m'che, sir in, manzana che' [Q'eqchi']

The trunk is used in **construction** to make house frame posts, beams, and rafters [A624, A972].

Arthrostemma parvifolium [FM].   Bellucia pentamera [GS].

Clidemia hirta (L.) D. Don

chigger nits [English]

To relieve **nervousness** or **anxiety**, 1 quart of leaves is added to 1 gallon of boiling water, steeped for 1 hour, and used warm as a bath once daily, at noon or bedtime, as needed [B2657]. For **anxiety**, 9 leaves and ½ cup of chopped bark are steeped in 1 quart of hot water for 1 hour, strained, and 1 cup consumed hot [B2657].

Clidemia hirta [MB].

Clidemia novemnervia (DC.) Triana

gwok shpím [Q'eqchi']; *pa sas* [Mayan]

The fruit is eaten as a **food** [A286].

The fruit is processed and used as a black **dye** [A286]. The soft leaves are used as "bush **toilet paper**" [A286]. The leaves are also used like **soap** with water to wash the hands and body [B3593].

Clidemia octona (Bonpl.) L.O. Williams

isb' i pim [Q'eqchi']

The fruit is eaten as a **food** by birds [B3528].

Clidemia petiolaris (Schltdl. & Cham.) Schltdl. ex Triana

masb'a it [Q'eqchi']

To stop **menstrual bleeding** when the "inside of the stomach is cold," 1 cup of leaves is boiled in 1 quart of water for 5 minutes and 1 cup consumed hot 3 times daily until the bleeding is a normal amount [B2561]. A "cold stomach" is when the arterial flow of blood is blocked and the veins are stagnant. This could also refer to **muscle cramps** [B2561].

Clidemia sericea D. Don

chiganettes [Creole]

To treat a persistent **cough** in a baby, 9 leaves are boiled in 1 cup of water for 5 minutes and sipped throughout the day [A347].

Conostegia xalapensis (Bonpl.) D. Don ex DC.

chigger nits [English]

The fruit is eaten as a **food** [B2699]. Standley and Record (1936) noted that the "berries are sweet and are of good flavor, suggestive of the huckleberries [*Gaylussacia* spp.] of the United States."

For **fever** in the absence of a rash, 1 quart of fresh leaves is boiled in 1 gallon of water for 10 minutes and used at room temperature as a bath once daily, at bedtime or noon [B2699]. To relieve **nervousness** or **anxiety**, 1 quart of leaves is added to 1 gallon of boiling water, steeped for 1 hour, and used warm as a bath twice daily, at noon and bedtime [B2699]. For **anxiety**, 9 leaves and ½ cup of chopped bark are steeped in 1 quart of hot water for 1 hour, strained, and 1 cup consumed hot [B2699].

Miconia affinis DC.

chigger nits [English]; *sirin* [Additional name recorded]

The ripe, purple fruits are sweet and eaten as a **food** [B2605, B2667].

To relieve **anxiety** and reduce **itching** in the absence of a rash, 1 double handful of leaves and branches is boiled in 1 gallon of water for 10 minutes, steeped for 20 minutes, and used warm as a bath over the affected area twice daily as needed [B2605, B2667].

(top) *Conostegia xalapensis* [RF].

(bottom) *Miconia affinis* inflorescence (left) and fruit (right) [FM].

Miconia albicans (Sw.) Steud.

flor azul [Spanish]; *shma mu cal* [Mayan]

As an **energy supplement**, nine 12-inch branches are soaked in 2 gallons of warm water for 10 minutes and poured over the body once daily, at noon or preferably at bedtime [B2257]. Alternatively, 1 small handful of chopped roots (preferably collected after 4 p.m.) is boiled in 3 cups of water for 10 minutes and 1 cup consumed at room temperature 3 times daily before meals [B2257].

To quickly clear **male urinary tract mucus** and to treat **prostate ailments**, 1 double handful of the entire plant with roots along with 1 small handful of the chopped wild yam tuber (*Dioscorea bartlettii*) and 1 small handful of balsam bark (*Myroxylon balsamum* var. *pereirae*) are boiled in 1½ quarts of water for 5 minutes and 1 cup consumed warm twice daily, before breakfast and at bedtime, for 7 days. Men will pass cloudy, thready urine [LR].

Miconia ciliata (Rich.) DC.

Maya seed [English]; *concha de la tortuga* [Spanish]; *ah sàr', ah sír, kak he a sir', sak he asàr, xoy pim* [Q'eqchi']; *chuk pa sas* [Mayan]; *maya, sirin* [Additional names recorded]

The fruit and seeds are eaten as a **food** [A287, A346].

For **headache**, 1 quart of leaves is boiled in 1 gallon of water for 10 minutes and poured directly over the head [A346]. For **fever**, 1 double handful of leaves is boiled in 2 gallons of water for 10 minutes and poured warm over the head and body once daily at bedtime [B3588, B3591]. For **dysentery**, 1 quart of fresh leaves and young stems is boiled in 1 quart of water for 10 minutes or mashed in cool water and placed to steep in the sun for 1 hour, after which 1 cup is consumed 3 times daily as needed [B3635].

Miconia holosericea (L.) DC.

la yerba de la mancha [Spanish]

To treat **skin fungus**, the leaves are rubbed directly against the affected area [B2256]. For **skin fungus**, 1 quart of leaves is boiled in 1 gallon of water for 5–10 minutes, steeped for 3 hours, reheated, and used as hot as possible to soak the affected area for 30 minutes [B2256].

Miconia hondurensis Donn. Sm.

chigger nits [English]; *ah sir, roq mukuy* [Q'eqchi']

For **chest pain**, 9 leaves are boiled in 1 quart of water for 10 minutes and 1 cup consumed warm 3 times daily before meals [B2476, B2507]. For **fever**, 1 quart of fresh leaves is boiled in 1 gallon of water for 10 minutes and poured, warm to hot, over the body once daily, at bedtime, as needed [B2708].

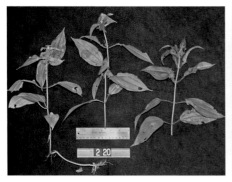

Miconia lacera [GS].

Miconia prasina [GS].

Miconia lacera (Bonpl.) Naudin
chigger nits [English]; *mas ba'it* [Q'eqchi']

The purple fruit is eaten as a **food** [B2599].

To treat **nervousness** or **anxiety**, 1 quart of leaves is added to 1 gallon of boiling water, steeped for 1 hour, and used warm as a bath twice daily, at noon and bedtime [B2599]. For **anxiety**, 9 leaves and ½ cup of chopped bark are steeped in 1 quart of hot water for 1 hour, strained, and 1 cup consumed hot [B2599].

To slow blood flow during **menstruation**, 1 handful of leaves is mashed in 1 cup of warm water, soaked for 30 minutes, and consumed [B2510].

Miconia prasina (Sw.) DC.
macho del eremul [Spanish]

The leaves are a component of a bath mixture used for **general ailments**, usually prepared in combination with 8 other plants. To prepare the bath, 1 quart of the mixed leaves is boiled in 1 gallon of water for 5 minutes and poured warm over the body or affected area once daily until better [B2255].

Miconia tomentosa (Rich.) D. Don ex. DC.
big chigger nits [English]

To treat **nervousness** or **anxiety**, 1 quart of leaves is added to 1 gallon of boiling water, steeped for 1 hour, and used warm as a bath twice daily, at noon and bedtime [B2608]. For **anxiety**, 9 leaves and ½ cup of the chopped bark are steeped in 1 quart of hot water for 1 hour, strained, and 1 cup consumed hot [B2608].

The flat leaves from this plant are used to **wrap** tortillas to keep them moist when traveling [B2608].

Mouriri exilis Gleason
juruch' ayin [Q'eqchi']

The stem is used as a **planting stick** because the wood is very hard and makes a good hole in the ground, especially when planting rice [B3632].

Pterolepis stenophylla Gleason

sa cha pàm, sa cha pím [Q'eqchi']

To stop the **bleeding** in **wounds**, 9 leaves and four 6-inch, young stems are mashed in ½ cup of water and applied as a poultice directly to the affected area [B3592].

Tococa guianensis Aubl.

ix k'a pim, ixq i pim [Q'eqchi']

As a **male contraceptive**, an unspecified amount of leaves is mashed, mixed with water, and consumed once as an effective treatment [B3554]. As a **female contraceptive**, the unspecified "male" counterpart of this plant is prepared and consumed [B3554]. To reverse the contraceptive effects, it is reported that another unspecified plant is used, but no further details were provided [B3554].

Topobea watsonii Cogn.

mountain sirin [English]

To treat **skin fungus** caused by evil magic (*obeah*), 9 leaves are ground, added to 1 quart of cold water, steeped in the sun for 1 hour, 1 cup consumed, and the remainder used warm or tepid as a bath [B2553].

MELIACEAE

Small or large trees, sometimes shrubs. Leaves alternate, sometimes spiral, pinnate. Inflorescences axillary or terminal, in panicles or thyrses. Flowers variously colored. Fruits a capsule, berry, or drupe. In Belize, consisting of 6 genera and 15 species. Uses for 5 genera and 9 species are reported here.

(top) *Tococa guianensis* [FM].

(bottom) *Carapa guianensis* [MB].

Carapa guianensis Aubl.

bastard mahogany [English]

Standley and Record (1936) reported that the wood was reddish-brown, easy to work, durable, and used extensively in general **construction** and in **carpentry** to make furniture. They also noted that the seeds were a rich source of **oil** "used in South America for making soap and for illuminating purposes."

Cedrela odorata L.

cedar, red cedar, Spanish cedar [English]; *cedro, cedro rojo* [Spanish]; *tyaw* [Q'eqchi']; *ku che* [Mopan]

As a **cough suppressant**, one 1-inch × 3-inch piece of fresh or dried bark is boiled in 1 quart of water for 10 minutes and ½ cup consumed hot 3 times daily for 3 days [B2309]. Alternatively, one 1-inch × 3-inch piece of bark along with one 1-inch × 3-inch piece of Billy Webb bark (*Acosmium panamense*) are boiled in 1 quart of water and ½ cup consumed twice daily [B2179]. To **loosen the mucus** from a chest cough, one 4-inch × 8-inch piece of bark is boiled in 1 quart of water for 5 minutes, a pinch of salt added to the mouth, and 1 cup consumed warm 3 times daily as needed [W115].

For **gastritis (*ciro*)**, non-insulin-dependent **diabetes** mellitus, **delayed menstruation**, **colds**, or **influenza**, refer to Billy Webb (*Acosmium panamense*) [P3]. To treat **toothache**, the leaf exudate is used in Maya traditional medicine [Flores and Ricalde, 1996].

To clear the lungs of **mucus** and to treat **bruises, falls, internal injuries, abdominal pain**, and **postoperative states,** 1 handful of grated bark is soaked in 3 cups of water for 6 hours and sipped all day [Arvigo and Balick, 1998]. To treat **bruises** and **sprains**, one 1-inch × 6-inch strip of fresh bark is soaked in 1 gallon of water for 2 hours and poured over or used to soak the affected area [A115, B1992]. For **internal injuries**, such as **bleeding**, and for treatment of **kidney ailments**, one 1-inch × 6-inch strip of fresh bark is soaked in 1 gallon of water for 2 hours and ¾ cup consumed

Cedrela odorata branches showing fruit (left) and bark (right) [MB].

warm 3 times daily for 7 days [A115, B1992, W115]. Following this 7-day treatment, a laxative is taken, followed by 7 days of rest and another 7 days of treatment [W115]. To treat **chronic nosebleed**, 1 piece of fresh bark is sniffed all day, which is believed to correct the condition permanently [Arvigo and Balick, 1998].

As a **female contraceptive**, one 10-inch piece of bark is boiled in 3 cups of water for 10 minutes and 3 cups consumed daily for 3 days before menstruation. This treatment is said to provide 1 month of protection and not lead to permanent sterility [Arvigo and Balick, 1998].

To treat **snakebite**, the root along with buttonwood root (*Piper amalago*) and *chicoloro* root (*Strychnos panamensis* var. *panamensis*) are boiled in water and consumed [Arnason et al., 1980].

To **repel fleas**, the leaves are spread under the affected bed [HR]. To prevent **bird lice**, which cause an itchy rash with pimple-like welts on humans who come in contact with the birds, the leaves are spread around the chicken coop [HR].

Steggerda (1943) reported that the sticky sap of this tree was used by the Yucatan Maya as **mucilage** and that local children made **toys** of the dried, open fruits. He also noted that "when the leaves appear in May, it is a **signal** for the Maya to plant corn. They tell that many trees begin to send forth their leaves with the first rains in April, but not the Cedro, for it does not put forth its leaves until the real rains begin."

The wood is used in **construction** for lumber and in **carpentry** to make altars and chairs for sacred ceremonies [A115, B1992, B2309, JB112]. Carved idols of Ix Chel, the Goddess of Childbirth, were placed under the birthing bed for assistance and speedy delivery without complications [RA]. The wood is used in **carpentry** to produce many household articles as well as for dories and cigar boxes [Horwich and Lyon, 1990]. Standley and Record (1936) noted that the wood was fragrant, durable, easily worked, and used in **construction** to make dugout canoes, boat planking, shingles, and chests ("as it is distasteful to insects"). Standley and Steyermark (1946) reported that the wood was durable and resistant to insect attack, presumably due to the presence of a volatile oil in the wood. Lundell (1938) suggested that "doubtless it was utilized by the ancient Maya." This species is one of the most important **timber** trees in the region [MB].

Guarea glabra Vahl.

wild orange [English]; *bâl ba*, *ból ba*, *cramantee* [Additional names recorded]

For **itchy and open foot sores**, the fresh bark is grated into a cloth, the juice squeezed out, 1 teaspoon of water added if necessary, and applied to the affected area 3 times daily [A745b]. As a treatment for **skin "cancer" (open sores)**, 1 ounce of the fresh inner bark is dried, powdered, mixed with 6 ounces of Vaseline or rum, and applied 3 times daily to the affected area [A745b].

To treat **snakebite**, 1 handful of bark along with 1 handful of Billy Webb bark (*Acosmium panamense*) and 1 handful of *copalchi* bark (*Ocotea veraguensis*) are boiled in 1½ quarts of water for 10 minutes, 4 ounces consumed warm every hour on the first day, 4 ounces consumed 3 times daily on the second day, and a laxative with 2 ounces of castor oil (*Ricinus communis*) consumed as soon as the condition has improved on the third day [LR]. Additionally, the usual poultice formula for snakebite wounds must also be used [LR]. This treatment is said to save the person's life more than just being a first-aid remedy. All snakebite formulas should always be followed by 2 ounces of a castor oil laxative (*Ricinus communis*) or "their guts will close up" from the venom, which could kill them, because the venom is "hot" [LR].

Lamb (1946) reported that the bark was used for **tanning** and as an **emetic**. The wood is used in **construction** for lumber [B2059]. Lamb (1946) noted that this plant resembles mahogany (*Swietenia macrophylla*) in its properties, and because the latter was so common, this tree was little utilized. He also noted that in other regions, it was valued in **carpentry** for making furniture and in **construction** as a general lumber.

Swietenia macrophylla King

broken ridge mahogany, mahogany [English]; *chacalte* [Mopan]; *sutz'uj* [Q'eqchi']
For **itchy and open foot sores** or **foot fungus**, especially bleeding cuts, 2 ounces of the inner bark are chopped, boiled in 4 ounces of water for 10 minutes, strained, and used as a wash directly over the affected area twice daily until better [B2488].

As a treatment for **skin "cancer" (open sores)**, the fresh inner bark is dried, powdered, 1 ounce mixed with 2 ounces of Vaseline, and applied directly over the affected area twice daily, or more frequently if the sore is very bad and rashy [B2672]. Alternatively, 1 pound of the bark is boiled in 6 cups of water, boiled down to 5 cups, 4 ounces of bay rum tincture (*Pimenta racemosa*) added to avoid spoilage, and used as a sponge bath twice daily, followed by an application of grease to the body after each treatment [B2672]. This treatment for **skin "cancer" (open sores)** is not to be poured over the head, as it causes hair loss [LR]. For **skin ailments** and **skin "cancer" (ulcer)** refer to Spanish elder (*Piper jacquemontianum*) [A326].

Thompson (1927) reported that this tree was used in **construction** to make dugout canoes. This is said to be the most important tree of Belize and is the "national tree" of the country [Marsh, Matola, and Pickard, 1995]. On the coat of arms of Belize, "the motto: '*Sub Umbra Floreo*'; under the tree we flourish," indicates its importance in Belizean history as the source for one of the earliest industries [Horwich and Lyon, 1990]. They also noted that the wood from this tree, because of its strength and long durability, was used in **carpentry** for making many household articles. Standley and Record (1936) noted that Belize was originally settled about the middle of the century by British logwood (*Haematoxylum campechianum*) and mahogany (*Swietenia macrophylla*) cutters, the value of the forests being such that the settlement was a continual source of strife, and it was not until 1798 that the British were left in undisputed possession. The export of forest produce was the sole reason for the original and continued settlement of the colony. It is an extremely important timber tree used in **carpentry** for making furniture and many other objects for local use and for export [MB]. Even today, it is not uncommon to find kitchen sinks constructed from this tree in the traditional thatched houses of Belize [MB, RA]. To make a wagon used to haul mahogany, refer to cabbage bark (*Lonchocarpus castilloi*) [Lamb, 1946].

(top) *Swietenia macrophylla* cut
timber ready for processing
at sawmill (left) and flowering
branch (right) [MB].

(bottom) Nineteenth-century
Belizean oxcart wheel made
from *Swietenia macrophylla*
[RBGKew].

Trichilia glabra L.

čabonče [Yucateca]

To **induce vomiting of "bad bile,"** the bark is soaked in water and consumed [Arnason et al., 1980].

Trichilia havanensis Jacq.

bastard lime, spoon tree [English]; *limoncillo, limoncillo macho, palo de cuchara* [Spanish]; *cot* [Additional name recorded]

To treat **skin rash** and **scabies**, the leaves are used to prepare a bath [A327]. **Caution** is used, as this treatment, it is said, may cause hemolysis, or the breaking apart of the red blood cells. For testing the reaction of the blood, 9 leaves are boiled in 1 cup of water for 5 minutes, steeped for 10 minutes, and 1 drop of blood added. If the blood spreads slowly, the decoction is good to use with that person; if the blood spreads too quickly, the boiled leaves will hemolyze the blood [A327].

Trichilia havanensis [GS].

To treat a **stubborn or infected sore**, tree resin (obtained from scraping the bark) is applied directly to the affected area once daily until better [A753].

To **repel fleas**, 1 handful of leaves is mashed, soaked in 1 quart of water, and applied all over the skin. This remedy is used by bushmasters and farmers [A812].

To treat **snakebite**, the bark is finely chopped, boiled in water for 10 minutes, and 2 teaspoons consumed warm 3 times daily before meals [A925]. Additionally, chopped bark and 1 "soft candle" are mixed and applied as a poultice directly to the bite once daily [A925].

The trunk and branches, which are prized because they are soft, pulpy, and cork-like, are used in **carpentry** for furniture, spoons, and other household utensils [A753, A812]. This wood is collected during the full moon, as it is believed that the wood will last longer [A812].

Trichilia hirta L.

red cedar, white ramon [English]; *sombra de carneiro* [Spanish]; *son* [Additional name recorded]

The wood is occasionally used as a **construction** material [B1831].

Trichilia minutiflora Standl.

white cabbage bark, wild lime [English]

The bark is a very **fragrant** component of bath mixtures, having a cedar-like smell, and is used to treat many **ailments** [B3214].

For preventing or controlling **high blood pressure**, 1½ handfuls of bark are boiled in 1 gallon of water for 15 minutes and 1 cup consumed 3 times daily. The blood pressure is checked every 3 days and the treatment continued as needed [A491].

The trunk is used in **construction** for house posts [A491].

Trichilia moschata Sw.

Trichilia minutiflora [MB].

Standley and Record (1936) reported that in Guatemala, the wood of this tree was used in **construction** to make marimba keys.

MENISPERMACEAE

Vines or lianas, sometimes small trees, shrubs, or herbs. Leaves alternate, usually simple, rarely palmately lobed or decompound. Inflorescences supra-axillary or cauliflorous, in spikes, racemes, or panicles. Flowers often small, greenish-white or dull reddish. Fruits a drupe-like monocarp. In Belize, consisting of 5 genera and 8 species. Uses for 1 genus and 1 species are reported here.

Cissampelos tropaeolifolia DC.

hoja preñada [Spanish]; *pik e tun* [Mayan]

To relieve **stomach pain** in young babies or to treat an **infected or protruding umbilicus**, 3–4 leaves are mashed, placed on the navel or umbilical area for 12 hours, and changed twice daily, or as needed if the leaves dry up too quickly [B2295]. To treat puhu (when a **baby's navel is oozing blood**), refer to *epasote* (*Chenopodium ambrosioides*) [HR, LR]. To relieve **headache**, 1 cup of mashed leaves and stems is soaked in 2 cups of water for 10 minutes, strained, and applied cool to moisten the head [B2531]. For **anxiety in pregnant women**, the leaves are prepared and used as a bath once daily, at bedtime, until better [A374; RA].

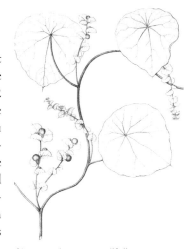

Cissampelos tropaeolifolia [NYBG].

MONIMIACEAE

Shrubs or small trees. Leaves opposite or rarely whorled, simple. Inflorescences axillary or borne on leafless nodes of older wood, in cymes. Flowers yellow or yellowish-green. Fruits a drupe. In Belize, consisting of 2 genera and 2 species. Uses for 2 genera and 2 species are reported here.

Mollinedia guatemalensis Perkins
Spanish ella [English]
To treat **bone pain**, an unspecified amount of leaves are soaked in a bucket of cool water and used as a bath over the affected areas until better [B2552]. For **anemia**, an unspecified amount of leaves are boiled in water, an unspecified amount consumed, and the remainder used as a bath [B2552].

Mollinedia guatemalensis showing flowers [MA].

Siparuna thecaphora (Poepp. & Endl.) A. DC.
bastard fig, wild coffee [English]; *chu' ché* [Q'eqchi']
To treat **headache**, 1 leaf is pressed to the head [B2548]. For **skin sores** and **measles**, 1 handful of leaves is boiled and used to wash the affected area [B2640]. To treat **swelling** of the legs and feet during pregnancy, the leaves are boiled in water, an unspecified amount consumed warm, and the remainder used as a bath [B3509].

To treat **hair loss caused by a mild fever**, the leaves are boiled in water, 1 cup consumed warm, and the remainder used cool as a bath [B2548].

MORACEAE

Trees, shrubs, sometimes lianas or herbs, with milky, sometimes watery latex. Leaves alternate, spiral, or distichous, sometimes opposite, simple. Inflorescences usually axillary, sometimes ramiflorous or cauliflorous, in racemes, spikes, or globose heads. Flowers green or greenish. Fruits a fleshy achene or drupe. In Belize, consisting of 9 genera and 39 species. Uses for 9 genera and 13 species are reported here.

Artocarpus altilis (Parkinson) Fosberg
breadfruit [English]; *arbol de pan, mazapán* [Spanish]
Standley and Record (1936) noted that this East Indian and Pacific plant, introduced many years ago to Tropical America, was cultivated for its fruit, which was eaten as a **food**. This fruit was an inexpensive way to feed slaves [RA]. Thompson (1927) was of the opinion that this species reached Belize via the West Indies and that there was no local Mayan name for the plant; only the Spanish name *mazapán* was used.

Brosimum alicastrum Sw. subsp. *alicastrum*
breadnut, red breadnut [English]; *ramón, ramón blanco, ramon rosa* [Spanish]; *capomo, chacox, macica, masicaran, ox, oš, ujushte* [Additional names recorded]
This fruit is eaten as a **food** [W1856; LR]. The outer shell of the ripe fruit is very sweet and tasty. During times of hardship, draught, or famine, the seeds are boiled and made into tortillas or combined with corn to make the corn supply last longer [LR, RA].

(left) *Artocarpus altilis* with inset showing leaves and flower [MB].

(right) *Brosimum alicastrum* subsp. *alicastrum* [FA].

During the *chicle* camp days, the leaves were used as **fodder** for horses and mules and said to be very nutritious for them [LR]. When in season, this fruit is the main **food** for monkeys, who also eat the leaves [LR]. Lundell (1938) reported that in addition to eating the boiled, starchy seeds, local people dried and ground them into flour, noting that "if we may judge by the extent and distribution of the *ramón* groves, we may assume that the ancients depended considerably upon this source of food." Thompson (1927), in addition to noting the palatability of the fruits and seeds, reported that the kernels, "after being steeped in water or lime, are ground and mixed with maize to make tortillas." He also provided a San Antonio (Toledo District) Mayan name for this species, *oš*. According to Lamb (1946), the boiled seeds had "a nutty flavor and somewhat resemble chestnuts." This tree was important to the ancient Maya and provided evidence for traditional management systems aimed at its improvement as a food source [Peters, 1983]. It has been reported that the leaves and young stems are soaked in water to make an emergency **substitute for a mother's milk** for newborn babies [RA]. It is also said that animals (e.g., cattle, horses) give more **milk** while nursing when they feed on the leaves [RA].

For **toothache**, the white sap is applied to a piece of cotton and stuffed into the tooth. This is said to shrink the gums so that the tooth comes out easily when a person cannot get to a dentist [LR]. To suffocate or stifle **beefworm**, the sap is applied directly to the larvae, covered with a piece of tape for 1 hour, and the worm extracted [LR].

(top left) *Castilla elastica* subsp. *elastica* with inset showing latex from cut stem [MB].

(top right) Sheet of rubber made from *Castilla elastica* subsp. *elastica* [RBGKew].

(bottom) *Dorstenia contrajerva* with inset showing inflorescence [MB].

Castilla elastica Sessé subsp. *elastica*

castilloa rubber, rubber tree [English]; *hule, ulé* [Spanish]

To treat **rheumatism**, cod liver oil is placed on the leaf and then applied as poultice to the affected area [Arnason et al., 1980].

Lundell (1938) reported that cloth was made impervious to water by covering it with the **latex** from this plant. Standley and Record (1936) cited an account by Major H.M. Heyder (1930) noting that the *chicle* tappers always carried small canvas bags covered with the latex of this tree, placing these bags under the cuts in the *chicle* tree where the sap was collected and harvested. Lundell (1938) also noted that the pre-Spanish Colonization Maya made **rubber balls** for playing the game known as *pelote* from this latex as well.

Dorstenia contrajerva L.

contrayerba, justo wes, lombrecina [Spanish]; *rah li ch'och, xwa b'aqnel* [Q'eqchi']; *x cambalhau* [Yucateca]

To expel **worms** in 2-year-old children, 1 root is boiled in 1 cup of milk for 10 minutes and 1 cup consumed warm once daily, before breakfast, for 3 days. If the child is frail, 1 cup is consumed warm every other day before breakfast during a 6-day period [B2303]. To expel **worms** in animals, the root is cut into small pieces and placed into the fodder [Arnason et al., 1980].

Steggerda (1943) noted that several Yucatan Maya traditional healers used toasted and ground roots to feed people to **increase their appetites**. Steggerda (1943) wrote:

> It has been used consistently over a long period to alleviate **disorders of the alimentary canal**, especially the stomach, and may be bought as a root or extract in the Merida drug stores. In early times X-cambalhau was prescribed for a great **variety of ailments**, including **colds**, **pain in the heart**, **insect bites**, **diarrhea**, **dysentery**, **indigestion**, **childbirth**, **irregular menses**, **blood-vomit**, **liver complaint**, **sores**, **gout**, **tumors**, **skin diseases**, and **infected gums** (Roys, 1931, p. 222). Today the plant is known to modern doctors in Merida as an **antidote for all poisons** and is employed as a **stimulant tonic** and **diaphoretic** in **fevers**, dysentery, **diarrhea**, and **indigestion**. Among the herb doctors the plant is used chiefly to cure **digestive disorders** and to treat poisonous **snakebites**. For digestive disorders the root is generally cooked with sugar or honey and the concoction taken by the spoonful. Sometimes the root is toasted and then ground into a powder and mixed with pozole or coffee. (p. 205)

For **drug withdrawal**, 5 leaves are soaked in 1 cup of water and consumed [A312]. For **marijuana withdrawal** specifically, the roots are smoked or made into tablets. To make the tablets, the roots of 3 plants are roasted until crispy dry, ground into a powder, used to fill 9 gelatin capsules (size 00), and 1 capsule consumed orally 3 times daily for 1 week [A312].

For vex (**anger**), 3 leaves are boiled in 1 cup of water for 10 minutes and sipped very warm until 1 cup is consumed by both parties. This is said to provide strength against the angry person. A second dose is sipped until the vexed person feels the anger has passed from the angry one [B2515]. For **fear** "in the bush," the leaves are boiled in water, an unspecified amount consumed, and the remainder used as a relaxing bath [B2515].

Standley and Record (1936) reported that this plant was widely used in traditional **medicine** in Belize and that the thick roots were used in Central America to **flavor cigarette tobacco**.

According to Steggerda and Korsch (1943), the leaves of this plant were stuffed into a **wound** for treatment.

Dorstenia lindeniana Bureau
rah li ch'och [Q'eqchi']

To treat **severe foot sores that result in sickness**, caused by getting toad feces or urine on the foot, 1 entire plant is mashed and applied as a poultice on the affected area [B2517].

Ficus americana Aubl.

fig, small fig matapalo [English]

The mature fruits are sweet, seedy, and eaten as a **food** by children, fish, and birds [B1900]. As these trees grow along the riverbanks, the fruits fall into the water, and the fish then eat them [RA].

Ficus insipida Willd.

red fig [English]

The fruit is eaten as a **food** by fish and forest creatures [B1919]. Standley and Record (1936) reported that the large fruits of this tree were sweet and "fairly good to eat." They also noted that birds and other animals fed on these fruits.

Lamb (1946) reported that the latex of this tree was "sometimes used to **adulterate *chicle*** (*Manilkara zapota*)" and that its properties were under investigation.

To treat **sores from leishmaniasis (baysore)**, an infection caused by a parasite transmitted through the bite of female sandflies, the white sap is collected and applied to completely cover the sore and the area around it twice daily until better. This application burns a bit [B1919].

(top) *Ficus americana* [MB].

(bottom) *Ficus maxima* [GS].

Ficus maxima Mill.

fig, fig tree, strangler fig, wild fig [English]; *higuero, mata palo* [Spanish]; *amate* [Mopan]

The dry and seedy fruit is eaten as a **food**, mostly by birds, forest animals, and fish, when the fruits fall into the water [Arvigo and Balick, 1998].

To relieve **headache** or **back pain**, the trunk is cut, the fresh latex collected, spread on a cotton gauze, another piece of cotton is spread over this, and the poultice applied to the affected area for 1 hour [A323]. To remove **rotten, painful teeth**, a small amount of white latex is placed on a cotton ball, stuffed inside the tooth to soften the flesh around the root, and within 1 hour, the tooth is said to break into pieces [Arvigo and Balick, 1998].

To heal **wounds**, 1 handful of fresh leaves is boiled in 1 gallon of water for 10 minutes and used cool as a bath twice daily [A901]. For **erysipelas**, the bark is powdered, mixed with baby oil, and applied to the skin twice daily [A901]. For **skin fungus, ringworm**, or **boils**, the latex from the stems and leaves is applied to the affected area

3 times daily [Arvigo and Balick, 1998]. Similarly, to suffocate and stifle **beefworm**, the resin is applied to cover the affected area [A901].

To **improve blood circulation** in children and adults, the leaves are used to prepare a bath [Arvigo and Balick, 1998]. For treatment of both early and later stages of **AIDS**, ½ pound of bark is soaked in 1 quart of water for 2 hours and then consumed all day long in ½ cup doses until finished [A323]. This treatment for AIDS is said not to be a cure but to make the person more comfortable, feel stronger, and improve the appetite [RA].

Arvigo and Balick (1998) reported that

> Maya people believe that the flower of the fig tree is visible only at noon by virgin children during Holy Week of Easter season. The flower is considered a **good luck charm**, but must be stolen from the "spirit" (*duende* in Spanish) of the tree by a child. To avoid the bad luck that occurs when an adult takes the flower, the adult has to struggle with the *duende* and must recite a secret prayer. However, if an adult happens upon the flower by accident and places it between the folds of a new white cloth, he or she will have good luck for life. (p. 105)

Ficus ovalis (Liebm.) Miq.

fig tree [English]; *matapalo* [Spanish]; *amate* [Mayan]

The green fruit, which is pinkish inside, is sweet like a fresh fig and eaten as a **food** by humans, fish, birds, and other mammals [A70].

Maclura tinctoria (L.) D. Don ex Steud.

fustic [English]

Lundell (1938) noted that this tree was the source of the **fustic wood** of commerce, which provided green, brown, and yellow **dye**. Record and Mell (1924) described the use of fustic as a dye as follows:

> Fustic is imported usually in the form of logs. These are chipped and the chips ground to a powder, which is put into bags ready for the dyer. In this state there is opportunity for adulteration which is difficult to detect on account of the small size of the particles. From the pulverized wood the dyer prepares an aqueous extract and paste or lake. A strong decoction of the wood is of a reddish-yellow color, the yellow becoming more pronounced upon dilution. The coloring principle, ma-

Maclura tinctoria [NYBG].

clurin, is readily soluble in water. The chips are often employed for **dyeing** by keeping them in bags at the bottom of the bath or of the dyeing vat. The natural color thus obtained, without mordants, is a fairly permanent dull yellowish-brown or khaki . . . Although the fustic industry has been very greatly reduced, except during the abnormal conditions created by the war, it is believed that the wood will always be a factor in the world's commerce. (p. 120)

This is one of the strong woods used in **carpentry** to make wooden tools, tool handles, and mortar sticks [Horwich and Lyon, 1990]. Standley and Record (1936) listed the wood as being useful in **carpentry** for inlaid cabinetwork, for making **fence posts**, and as **fuel**. Steggerda (1943) noted that in the Yucatan, the wood of this tree was used in **carpentry** to make the pestles of chocolate mixers and the sap applied to cotton and placed into a decayed tooth to treat a **toothache**.

Poulsenia armata (Miq.) Standl.
chi chi costé [Mayan]

The raw fruit is eaten as a **food**. Seeds are spit out after the fruit is eaten, and the fruit leaves a gummy substance on the lips [R50].

Standley and Record (1936) reported that indigenous people in Panama soaked the inner bark of this tree in water, pounded it into thin sheets, and made a "**coarse cloth**" from it, and they noted that this type of **fiber** was sometimes made elsewhere in Central America from other trees in this family.

Pseudolmedia spuria (Sw.) Griseb.
bastard cherry, cherry, wild cherry [English]; *mambas* [Spanish]; *manash, manax, manga* [Additional names recorded]

(left) *Poulsenia armata* branch with fruit [RF].

(right) *Pseudolmedia spuria* [MB].

The fruits are juicy, deliciously sweet, slightly tart, somewhat like fully ripe prunes (*Prunus domestica*), and eaten as a **food** [M201, N46831].

For **fever**, the bark along with Billy Webb bark (*Acosmium panamense*) and *tres puntas* leaves (*Neurolaena lobata*) are combined, steeped in water, and 1 teaspoon consumed 3 times daily for a child or 1 tablespoon consumed 3 times daily for an adult [B3733]. For a person who is **tired**, 4 ounces of bark are boiled in 1 quart of water for 10–15 minutes and consumed cool [B3733]. For **thrush** in a baby, 8 ounces of bark are chopped, soaked in water, strained, 1 ounce of honey added, steeped in ⅓ cup of water until color appears, and applied on a piece of cotton wrapped around a finger with gloves to clean the affected area. When a lot of thrush is collected on the cotton, the cotton ball is changed and the treatment repeated [B3733].

To make **wine**, 3 pounds of fruit are washed, the seeds squeezed out, the fruit flesh placed in 2½ gallons of water, and soaked overnight. The following day, 2½ pounds of sugar and ½ teaspoon of bread yeast are added and soaked for 3–4 days. If the infusion turns into vinegar, more sugar is added [B3733; RA].

This is considered a good **timber** wood [Robinson and Furley, 1983].

Trophis racemosa (L.) Urb.

cherry, white ramon (female), red ramon, white breadnut, white ramón, wild waya [English]; *el dorado, raman verde, ramón, ramon blanco, ramon colorado* [Spanish]; *cha cox, sacua yun, yak ox, yaxox* [Mayan]

The fruit is eaten as a food by humans, horses, cows, birds, and other animals [A264, A762, B2070, B3263, JB111]. The leaves are used as **fodder** for cattle, horses, and sheep and fed to these animals especially during gestation [A264, B2070, B2450].

For **diarrhea**, two 2-inch pieces of bark are boiled in 1 quart of water for 20 minutes, cooled, and ½ cup consumed every hour until the stools return to normal [B3262].

To "drive off" the poison of a **snakebite**, the root of this plant, along with 1 ounce of root from rat root (an undetermined species), 1 ounce of snake root (root) (*Pentalinon andrieuxii*), and 1 ounce of skunk root (root) (*Chiococca alba*) are grated together, mixed with the toasted head of the snake, and all of this burned down to ashes. The ashes are then applied as a poultice directly to the bite and changed every 2 hours [B3262].

Trophis racemosa [MB].

MYRICACEAE

Small trees and shrubs, often aromatic. Leaves alternate, simple. Inflorescences axillary, in catkins. Flowers small. Fruits a drupe or nutlet. In Belize, consisting of 1 genus and 1 species. Uses for 1 genus and 1 species are reported here.

Myrica cerifera [NYBG].

Myrica cerifera L.

tea bark, tea box [English]; *myrtle* [Additional name recorded]

To treat **kidney and bladder ailments**, especially to soothe the pain of **kidney stones** and to **cleanse the urinary tract**, 1 small handful of leaves is boiled in 3 cups of water for 5 minutes and sipped all day until finished [A343]. For **bloody urine**, especially in men, refer to black cockspur (*Acacia collinsii*) [A342; RA]. To overcome an **addiction** to black tea (*Camellia sinensis*), the decoction described for kidney and bladder ailments is consumed instead [A343]. As a **diuretic**, 2–3 young to just-developed leaves are boiled in water and consumed 2–3 times daily [Mallory, 1991].

According to Standley and Record (1936), "the greenish **wax** obtained by boiling the fruits in water is employed in the United States for making candles that burn with an agreeable fragrance."

MYRISTICACEAE

Trees, sometimes shrubs, aromatic. Leaves alternate, sometimes distichous, simple. Inflorescences terminal, sometimes cauliflorous, in racemes, corymbs, or very complex. Flowers white to creamy green to yellow, sometimes greenish. Fruits a capsule or dehiscent berry. In Belize, consisting of 2 genera and 3 species. Uses for 1 genus and 1 species are reported here.

Virola koschnyi Warb.

black banak [English]; *palo de sangre, sangre* [Spanish]; *banak* [Mayan]

The trunk and branches are used to make **veneer** [A454]. Standley and Record (1936) noted that the seeds of this tree had a high amount of **oil** and that the **timber** was "the most important secondary timber now being exploited in British Honduras."

MYRTACEAE

Subshrubs to large trees. Leaves opposite, sometimes alternate, simple. Inflorescences axillary or terminal, in panicles, dichasia, racemes, or with solitary flowers. Flowers

(top) *Virola koschnyi* germinating seed (upper left), seedling (center), and mature fruit showing red aril (lower right) [CV].

(bottom) *Calyptranthes bartlettii* [MB].

usually white, sometimes reddish. Fruits a berry, rarely a capsule. In Belize, consisting of 13 genera and 61 species. Uses for 9 genera and 19 species are reported here.

Calyptranthes bartlettii Standl.

contramor [Spanish]

For **diarrhea**, 1 large double handful of leaves is boiled in 2 gallons of water for 10 minutes, steeped until warm, and used as a sitz bath for 20 minutes once daily, at bedtime, until better [B1894]. To treat children suffering from **diarrhea** or **fright** (*susto*), 1 small handful of leaves is steeped in 3 cups of boiling water for 20 minutes and 3 teaspoons consumed warm once daily for infants, 6 teaspoons consumed warm once daily for children ages 2–6, and ½ cup consumed warm 3 times daily for children ages 7–12 until better [B1894].

Calyptranthes calderonii Standl.
blossom berry [English]; *chiri mis* [Mayan]
The wood is used in **construction** to make house walls [JB118].

Calyptranthes chytraculia (L.) Sw. var. *americana* McVaugh
blossom berry [English]; *ginda, guayabillo, pimiento* [Spanish]
To soothe **anxiety**, 1 large double handful of leaves is boiled in 2 gallons of water for 10 minutes, steeped until warm, and poured over the body or used for soaking in the tub once daily, at noon or bedtime, as needed [A398; RA]. For **nervousness**, refer to timber sweet (*Nectandra salicifolia*) [A399].

Calyptranthes lindeniana O. Berg
blossom berry [English]
The fruit is tasty and eaten as a **food** [A68].

Calyptranthes pallens Griseb.
lemon tree (not the standard lemon) [English]
To **cleanse the body**, the leaves are boiled in water and consumed as a tea [R93].

Chamguava gentlei (Lundell) Landrum var. *gentlei*
Pine Ridge spice [English]
To treat **exhaustion**, six 6-inch stems are boiled in 2 cups of water for 10 minutes and ⅓ cup consumed 3 times daily [A374]. Alternatively, 1 large double handful of leaves and stems is boiled in 2 gallons of water for 10 minutes, steeped until warm, and poured over the body or used for soaking in the tub once daily, at noon or bedtime, until better [A374; RA].

Eucalyptus sp.
eucalyptus [English]; *melina* [Spanish]
To relieve **cough** and **congestion**, 12 leaves are boiled in 1 quart of water for 5 minutes and used as a steam bath for the head. A tent is made with towels around the head, and the hot decoction and the steam is breathed in deeply [A608; HR].

The trunk is used in **construction** for making lampposts and telephone posts [A605].

Eucalyptus sp. [GS].

Eugenia biflora (L.) DC.

To treat **sick babies**, especially newborns, 1 handful of leaves is boiled in 1 gallon of water for 10 minutes, cooled to warm, and used as a bath for 20 minutes once daily until better [A369]. To prevent **fever** after childbirth, 1 large double handful of leaves is boiled in 2 gallons of water for 10 minutes and used as a bath once daily for 7 days following childbirth [A369].

Eugenia capuli (Schltdl. & Cham.) Hook. & Arn.

blossom berry, walk naked [English]; *cacho de venado, granada cimarrona, Indio desnudo, limoncillo, vaina de espada* [Spanish]; *chi lun che* (female) [Mayan]; *chrimis embra* [Additional name recorded]

The ripe fruit is sour and eaten as a **food** [B3265].

To treat **stomach pain** and **diarrhea**, 1 ounce of bark is boiled in 2 cups of water for 10 minutes and ¼ cup consumed cool 3 times daily until better [A235]. For **snakebite**, 1 ounce of root along with 1 ounce of Billy Webb root (*Acosmium panamense*) and 1 ounce of *chicoloro* vine (*Strychnos panamensis* var. *panamensis*) are boiled in 2 cups of water for 10 minutes, ½ cup consumed the first hour, ¼ cup consumed the second hour, and ⅛ cup consumed the third hour [A911].

The branches are used in **construction** to make broom handles [A235]. The straight-growing trunk is used in **construction** for house posts [R87].

Eugenia riograndis Lundell

The leaves are a component of bath mixtures [A360].

Eugenia trikii Lundell

chi lon che macho [Mayan]

The mature fruits are sour and eaten as a **food** by humans and birds [A246].

(left) *Eugenia capuli* [GS].

(right) *Eugenia trikii* [GS].

(top left) *Eugenia venezuelensis* [GS].

(top right) *Melaleuca leucadendra* [NYBG].

(bottom) *Myrcia splendens* [RF].

Eugenia uniflora L.

Surinam cherry [English]; *pitanga*
[Additional name recorded]
The red fruit is eaten raw as a **food** and said to
be delicious; the first bite is sour, followed by
a very sweet flavor that remains in the mouth
[R4].

Eugenia venezuelensis O. Berg

molb' i ba' [Q'eqchi']
The ripe fruits provide **food** for birds; hunters know this and seek their quarry by this
plant [A618].
    The wood is used as **fuel** and **firewood**
[A618].

Melaleuca leucadendra (L.) L.

fever tree [English]
To treat **fever**, the bark is used to prepare a bath treatment [B2131].

Myrcia splendens (Sw.) DC.

Standley and Record (1936) reported that "the fruit is **edible**, with a somewhat acid,
spicy, agreeable flavor."

Pimenta dioica leaves and fruits with inset showing bark [MB].

Pimenta dioica (L.) Merrill

allspice [English]; *pimienta gorda, pimiento* [Spanish]; *bosh pol pimienta, naba cuc* [Mopan]; *pens, q'ehen* [Q'eqchi']

The berries are used as a **spice** in seasoning meat stews, hot cereals, and liquored beverages and, at the time of this study (1990s), were collected from wild trees in the forest to be sold at a rate of BZ$120 per quintal (100-pound sack), then valued at US$60, to wholesalers for the **international spice trade** [B2046, B2404, JB84; Arvigo and Balick, 1998]. To enhance **flavor**, 3–6 leaves are boiled in fish soup [B2404]. As a **coffee substitute**, 3 leaves are boiled in 3 cups of water, milk added to taste, and consumed as a beverage [B3135, JB21]. For **thirst** when there is no supply of drinkable water, the leaves are chewed [Mallory, 1991]. Standley and Record (1936) reported that the leaves were used to make a hot **beverage**.

For temporary and quick relief of **toothache**, 1 leaf or a few berries are chewed into a paste and applied as a poultice to the gums every hour as needed [A107; Arvigo and Balick, 1998]. Using green, unripe berries is the most effective **toothache** treatment [RA]. To treat **foot fungus**, 1 handful of berries and 2 ounces of cow fat or a "soft candle" are mashed and applied as a poultice or ointment to the affected area in the nighttime [A107, B1833]. For **rheumatism**, the berries are crushed, boiled down to a paste, spread onto a piece of cotton, and used as a plaster on the affected area [Arvigo and Balick, 1998].

To treat **infantile colic** or **minor stomach pain**, such as **menstrual cramps, nausea, indigestion**, or **intestinal gas**, 3 leaves are boiled in 3 cups of water for 5 minutes and sipped as hot as possible all day as needed [A107, A818, B1833, B3135]. For **intestinal gas, loose stool, indigestion**, or a **bad belly**, 1 seed along with 1 garlic clove and one 3-inch piece of a dried sweet orange peel (*Citrus sinensis*) are boiled in 1 quart of water for 10 minutes and ¼ cup consumed every hour until better [B2404]. To treat **gastritis** (ciro), 9 berries along with one 1-inch ginger root, one 3-inch piece of a dried sweet orange peel, and 1 garlic clove are boiled in 1 quart of water for 10 minutes and ¼ cup consumed every hour [A818]. To treat **gastritis** (ciro) in infants, 9 drops of the decoction described for gastritis (*ciro*) are put into the mouth every hour [A818].

For **diarrhea**, one 1-inch × 2-inch piece of bark is boiled in 3 cups of water for 10 minutes, 3 leaves added, and 1 cup consumed cool 3 times daily before meals [B2046]. For **chronic diarrhea**, refer to guava (*Psidium guajava*) [B2138]. To treat **dysentery**, the seed along with the *chicle* seed (*Manilkara zapota*) is macerated, boiled in water, and consumed. Alternatively, this treatment can be prepared with the omission of the *chicle* seed [Arnason et al., 1980].

To treat **painful menstruation**, 1 large double handful of leaves is boiled in 2 gallons of water for 10 minutes, 1 cup consumed as hot as possible, and the remainder used hot as a sitz bath for 20 minutes [B1833].

To treat **exhaustion** or **low energy**, the leaves are boiled in water and used as a bath [Arvigo and Balick, 1998]. For **hangover** or **nervousness**, 3 leaves are boiled in 2 cups of water for 5 minutes, cooled to lukewarm, and ½ cup consumed 4 times daily or sipped slowly all day until finished [LR]. **Caution** is advised against daily consumption of the tea over a long period of time, as it is said this will lower sexual function [Arvigo and Balick, 1998]. This is considered a "hot" remedy. **Caution** is also advised against using it too often in cooking, as rashes and excessive weight loss can result [Arvigo and Balick, 1998].

Psidium guajava L.

guava [English]; *guajava, guayaba* [Spanish]; *pi chi, pici, pu tá* [Mopan]; *patah* [Q'eqchi']; *coloc, put* [Additional names recorded]
The fruit is considered sweet and is eaten fresh as a **food** or made into a jelly [B2273, B3189, B3577, W122].

To treat any type of **pain**, 1 large double handful of leaves along with 1 large double handful of mistletoe leaves and vines (*Phoradendron* sp.) are boiled in 2 gallons of water for 10 minutes, steeped to warm, and used warm as a bath twice daily, in the morning and in the evening, until better [B2138; RA]. To prevent **scabs** from forming as the result of **measles** or **chicken pox**, 1 large handful of leaves is boiled in 2 gallons of water for 10 minutes and used lukewarm to soak the body for 20 minutes [B2652; HR]. To treat **measles, itching**, and **sores**, refer to red vervain (*Stachytarpheta miniacea*) [A378].

Steggerda (1943) reported that the Yucatan Maya used the leaves to **promote**

Psidium guajava with inset showing flower [MB].

sweat by soaking the leaves in a person's bath water. He also noted that the water was said to soothe **irritated skin**.

To treat **eye pain** from sun strain, conjunctivitis (pinkeye), or accident, the flowers are mashed or pounded and applied as a poultice directly to the eye [Arvigo and Balick, 1998]. To treat **bleeding gums**, **bad breath**, and to **prevent hangovers**, the tender leaves are chewed [Arvigo and Balick, 1998]. To treat **bleeding gums** or **mouth soreness**, 1 cup of green leaves is steeped in 3 cups of boiling water for 20 minutes and used as a gargle twice daily [Arvigo and Balick, 1998]. For **mouth sores**, 1 small handful of bark is mashed, soaked for 30 minutes in 1 cup of water, 1 teaspoon of honey added, and used twice daily as a mouthwash [A830, W122; RA]. As a general **mouthwash**, the decoction described above for mouth sores is used without the honey [A830, W122; RA]. For **skin sores**, following a regular bath of the area with soap and water, 1 small handful of bark is mashed, soaked for 30 minutes in 1 cup of water, applied to the affected area twice daily, allowed to air dry, and repeated as needed [A830, W122; RA]. Alternatively, the young leaves and flowers are boiled in water and used as a wash over the affected area [Arvigo and Balick, 1998]. For **stubborn, infected, or chronic skin sores**, the leaves and flowers are toasted, powdered, and sprinkled over the affected area following the wash described above [Arvigo and Balick, 1998]. To treat **skin wounds** and **skin ulcers**, one 1-inch × 2-inch piece of bark is boiled in 3 cups of water for 10 minutes and used as a wash over the affected area [Arvigo and Balick, 1998]. To treat **athlete's foot fungus**, the bark is boiled in water and used to bathe the feet [Arnason et al., 1980]. To **cleanse wounds**, the bark along with *nanci* bark (*Byrsonima crassifolia*) are boiled in water for 10 minutes and used as a wash over the affected area [B2273].

For **dysentery**, **upset stomach**, **diarrhea**, or **colds**, 9 leaves and 9 young fruits are boiled in 3 cups of water for 10 minutes and 1 cup consumed warm 3 times daily before meals [Arvigo and Balick, 1998]. To treat **dysentery**, **upset stomach**, **diarrhea**, **vomiting**, or **sore throat**, one 1-inch × 2-inch piece of bark is boiled in 3 cups of water for 10 minutes and 1 cup consumed 3 times daily before meals [Arvigo and Balick, 1998]. For **dysentery**, 2 ounces of bark are boiled in 2 cups of water for 10 minutes and ½ cup consumed cool 3 times daily until better [B2652]. To treat **infantile dysentery**, 1 ounce of bark is boiled in 2 cups of water for 10 minutes and ¼ cup consumed cool 3 times daily until better [B2652].

For **stomach pain** or **intestinal discomfort**, 2 ounces of bark and 1 small handful of leaves along with one 6-inch piece of a dried sweet orange peel (*Citrus sinensis*) and two 3-inch pieces of cinnamon bark (*Cinnamomum verum*) are boiled in 1 quart of water for 10 minutes and ½ cup consumed cool 3 times daily until better [B2652]. Alternatively, 4 ounces of bark and 2 small handfuls of leaves along with two 6-inch pieces of a dried sweet orange peel and four 3-inch pieces of cinnamon bark are boiled in 1 gallon of water for 10 minutes and used warm as a bath once daily at bedtime [B2652]. For **stomach pain** or **diarrhea**, 1 handful of bark or leaves is boiled in 2 quarts of water for 10 minutes and consumed cool [A239, A830, B1878, W122].

To treat **diarrhea** (**"bad stomach"**) in young children, 1 green fruit is boiled in 1 cup of water and eaten [B1878]. Alternatively, the bark along with *nanci* bark are boiled in water for 10 minutes and consumed [B2273]. In a third reported treatment, one 3-inch × 1-inch piece of bark, 9 leaves, and 1 young fruit are boiled in 1 quart of water for 10 minutes, steeped until tepid, and ½ cup consumed 6 times daily until better [B2273]. For a fourth treatment, refer to sweet orange (*Citrus sinensis*) [B2279].

To treat **diarrhea** in children, the young fruits and leaves are boiled and mashed in water as part of an unspecified treatment [B2273]. Alternatively, 3 leaves are mashed in ½ cup of water and 1 tablespoon consumed cold every hour for children 1–6 years old, 1 ounce consumed cold every hour for children 6–12 years old, and ½ cup consumed cold every hour for adults [B3189]. In another reported treatment, three 3-inch young leaf shoots are boiled in 2 cups of water for 10 minutes and ½ cup consumed cool 3 times daily until better [B3577]. For **chronic diarrhea**, 2 ounces of chopped root along with 2 ounces of chopped *kibix* vine (*Bauhinia herrerae*) and 9 allspice seeds (*Pimenta dioica*) are boiled in 1 quart of water for 10 minutes and ⅓ cup consumed cool 3–4 times daily for 2–3 days or until better [B2138]. For **leukorrhea** (**vaginal discharge**) or **relaxed vagina walls** after childbirth, 1 cup of green leaves is steeped in 3 cups of boiling water for 20 minutes and used as a douche twice daily [Arvigo and Balick, 1998]. For a **postpartum treatment**, refer to amaranth (*Amaranthus dubius*) [BW].

Steggerda (1943) reported that the wood was used by the Maya in the Yucatan region for house **construction**. Thompson (1927) noted that the leaves of this plant were used to **wrap** locally made cigarettes in the highlands of Guatemala.

Psidium sartorianum (O. Berg.) Nied.

half crown [English]; *pichiche* [Yucateca]

Standley and Record (1936) noted that the **edible** fruit "is reported to have a rich spicy subacid flavor." Steggerda (1943) noted that the fruit was not eaten by humans but was eaten by wild pigs (*Sus* spp.).

Steggerda and Korsch (1943) reported that the Yucatan Maya treated **bruises** by placing olive oil and this leaf at the site to change the skin back to its natural color. Steggerda (1943) stated that a decoction of the leaves was taken as a bath, was used to treat **pimples** and **skin eruptions**, and was used to treat **night sweats**. He also reported that the roots were cooked in water and the liquid consumed to treat **dysentery**.

Steggerda (1943) noted that the wood was hard and used in **construction** to make machete handles and house poles.

Syzygium jambos (L.) Alston

ma sa'an [Q'eqchi']

The mature fruit is eaten as a **food** [B3576].

NYCTAGINACEAE

Herbs, shrubs, or trees. Leaves opposite, subopposite, rarely whorled or alternate, simple. Inflorescences terminal or axillary, variously branched, ultimate branches usually in cymes. Flowers showy or inconspicuous. Fruits an achene. In Belize, consisting of 6 genera and 12 species. Uses for 3 genera and 3 species are reported here.

Mirabilis jalapa L.

maravilla [Spanish]; *tza kel bák* [Mayan]

This plant is used as an **ornamental** [B2388].

To treat **cough** and **colds**, 9 leaves along with 9 *chayote* leaves (*Sechium edule*), the juice from 1 lime fruit (*Citrus* sp.), 2 *guanábana* leaves (*Annona muricata*), 2 avocado leaves (*Persea americana*),

(top) *Syzygium jambos* [NYBG].

(bottom) *Mirabilis jalapa* [NYBG].

and 2 *gumbolimbo* leaves (*Bursera sima-ruba*) are boiled in 1 liter of water for 5 minutes and ½ cup consumed warm 6–8 times daily until better [B2283].

To treat **vitiligo** (a **skin pigmentation disorder**), the flowers are rubbed onto the affected areas [Arnason et al., 1980].

Neea psychotrioides Donn. Sm.

pigeon plum [English]; *hoja de salat*, *salat* [Spanish]

The mature fruit is eaten as a **food** [B2659].

To treat **fever** in an infant, fresh leaf juice is used as a bath or rubbed over the body [A749]. To treat **babies or very young children who cannot sleep**, often accompanied with **crying**, especially "when they have been frightened by chickens," 3–6 leaves are mashed, fried, or heated in ½ cup of oil and rubbed all over the body twice daily, at noon and bedtime, until better [B1937, B2298].

Neea psychotrioides showing leaves and fruits (top) and close-up of leaves (bottom) [HW].

To treat dependent **edema**, specifically when a pregnant women suffer from **swollen legs**, 1 handful of leaves is boiled in 2 quarts of water for 5 minutes, ½ cup consumed 3 times daily, and the remainder used as a bath over the affected area [B2492].

Pisonia aculeata L.

tiger nail embra [English]; *hierba del malaire* [Spanish]; *beeb* [Yucateca]

To **promote sweat** and to treat **fever**, **nervousness**, and *aires* (a locally recognized condition caused by exposure to a cold or wet wind, particularly when a person is overheated, resulting in muscle cramps or joint pain), 1 entire plant is boiled in 1 gallon of water for 10 minutes, ½ cup consumed 3 times daily, and the remainder used hot as a bath once daily at bedtime [A361]. Steggerda (1943) reported that the Yucatan Maya boiled young plants in water and used it as a bath to treat **fever** in people suffering from **malaria**. He also noted that "a use for Beeb branches is to place them over the open parts of chicken coops so that **bats** become entangled in the thorns, which are so strong and peculiarly hooked that the Maya also use them to retrieve buckets that have fallen into wells."

(top) *Pisonia aculeata* [RH].

(bottom) *Ximenia americana* var. *americana* [NYBG].

OLACACEAE

Small to medium trees, sometimes shrubs, rarely lianas, or sometimes root parasites. Leaves alternate, sometimes spiral or distichous, simple. Inflorescences axillary, in racemes, panicles, umbels, spikes, or fascicles. In Belize, consisting of 3 genera and 3 species. Uses for 1 genus and 1 species are reported here.

Ximenia americana L. var. *americana*
Standley and Record (1936) reported that the fruit had a juicy, acidic flesh; that it was eaten as a **food**; and that the **astringent** bark was used in **tanning**.

ONAGRACEAE

Herbs, shrubs, lianas, or trees. Leaves alternate, opposite or whorled, simple. Inflorescences usually axillary, in spikes, panicles, or of solitary flowers. Flowers often showy. Fruits a capsule, sometimes a berry or nut-like. In Belize, consisting of 2 genera and 10 species. Uses for 1 genus and 2 species are reported here.

Ludwigia affinis (DC.) H. Hara

swamp clove, wild clove [English]; *clavito*, *clavo* [Spanish]

For **hemorrhage**, 1 entire plant is boiled in 2 quarts of water for 10 minutes, cooled, and consumed as a tea [A337]. To stop **bleeding** on a **wound** or an **abscess**, fresh leaves are mashed in water to form a thick mash that is applied as a poultice to the affected area [A337].

For **toothache**, 1 entire plant with roots is boiled in 1 quart of water for 5 minutes, strained, a pinch of salt added, and used lukewarm as a mouthwash once every 30 minutes. Using the root adds a **numbing effect** [LR]. For non-insulin-dependent **diabetes** mellitus, 1 entire plant with roots is boiled in 1 gallon of water for 10 minutes and 1 cup consumed 3 times daily, after meals, until finished. Following the preparation and consumption of a second gallon, the blood sugar is checked. This entire treatment is repeated, if necessary [HR].

Ludwigia octovalvis (Jacq.) P.H. Raven

wild clove [English]; *clavito*, *clavo* [Spanish]; *klaux pim* [Q'eqchi']

This plant is cultivated as an **ornamental** [B2456].

To treat an **abscess**, 2 handfuls of the entire plant and 1 teaspoon of salt are boiled in water and applied warm to the affected area [C20]. To treat **fever**, 1 quart of leaves and young stems is mashed in 1 quart of warm water and used as a bath over the entire body once daily, at noon or bedtime [B3630]. To lower the glucose level in the blood at the onset of non-insulin-dependent **diabetes** mellitus in adults over 30 years old, 1½ pounds of the entire plant are boiled in water and consumed all day for 14 days [C20].

OXALIDACEAE

Herbs or subshrubs, rarely shrubs, trees, vines, or lianas. Leaves alternate, sometimes basal or pinnate. Inflorescences axillary, sometimes cauliflorous, in cymes, umbels, racemes, or of solitary flowers. Flowers small and inconspicuous to large and showy, variously colored. Fruits a capsule or berry. In Belize, consisting of 2 genera and 3 species. Uses for 2 genera and 2 species are reported here.

Ludwigia octovalvis [NYBG].

(top) *Biophytum dendroides* showing close-up of fruit [JDL].

(bottom) *Oxalis latifolia* [GAC].

Biophytum dendroides (Kunth) DC.

hierba para culebra [Spanish]; *pim re k'anti* [Q'eqchi']; *xiv can* [Mayan]

For **snakebite**, 1 handful of roots is boiled in 3 cups of water for 20 minutes and ½ cup consumed every 30 minutes [B1980]. To treat the swelling of **snakebite**, the leaves are parched, powdered, and applied externally to the bite [B1890].

Oxalis latifolia Kunth

This plant is used as an **ornamental** [B2336].

Argemone
mexicana [MB].

PAPAVERACEAE

Herbs, small trees, sometimes subshrubs, with orange, yellow, or white latex. Leaves usually alternate, sometimes subopposite, simple. Inflorescences axillary or terminal, in racemes, panicles, or of solitary flowers. Flowers usually showy. Fruits a capsule. In Belize, consisting of 2 genera and 2 species. Uses for 1 genus and 1 species are reported here.

Argemone mexicana L.
To treat **toothache** and **halitosis**, the stem and leaf latex are used by the Yucatan Maya, but no further details were provided [Flores and Ricalde, 1996]. As a **sedative** for **insomnia**, **pain**, and **cough**, the white, milky sap is used in an unspecified way [Arvigo and Epstein, 1995].

PASSIFLORACEAE (INCLUDING TURNERACEAE)

Vines or lianas, herbs, shrubs, or sometimes trees, often with tendrils. Leaves alternate, simple or rarely compound. Inflorescences mostly axillary and sessile, sometimes terminal or cauliflorous, in cymes, rarely racemes or fascicles. Flowers small to large, often showy, variously colored. Fruits a berry or capsule. In Belize, consisting of 4 genera and 33 species. Uses for 2 genera and 9 species are reported here.

Passiflora biflora Lam.
granadillo, melon de ratón [Spanish]
To treat **toothache**, the leaves are placed in the mouth cavity, but the effect was unspecified [B2511].

To treat **snakebite** when "in the far bush," the leaves are soaked in cool water, squeezed by hand, and the water consumed. It is said that this infusion "will hold the poison for 3 days" [B2511].

Passiflora mayarum [GS].

Passiflora foetida L.
granadilla del monte [Spanish]; *sa yèp* [Q'eqchi']
The fruit is eaten as a **food** [A744, B3589].

Passiflora mayarum J.M. MacDougal
wild passion fruit [English]; *granadillo* [Spanish]
The mature fruit is eaten as a **food** [A498].

Passiflora urbaniana Killip
grandpa's balls [English]
For **fever**, the vine is boiled in water and used cool to bathe the head [A354]. To treat **headache**, 1 handful is boiled in 1 gallon of water for 5 minutes, steeped until cooled, and poured over the head twice daily, at noon and bedtime, until better [A354]. To treat **thrush**, 9 fresh leaves and one 3-inch piece of fresh vine are boiled in 2 cups of water, strained, applied to the inner mouth, and used as a mouthwash gargle 3 times daily with warm tea until better [A354].

Passiflora sp.
The fruit, a purple berry, is said to be eaten as a **food** [R90].
For **wart** removal, the ripe fruit is mashed and applied topically to the skin twice daily until better [R90].

Turnera aromatica Arbo
wild damiana [English]; *skabon che* [Additional name recorded]
To treat **cough** and **nervousness**, three 10-inch pieces of branchlets are boiled in 3 cups of water for 15 minutes and 1 cup consumed 3 times daily before meals [A384].

Turnera diffusa Willd. ex Schult.

swamp bush [English]

For **general pain**, 1 entire plant is boiled in 1 quart of water for 10 minutes and 1 cup consumed lukewarm 3 times daily, before meals, for 3 days [A350]. For **toothache**, 9 leaves are chewed [A350].

Turnera scabra Millsp.

wild damiana [English]; *claudioso* [Spanish]

As an **expectorant** or a **sedative**, 1 entire plant is boiled in 1 quart of water for 10 minutes and 1 cup consumed warm 3 times daily, before meals, for 3 days [A649]. As a treatment for non-insulin-dependent **diabetes** mellitus, 3 twigs along with one 1-inch × 6-inch piece of bay cedar bark (*Guazuma ulmifolia*) are boiled in 3 cups of water for 10 minutes and 1 cup consumed 3 times daily, after meals, for 3 days, following which the blood sugar is checked [A674].

Turnera ulmifolia L.

Jamaica herb [English]; *claudioso, clavo de Cristo, clavo de oro* [Spanish]

For **fever**, the leaves are used to prepare a bath [B1959]. To treat non-insulin-dependent **diabetes** mellitus, one 12-inch branch with leaves is boiled in 1 quart of water for 10 minutes and 1 cup consumed 3 times daily, before meals, until the blood sugar is normal [HR]. To treat a **cough**, 1 cup of the decoction described for diabetes is consumed 3 times daily, before meals, until better

Turnera ulmifolia [MB].

[HR]. To **cleanse the blood**, the roots of 3 plants are boiled in 1 quart of water for 10 minutes and 1 cup consumed 3 times daily, before meals, for 4 days [B1959].

PEDALIACEAE

Herbs. Leaves opposite, subopposite or alternate, simple. Inflorescences axillary, of solitary flowers or in cymes. Flowers pink, bluish, white or cream. Fruits a capsule. In Belize, consisting of 1 genus and 1 species. Uses for 1 genus and 1 species are reported here.

Sesamum indicum L.

sesame [English]; *wangala* [Garífuna]

The seeds are rich in calcium and are a preferred **food** for nursing mothers and menopausal women. The seeds are roasted, ground, and mixed with garlic, lime, and salt to make a tasty sauce or salsa and given especially to people who need to consume more calcium [RA]. The seeds are made into a sweetened **candy** known as "*wangala*," and sold in local markets [Mallory, 1991].

(top left) *Sesamum indicum* [MB].

(top right) Carrying harvested *Sesamum indicum* [MB].

(bottom) *Petiveria alliacea* [MB].

As a **laxative**, oil is extracted from the seed and used in an unspecified way [Mallory, 1991]. To treat **diarrhea**, the leaves are boiled in water and consumed [Mallory, 1991]. Steggerda (1943) reported that the Yucatan Maya ground the seeds, mixed them with *masa* (corn meal), and gave them to nursing mothers to **increase lactation**.

PHYTOLACCACEAE

Herbs, shrubs, lianas, or trees. Leaves alternate, simple. Inflorescences axillary or rarely terminal, in spikes, racemes, or panicles. Flowers small. Fruits a berry, drupe, drupelet, or capsule. In Belize, consisting of 4 genera and 5 species. Uses for 3 genera and 4 species are reported here.

Petiveria alliacea L.

guinea-hen plant, guinea-hen root, skunk root, skunkweed [English]; *zorrillo* (female) [Spanish]; *pa'har pim* [Q'eqchi']

As a treatment for **dizziness** caused by "bad air in the head," one 10-inch piece of mashed root along with three 3-inch pieces of *ruda* leaves (*Ruta graveolens*), 1 piece of *alcanfor* (camphor block, an over-the-counter product), and 1 piece of assafoetida resin (*Ferula assafoetida*) are mixed with 1 pint of strong rum and used to bathe the head [B2122].

For **measles**, 1 handful of fresh leaves is boiled in water for 10 minutes, an unspecified amount consumed warm 10 times daily, and the remainder used as a bath once daily. This treatment is repeated for 5 days [C52].

For **sinus headache**, refer to warrie wood (*Diphysa americana*) [B2155].

To **cleanse the uterus** and assist the flow of **menstruation**, 2 handfuls of leaves are boiled in 3 cups of water for 5 minutes and consumed warm as a strong dosage [B1861]. Alternatively, 1 handful of roots is boiled in 3 cups of water for 5 minutes and consumed warm [B1861].

To **dispel witchcraft**, 1 handful of roots is boiled in 3 cups of water for 5 minutes and 1 cup consumed once daily for 3 days [B1861].

Caution is used, as an overdose can be toxic and causes pain, burning, and blood in the urine. A person should not exceed 3 cups as a complete dosage. It should also not be taken during pregnancy, as it may have abortifacient effects [HR, LR, RA].

To improve a dog's ability to follow a scent, the leaves are crushed and put on the nose; "it acts as a **stimulant**" for the dog [Arnason et al., 1980].

Phytolacca icosandra L.

burr vine, scorpion tail [English]; *calaloo* [Garífuna]; *telcox* [Yucateca]

Steggerda (1943) reported that the fruits of this plant were boiled in water and the resulting beverage used to treat **smallpox.** "The informant related that this medicine caused the pox to 'break out,' for if it remained 'inside,' the patient died. The raw leaves of this plant are crushed and rubbed on **pimples** and **sores.**" To treat **burns**, the stems and leaves are applied as a poultice to the affected area [Mallory, 1991]. As a tonic "to **clear and build blood**," the young leaves and shoots are boiled in water and eaten [Mallory, 1991]. To treat **eye inflammation**, the juice from the stems and leaves is squeezed out and used as eye drops [Mallory, 1991].

Phytolacca rivinoides Kunth & C.D. Bouché

pigeon berry, pokeweed [English]; *jocato* [Creole]; *coch otón, yaki* [Q'eqchi']; *telcox* [Mayan]

The leaves are eaten as a food [B3670]. To treat **anemia** to build up the red blood cell count, the leaves along with red cayenne pepper (*Capsicum annuum*), onion, and seasonings are cooked together to prepare a soup [B3670]. Standley and Record (1936) reported that the leaves were boiled and used as a hot **beverage.**

Some healers consider the root tea to be a **panacea** to treat several ailments [RA]. To treat *arthritis*, "dissolve" **tumors**, and "expel **viruses** from the blood," the roots or leaves are dried, powdered, 1 teaspoon added to 1 pint of boiling water, and 1 teaspoon consumed every 2 hours. Alternatively, the fruits are harvested when ripe (a black or deep purple color), and 4 berries are taken every day. Both treatments often cause dark urine while cleansing, and treatment is continued "until the urine is clear" [A926].

(left) Polo Romero collecting *Phytolacca icosandra* [MB].

(right) *Rivina humilis* [MB].

Rivina humilis L.

coqueta, coralillo, tomatillo, yierba mora [Spanish]; *cusucán* [Additional name recorded]

For a **sore** or **swelling**, an unspecified part of the plant is used to prepare a bath [A313]. To treat **erysipelas**, 1 entire plant is ground in a mill, 9 Spanish thyme leaves (*Lippia graveolens*) and three 3-inch branches with leaves of *epasote* (*Chenopodium ambrosioides*) are mashed in, and this is applied as a plaster to the skin twice daily for 2 hours at a time until better [B2092, B2123]. To treat **erysipelas**, **sores**, or **swelling**, *dama de noche* leaves (*Cestrum nocturnum*) and little bamboo leaves (*Commelina erecta*) are added to the mixture described above for erysipelas and applied as a plaster with either white lime (an over-the-counter product) or *tricofero* juice (the "pink juice," also an over-the-counter product) to the affected area [B2123]. To treat **swelling in the feet**, a mixture of white lime (an over-the-counter product) and warm water is used to bathe the affected area followed by an application of the plaster described above in the second treatment for erysipelas [B2123; MB].

For **kidney infection**, 4 ounces of bark along with 8 ounces of *gumbolimbo* bark (*Bursera simaruba*) are boiled in 1 quart of water for 10 minutes and 1 cup consumed 3 times daily until better [B2123].

PIPERACEAE

Herbs, shrubs, subshrubs, small trees, or occasionally vines. Leaves alternate, sometimes opposite or whorled, simple. Inflorescences opposite the leaves, variously terminal or axillary, in dense spikes or racemes. Flowers very small. Fruits berry-like or drupes. In Belize, consisting of 2 genera and 46 species. Uses for 2 genera and 22 species are reported here.

Peperomia costaricensis C. DC.
  teb' pim [Q'eqchi']
  To treat **influenza**, refer to *pa ulul* (*Begonia sericoneura*) [C41, C42, C43]

Peperomia macrostachyos (Vahl) A. Dietr.
  To treat **skin sores** and **itching**, 1 handful of leaves is mashed in ¼ cup of water and applied as a poultice to the affected area [B2556]. To treat **itching feet**, the leaves are mashed in cold water and applied to the affected area [R25].
    To treat **influenza**, refer to *pa ulul* (*Begonia sericoneura*) [C41, C42, C43].

Peperomia obtusifolia (L.) A. Dietr.
  climbing piper [English]; *teb' pim* [Q'eqchi']
  To "take away the poison" from **snakebite** "when bit in the jungle and far from home," 1 handful of leaves is mashed and applied as a poultice to the bite [B2513].

Peperomia pellucida (L.) Kunth
  wild mint [English]
  For treating **foot fungus**, such as **athlete's foot**, the leaves are mashed and rubbed on and between the toes [B2334].

Piper aduncum L.
  cow's foot, Spanish elder [English]; *puchuch, ob'el* [Q'eqchi']
  The leaves are used like a **scrub brush** to wash plates and pots [B3523].

(top) *Peperomia obtusifolia* [MB].

(bottom) *Peperomia pellucida* [DA].

Piper amalago L.

buttonwood, Spanish elder [English]; *cordoncillo, cordoncillo chico* [Spanish]; *tzak al bak* [Mopan]; *lum pom, puchuch* [Q'eqchi']

To make a bath mixture, 1 handful of leaves from a 9-plant mixture, including this plant, is added to 2 quarts of water and used tepid as a bath once daily at bedtime [A60, A123, A256; Arvigo and Balick, 1998].

For **headache**, refer to *annona* (*Annona reticulata*) [B2355, W1832]. To treat **headache**, **constipation**, and for use as a **sedative**, the leaves are mashed in water and consumed cold [Arvigo and Balick, 1998]. To treat **toothache**, the root is mashed and applied as a poultice over the gum or affected area [Arvigo and Balick, 1998]. Alternatively, the root is chewed to numb or "freeze" the tooth in Cayo District [Mallory, 1991].

For **pain**, **anemia**, or **fever**, nine 12-inch branches with leaves are mashed in a bucket of warm water until green, three 9-inch branches with leaves of *gumbolimbo* (*Bursera simaruba*) and 9 *ob'el* leaves (*Piper auritum*) are mashed into the infusion, and the infusion is poured warm over the body as needed [B3673, B3674, B3677]. To treat **pain**, **rheumatism**, **swelling**, **skin ailments**, **fatigue**, and **sleeplessness**, 1 large double handful of freshly picked leaves is boiled in 2 gallons of water, cooled to very warm, and poured over the body as needed [Arvigo and Balick, 1998]. To treat an **abscess**, which sometimes causes fever or swelling of the lymph nodes, 1 handful of fresh leaves is mashed in 1 liter of cold water and used as a bath over the affected area 3 times daily for 5 days [C75]. To ease body **pain** from influenza, refer to red vervain (*Stachytarpheta miniacea*) [B2290]. For **menstrual cramps** or **delayed menstruation**, the leaves are used to prepare a sitz bath given for 20 minutes once daily, before bedtime, for 3 consecutive nights [Arvigo and Balick, 1998].

As a first-aid treatment for **snakebite**, one 12-inch piece of root is chewed [LR]. Alternatively, 9 mature leaves are boiled in 3 cups of water for 5 minutes and sipped continuously during the day until proper help is found [Arvigo and Balick, 1998]. For

Piper amalago [MB].

snakebite, 1 handful of leaves is boiled in 1 quart of water for 5 minutes and used lukewarm to bathe the bite every 2 hours. Following the bath, 7 or 9 roots and 7 or 9 leaves are mashed, applied as a poultice to the bite, and changed after each bath [LR]. Alternatively, 1 piece of root, equal in length to the person's forearm, is boiled in 3 cups of water for 10 minutes and consumed while additional treatment is being sought [Arvigo and Balick, 1998]. For another reported treatment, refer to Spanish cedar (*Cedrela odorata*) [Arnason et al., 1980].

To make **snakes** run away from you, the stems are used to beat them [A256].

To soothe a spiritual disease called **wind (*viento*)**, the leaves and branches are used to make a warm bath [B2297; Arvigo and Balick, 1998]. To treat babies with **grief (*pesar*)**, refer to cross vine (*Paullinia tomentosa*) [Arvigo and Balick, 1998].

Piper auritum Kunth

cow foot, bullhoof [English]; *Santa María* [Spanish]; *puchuch, ob'el* [Q'eqchi'];
maculán, shma culan [Mayan]

This plant has several uses [B3230]. The leaves are a component of a bath mixture [A810; Arvigo and Balick, 1998]. The Mennonites of Belize use this plant to prepare a **tea** [B1799]. This root was mixed with "Wizard oil," a locally produced medicinal oil more commonly used in the past, to treat **cuts** [B2143]. For **cough**, refer to *pega ropa* (*Priva lappulacea*) [A839].

To treat **pain** and **swelling**, 3 leaves are boiled in 1 quart of water for 5 minutes and used cool as a bath over the affected area [A810]. For **pain**, **anemia**, or **fever**, refer to buttonwood (*Piper amalago*) [B3673, B3674, B3677]. To soothe a **toothache** or **headache**, the leaves or roots are chewed and applied as a poultice to the affected area [A810, B3230].

For **rheumatism** or **arthritis**, the leaves are heated, "Vicks" (Vicks VapoRub® topical ointment) rubbed on the affected area, the leaves added on top, and left on for 1–2 hours [B2396]. Alternatively, the leaves are heated and applied warm as a poultice directly to the affected area [B3230]. To treat **arthritis**, 3 leaves are boiled in 1 quart of water for 5 minutes and ½ cup consumed until better, usually for 9 days [B3230].

To treat non-insulin-dependent **diabetes** mellitus, 7 fresh leaves are boiled in 1½ quarts of water for 10 minutes and consumed cold as a beverage 10 times daily for 15 days [C88]. To treat **parasitic intestinal worms**, the decoction described for diabetes is sipped all day long in place of water for 9 days during the new moon phase [A810]. Alternatively, the young green stems are chewed [G17].

As a **female contraceptive**, 6 leaves are boiled in 1 quart of water for 5 minutes and 1 cup consumed warm twice daily, in the morning and in the evening, for 1 month [B3230]. For **swelling** of the genital organs, 1 leaf is heated in kitchen oil and applied as a poultice to the affected area [B2143]. For **women's ailments**, refer to castor plant (*Ricinus communis*) [B2176].

Piper auritum [MB].

The leaves are used to **wrap** fish before being baked [A810]. The leaves of this plant are "used as a **food flavoring** throughout Central America" [Horwich and Lyon, 1990].

Piper dilatatum Rich.
puchuch [Q'eqchi']
To treat **colds** or **influenza**, 1 double handful of fresh leaves is boiled in water for 15 minutes and used as a bath. Additionally, the same decoction of leaves is boiled a second time and consumed cold. This 2-part treatment is repeated twice daily for 7 days [C71].

Piper cf. *guazacapanense* Trelease & Standl.
To treat **mouth, tongue, and lip sores**, in people of all ages but especially in children, 1 handful of fresh leaves and roots is mashed in cold water and consumed 3 times daily for 7 days [C28].

Piper hispidum Sw.
cordoncillo [Spanish]; *xna' q'ehen, po ch'ueh* [Q'eqchi']
As an all-purpose medicament for **general sicknesses**, 1 double handful of leaves is mashed, boiled in water, applied as a poultice directly to the affected area, and the remainder used as a bath [B3654]. For a bath mixture for **general ailments**, refer to Spanish elder (*Piper schiedeanum*) [B3537].

To relieve **coughing** or to treat **muscle pain**, 9 leaves are boiled in 1 quart of water for 5 minutes and 1 cup consumed as hot as possible 3 times daily. Alternatively, 1 handful of leaves is boiled in 1 gallon of water for 5 minutes and poured warm as a bath over the body once daily at noon [B2676, B2712].

To treat **water retention** in pregnant women, or "when they begin to swell," the leaves along with Spanish elder leaves (*Piper jacquemontianum*) are mixed in water and consumed [B3536]. Alternatively, refer to Spanish elder (*Piper schiedeanum*) [B3537].

Piper jacquemontianum Kunth
buttonwood, Spanish elder (female) [English]; *cordoncillo* [Spanish]; *kam pom* [Q'eqchi']
To prepare a bath for **general ailments**, 1 handful of a 9-leaf mixture including this plant is boiled in 2 quarts of water for 5 minutes and poured very warm over the body once daily at bedtime [A58, A636; Arvigo and Balick, 1998]. Alternatively, refer to Spanish elder (*Piper schiedeanum*) [B3537].

Piper jacquemontianum [MB].

For treatment of **skin ailments** and **skin "cancer" (open sores)**, 1 handful of fresh leaves along with one 1-inch × 6-inch piece of mahogany bark (*Swietenia macrophylla*) are boiled in 1 gallon of water for 10 minutes and used as hot as possible as a bath twice daily until better [A326].

To treat **skin rash** or to "**preserve the skin**," the leaves are blended, juiced, mixed with olive oil and bay rum, and applied topically as a lotion to the affected area [B2604]. Fresh plant juice is blended with oil or petroleum jelly and used as a **beauty aid** [A326].

For **stomach pain**, **cough**, or **pain**, 1 teaspoon of freshly extracted leaf juice is consumed 3 times daily [A326]. For **stomach pain**, 9 leaves are steeped in 2 cups of boiling water and consumed as needed [B2604]. For **pain**, **tiredness**, or **snakebite**, the leaves are used to prepare a bath [B2133, B2604]. Alternatively, the leaves along with the leaves of an unidentified species referred to as "I'll not leave" and sour orange leaves (*Citrus aurantium*) are mixed, boiled in water, and used as a bath [B2133].

For **snakebite**, 2 ounces of the root along with the grated seeds of *bisseemorro* (possibly *Cola acuminata* (P. Beauv.) Schott & Endl.) and the grated seeds of *cedron* (*Simaba cedron* Planch.) are mixed well with a "soft candle" and applied as a poultice over the bite following a hot bath twice daily, in the morning and in the evening [B1999; HR]. To prepare this bath, 1 handful of leaves and inflorescences is boiled in 1 gallon of water for 10 minutes and poured as hot as can be tolerated over the bite twice daily [B1782].

To treat **water retention** in pregnant women, this plant is one of the 3 *Piper* plants boiled to prepare a decoction given as a tea "when they begin to swell" [B3535]. Alternatively, refer to *cordoncillo* (*Piper hispidum*) [B3536]. For a third reported treatment, refer to Spanish elder (*Piper schiedeanum*)[B3537].

For **ritual cleansing** before certain Maya ceremonies or to **dispel witchcraft**, the leaves and inflorescence are used to prepare a bath [B1782].

Piper marginatum Jacq.
buttonwood, Spanish elder [English]; *cordoncillo* (female) [Spanish]; *puchuch* [Q'eqchi']

For **general ailments** and to soothe **wind (*viento*)**, 1 handful of a leaf mixture including this plant is boiled in 2 quarts of water for 5 minutes and used tepid as a bath once daily at bedtime [A59, B2340]. Alternatively, this plant is used to prepare a steam bath for 20 minutes once daily, at noon or bedtime [B2383].

To treat **edema** in pregnant women, characterized by **swollen legs** from the knee to the feet occurring after 6 months of pregnancy, especially if the woman is weak, has improper nutrition, or eats a lot of salt, 1 full handful of fresh leaves is boiled in 1 cup of water for 10 minutes and consumed 3 times daily for 7 days [C26]. Additionally, this same decoction is used as a bath once daily, in the evening, for 7 days [C26]. To **keep the body warm**, 1 quart of fresh leaves and flowers is boiled in 1 gallon of water for 5 minutes and used as a bath once daily, at noon or bedtime [B2340].

For **snakebite**, 1 handful of leaves and inflorescences is boiled in 1 gallon of water for 10 minutes and poured as hot as can be tolerated over the bite twice daily [B1781].

To **dispel witchcraft** or for **ritual cleansing** before certain Maya ceremonies, the leaves and inflorescences are used to prepare a bath [B1781].

Piper neesianum C. DC.
buttonwood, Spanish elder [English]; *cordoncillo* [Spanish]; *bak er*, *mam* [Mayan]
The leaves of this plant are considered an important ingredient of bath mixtures [A52].

The roots, leaves, and bark are used for treating **snakebite**. Two ounces of the root are toasted along with grated seeds referred to as *bisseemorro* (possibly *Cola acuminata* (P. Beauv.) Schott & Endl.) and *cedron* (*Simaba cedron* Planch.). This is mixed together with a "soft candle" (made from animal fat) to make a poultice. The bite is bathed with a hot leaf decoction and the poultice then

(top) *Piper marginatum* [MB].

(bottom) *Piper neesianum* [MB].

placed over the snakebite twice daily after each of 2 baths daily [B1805; HR]. To prepare the baths, 3 leaves are boiled in 1 gallon of water for 10 minutes, and the bite is bathed in the decoction when it is lukewarm by pouring the entire gallon over the area twice daily, in the morning and in the evening [RA]. As a first-aid treatment for **snakebite** in the field or far from home, the bark and roots are chewed [B1805].

Piper peltatum L.
cow foot, piper [English]; *Santiago, Santa María, San Diego* [Spanish]; *shu tu it, u tu it* [Mopan]; *ch'ut it* [Q'eqchi']
To **stimulate lactation** in women, 1 quart of chopped leaves is boiled in 1 gallon of water for 10 minutes and used very warm to wash the back and breasts twice daily, in the morning and in the evening [B1784; Arvigo and Balick, 1998].

To soothe **muscle pain**, 9 leaves are boiled in 1½ gallons of water for 10 minutes and used hot as a bath or a steam bath [A154, A584]. For **muscle pain** and **open sores**, the leaves are warmed and applied directly to the affected area [B2655]. To

Piper peltatum [MB].

Piper pseudofuligineum [MB].

soothe **back pain**, 4–5 warmed leaves are applied directly to the affected area or mashed against the back once daily, in the evening, until better [B1784, B2655, B2677, B3533, B3659]. For **back pain**, **sores**, **boils**, and **wounds**, the leaves are heated in oil and applied as a plaster to the affected area [A154, B2395].

To treat an infant's **umbilicus** that is not healing properly following delivery, the leaves are heated in oil and placed on the navel for 1–2 hours and repeated every 2 hours until better [B2395]. To treat **gas pain** in infants, the leaves are heated in oil and placed on the navel for 1–2 hours and repeated every 2 hours until better [B2395].

To treat a **prolapsed anus** of children who have fallen so hard that their anus protrudes, the leaves are mashed and inserted directly into the anus [A584]. To treat **diarrhea** in babies, one 6-inch piece of the root is boiled in 1 cup of water for 10 minutes and 2 teaspoons consumed cool 3 times daily until better [B2523].

To relieve pain of **muscle spasms**, **headache**, and **stomach pain**, 1 large leaf is warmed and applied to the affected area and a towel wrapped over the leaf, with the person resting during the treatment [Arvigo and Balick, 1998]. To treat **back pain**, 1 large leaf is heated in a small amount of cooking oil and applied to the affected area, a towel wrapped around the back to keep it warm, and left on overnight [Arvigo and Balick, 1998]. For **rheumatism** and **arthritis**, the leaves are utilized to prepare a bath used over the affected area [Arvigo and Balick, 1998].

The leaves are used for **wrapping** foods, such as onions (*Allium cepa*), fish, or potatoes (*Solanum tuberosum*), before baking [B3533].

Piper pseudofuligineum C. DC.

buttonwood [English]; *cordoncillo* [Spanish]; *puchuch* [Mayan]

For **snakebite**, 1 handful of leaves and inflorescences is boiled in 1 gallon of water for 10 minutes and poured as hot as can be tolerated over the bite twice daily [B1783]. To **dispel witchcraft** or for **ritual cleansing** before certain Maya ceremonies, the leaves and inflorescence are used to prepare a bath [B1783].

Piper pseudolindenii C. DC.

Spanish elder [English]; *cordoncillo chico*, *cordoncillo-che-che* [Spanish]

Piper pseudolindenii [GS].

To make a **healing bath**, 1 large handful of leaves is boiled for 10 minutes in 1 gallon of water [A247]. To treat **toothache** or **stomach pain**, one 1-inch piece of the root is chewed as needed [A247].

The roots and leaves are part of a **snakebite** formula. Two ounces of the roots are toasted along with the grated seeds of *bisseemorro* (possibly *Cola acuminata* (P. Beauv.) Schott & Endl.) and *cedron* (*Simaba cedron* Planch.). This is mixed well together with a "soft candle" to make a poultice. The bite is bathed with a hot leaf decoction and the poultice then placed over the snakebite twice daily after each bath [A247; HR]. To prepare the bath, 3 leaves are boiled in 1 gallon of water for 10 minutes and the bite bathed in the decoction when it is lukewarm by pouring the entire gallon over the area in the morning and in the evening [RA].

Piper psilorhachis C. DC.

buttonwood, piper, Spanish elder [English]; *cordoncillo*, *Santa María* [Spanish]; *kaq puchuch* (female) [Q'eqchi']; *chucsuc* [Additional name recorded]

To treat **general ailments** or, in particular, for "when the **bones ache**," 1 handful of an unspecified bath leaf mixture including this leaf is boiled in water and used cool as a bath once daily at bedtime [A126, B2480]. To treat **fever** and **colds**, a bath is prepared using this plant [W1831]. For **colds**, 1 handful of fresh leaves is boiled in 3 cups of water for 15 minutes and consumed warm 3 times daily for 5 days [C32].

This plant is used as an antidote for **snakebite** when other treatments are not available [W1831]. For **snakebite**, 2 ounces of the root are toasted along with grated seeds of *bisseemorro* (possibly *Cola acuminata* (P. Beauv.) Schott & Endl.) and *cedron* (*Simaba cedron* Planch.). This is mixed with a "soft candle" to make a poultice. The bite is bathed with a hot leaf decoction and the poultice then placed over the snakebite twice daily after each of 2 daily baths. To prepare the baths, 3 leaves are boiled in 1 gallon of water for 10 minutes and the bite bathed in the decoction when it is lukewarm by pouring the entire gallon over the area in the morning and in the evening. If the **snakebite** is received "in the field," the bark and root are chewed as part of the treatment [HR, RA].

Piper schiedeanum Steud.

Spanish elder [English]; *cordoncillo, Santa María* [Spanish]; *puchuch* [Q'eqchi']

A bath mixture for **general ailments** is prepared from a mixture of these leaves along with Spanish elder leaves (*Piper jacquemontianum*) and *cordoncillo* leaves (*Piper hispidum*) [B3537].

To treat **fever**, 1 quart of leaves is mashed in 1 gallon of cool water and used as a bath over the entire body [B3505]. To treat **arthritis ("burning bones")** or **bone pain**, refer to *rah li ch'och* (*Tectaria panamensis*) [R30, R31, R34]. For **water retention** in pregnant women when they "begin to swell," 3 leaves along with 3 *Piper hispidum* leaves and 3 *Piper jacquemontianum* leaves are boiled in 2 quarts of water for 10 minutes and 1 cup consumed 3 times daily for 7 days [B3537].

To treat **evil magic (*obeah*)** or **bad spirits**, 1 cup of fresh bark and 1 quart of leaves are boiled in 2 gallons of water for 10 minutes and used as a bath once daily for 7 days [B2475].

Piper tuberculatum Jacq.

buttonwood, Spanish elder [English]; *cordoncillo* [Spanish]

A "rare species," the leaves are boiled in water and used as a component of baths for **general ailments** [A121].

Piper tuerckheimii C. DC.

piper [English]; *puchuch, uta et pe, sák e pú chuch* [Q'eqchi']

To treat **arthritis ("burning bones")** or **bone pain**, refer to *rah li ch'och* (*Tectaria panamensis*) [R30, R31, R34].

For women during pregnancy who "feel a **hurting in the belly**," such as from an overweight baby or a wrong positioning or turning of the baby, 1 large double handful of leaves is squeezed by hand in 4 gallons of cold or cool water, ½ cup consumed daily, and the remainder used cool as a sitz bath for 20 minutes, repeated for 3 days [B2549].

As a **hemorrhoid** treatment, 1 ounce of the root and 9 leaves are boiled in 1 quart of water for 10 minutes and 1 cup consumed cool 3 times daily until better [B3563]. To treat **epileptic seizures**, refer to *Tectaria heracleifolia* [C33, C34].

Piper umbellatum L.

puchuch, tush pim (female) [Q'eqchi']

To treat **edema** in women after 6 months of pregnancy who do not eat properly, are weak, eat a lot of salt, and whose feet swell, 2 small handfuls of fresh leaves are boiled in 1 liter of water for 5 minutes and an unspecified amount consumed warm 4 times daily for 3 days [C35]. Additionally, the fresh leaves are chewed, and the decoction described above is used warm as a bath [C35].

Piper yucatanense C. DC.

Spanish elder [English]; *cordoncillo* [Spanish]

To treat small **cuts** and **bruises**, one 6-inch piece of fresh root is crushed and applied as a poultice to the affected area, covered with a bandage, left on for 3 hours, and reapplied as needed until better [R49].

PLANTAGINACEAE

Usually herbs, sometimes small shrubs. Leaves alternate, often in basal rosettes, simple. Inflorescences axillary, in spikes or appearing head-like. Flowers small. Fruits a capsule or nut. In Belize, consisting of 1 genus and 1 species. Uses for 1 genus and 1 species are reported here.

Plantago major L.

Steggerda (1943) reported that the Yucatan Maya boiled the young leaves of this plant and used the tea as a therapy for **diarrhea**, particularly for babies.

Steggerda and Korsch (1943) reported that a poultice of this plant was placed over a **dog bite** to treat it.

POLYGALACEAE

Herbs, subshrubs, shrubs, trees, or lianas. Leaves usually alternate, rarely opposite or verticillate. Inflorescences terminal or axillary, usually in racemes or panicles, rarely spikes, heads, or of solitary flowers. Flowers showy, variously colored. Fruits usually a capsule, sometimes a berry, drupe, or samara. In Belize, consisting of 3 genera and 16 species. Uses for 2 genera and 3 species are reported here.

Polygala paniculata L.

wintergreen [English]; *pepamento* [Spanish]

For treatment of **stomach gas**, or "**wind**," 5 entire plants with roots are boiled in 3 cups of water for 5 minutes and 1 cup consumed cool 3 times daily. This treatment is not given to pregnant women [B3758].

(top) *Plantago major* [MB].

(bottom) *Polygala paniculata* [MB].

Securidaca diversifolia (L.) S.F. Blake
man vine [English]; *bejuco de hombre,
bejuco verde* [Spanish]; *ya ax ak* [Mopan];
teelom k'aham [Q'eqchi']

As a treatment for **anxiety, back pain, neckache, headache, muscle spasms, mucus in male urine, mucus in the stool, constipation, gastritis (*ciro*), intestinal gas, indigestion,** an **inability to eat** even a small portion of food, or other **digestive or alimentary tract ailments**, 1 handful of the chopped, woody vine is boiled in 1 quart of water for 10 minutes and 1 cup consumed warm 3 times daily, before meals, until better, usually for 9 days [B1857; Arvigo and Balick, 1998].

To treat male **impotence**, 1 handful of the fresh or dried chopped roots is boiled in 1 quart of water for 10 minutes and 1 cup consumed 3 times daily, before meals, until better, usually for 9 days [B1857; Arvigo and Balick, 1998].

Caution is urged when using these treatments. It is advised to abstain from all acidic foods, cold drinks, beef, and chili because acidic substances reduce the effect of the medicine [Arvigo and Balick, 1998].

Securidaca sylvestris Schltdl.
man vine [English]; *behuco de hombre, behuco verde* [Spanish]; *k'aham pim, teelom k'aham* [Q'eqchi']

To treat **gastritis (*ciro*)**, 1 handful of chopped root is boiled in 1 quart of water for 10 minutes and ½ cup consumed cool 3 times daily [B3650].

POLYGONACEAE

(top) *Securidaca diversifolia* [RH].

(bottom) *Securidaca sylvestris* [RF].

Herbs, shrubs, trees, lianas, or vines. Leaves alternate, occasionally whorled or opposite, simple. Inflorescences axillary or terminal, in cymes, spikes, panicles, racemes, or heads. Flowers white, greenish, pinkish, or reddish. Fruits an achene. In Belize, consisting of 5 genera and 20 species. Uses for 2 genera and 4 species are reported here.

Coccoloba barbadensis Jacq.

fresh water grape, wild grape [English]

Standley and Record (1936) noted that the fruits of this tree were eaten as a **food**. Steggerda (1943) reported that the fruit was eaten as a **food** by birds.

To staunch bleeding from **snakebite**, 2 ounces of the root are toasted along with grated seeds known as *bisseemorro* (possibly *Cola acuminata* (P. Beauv.) Schott & Endl.) and *cedron* (*Simaba cedron* Planch.). This is mixed with a "soft candle" to make a poultice. The bite is bathed with a hot leaf decoction, and the poultice is then placed over the snakebite twice daily after each of 2 daily baths. To prepare the bath, 3 leaves are boiled in 1 gallon of water for 10 minutes and the bite bathed in the decoction when it is lukewarm by pouring the entire gallon over the area in the morning and in the evening [HR, RA].

To treat **hemorrhaging** and the constant **break-through bleeding** during menopause, 4 ounces of the root are boiled in 1 quart of water for 10 minutes, steeped until cool, and sipped all day until finished [A672].

Steggerda (1943) reported that the Yucatan Maya used the leaves of this tree to **wrap** tortillas and candy. That author also reported that once the bark was removed from the stem, the hard wood was used in house **construction**, especially for center beams.

Coccoloba belizensis Standl.

berry tree, bob, wild grape [English]; *papa turro, uva montes, uva silvestre* [Spanish]; *niiche* [Additional name recorded]

The ripe, purple fruits are sweet and eaten as a **food** by humans and birds [B2030, B2153].

The wood of the trunk is used in **construction** for house frames and poles [A770].

Coccoloba uvifera L.

grape, sea grape [English]; *uva* [Spanish]; *niche* [Additional name recorded]

The ripe fruit is eaten as a **food** by humans and birds [A411]. To make **wine**, 1 quart of the fruits is soaked in 1 quart of water with either 1 pound of brown sugar or 1½ pounds of white sugar for 15 days, strained, and consumed.

(top) *Coccoloba barbadensis* [GS].

(bottom) *Coccoloba uvifera* [MB].

This makes about 2 quarts of wine [A411]. Marsh, Matola, and Pickard (1995) noted that the tart, **edible** fruits were made into **preserves**, **syrup**, and **wine** and that the leaves were used as a form of **paper** that could be written on using a sharp object.

To treat **diarrhea**, the bark is boiled in water and consumed until better [B1961]. For **thrush**, one 12-inch piece of bark is beaten, soaked in 3 cups of water, and used as a mouthwash 3 times daily [B1961].

The wood is used for **fuel** [A411].

Gymnopodium floribundum Rolfe
bastard logwood [English]
The wood of this tree is used to make **charcoal** [Horwich and Lyon, 1990].

PORTULACACEAE

Herbs, low shrubs or subshrubs, mostly succulent. Leaves alternate, spiral or apparently opposite, simple. Inflorescences of solitary flowers or in panicles. Flowers small and showy, variously colored. Fruits a capsule. In Belize, consisting of 1 genus and 3 species. Uses for 1 genus and 3 species are reported here.

Portulaca grandiflora Hook.
This plant is used as an **ornamental** [B2378].

Portulaca oleracea L.
purslane [English]; *verdolaga* [Spanish]; *xucul* [Mopan]
The leaves and stems are potherbs eaten as a **food** either raw, in salads, boiled, or fried, and they are especially good for people recovering from illness [A94; Arvigo and

Gymnopodium floribundum [NYBG].

Portulaca grandiflora [MB].

Balick, 1998]. "The thick, older stems are pickled in salt and vinegar" [Arvigo and Balick, 1998]. The leaves are said to be rich in vitamins, protein, and minerals [A94, B1979; Arvigo and Balick, 1998]. Standley and Steyermark (1946) noted that it was commonly sold as a **vegetable** in local Guatemalan markets. Steggerda (1943) noted that this plant was an accepted **fodder** for chickens, turkeys, and pigs and was sold in the Merida market for this purpose at that time.

As a **diuretic**, a **blood cleanser**, and to **nourish the body**, the leaves and stem from 1 large plant are boiled in 3 cups of water for 5 minutes and consumed [Arvigo and Balick, 1998]. To relieve **dry cough**, fresh juice from the leaves and stems is mixed with sugar or honey and consumed as needed [Arvigo and Balick, 1998].

As a treatment for **urinary ailments** and **anemia**, 4 ounces of the leaf and stem are boiled in 1 quart of water for 10 minutes and 1 cup consumed 3 times daily, before meals, for 2 days [A94].

Portulaca oleracea [MB].

To stop **external bleeding**, to treat **headache** resulting from overexposure to the sun, and to heal **ulcers**, **wounds**, and **sores**, the leaves and stem are crushed and applied as a poultice to the affected area [Arvigo and Balick, 1998].

Steggerda (1943) reported that the Yucatan Maya used a decoction of this plant to treat **worms**.

Portulaca pilosa L.

mañanita [Spanish]; *tsayoch* [Additional name recorded]

As a medicament for a **rash**, 9 leaves along with 9 arnica leaves (*Montanoa speciosa*), 9 pomegranate leaves (*Punica granatum*), and 9 polly red head leaves (*Hamelia patens* var. *patens*) are crushed, soaked in a bit of water, mashed to a pulp, rubbed on the affected area, and left on for 1 hour [B2099, B2100].

PRIMULACEAE (INCLUDING MYRSINACEAE AND THEOPHRASTACEAE)

Subshrubs to trees. Leaves alternate, pseudoverticillate, simple. Inflorescences axillary, terminal or subterminal, in panicles, racemes, fascicles, corymbs, umbels, or rarely of solitary flowers. Flowers orange, pink, green, or white. Fruits a berry or drupe. In Belize, consisting of 10 genera and 29 species. Uses for 4 genera and 6 species are reported here.

Ardisia compressa Kunth
Standley and Record (1936) reported that the fruit had an agreeable flavor and was eaten as a **food**.

Ardisia escallonioides Schltdl. & Cham.
hullaba, *residan* [Spanish]

The fruit, reportedly sweet, is eaten as a **food** by birds [B2192].

For unspecified **discomfort**, 1 handful of leaves is steeped in 1 gallon of hot water for 30 minutes and poured warm as a bath over the affected area [B2192].

Deherainia smaragdina (Planch. ex Linden) Decne. subsp. *smaragdina*
flower of death [English]; *flor de la muerte* [Spanish]

As treatment for people who have had a **stroke** or those who have a **terminal illness** but "are afraid to die," 1 handful of leaves is boiled in 2 quarts of water for 10 minutes and used tepid as a bath once daily at bedtime [A87]. This formula is said to "release the stiffness" from stroke [HR].

To treat "**people who hear voices**," these plants are grown all around them in pots or transplanted into the ground from the forest [HR].

Ardisia escallonioides [NYBG].

Deherainia smaragdina subsp. *smaragdina* [MB].

Jacquinia macrocarpa Cav. subsp.
macrocarpa

To treat **toothache**, the seeds and roots are ground to extract a resin used by the Yucatan Maya [Flores and Ricalde, 1996].

This species may be the one referred to by Lundell (1938) for making **necklaces**, **personal adornments**, and **festival decorations** from the corollas of the flower.

Parathesis cubana (A. DC.) Molinet & M. Gómez

pigeon grape [English]; *residan* [Spanish]; *ix pam bul* [Additional name recorded]

The sweet fruit is considered a high source of vitamin C and eaten as a **food** by humans and birds [A654, B2130].

To treat unspecified **discomfort**, 1 handful of leaves is steeped in 1 gallon of hot water for 30 minutes and used warm as a bath as needed [B2130].

Parathesis sessilifolia Donn. Sm.

gua ca tèn, *gua ca tún* [Q'eqchi']

The fruit is eaten as a **food** [B2505].

To relieve **diarrhea**, the leaves are boiled in water and consumed frequently until better [B2491].

PROTEACEAE

Prostrate shrubs to large trees. Leaves usually alternate, less frequently opposite or whorled, compound or simple. Inflorescences terminal or lateral, in racemes. Flowers variously colored. Fruits a follicle. In Belize, consisting of 2 genera and 2 species. Uses for 1 genus and 1 species are reported here.

(top) *Jacquinia macrocarpa* subsp. *macrocarpa* [NYBG].

(bottom) *Roupala montana* [MN].

Roupala montana Aubl.

Standley and Record (1936) reported that this shrub or small tree yielded an attractive **lumber** but also that its uses were few due to the small size of the plant.

RHAMNACEAE

Trees, shrubs, or lianas. Leaves usually alternate, occasionally simple. Inflorescences axillary or terminal, usually in cymes, sometimes racemes. Flowers small. Fruits a drupe, capsule, or schizocarp. In Belize, consisting of 5 genera and 6 species. Uses for 4 genera and 5 species are reported here.

Gouania lupuloides (L.) Urb.

chew-stick [English]; *bejuco, botón de Indio* [Spanish]; *waika chustick* [Creole]; *chu wis* [Additional name recorded]

In ancient times, the foamy stick was chewed to **clean the teeth** [B2121]. "The stem is drawn between the teeth to cleanse them and give a pleasant persistent taste" [Robinson and Furley, 1983].

To treat **neck pain**, 1 large handful of leaves is mashed in 1 gallon of water, soaked overnight in the night dew, and used as a bath [A733].

To treat **snakebite** when it causes the spitting up of blood, one 12-inch piece of vine is mashed, soaked in 1 quart of water for 20 minutes, and ½ cup consumed hot every 10 minutes until the bleeding stops [B2121].

According to Standley and Steyermark (1949):

Gouania lupuloides [MB].

The stems of this and other species probably contain saponin, and when they are chewed large quantities of lather are produced. They have been dried and exported in large amounts from Tropical America to the United States and Europe for use in preparation of **dentifrices**. In Central America the stems are often chewed to **clean the teeth** and **harden the gums**. A decoction of the root is used in Yucatan as a gargle for **sores** in the mouth and throat. (p. 285)

Gouania polygama (Jacq.) Urb.

To treat **gastritis (*ciro*)**, 1 small handful of the top part of the root is collected, dried, boiled in 3 cups of water for 10 minutes, and 1 cup consumed warm 3 times daily, before meals, until better [B3197].

Krugiodendron ferreum (Vahl) Urb.

axe master [English]; *quebracho* [Spanish]; *exmasta bark* [Creole]

As an **astringent** after childbirth, one 1-inch × 12-inch piece of bark is boiled in 3 cups of water for 10 minutes, strained through a cloth, and used as a douche once daily for 7 days or as needed [Arvigo and Balick, 1998]. As a **blood tonic**, this plant is used in an unspecified treatment [Arvigo and Balick, 1998].

For **anemia**, **pleurisy**, **emphysema**, and **malaria**, this plant is used as part of an unspecified plant mixture consumed as a tea [Fernandez, 1990]. Steggerda (1943) noted that the bark and roots have been used since early days as a mouthwash for **toothache** and **gum ailments**. He concluded, "One yerbatero adds that the roots of this tree can be boiled and the liquid drunk as a purgative. He warns, however, that while using this medicine, the patient should not eat chile, pork, or any form of lard."

Steggerda (1943) also reported that the Yucatan Maya used the hard wood of this plant in house **construction** and that the tree was often left standing when farmers cleared their fields.

Sageretia elegans (Kunth) Brongn.
cherry, wild plum [English]; *ciruelillo* [Spanish]
The mature, red fruit is eaten as a **food** [A662, B2136].

Ziziphus mauritiana Lam.
Belize plum, governor plum, hog plum [English]
This mature fruit is eaten as a **food** and can be used in making **jams** and **juices** [A669]. When mature, the fruits are brown and have a "plum-like" flavor [B1843]. If the fruit is eaten green, the taste is somewhat **astringent** [B1965].

(top) *Krugiodendron ferreum* with inset showing flower [PA].

(bottom) *Ziziphus mauritiana* [MB].

RHIZOPHORACEAE

Shrubs or trees. Leaves opposite or verticillate, simple. Inflorescences axillary, in cymes, fascicles, or dichotomously branched. Flowers white or yellowish-white. Fruits a berry or capsule. In Belize, consisting of 2 genera and 2 species. Uses for 2 genera and 2 species are reported here.

Cassipourea guianensis Aubl.

water wood [English]

Standley and Record (1936) reported that this heavy, tough, strong, durable wood was used locally in the **construction** of railway ties and house frames.

Rhizophora mangle L.

red mangrove [English]; *colorado*, *mangle*, *mangle colorado* [Spanish]; *tapche* [Additional name recorded]

The leaves of this plant are used "to make a refreshing **tea**" [Marsh, Matola, and Pickard, 1995].

To treat **toothache** and **diarrhea**, the leaf juice and stem resin are used by the Yucatan Maya, but no further details were provided [Flores and Ricalde, 1996].

To treat **skin ailments**, **stubborn or serious sores**, **swelling**, or **leprosy**, 1 handful of chopped bark is boiled in 1 gallon of water for 10 minutes and used hot as a bath [Arvigo and Balick, 1998].

(top) *Cassipourea guianensis* with inset showing flowers [PA].

(bottom) *Rhizophora mangle* [MB] with inset showing leaves and buds [GS].

The bark contains tannins used as a **dye** and for **tanning** [B1948]. For **tanning**, 10 pounds of bark and 1 normal-sized cowhide are soaked or boiled in enough water to cover the hide until the color has penetrated through the skin. The hide is then removed, stretched to dry, and oiled well until flexible and soft [HR].

Standley and Record (1936) reported that the wood of this plant was used for **fuel**, **charcoal**, and sometimes, **construction**.

ROSACEAE

Herbs, trees, or shrubs. Leaves usually alternate, rarely opposite, simple, pinnately or palmately compound. Inflorescences terminal or axillary, in racemes, cymes, panicles, or of solitary flowers. Flowers small and inconspicuous or large and showy, variously colored. Fruits a drupe, pome, capsule, or an aggregate of follicles, achenes, or drupelets. In Belize, consisting of 4 genera and 4 species. Uses for 1 genus and 1 species are reported here.

Rosa chinensis Jacq.

rose [English]; *rosa* [Spanish]; *nikte* [Additional name recorded]

The following treatments apply only to the red or pink varieties.

For **headache**, 1 medium-sized rose along with 2 blades of lemon grass (*Cymbopogon citratus*) are boiled in 1 cup of water for 5 minutes and 1 cup consumed warm 3 times daily before meals [B2333]. To treat **red**, **inflamed eyes** in infants, children, and adults, 1 flower is steeped in 1 cup of boiling water, cooled, strained through a cloth, and 3 drops placed in the affected eye 3 times daily until better [Arvigo and Balick, 1998].

To treat **diarrhea** in infants and children or **fever**, 1 flower and 9 leaves are steeped in 1 cup of boiling water for 15 minutes, strained, cooled, and consumed [Arvigo and Balick, 1998]. For **diarrhea** in adults and **uterine hemorrhage**, 3 flowers and 1 handful of leaves are steeped in 1 cup of hot water for 15 minutes, strained, and consumed cool [Arvigo and Balick, 1998].

For when a person is feeling **uncertain** and **indecisive**, refer to *alamanda* (*Allamanda cathartica*) [B2651]. To improve **luck**, **fortune**, and **love**, refer to *alamanda* [BW]. To treat someone possessed by **evil spirits**, refer to pine (*Casuarina equisetifolia*) [BW].

Rosa chinensis [MB].

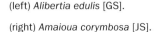
(left) *Alibertia edulis* [GS].

(right) *Amaioua corymbosa* [JS].

RUBIACEAE

Trees, shrubs, sometimes lianas, vines, and herbs. Leaves usually opposite, sometimes whorled, simple. Inflorescences terminal or axillary, in cymes, panicles, umbels, racemes, spikes, heads, or of solitary flowers. Flowers variously colored. Fruits a berry, drupe, capsule, samara, pseudosamara, schizocarp, or syncarp. In Belize, consisting of 53 genera and 148 species. Uses for 24 genera and 53 species are reported here.

Alibertia edulis (Rich.) A. Rich. ex DC.
> gibnut fruit, wild guava [English]; *guayaba de monte* [Spanish]; *tun tun, q'ani che'* [Q'eqchi']
> The mature fruit is eaten as a **food** [A567]. Standley and Record (1936) noted that the fruit was "of poor flavor."
> The stems are used for **fishing rods** [A567].

Amaioua corymbosa Kunth
> bastard coffee, wild coffee [English]
> The wood is prized as a **hardwood** used for lumber [B2003].

Chiococca alba (L.) Hitchc.
> rat root, skunk root [English]; *dama de la noche de Pine Ridge, zorrillo* [Spanish]; *pay che* [Mopan]; *pahar che'* [Q'eqchi']
> This plant is said to be the "**thinking herb**" of the Maya and is used to give relief to a great **variety of ailments** [Arvigo and Balick, 1998]. It is used when all else fails or the practitioner is unsure of the ailment [Arvigo and Balick, 1998]. As a general **tonic** and to treat **colds**, the stems and leaves along with the *contribo* vine (*Aristolochia trilobata*) are boiled in water and consumed [RA; Mallory, 1991].

To **improve blood circulation** and to treat **insomnia**, 1 large handful of leaves is boiled in 1 gallon of water for 10 minutes and used warm as a bath once daily at bedtime [A652, A653; HR]. If the insomnia is caused by bad dreams, 1 teaspoon of chopped, dried roots is boiled in 1 cup of water for 5 minutes and consumed warm once daily at bedtime [A652; HR]. To treat **snakebite**, refer to white breadnut (*Trophis racemosa*) [B3262].

For **stubborn sores**, **rash**, or **ulcers**, 1 small handful of chopped roots is boiled in 3 cups of water for 10 minutes, strained, and used as a wash over the affected area [Arvigo and Balick, 1998]. To treat **constipation, stomach or intestinal ulcers, delayed menstruation, obstructed bowels, colitis, endometriosis, pain,**

Chiococca alba root (top) and section of vine (bottom) [MB].

nervousness, dementia, or **depression,** 1 cup of the decoction described above for sores is consumed 3 times daily, before meals, for adults and 1 cup sipped throughout the day for children [Arvigo and Balick, 1998]. To treat **constipation, ulcers, skin "cancer" (open sores), organ congestion,** or **gastritis** (ciro), 1 small handful of the roots or the bark from the vine is boiled in 3 cups of water for 10 minutes, and 1 cup consumed warm 3 times daily, before meals, for 9 days, after which treatment is stopped completely for 3 days, following which the 9-day treatment is repeated. This is continued as needed [A745a]. With this treatment, **gastritis (*ciro*), constipation,** and **organ congestion** may need only 1 treatment. **Skin "cancer" (open sores)** may require more time, and exactly how many treatments is unknown [A745a].

For **uterine tumors**, refer to *kibix* (*Bauhinia herrerae*) [B1898, B2220]. To flush out the uterus before the onset of **menstruation**, 1 small handful of the root is boiled in 3 cups of water for 10 minutes and 1 cup consumed 3 times daily, before meals, for 5 days [A38]. To flush out a **congested uterus**, 1 cup of chopped, dried root is boiled in 3½ cups of water for 20 minutes and 1 cup consumed 3 times daily for 10 days. This dose should not be exceeded, as it may be **toxic** to the system [A38, A745a; HR].

As **a uterine cleanser** to counteract the papaya seed (*Carica papaya*) sterility treatment used by women for less than 2 years, 2 tablespoons of root along with 1 garlic clove are boiled in 1 pint of water for 10 minutes and 1 cup consumed twice weekly, at bedtime, until conception, which is said to occur within 3 months [Arvigo and Balick, 1998]. To **purge, reverse, or "wash out" a male contraceptive**, refer to Billy

Webb (*Acosmium panamense*) [A38, B1848, B2220; HR]. As a **male aphrodisiac**, refer to *contribo* (*Aristolochia trilobata*) [Mallory, 1991]. This plant should not be taken during pregnancy, as it may have **abortifacient effects**.

To treat **alcoholism**, ⅓ of an empty rum bottle is filled with chopped, dried roots and the remainder filled with rum, and this is set to steep for 10 days. If the person is willing to stop drinking, 1 shot glass is consumed daily until finished. If the person is not willing to stop drinking, the whole bottle is consumed at once when they are drunk and looking for more alcohol; this dosage is said to cause severe nausea whenever alcohol is smelled in the future [A38, A745a; HR]. Alternatively, 1 handful of chopped root is placed in about 1 quart of rum, vodka, or gin and then steeped in the sun for 5 days, strained, and consumed in the same manner as described above for alcoholism [Arvigo and Balick, 1998].

To strengthen the **spiritual powers** of a shaman, 1 tablespoon of the roots and the bark from the vine is boiled in 1½ cups of water for 10 minutes and consumed once daily for 9 consecutive Fridays [RA; Arvigo and Balick, 1998]. To **dispel witchcraft, evil magic, and the evil eye (*mal ojo*)**, 1 small handful of chopped, dried roots is boiled in 3½ cups of water for 10 minutes, strained, and 1 cup consumed 3 times daily, before meals, for 9 days [A38; HR; Arvigo and Balick, 1998]. To **dispel witchcraft**, 1 small branch of *ruda* (either *Ruta graveolens* or *R. chalapensis*) is mashed, the leaf juice and a pinch of powder from the Guatemalan white stone *Esquipulas* (bought in the market) added to the decoction described above to dispel witchcraft, and 1 cup consumed 3 times daily for 9 days accompanied by saying the "Our Father" prayer 9 times daily [HR, RA].

Caution is used with this plant. It is reported that the vine has a cross pattern said to indicate a medicinal use [A765]. It is reported that this plant has a potential toxicity, in that after 9–10 days of 3 cups consumed daily, diarrhea may develop [A765]. Additionally, this plant often proves to be too potent for the elderly and the very weak [Arvigo and Balick, 1998].

Coccocypselum herbaceum P. Browne

To treat **swelling**, the young leaves and stems are mashed with a bit of castor oil (*Ricinus communis*), applied as a poultice to the affected area, and changed twice daily until better [B2613].

Coccocypselum hirsutum Bartl. ex DC.

oregano del monte [Spanish]; *unicte tunich* [Mayan]

For **headache**, 1 handful of leaves is mashed, wrapped in a cloth, and applied as a poultice to the forehead for 2–3 hours [A267]. Alternatively, 1 handful of leaves is boiled in 1 quart of water for 10 minutes and used cool to wash the head for a headache that comes in the daytime or used warm to wash the head at bedtime for a headache that comes in the nighttime [B2261].

To lessen the **pain** from a scorpion bite or snakebite, 9 leaves are mashed finely and about 1 teaspoon of the juice squeezed onto the sores as needed [B3601]. Ad-

Coccocypselum herbaceum showing buds (left) and fruit (right) [PA].

ditionally, the leaves are mashed, applied as a poultice to the bite, left on for 1 hour, and then the replaced by a new poultice [B3601].

Coffea arabica L.

coffee [English]; *café* [Spanish]; *cape* [Q'eqchi']; *café xiv* [Mayan]

The ripe fruit flesh is eaten as a **food** [Arvigo and Balick, 1998]. To prepare a **beverage**, 1 teaspoon of the dried, roasted seeds is steeped in 1 cup of hot water for 10 minutes and consumed [B2405]. This is considered a "hot" plant in local medical systems [Arnason et al., 1980]. As a **stimulant** and a **diuretic**, 3 leaves are boiled in 1 cup of water for 10 minutes and consumed hot [Arvigo and Balick, 1998].

To treat infants with **fever**, the shells of green, unroasted seeds are mashed, and the exuding resin is placed on a piece of cloth and applied as a poultice to the forehead and soles of the feet [Arvigo and Balick, 1998]. Alternatively, refer to *Santa Mariá* (*Pluchea odorata*) [Steggerda, 1943]. To **improve blood circulation**, 1 handful of large leaves is boiled in 1 gallon of water for 10 minutes and used warm as a bath once daily at bedtime [A204]. To **stimulate urination**, 1 handful of green seed shells is boiled in 3 cups of water for 10 minutes and sipped all day [Arvigo and Balick, 1998].

To prepare a long-lasting **contraceptive** said to work for up to 1 year, 1 handful of freshly picked green coffee beans (*Coffea arabica*) is boiled in 3 cups of water for 10 minutes and consumed throughout the day for 3 consecutive days. To be effective, this treatment is begun immediately following childbirth [Arvigo and Balick, 1998].

Coffea arabica [MB].

To facilitate **de-shelling** coffee beans (*Coffea arabica*), refer to *gumbolimbo* (*Bursera simaruba*) [HR]. For **coffee substitutes**, refer to ginger (*Zingiber officinale*) [Arvigo and Balick, 1998], lemon grass (*Cymbopogon citratus*) [Arvigo and Balick, 1998], *bu kút* (*Cassia grandis*) [B3203], *frijolillo* (*Senna occidentalis*) [BW; Arvigo and Balick, 1998], *Senna sophera* [B3096], and allspice (*Pimenta dioica*) [B3135, JB21]. For a tree that is commonly inter-cropped with coffee, refer to mother of cacao (*Muellera frutescens*) [R8].

Coffea liberica W. Bull ex Hiern
American coffee [English]
To prepare a **beverage**, the fruits are cleansed, dried, roasted, ground, and boiled in water [R18]. The husks on this species are said to be more difficult to remove [R18].

Coussarea cf. *paniculata* (Vahl) Standl.
For **anemia**, the leaves are boiled in water, 1 cup consumed, and the remainder used as a bath 3 times daily for 7 days [B2487].

Coutarea hexandra (Jacq.) K. Schum.
Standley and Record (1936) noted that in El Salvador local healers used the bark of this tree as a **quinine substitute**, presumably for the treatment of **fever** and **malaria**.

(top left) *Coutarea hexandra* [NYBG].

(top right) *Erithalis fruticosa* [MB].

(bottom) *Ernodea littoralis* [IA].

Diodia brasiliensis Spreng.
la muerte [Spanish]
For a shaman to **ward off evil**, 9 leaves are boiled in 2 cups of water for 5 minutes and 1 cup consumed warm twice daily, in the morning and in the evening, on a Friday 1–2 times yearly, depending on the need and the danger [A152].

Erithalis fruticosa L.
botoncillo [Spanish]
The fruit is eaten as a **food** by birds and is especially favored by the gray catbird (*Dumetella carolinensis*), the black catbird (*Melanoptila glabrirostris*), and the Yucatan vireo (*Vireo magister*) [A409].

Children use the fruits as a source of **dyes** for making temporary tattoos on their skin [A409].

Ernodea littoralis Sw.
wild cherry, yellow jugs [English]
The fruit is sweet and eaten as a **food** [B1952]. The fruit and seeds are eaten as a **food** by hummingbirds and warblers [A405].

Geophila repens (L.) I.M. Johnst.

To treat a child with **measles**, 1 large handful of the entire plant is boiled in 2 gallons of water for 10 minutes, cooled, and used cool as a bath twice daily until better [B2729].

Geophila sp.

This is one of the many plants used in a bath mixture for **general ailments**, including **tiredness**, **pain**, **paralysis**, **swelling**, **skin ailments**, and **burns** [B2347].

Geophila repens [MB].

Gonzalagunia panamensis (Cav.) K. Schum.

monkey tail [English]; *saca tinta* [Spanish]

The fruit is eaten as a **food** by birds [B2007].

To treat **fever**, 1 double handful of leaves is mashed in 2 gallons of water, steeped in the sun for 2–3 hours, and used as a bath [A791].

The fruits are used as a **dye** and yield a purple or a blue color [B2007, B2071]. To prepare the **dye**, 1 quart of berries is mashed in ½ gallon of water, steeped for 3 hours, and strained [HR].

Guettarda combsii Urb.

glassy wood, green star, pine-ridge glossy wood [English]; *arepa, verde lucero* [Spanish]; *cas cab, tas tas, tas tab* [Mayan]

To treat **leishmaniasis (baysore)**, the leaves are toasted, powdered, and applied directly to the affected area [A661]. Alternatively, the bark is steeped in water and used hot as a wash over the affected area [A661]. For **painful urination**, 9 leaves are boiled in 3 cups of water for 10 minutes and sipped all day [A390]. To treat **diarrhea**, one 12-inch piece of fresh root is mashed in 1 quart of water, steeped for ½ day, and ½ cup consumed cool 6 times daily for 2 days [A905; HR].

The leaves are used as a detergent **scouring pad** to wash greasy dishes [A905]. Leaves are also used as a **baking sheet** when making corn cakes [A390].

The wood is used in **construction**, particularly in building house frames and house

Guettarda combsii [GS].

poles, and is valued for its durability [A764, A817, A905]. Lamb (1946) reported that the wood was occasionally used in **construction** for inlay work and turnery.

Guettarda elliptica Sw.

prickle wood [English]; *tzib che* [Mayan]

To treat infants with **skin sores** or **bad winds** (mal viento), 1 large double handful of leaves is steeped in 1 gallon of water all afternoon and used warm as a bath once daily, at bedtime only [A668]. "Bad winds" would occur when an errant, evil spirit is around and the baby is suddenly cranky and frightened [A668].

Steggerda (1943) reported that the Yucatan Maya, during Colonial times, used this plant as an antidote for **spider bites**, **snakebites**, and to treat **dysentery**. He noted that the plant was no longer used according to the healers he interviewed.

To prevent against harmful effects from the powerful presence of spiritual forces, the branches and leaves are used to brush the bodies of participants at sacred Maya **ceremonies** [RA].

Guettarda macrosperma Donn. Sm.

wild grape [English]; *pimentilla* [Spanish]; *kruz k'ix* [Q'eqchi']

The fruit is eaten as a **food** [A629].

The wood is used for **fuel** [A629].

Hamelia patens Jacq. var. *patens*

polly red head, red head, scarlet bush [English]; *Indios, sanalo todo* [Spanish]; *ix canan, sac te much* [Mopan]; *klaux pim* [Q'eqchi']; *ax canaan, canaan* [Mayan]; *chactoc, corallilo* [Additional names recorded]

The fruit is eaten as a **food** [Arvigo and Balick, 1998]. Standley and Record (1936) noted that the fruits were of "poor flavor." However, the fruits are a favorite of birds, and butterflies like the flowers [RA].

To prepare a bath, 1 handful of leaves is mashed and soaked in 1 gallon of water and used tepid as a bath once daily at bedtime [A71].

To treat all types of **skin ailments** and **skin infections**, especially **red rash, sores, burns, pimples, itching, cuts, fungus**, and **insect bites**,

(top) *Guettarda macrosperma* [GS].

(bottom) *Hamelia patens* var. *patens* [MB].

1 double handful of leaves, flowers, and fruits is boiled in 1 gallon of water for 10 minutes, cooled, and used as a bath 6 times daily [A893, B1777; Arvigo and Balick, 1998]. To treat **skin infections**, 1 cup of fresh leaves, stems, and flowers and ½ cup of cooking oil or lard are steeped over a low flame or on a warm stove, brought almost to a boil and them shut off several times daily for 3 days, and one 5-inch piece of candle is added and allowed to melt, the mixture poured off through a strainer, stored in a clean jar, and used as a salve or cream over the affected area as needed [A931]. Steggerda (1943) noted that the Yucatan Maya toasted the leaves of this plant, crushed them, and applied them on the hands to harden **blisters**. This treatment was also said to be used for **sores**, **wounds**, and **eczema**.

An extract of this plant was tested on **wounds** in mice, and the results showed that the topical treatment increased the breaking strength of the wounds, showing that the plant promoted wound healing [Gomez-Beloz et al., 2003].

For **skin rash**, 9 leaves either alone or along with 9 plum leaves (*Spondias purpurea*) are mashed in ½ cup of water and applied as a poultice to the affected area [B3232]. Alternatively, refer to *mañanita* (*Portulaca pilosa*) [B2099, B2100] and stinkin' bush (*Cornutia pyramidata*) [B2104]. For **itchy skin**, refer to wire wis (*Lygodium venustum*) [B2190].

For **fever**, **headache**, **bad winds** (mal viento), **spider bites**, **skin sores**, **scabies**, **leishmaniasis (baysore)**, **itching**, and other **skin ailments**, 1 large handful of leaves and flowers (if available) is boiled in 1 gallon of water for 10 minutes, 1 cup consumed twice daily, and the remainder used tepid as a bath twice daily as needed [A316, A811, B2293, B2533, B3232, JB104]. To treat **cradle cap sores** and **dandruff**, the decoction described for fever is used to bathe the head, followed by a massage with a powder made from the dried leaves [B2409].

To relieve **itching**, the leaves are rubbed over the affected area [A811]. For **wounds**, **skin ulcers**, **leishmaniasis (baysore)**, **stubborn skin sores**, **ulcers**, **scaly skin**, **irritations**, or **fungus**, 1 handful of leaves is parched over low heat, powdered, sprinkled lightly on the affected area after it has been cleansed, covered in gauze, and kept clean [A71, A931, B1777, B2293; Arvigo and Balick, 1998]. For **skin fungus**, the leaves are heated and rubbed directly on the affected area [B2055]. To treat **athlete's foot fungus**, the leaves are mashed and placed between the toes [B2735]. To treat **bee stings**, **wasp stings**, or **"doctor fly" stings**, the leaves are warmed and applied as a poultice to the affected area [Arvigo and Balick, 1998]. Alternatively, the fresh leaves are crushed and the leaf juice rubbed on the sting [Arvigo and Balick, 1998]. To treat **ringworm**, 1 large handful of leaves is boiled in 1½ quarts of water until reduced to 1 quart, kept for several days, and used as a wash over the affected area 3 times daily [A931]. To treat all the **skin ailments** mentioned above, the leaves are squeezed in the hands during the rainy season, and the expelled red juice is used over the affected areas [RA].

To treat non-insulin-dependent **diabetes** mellitus, nine 3-inch young leaf tops and young buds are steeped in 1 quart of water for 10–15 minutes, and 1 cup is consumed 3 times daily after each meal for adults and ½ cup consumed 3 times daily after each meal for children [A931].

To alleviate **menstrual cramps** or **pain**, 1 entire young plant about 2 feet tall is prepared in water and used as a sitz bath or 3 cups consumed daily [Arvigo and Balick, 1998]. Alternatively, 1 large double handful of leaves and young stems is boiled in 2 gallons of water for 10 minutes and used as a sitz bath [RA]. To reduce **pain** and **swelling** following a miscarriage, the leaves are boiled in water and used as a vaginal steam bath [Arvigo and Balick, 1998].

To make a household "**iodine**," three 10-inch stems are boiled in 3 cups of water for 10 minutes, a rusty nail added for 15 minutes, the mixture strained off, bottled, and used as needed [Arvigo and Balick, 1998].

This species is considered the "female" counterpart to butterfly weed (*Asclepias curassavica*) [B3694].

Hamelia rovirosae Wernham

To treat **skin ulcers** and **skin irritations**, 1 handful of leaves is boiled in 1 gallon of water for 10 minutes and used cool as a wash over the affected area twice daily until better [B2671].

Hoffmannia bullata L.O. Williams

To treat **hepatitis**, 1 double handful of fresh leaves along with 1 large handful of fresh leaves of *Psychotria pleuropoda* are mashed in cold water and consumed cold 3 times daily for 7 days [C72, C73].

Hoffmannia ghiesbreghtii (Lem.) Hemsl.

crazy head [English]; *oh câehp, oh cóehp, map pim* [Q'eqchi']; *can um puc* [Mayan]

To relieve **headache** and **irritations** when the eyes hurt, 1 handful of leaves is added

Hoffmannia ghiesbreghtii [IA].

to 1 gallon of boiling water for 10 minutes, cooled, and poured over the head and hair [B2518]. To treat "**crazy head**," which is a mental condition of **insanity**, **madness**, or **hysteria**, 9 fresh leaves are mashed in 1 cup of cold water and 1 cup consumed 3 times daily until calm [R36].

Manettia reclinata L.

klaux q'ehen, kolaars [Q'eqchi']

As a treatment for **swelling** or **boils**, the plant is mashed, soaked in water, and used as a bath over the affected area [B3550]. To treat **vomiting of blood**, the entire vine is mashed by hand in tepid water and consumed 3 times daily [B2566].

Morinda panamensis Seem.

turkey victuals, yellow-wood [English]; *palo amarillo* [Spanish]; *q'an i che'* [Q'eqchi']

For an unspecified treatment, one 6-inch piece of root dug from the "east" and one 6-inch piece of root dug from the "west" from the same plant are used to prepare a beverage that is consumed lukewarm twice daily [A973].

To treat **yellow jaundice**, the stem and leaves are used to prepare a decoction and 1 cup consumed 3 times daily. The first cup is consumed in the morning, followed by a bath with cool dew water that allows all the heat to leave the body [A599]. To treat **hepatitis**, refer to *frijolillo* (*Senna occidentalis*) [C30, C50].

Morinda panamensis [GS].

Morinda royoc L.

wart vine, wild pine, wort vine [English]; *ax, ax ak* [Mayan]

For removal of **warts**, the fruit is cut, heated over a flame, and rubbed frequently over the affected area [B2093, B2151]. To treat yellow **jaundice**, 3 teaspoons of the chopped yellow roots are boiled in 3 cups of water for 10 minutes and 1 cup consumed 3 times daily or used warm as a bath once daily until better [B2093, B2151].

Morinda yucatanensis Greenm.

wart vine, wild pine [English]; *ax ak* [Mayan]

For removal of **warts**, the fruit is sliced, heated over a flame, and the juice rubbed on the affected area as needed [A47]. Alternatively, the fruit is sliced, heated, applied to the affected area, secured with a cloth, kept on overnight, changed in the morning, and repeated until better [B1820]. Steggerda (1943) reported that the juice of the fruit was used to treat **granulated eyelids**.

Standley and Record (1936) reported that this plant was once employed by the Maya as a **dye**, but no details were provided.

(top) *Morinda royoc* with ripe fruit [MB].

(bottom left) Inflorescence of *Pentas lanceolata* [GAC].

(bottom right) *Posoqueria latifolia* [GS].

Pentas lanceolata (Forssk.) Deflers
This shrub is used as an **ornamental** [B2335].

Posoqueria latifolia (Rudge) Roem. & Schult.
mountain guava, snake-seed [English]; *cacalichuche, chholom k'oy* [Additional names recorded]
The wood is used in **construction**, especially for house posts [A977].

Psychotria acuminata [MB].

Psychotria brachiata [MB].

Psychotria acuminata Benth.

anal, ix anal (female) [Mayan]

To treat **skin ailments, rheumatism, nervousness, insomnia, headache, swelling, bruises**, and **infantile diseases**, 1 large double handful of leaves is boiled in 1 gallon of water for 10 minutes and used as a bath [Arvigo and Balick, 1998]. Alternatively, the leaves are used in combination with other leaves to prepare a bath [Arvigo and Balick, 1998].

To treat **back pain**, 1 large double handful of leaves along with 1 large double handful of marigold leaves (*Tagetes erecta*) are used to prepare a bath used once daily, at bedtime, for 3 consecutive nights [Arvigo and Balick, 1998]. To treat **wind** (viento), 1–3 entire small plants about 2 feet tall are boiled in 2 gallons of water for 10 minutes and used on warm days either as a warm bath at noon or a cool bath at bedtime as needed [B1775].

Following a bath treatment, it is considered **unwise** to be exposed to the cold and windy night air or daytime breezes [Arvigo and Balick, 1998].

This is the "female" counterpart to dog's tongue (*Psychotria tenuifolia*) [Arvigo and Balick, 1998].

Psychotria berteriana DC.

As a **female contraceptive**, 1 handful of leaves and 2 cups of water are mashed in a calabash gourd (*Crescentia cujete*) and 2 cups consumed once. This is said to work for 3 years [B2493].

Psychotria brachiata Sw.

To ease **joint pain**, 1 large handful of young leaves is boiled in 2 gallons of water for 10 minutes and used warm as a bath once daily at bedtime [B2680]. Alternatively, about 1 cup of fresh leaves is mashed and applied as a poultice to the affected area [B2680].

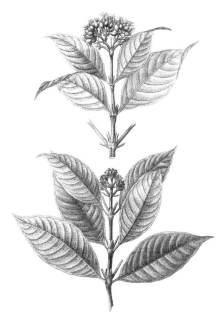

Psychotria capitata [NYBG].

Psychotria fruticetorum [GS].

Psychotria capitata Ruiz & Pav.
To **calm** and **relax** babies when they are "frightened," 1 double handful of leaves and stems is boiled in 1 gallon of water for 10 minutes and used warm as a bath once daily at bedtime [B2646].

Psychotria costivenia Griseb. var. *costivenia*
The fruit is eaten as a **food** by birds, especially pigeons [B2452].

Psychotria elata (Sw.) Hammel
To treat a **loss of energy**, 1 large handful of leaves and stems is boiled in 2 gallons of water for 10 minutes and used lukewarm as a bath once daily, at noon, as needed [B2666]. For **knee dislocations** and **weak knees**, the leaves are mashed with salt, applied as a poultice to the affected area, and changed twice daily [B2666].

Psychotria fruticetorum Standl.
bird berries, mountain lime, turtle bones [English]; *gueso de tortuga*, *lemonaria* (male), *lima del monte* [Spanish]; *bak y ak* [Mayan]
This leaves are used to prepare a bath for **general ailments** [B2635]. To treat **headache**, the leaves are prepared into a poultice applied directly to the head [B2635].
For **stomach and liver disorders** or **general pain**, 1 root is boiled in 3 cups of water and consumed hot as needed [B1795]. To treat **chronic gastritis** (ciro) and **constipation**, the roots of this plant are 1 of 5–6 different plants used in an un-

specified treatment [B1795; RA]. To treat **gonorrhea** and **bladder ailments**, three 6-inch, small stems with leaves are boiled in 3 cups of water for 10 minutes and 1 cup consumed 3 times daily for 10 days [A167, A671].

Psychotria glomerulata (Donn. Sm.) Steyerm.

rash shon ch [Mayan]

To treat **heart palpitations**, the flower along with sweet orange blossoms (*Citrus sinensis*) are steeped in a bottle of white rum, 3 drops diluted into 1 tablespoon of water, and consumed 3 times daily [B2617, B2643].

Psychotria grandis Sw.

Shkua tàsh, shkua tísh, rax ik che' [Q'eqchi']; *rash a wa* [Mayan]

The leaves are eaten as a **food** by tapirs [B3651].

The leaves and the stems are used to make **toys** [B3651].

Psychotria limonensis K. Krause

To treat **fever** caused by evil magic (*obeah*) or the voodoo man burning too many candles, 1 large handful of leaves is boiled in 1 gallon of water and a "hot, hot" bath poured over the body once daily, at noon or in the evening, every other day for a total of 3 baths [B2479].

Psychotria marginata Sw.

mountain bill bird potter [English]; *anal* [Mayan]

To soothe **fever**, **sores**, or **rash**, 1 handful of leaves is boiled in 1 gallon of water for 10 minutes and used cool as a bath twice daily [B2714]. To treat **fever** in a baby, 3 teaspoons of the decoction described above for fever are consumed and the remainder used warm as a bath once daily at bedtime [B2565].

Psychotria microdon (DC.) Urb.

Steggerda (1943) reported that the leaves of this shrub were boiled in water and used to bathe babies with **diarrhea**.

Psychotria glomerulata [MB].   *Psychotria grandis* [GS].

Psychotria nervosa [MB].

Psychotria poeppigiana [MB].

Psychotria nervosa Sw.

contra hierba [Spanish]; *anal* (male), *anal xiv*, *canaan* [Mayan]

This is one of the plants used to prepare a bath for **general ailments** [B1759, B2444].

To soothe **headache**, 1 handful of leaves is mashed and soaked in 2 quarts of warm water for 2 hours and used as a wash over the face [B2299]. For **toothache**, the leaves are warmed, slightly chewed to mash, and applied as a poultice in the mouth against the gums [B1759].

To treat **leishmaniasis (baysore)**, 1 handful of leaves is burned and the ashes applied directly to the affected area [B2299]. To treat **skin ailments**, the leaves are boiled in water and used cool as a wash over the affected area twice daily [B1759].

As a treatment for **snakebite**, equal parts of the leaves along with green *calalou* leaves (*Amaranthus retroflexus*) are boiled in 1 quart of water for 10 minutes and poured hot as a wash over the bite [W1360].

Psychotria pleuropoda Donn. Sm.

To treat **hepatitis**, refer to *Hoffmannia bullata* [C73, C72].

Psychotria poeppigiana Müll. Arg.

hot lips, womb bush [English]; *sclavel* [Spanish]; *xjolom aj tz'o'* [Q'eqchi']

To treat **gingivitis** or **bleeding gums**, 1 handful of leaves is mashed in 1 quart of cool water, steeped in the sun for 2 hours, 1 cup consumed, and the remainder used to wash the head [B3545].

To prepare a **vaginal douche**, 1 entire plant with roots is boiled in 2 quarts of water for 10 minutes, steeped until tepid, strained, and used twice daily for 3 days [B2598]. For **irregular and painful menstruation** or **infertility** in women, 1 large double handful of leaves is boiled in 2 gallons of water for 10 minutes, used hot as a sitz bath once daily for 20 minutes, and repeated 2 days later. This should be used during the week before the onset of menstruation [HR].

For **delayed parturition** or **delayed delivery** of the fetus, the root is boiled and consumed in an unspecified treatment [A217]. For **postpartum hemorrhaging**, 1 entire plant with roots is boiled in 1 quart of water for 10 minutes, steeped until tepid, strained, and sipped cool as needed [A217]. For **miscarriage**, 3 roots are boiled in 1 quart of water for 10 minutes and consumed hot until everything passes out [HR].

The bracts of the flowers are said to resemble female reproductive parts [A217].

Psychotria pubescens Sw.

hog bush, wild skanan [English]; *anal hembra* (female), *hierba* [Spanish]; *anal, anal xiv* (male), *ix anal, ix anal xiv* [Mayan]

The leaves are used as **forage** for pigs and wild animals [A813].

For **wind** (viento) and for **general ailments**, the leaves are part of a 9-plant bath formula [A57, A171, A488, A634, A767, A813, B1880].

To treat **snakebite**, 1 root from 1 plant is roasted until very dry, powdered, mixed with a piece of "soft candle" (just enough to make it stick together), and applied as a poultice over the bite after it has been bathed. Following this, 1 snake plant leaf (*Sansevieria trifasciata*) is pounded until soft and fibrous, applied as a poultice to wrap up the snakebite, and replaced twice daily until the "tooth scale" comes out [A488].

The dry leaves are a source of kindling **fuel** [JB85].

Psychotria tenuifolia Sw.

dog's tongue [English]; *lengua del perro* [Spanish]; *anal* (male) [Mopan]; *anal pik* [Mayan]

The leaves are part of a 9-plant bath formula used for **general ailments** [A51, B1771].

To treat **wounds**, **rash**, **swelling**, **burns**, **nervousness**, or **sleeplessness**, the leaves are added to an unspecified mixture of medicinal leaves (usually 9 kinds) to prepare a bath [Arvigo and Balick, 1998]. For **infected sores**, the leaves and flowers are mashed with about 1 teaspoon of water and applied as a poultice to the affected area [Arvigo and Balick, 1998]. To reduce **stubborn swelling**, the leaves along with cancer herb leaves (*Acalypha arvensis*) are boiled in water, cooled, and used as a wash over the affected area [Arvigo and Balick, 1998]. For **burns**, 1 handful of leaves is crushed, heated in 4 ounces of sweet almond oil until very warm, soaked until tepid, and the liquid squeezed out through a cloth and applied as a lotion twice daily to the affected area [A827; RA]. One healer in Belize reported being shown in a dream vision the use of this plant for all manner of **skin burns** [RA].

This is the "male" counterpart to *ix anal* (*Psychotria acuminata*) [Arvigo and Balick, 1998].

Psychotria tenuifolia [MB].

Psychotria sp.
canaan [Mayan]

To treat **headache**, the leaves are used to prepare a facial wash [B2349]. To treat **leishmaniasis (baysore)**, the leaves are burned, the ashes placed directly on the affected area, and changed twice daily [B2349].

Randia aculeata L.
bastard lime, wild lime [English]

To treat **diarrhea**, **dysentery**, or **intestinal disorders**, 1 small handful of leaves and one 1-inch × 6-inch piece of fresh bark are boiled in 1 quart of water for 10 minutes, steeped until cool, and 1 cup consumed 3 times daily, before meals, until better [A752]. For **migraine**, 1 large handful of leaves and stems is boiled in 2 quarts of water for 10 minutes, steeped until warm, and used as a bath over the head 3 times daily [A339].

Randia lundelliana Standl.
Pine Ridge guava, wild lime [English]

The immature and mature fruit is eaten as a **food** by humans and birds [A648].

To treat **headache**, the thorns are used to pierce the temple region of the head [A648]. To treat **fever** and **general pain**, the leaves are used to prepare a warm bath [A648]. To treat **stubborn sores** and **skin irritations**, the leaves are used to prepare a cool bath following which additional leaves are roasted, powdered, and sprinkled over the affected area [A648].

Sabicea panamensis Wernham
This plant is used to prepare a bath treatment for **general ailments** [B2629].

Simira salvadorensis (Standl.) Steyerm.
high ridge redwood, John Crow redwood, redwood [English]; *palo colorado* [Spanish]; *sac te m'ooch* [Mayan]

This wood is of a deep red color and recommended as a long-lasting lumber often used in house **construction** [A88].

(top) *Randia lundelliana* [GS].

(bottom) *Sabicea panamensis* [MB].

Spermacoce ocymifolia [MB].

Spermacoce ocymifolia Willd. ex Roem. & Schult.

yierba del crucita [Spanish]

To treat **cuts** on the foot, 1 large handful of fresh leaves and stem is boiled in water and poured hot over the affected area [B2563]. To treat **measles**, 1 large handful of fresh leaves and stems is boiled in 1 quart of water for 10 minutes and consumed warm 3 times daily for 10 days [C68].

To treat a **nervous breakdown** or when **voices are heard** or **hallucinations** are present, 1 entire plant is boiled in water and an unspecified amount consumed 3 times daily until better [B2618]. Alternatively, the plant's purple-flowered variety is used for this treatment [B2618].

It was noted that this plant is used to treat a person "possessed by **evil spirits**," but no specific information was provided [C68].

Spermacoce suaveolens (G. Mey.) Kuntze

The entire plant is used to prepare a bath used to treat **any ailment** [A372]. For **dysentery** or **intestinal disorders**, 3 entire plants are boiled in 3 cups of water for 20 minutes and 1 cup consumed cool 3 times daily, before meals, until better [A372].

Spermacoce verticillata L.

The leaves are eaten as a **food** by rabbits, goats, and pigs [A410].

RUTACEAE

Trees, shrubs, or sometimes herbs. Leaves alternate, opposite, rarely whorled, simple, pinnately or palmately compound. Inflorescences terminal, lateral, or axillary, sometimes cauline, in thyrses, cymes, racemes, spikes, or cincinni. Flowers usually showy. Fruits a capsule, mericarp, berry, hesperidium, drupe, or sometimes a samara. In Belize, consisting of 9 genera and 22 species. Uses for 5 genera and 11 species are reported here.

Amyris sylvatica Jacq.

hoja de espada [Spanish]; *ix coc* [Mayan]

As an antivenom first-aid treatment for **snakebites**, 1 piece of fresh bark equal to the size of the person's forearm is cut and chewed constantly, with the juices swallowed, until help is obtained or the person arrives at the hospital [A834]. As part of a **snake-**

(top left) *Amyris sylvatica* [NYBG].

(top right) *Casimiroa tetrameria* [WJH].

(bottom) *Citrus aurantiifolia* [IA].

bite treatment following a first-aid treatment, 1½ handfuls of fresh roots are boiled in 2 quarts of water for 10 minutes and 1 cup consumed tepid 3 times daily [A834].

Casimiroa tetrameria Millsp.

white sapote [English]; *mata sano* [Spanish]; *yuy* [Yucateca]

Standley and Record (1936) reported that the white flesh of this fruit was sweet and eaten as a **food**, but also that it was little used because of the belief it was not healthy. "As a matter of fact, there has been extracted from the seeds and leaves a glucoside having a soporific effect, hence there is probably some basis for another belief that eating the fruit **induces drowsiness**." Steggerda (1943) reported that the wood of this tree was **burned** in lime kilns and that in addition to humans, birds and deer ate the fruit as a **food**.

Citrus aurantiifolia (Christm.) Swingle

lime, sweet lime [English]

To prepare a **food** for nursing mothers and menopausal women, refer to sesame (*Sesamum indicum*) [RA].

The fruit is chewed to help get rid of **sputum** [Arnason et al., 1980]. To treat **cough** in a dog, refer to belly full tie tie (*Philodendron radiatum* var. *radiatum*) [BW]. To treat **cough** and **colds**, refer to *maravilla* (*Mirabilis jalapa*) [B2283]. For

dehydration, refer to coconut (*Cocos nucifera*) [Arvigo and Balick, 1998]. To treat **dysentery**, refer to coriander (*Eryngium foetidum*) [BW]. To treat *pasmo* (**congested blood; more than the usual amount in an area of the body**), refer to *Mandevilla hirsuta* [B2649].

For **nervousness**, refer to *ich wâ* (*Lepidaploa tortuosa*) [B2633] and wild sage (*Lantana trifolia*) [B2096]. To treat **high blood pressure**, refer to prickly pear (*Opuntia cochenillifera*) [BW]. To treat **spleen ailments**, refer to physic nut (*Jatropha curcas*) [Arvigo and Balick, 1998].

Citrus aurantium L.

sour orange tree [English]; *naranja agria* [Spanish]; *zutz pakal* [Mopan]

The fruit is eaten as a **food** [B2274]. The rind may be dried and preserved [Arvigo and Balick, 1998]. Steggerda (1943) reported that the fruits were sometimes used to wash wild meat and fowl to remove the gamey taste and also noted that the juice was utilized as **vinegar**.

To **promote sweat** to treat **fever** and certain **skin ailments**, 1 handful of leaves is boiled in 2 gallons of water for 10 minutes and used warm as a bath once daily until better [A147].

As a general **tonic**, to **stimulate the appetite**, and for **liver ailments**, fresh fruit juice is consumed [Arvigo and Balick, 1998]. To treat **colds, influenza, fever, blood clots, diarrhea, indigestion, vomiting,** or **infantile colic**, 9 leaves are boiled in 3 cups of water for 2 minutes, steeped for 10 minutes, 1 cup consumed 3 times daily before meals for adults and 1 cup consumed through small doses per day for infants [Arvigo and Balick, 1998]. For **coughs, colds, bronchitis, mucous congestion,** or **indigestion**, the rind of 1 fruit is steeped in 3 cups of hot water for 20 minutes, strained, and 1 cup consumed 3 times daily before meals [Arvigo and Balick, 1998]. To treat **stomach and chest pain** or to regulate

Citrus aurantium with inset showing leaf and thorns [IA].

menstrual bleeding, refer to *Santa Mariá* (*Pluchea odorata*) [Steggerda, 1943]. For **stomach pain**, refer to cowitch (*Urera baccifera*) [Steggerda and Korsch, 1943].

To treat **digestive difficulties**, **gas**, or **vomiting**, 9 leaves along with 9 spearmint leaves (*Mentha spicata*) are boiled in 2 cups of water for 5 minutes, steeped for 10 minutes, and ½ cup consumed warm 3–4 times daily, before meals, until better [A147, B2274]. To provide quick relief for acute attacks of **indigestion**, 1 leaf is chewed and the juices swallowed [Arvigo and Balick, 1998].

To treat **high blood pressure**, 2–3 tablespoons of fresh fruit juice are consumed once daily, in the morning, for 10 days [Arvigo and Balick, 1998]. For **nervousness**, **hysteria**, or **insomnia**, 1 small handful of flowers is steeped in 2 cups of boiling water for 20 minutes, strained, and consumed slowly [Arvigo and Balick, 1998]. To treat **pain**, **tiredness**, or **snakebite**, refer to Spanish elder (*Piper jacquemontianum*) [B2133].

The entire plant is used as a grafting **rootstock** for sweet orange (*Citrus sinensis*) [B2274].

It is believed that to **ward off evil magic**, **envy (*envidia*)**, and the **evil eye (*mal ojo*)**, 1 very young, green fruit is carried in one's pocket [Arvigo and Balick, 1998]. To treat **bad winds** (mal viento), 3 young leaves are gathered and placed in a cross formation over the person's wrists while reciting prayers [Arvigo and Balick, 1998].

Citrus limon (L.) Osbeck

lemon, lemon tree [English]; *limón* [Spanish]
The fruit is eaten as a **food** [B2277].

To treat **high blood pressure**, the fruit is eaten or prepared and consumed as a beverage [B2277]. To treat **heat exhaustion ("overheating")**, the fruit is juiced and 2 cups consumed with aspirin 3 times daily until better [B2277]. For use as a "cooling" medicine to treat **excessive body heat**, refer to coconut (*Cocos nucifera*) [Arvigo and Balick, 1998]. To ease **cold sweats**, usually occurring with **asthma**, **exhaustion**, **excessive coughing**, or with **anemic babies**, refer to *helecho* (*Pteridium caudatum*) [B2142; BW]. To halt **vomiting**, 9 leaves are boiled in 1½ cups of water for 5 minutes, steeped for 10 minutes, and consumed warm until better [B2277]. For two **cough** treatments, refer to pine (*Pinus caribaea* var. *hondurensis*) [A401]. To treat **fever** or **painful "wind spasms" (muscle pain due to being in a "draft")**, refer to grande Betty (*Cupania belizensis*) [A395].

Citrus limon [NYBG].

Citrus × paradisi Macfad.

grapefruit [English]; *toronja* [Spanish]

The fruit is eaten as a **food** [A149, B2353].

As a general household **tonic** given by mothers to their children, refer to *sorosi* (*Momordica charantia*) [A63, B1760]. To treat **fever** and **malaria**, the inner white flesh (the bitter part encasing the juicy fruits) of 1 fruit is boiled in 2½ cups of water for 5 minutes, steeped for 15 minutes, and ½ cup consumed 4 times daily until better [A149]. For **malaria**, refer to *sorosi* (*Momordica charantia*) [B2124]. To treat non-insulin-dependent **diabetes** mellitus, 5 entire fruits and one 2-inch piece of root are boiled in 1 gallon of water, reduced to ¾ gallon, cooled, and consumed as a beverage 12 times daily for 30 days [C93].

Citrus × paradisi [MB].

Citrus sinensis (L.) Osbeck

orange, sweet orange, Washington orange [English]; *naranja dulce* [Spanish]; *chiin* [Q'eqchi']; *paàk'al* [Yucateca]

The delicious fruit is eaten as a **food** and often made into **beverage** [B2279, R48]. Steggerda (1943) reported that the Yucatan Maya distilled an oil from the flowers, which was used to **flavor** refreshments, and that a **beverage** was also made from the leaves.

To treat **diarrhea ("bad stomach")**, a 1:1:1 ratio of the bark along with *nanci* bark (*Byrsonima crassifolia*) and guava bark (*Psidium guajava*) is boiled in 3½ cups of water for 10 minutes, steeped for 20 minutes, and ½ cup consumed warm every 3 hours all day long until better [B2279]. For two treatments for **stomach pain** and **intestinal disorders**, refer to guava (*Psidium guajava*) [B2652]. To treat **diarrhea**, refer to scorpion tail (*Heliotropium angiospermum*) [B2294]. For **stomach pain** due to food poisoning, refer to scorpion tail (*Heliotropium angiospermum*) [B1939]. For **intestinal gas**, **loose stool**, **indigestion**, or a **bad belly**, refer to allspice (*Pimenta dioica*) [B2404].To treat **adult colic** and **stomach pain**, refer to turmeric (*Curcuma longa*) [LR]. To treat **colic** pain in babies, refer to *artemisia* (*Egletes viscosa*) [LR]. To relieve **spasms** in babies with colic, refer to scorpion tail (*Heliotropium angiospermum*) [B1901]. For **gastritis** (ciro), refer to *tip te ák* (*Aristolochia arborea*) [B2199], *guaco* (*Aristolochia grandiflora*) [B3669], and allspice (*Pimenta dioica*) [A818].

For **women's ailments**, refer to castor plant (*Ricinus communis*) [B2176]. To treat **heart palpitations**, refer to *rash shon ch* (*Psychotria glomerulata*) [B2617, B2643]. To treat "**pain in the heart**," the leaves are used to make a tea [Arnason et al., 1980]. To treat *espantajo* (a **baby's illness caused by shock or sudden surprise**) or *susto* (**fright**), 9 leaves are placed on the baby's wrists and forehead while repeating the 9 prayers specific to this ailment [RA; Arnason et al., 1980].

To treat **nausea** and **hangover**, the leaves are used to make a tea [Arnason et al., 1980].

To **dispel evil spirits**, refer to marigold (*Tagetes erecta*) [Arvigo and Balick, 1998].

When a grafting **rootstock** is needed for this plant, the entire sour orange tree (*Citrus aurantium*) is used [B2274].

Ruta chalepensis L.

rue [English]; *ruda* [Spanish]; *sink in* [Mopan]; *sik'ij* [Q'eqchi']

Steggerda and Korsch (1943) reported that an ancient remedy for treating **nosebleed** was to inhale the juice of this plant through the nose. They also noted that "roasted horse dung, or black dog's excrement made into a powder and blown into the nose with a little reed are also supposed to stop a nosebleed." Steggerda and Korsch (1943) noted that the Maya of the Yucatan made a poultice of the leaves of this plant, mixed with salt and egg white, to treat **toothache**.

For **sunburn**, **heat exhaustion**, **sun stroke**, or **headache**, 1 cup of leaves is soaked in 1 cup of wintergreen alcohol for 9 days and then rubbed on the head, neck, and forehead every hour until better [RA]. To make a person come back from **fainting spells**, the same decoction described for sunburn is placed under the nose and smelled [RA].

To disinfect **cuts**, **infections**, and reduce **swelling** (*hinchazón*) in the body, ½ cup of leaves along with ½ cup of *epasote* leaves (*Chenopodium ambrosioides*) are steeped for 9 days in 2 cups of alcohol and applied as a wash [B2316]. For **swelling**, a cotton cloth is soaked in this tincture, applied as a dressing over the area, and changed 3 times daily or as needed [RA].

To treat **stomach pain** or **epilepsy**, three 6-inch sprigs are squeezed into 1 cup of water until the water turns green and the leaf parts are well dissolved, strained, and consumed at room temperature 3 times daily before meals. For stomach pain, treatment is continued as needed, and for epilepsy, this treatment is only used when medication is not available [A151]. For **spasms**, refer to *Indigofera suffruticosa*

Ruta chalepensis [MB].

[Steggerda and Korsch, 1943]. This plant should not be taken during pregnancy, as it may have **abortifacient effects**.

To stop **nightmares**, four 6-inch sprigs are placed under the pillowcase in a cross pattern [RA]. To **ward off evil**, **envy (*envidia*)**, and to **dispel witchcraft**, three 6-inch sprigs are squeezed into 1 cup of water until the water turns green and the leaf parts are well dissolved, ¼ teaspoon of the powder from the Guatemalan white stone *Esquipulas* (bought in the market) and 1 tablespoon of the holy water from the Catholic Church added, the decoction strained, and 1 cup consumed 3 times daily as needed [A151]. Alternatively, one 1-inch sprig is eaten whenever one is in the presence of the threat or feels there is an immediate threat [RA]. To **dispel witchcraft**, refer to skunk root (*Chiococca alba*) [HR, RA]. Gann (1918) reported that the Maya of this region utilized the leaves and applied them externally to children suffering from **convulsions** as well as in the same fashion to treat "almost any **nervous complaint** in adults."

Ruta graveolens L.

rue [English]; *ruda* [Spanish]; *sink in* [Mayan]

For **stomach cramps**, **delayed menstruation**, **intestinal worms**, **epilepsy**, **vomiting**, or **nervousness**, 9 small branches with leaves are squeezed into 1 cup of water, strained, and consumed twice daily before meals [Arvigo and Balick, 1998]. To ease **childbirth** and help with **contractions**, the decoction described for stomach cramps is sipped during delivery [Arvigo and Balick, 1998].

To treat **muscle soreness**, **back pain**, **headache**, and **muscle spasms**, the leaves are soaked in alcohol and used as a liniment [Arvigo and Balick, 1998]. For **fever**, **exhaustion**, or **fainting spells**, the leaves are soaked in alcohol and used to massage the body [Arvigo and Balick, 1998]. As a treatment for **dizziness** caused by "bad air in the head," refer to skunk root (*Petiveria alliacea*) [B2122]. As a **facial wash**, refer to *saramuya* (*Annona squamosa*) [Arnason et al., 1980].

To treat **bad winds (*mal viento*)**, refer to warrie wood (*Diphysa carthaginensis*) [A647]. For when a person is feeling **uncertain** and **indecisive**, refer to *alamanda* (*Allamanda cathartica*) [B2651]. To treat **evil magic (*obeah*)**, **envy (*envidia*)**, **evil eye (*mal ojo*)**, **fright (*susto*)**, or **grief (*pesar*)**, the fresh leaves are either eaten, prepared as a beverage, or used in a bath [Arvigo and Balick, 1998]. For **evil eye (*mal ojo*)**, refer to *flor de la virgen* (*Caesalpinia pulcherrima*) [Arnason et al.,

Ruta graveolens [NYBG].

1980]. To treat **evil magic (*obeah*)**, refer to *begonia* (*Begonia sericoneura*) [BW]. For people who suspect they have been affected by **evil magic (*obeah*)**, refer to *dormilón* (*Mimosa pudica* var. *unijuga*) [HR]. To **ward off evil**, **envy (*envidia*)**, and **bad spirits**, fresh sprigs are added to protective amulets [Arvigo and Balick, 1998]. Additionally to **ward off envy (*envidia*)**, **jealousy**, **anger**, or **bad intentions** toward the family, 3 sprigs are made into a cross and placed over the door so that all who enter the house are cleansed of these emotions [Arvigo and Balick, 1998]. To immediately cancel all negative effects of **envy (*envidia*)**, when someone has recently been "envied," 1 small fresh sprig is placed under the tongue and allowed to slowly dissolve [Arvigo and Balick, 1998]. To **keep the devil away**, refer to morning glory (*Ipomoea carnea* subsp. *fistulosa*) [RA]. To **dispel witchcraft**, refer to skunk root (*Chiococca alba*) [HR]. To treat **spiritual ailments**, such as **fright (*susto*)**, **envy (*envidia*)**, **grief (*pesar*)**, and **evil magic (*obeah*)**, refer to bay cedar (*Guazuma ulmifolia*) [RA].

To treat **head lice** (*Pediculus humanus*), refer to *annona* (*Annona reticulata*) [A947, B1817].

Zanthoxylum caribaeum Lam.

bastard prickly yellow, prickly allah, prickly yellow, prickly wood [English]

To treat **fever** in an infant, 1 handful of leaves is soaked in 1 gallon of water and used as a bath once daily, at bedtime, until better [A826]. For **migraine**, the thorns are used to pierce the temple region of the head. This is repeated for 3 days; however, often only 1 piercing is needed [A826].

Zanthoxylum juniperinum Poepp.

black prickly yellow, prickly yellow macho [English]; *si lon che* [Mayan]; *kuxche* [Additional name recorded]

To treat **headache**, 9 thorns are used to puncture the skin of the forehead with 3 thorns forming a triangle in 3 distinct places: over the right temple, over the left temple, and in the middle of the forehead. This is repeated as needed until better, usually for 1–3 days [A531]. To treat **wind (*viento*)**, 1 large double handful of leaves is boiled in 2 gallons of water for 10 minutes and used hot as a bath once daily, at noon or bedtime, as needed [A531].

The wood is used as a source of **fuel** [A252].

Zanthoxylum caribaeum [NYBG].

SALICACEAE (INCLUDING FLACOURTIACEAE)

Trees or shrubs. Leaves usually alternate, distichous or sometimes spiral, rarely opposite, simple. Inflorescences terminal, subterminal or mostly axillary, in racemes, panicles, corymbs, or cymes. Flowers small, white or yellowish-white. Fruits a berry, capsule, or drupe. In Belize, consisting of 13 genera and 25 species. Uses for 5 genera and 7 species are reported here.

Casearia corymbosa Kunth

Billy hop, blossom berry [English]; *balelac de aquada, café de montaña, café de monte, paletillo* [Spanish]; *kape che'* [Q'eqchi']; *canjuro, sak tan té* [Additional names recorded]
The fruit is eaten as a **food** [B1891].

For **gastritis (*ciro*)**, 9 fruits are mashed, boiled in 3 cups of water for 5 minutes, and 1 cup consumed warm 3 times daily, before each meal [B2246]. To treat **heart pain**, an unspecified treatment is prepared [A804]. Steggerda (1943) reported that to treat **bile disorders** and **diseases of the spleen**, the Yucatan Maya were said to have boiled the leaves in water and used the decoction as a bath. For **chills**, 1 large handful of leaves is boiled in 2 gallons of water for 5 minutes, cooled, 3 teaspoons consumed, and the remainder used as a warm bath [B2522].

The wood is used in **construction** for house lumber [A90]. The smaller branches are prized for their flexibility and are used as roof thatch [A82].

Casearia sylvestris Sw. var. *sylvestris*

worm plant [English]; *mool bá, mulbá, wuqub' q'ehen* [Q'eqchi']
To treat a **low fever in an inactive, sweating baby**, 1 handful of leaves is boiled in 1 gallon of water for 5 minutes, 3 teaspoons consumed warm once daily, at bedtime or naptime, and the remainder used as a bath until better [B2544].

Flacourtia cataphracta Roxb. ex. Willd.

cherry [English]
The dark purple fruit is eaten as a **food** [B2345].

(top) *Casearia sylvestris* var. *sylvestris* [NYBG].

(bottom) *Flacourtia cataphracta* [NYBG].

Laetia procera (Poepp.) Eichler

drunken bayman wood [English]; *may té* [Q'eqchi']

The fruit is eaten as a **food** by birds [B3640].

To **treat wounds**, the resin or mashed leaves are applied with a cloth as a poultice twice daily to the affected area until better [B3640].

The wood **breaks easily**; therefore, it is not used in construction for making houses [B3640].

Laetia thamnia L.

bully hop, bullyhob, night perfume [English]; *perfume de la noche* [Spanish]

The wood is used in **carpentry** for making furniture [A832].

Xylosma cinerea (Clos) Hemsl.

To treat **chicken pox**, 1 large handful of leaves is boiled in 2 gallons of water for 5 minutes, cooled, and poured over the head as a bath once daily, at bedtime or naptime [A382].

Zuelania guidonia (Sw.) Britton & Millsp.

drunken bayman, drunken bayman wood, water-wood [English]; *ta mai, tamai, tamay* [Spanish]; *chu ya ak* [Additional name recorded]

The fruit is eaten as a **food** by humans and birds [B1926, B3122]. Stingless bees, locally called "drunken baymen," collect the sweet, brown **resin** [B3736].

This plant is 1 of 90 species whose leaves are used in Maya bath mixtures [B1872].

To **lose weight**, one 12-inch piece of bark is boiled in 2 quarts of water for 20 minutes and ½ cup consumed 6 times daily for 2 days of each week when the moon is full or waxing [B3122].

The wood is used as **firewood** [B1872].

(top) *Laetia thamnia* [GS].

(middle) *Xylosma cinerea* [GS].

(bottom) *Zuelania guidonia* [MB].

Steggerda (1943) noted that the Yucatan Maya were said to have used the leaves of this tree, boiled with another unspecified plant, as a bath to reduce **fever**.

SANTALACEAE (INCLUDING VISCACEAE)

Subshrubs or shrubs, hemiparasitic on woody plants. Leaves opposite, simple, well developed or scale-like, often both on same plant. Inflorescences axillary and/or terminal, in spikes. Flowers small, usually greenish or yellow. Fruits a berry. In Belize, consisting of 2 genera and 10 species. Uses for 1 genus and 2 species are reported here.

Phoradendron piperoides (Kunth) Trel.

God Almighty bush, scorn the earth [English]; *matapalo* [Spanish]
To treat the onset of **asthma** attacks, 9 leaves are boiled in 1 cup of water and sipped hot [Arvigo and Balick, 1998]. To treat **skin ailments**, **skin infections**, **swelling**, and **bruises**, 1 entire plant is boiled in water for 10 minutes and used warm as a wash over the affected area [Arvigo and Balick, 1998]. For **dog bite**, 1 entire plant and an unspecified amount of garlic are mashed together and applied as a poultice to the affected area [Arvigo and Balick, 1998].

Phoradendron piperoides [PA].

Phoradendron quadrangulare (Kunth) Krug & Urb.

God Almighty bush, scorn the earth [English]; *mata palo* [Spanish]
To treat **"cancer"** (**internal growths**), two 12-inch pieces of the vine with leaves are boiled in 1 quart of water for 10 minutes and 1 cup consumed tepid 3 times daily. This is said to shrink **tumors** [W120; RA].

After a **dog bite** has been washed out, 1 fresh vine with leaves and 1 fresh garlic clove are mashed, applied as a poultice to the bite, and changed 3 times daily [W120; RA].

SAPINDACEAE

Lianas, trees, or sometimes shrubs. Leaves usually alternate, very rarely opposite, usually pinnately compound, or sometimes palmately compound. Inflorescences terminal, axillary, or sometimes cauliflorous in lianas, in cymes, panicles, racemes, corymbs, umbels, or fascicles. Flowers greenish-white or yellow. Fruits a capsule, drupe, schizocarp, or berry, often winged. In Belize, consisting of 16 genera and 40 species. Uses for 9 genera and 15 species are reported here.

Allophylus cominia (L.) Sw.

bastard cherry, cherry, red seed [English]; *belo, huesillo, palo de caja* [Spanish]; *bichach, bikbach* [Additional names recorded]

(top left) *Allophylus cominia* [MB].

(top right) *Cardiospermum grandiflorum* [DA].

(bottom) *Cupania belizensis* [MB].

The fruit is eaten as a **food** by humans and birds [B2129].

This plant is used as a **decoration** in the home at Christmas time for flower arrangements because of its red berries [A161].

Cardiospermum grandiflorum Sw.

cruxi [Mayan]

The leaves are boiled and used in baths for **general ailments** [A136].

Cupania belizensis Standl.

bastard grandy Betty, grande Betty, grandy Betty [English]; *palo carbón, palo de carbon* [Spanish]; *chac pom* [Mopan]

To treat **swelling**, **sprains**, **fractures**, and **bruises**, 1 handful of leaves is boiled in 2 quarts of water for 10 minutes, cooled, and used as a bath 3 times daily [A86, B1814, JB16; Arvigo and Balick, 1998]. To treat **fever** or **painful "wind spasms" (a muscle pain due to being in a "draft")**, the bath described above for swelling is used while concurrently consuming a tea made from the lemon tree leaf (*Citrus limon*) [A395]. To treat **skin swelling**, **irritations**, or **infections**, 1 handful of leaves and fruits is boiled in 1 quart of water, reduced to 1 pint, cooled, and used as a wash over the affected area until better [A930].

For **diarrhea** or a "**bad belly**," one 1-inch × 3-inch piece of bark is boiled in 3 cups of water for 10 minutes and sipped all day. During this treatment, cold foods and cold drinks are avoided [Arvigo and Balick, 1998].

The wood is used to make **charcoal** [A86, B1814, B2464, B3283, JB16; Arvigo and Balick, 1998]. The wood is used in **construction** to make house posts, roof supports, and walking sticks [A96, B3283]. Small trees and branches are used as **canoe poles** to pull dugout canoes up river due to the right combination of strength and flexibility of the wood [R67].

Cupania rufescens Triana & Planch.

white grandy Betty [English]; *xucuroi* [Mayan]
The wood of this tree is used as **lumber** [B3331].

Matayba apetala Radlk.

bastard willow, boy job [English]; *mabehu, zacuayum* [Additional names recorded]
The wood of this tree is used as a material for **construction**; specifically, the trunks are used for house posts [A532]. Standley and Record (1936) reported that the wood was used in **construction** to make house beams and frames.

Paullinia clavigera Schltdl.

cross vine [English]; *k'atsucutche, kurus k'ix* [Q'eqchi']; *cruxi, puk sak an* [Mayan]
The fruit, a berry, is eaten as a **food** [A568].

The leaves are used in bath mixtures for **general ailments** [A50, B2193]. This plant is said to be a medicinal plant because of the cross formation of the leaves [A50].

To treat **colds**, 1 large handful of fresh leaves is boiled in 1 cup of water for 10 minutes or until the water is red colored, and 1 cup is consumed warm 3 times daily for 7 days [C37].

This is considered a **sacred** plant because the leaves form a cross [RA].

Paullinia clavigera [GS].

Paullinia costata Schltdl. & Cham.

The fruit is eaten as a **food** for birds, especially pheasants [B2664].

This is considered a **sacred** plant because the leaves form a cross [RA].

Paullinia fuscescens Kunth

As a **fish poison**, the vine is mashed up in the water of a dammed stream and used to stun fish. The quantity of plant material used in a 6-foot × 6-foot area of stream is approaching the number of chopped up branches that could be packed into a large flour sack [B2185].

This is considered a **sacred** plant because the leaves form a cross [RA].

Paullinia pinnata L.

fish poison [English]; *tie tie* [Creole]; *macalte ik* [Mayan]

Standley and Record (1936) reported that this as well as other species of *Paullinia* and *Serjania* are commonly used in Central America as **fish poison**. They noted:

> The stems and leaves are macerated and thrown into ponds or quiet streams, whereupon after a short time the fish become stupefied and float on the surface of the water, so that they may be collected easily. The poisonous properties of the plants are not deleterious to the fish as human food, and it is stated that if the fish are left in the water they recover after a while and swim away. (pp. 234–5)

Paullinia tomentosa Jacq.

cross vine, skipping rope vine [English]; *hierba del pensamiento*, *yerba del pensamiento* [Spanish]; *cruxi* [Mopan]; *ix cruxi* (female) [Mayan]

The leaves are boiled alone or in combination with 8 other medicinal plants and used in baths for **general ailments**, including **nervousness**, **sleeplessness**, **skin ailments**, **swelling**, **bruises**, and **burns** [A137, A156; Arvigo and Balick, 1998]. About 30 different leaves are used for bathing sick people; the Maya always used 9 leaves in each bath, with the choice of leaves based on the problem [B1766].

To treat **pain** in a baby from sleeping on a cold cement floor, the leaves and stems are mashed, soaked in water for 1 hour, heated, and used as a bath [Arvigo and Balick, 1998]. To treat babies with **grief (*pesar*)**, the leaves along with buttonwood leaves (*Piper amalago*) and 9 black beans (*Phaseolus vulgaris*) are

Paullinia tomentosa [MB].

Sapindus saponaria [MB].

boiled in water and used as a bath once daily, every other day, for a total of 3 baths [Arvigo and Balick, 1998]. For treatment of **insanity**, the leaves are boiled and used as a tea and a bath [B2253].

This is considered a **sacred** plant because the stem and leaf petioles form a cross pattern [RA].

Sapindus saponaria L.

mountain cherry, soap berry, soap seed, soap-seed tree, soap tree [English]; *seahone té* [Mopan]; *xab'on che'* [Q'eqchi']; *guiril, jabón che, zubul* [Additional names recorded]

The skin of the fruit is crushed and used as a **soap** [A685]. When ripe, the greenish-brown fruits are mashed, and they exude a slimy substance that feels like dish soap, creating suds in hot water [R16]. The ripe fruits may also be rubbed to get the skins off, producing a foamy residue, but the fruits cannot be made into a bar of soap [HR]. Lundell (1938) noted that the "pulp of the fruits, rubbed in water, gives a soapy lather, which is substituted for soap."

Serjania lundellii Croat

barbasco, santa culebrilla [Spanish]; *hab* [Additional name recorded]

For babies who like to "sleep on cold floors and wake up sick and crabby with **aches** and **pains**," 1 double handful of the entire plant is boiled in 1 gallon of water for 10 minutes and used warm as a bath once daily or as often as needed [A385, B1770]. For babies and adults with frieldad (a **stiffness** in the body), 1 entire plant is used to prepare a bath [A501].

As a **fish poison**, the entire vine is chopped up and added to a dammed stream, and the stunned fish are easily harvested [B2028]. As with *Paullinia fuscescens*, for a 6-foot × 6-foot area of stream, the equivalent of a 100-pound flour sack stuffed with chopped up

Serjania lundellii [MB].

branches of the plant material is used [RA]. Reporting on this practice by the Maya in the area of San Antonio (Toledo District), Thompson (1930) wrote:

> As many as sixteen or twenty men may take part in fish-poisoning. They pass the night in vigil at the hut of any one of the party, partaking of chicken and rice. Candles and copal are lighted, and an offering of cacao is set aside during the course of the night. At dawn all proceed to the river, taking copal, one large candle and four small ones of black beeswax. A section of the river has already been decided upon. The candles are first lit, and subsequently two light logs of wood are placed in the river, one at each bank at the uppermost point of the section to be poisoned. On each of these logs a small fire is kindled, and lumps of copal placed on top of them. The two logs are then floated down stream as soon as the smoke of the copal begins to wreathe upward. An attempt is made to keep each log to its own side of the river. As the logs float down stream the following prayer is recited:
>
> [Free Translation] "O God, Lord Ha, spirit of the water, I am about to molest your very heart, I am going to stir up the mud in your limpid pools because of the fish in your keeping. I do this on account of my great need. See, I pray you, that the crocodile harm me not, bid him seek his lair, that he frighten me not. May the snake not bite me, so that I may work in peace. Permit that I kill all the fish that are here in your keeping."
>
> On the conclusion of this ceremony nets of henequen fiber are stretched across the river at the top and bottom of the reach that is to be worked. The pieces of the poisonous liana wood [*Serjania lundellii*], four in number, are thrown into the water. Soon all the fish in the reach are floating dead upon the surface.
>
> The capture of fish by poisoning is widely practiced in the Amazon area, and possibly was one of the cultural traits introduced from South America via Panama, where it is a common practice. (pp. 90–1)

Serjania mexicana (L.) Willd.
The fruit is eaten as a **food** by birds [B2665].

Talisia oliviformis (Kunth) Radlk.
guaya [Spanish]; *guinep, kinep, uayum* [Additional names recorded]
The fruit is eaten as a **food** by birds and other animals [A549].

Thouinia paucidentata Radkl.
dzol [Additional name recorded]
The stem is used in **construction** for making posts [B2296].

Steggerda (1943) noted that "the Maya believe that the tree contains a charm against **bad or evil winds**." He also discussed boiling its bark for treating **snakebites** and the treatment of **severe cough** by boiling the bark in saltwater and drinking ½ cup at night. Finally, he reported that the leaves of this plant were also said to be used by the Yucatan Maya for healing the "chiclero **ulcer**" (a form of **leishmaniasis**) by applying this sap directly to the wound.

(top) *Serjania mexicana* [MB].

(bottom) *Chrysophyllum mexicanum* [MB].

SAPOTACEAE

Small or large trees with white or rarely yellow latex. Leaves alternate, spiral, or less frequently distichous. Inflorescences axillary, ramiflorous or cauliflorous, in fascicles. Flowers small. Fruits a berry or rarely a drupe. In Belize, consisting of 5 genera and 24 species. Uses for 4 genera and 11 species are reported here.

Chrysophyllum cainito L.
star apple [English]; *caimito* [Spanish]
The fruit is eaten as a **food** [JB99].
This plant is used in an unspecified way to treat a **sore mouth** [W1869].
Standley and Record (1936) reported that the wood of this tree was hard, heavy, strong, fairly durable, and used for heavy **construction**.

Chrysophyllum mexicanum Brandegee ex Standl.
damsel, wild coco plum [English]; *sikiya* [Q'eqchi']; *agya, chike, chi ke, chi que, siciya, sik ey yáh, tzik i ya* [Mayan]

The fruit is eaten as a **food** by humans and birds [A585, B2186, JB103].

To bathe young babies, a leaf decoction is used [B2186]. To treat the "**culebrilla worm**" in babies, 1 handful of leaves along with 1 handful of avocado leaves (*Persea americana*) and 1 handful of Jack in the bush leaves (*Chromolaena odorata*) are boiled in 1 gallon of water for 10 minutes and used warm as a bath once daily, at noon, for 3 days [B2115]. Additionally, garlic and anise seeds (*Pimpinella anisum*) are eaten. This condition is reportedly hard to diagnose [B2115].

Chewing gum is made from the resin and the ripe fruits [A585, B2451].

The trunk is used in **construction** for making house frames, animal trap posts, and fence posts [A585, B1773].

Manilkara chicle (Pittier) Gilly
chicle macho [Spanish]
Standley and Record (1936) reported that there might have been 2 varieties of this plant, one with high-quality white **latex** exported as a form of ***chicle*** under the name of "crown gum" and another that produced poor-quality latex used to adulterate *chicle* sap (*Manilkara zapota*) in the *chicle* industry.

Manilkara zapota (L.) P. Royen
chicle tree, red sapodilla [English]; *chicle, chico zapote, sapadilla, sapote, zapote, zapote colorado, zapote morado, zapotillo* [Spanish]; *muuy* [Q'eqchi']; *ya* [Yucateca]
The mature, red-brown fruit is eaten as a **food** [A484, B3754].

Steggerda (1943) reported that the latex was mixed with salt and used to treat **toothache**.

To treat **diarrhea**, 2 ounces of bark are boiled in 3 cups of water for 10 minutes, strained off quickly, and 1 ounce diluted into 2 ounces of water, with ½ ounce consumed every 2 hours until better for babies; 1 undiluted tablespoon consumed every 2 hours as needed up to a maximum dosage of 3 tablespoons in 1 day for children, and ¼ undiluted cup consumed every 2 hours as needed for adults [A484; RA]. Alternatively, one 2-inch × 4-inch piece of the red inner bark is boiled in 1 cup of water and several cups consumed daily until better [B3754]. To

(top) Axe handle made of *Manilkara zapota* wood [RBGKew].

(bottom) *Manilkara zapota* trunk showing zigzag tapping pattern and white latex [MB].

treat **bloody diarrhea**, refer to *Sagittaria lancifolia* subsp. *media* [B2171]. To treat **dysentery**, the seeds of this plant, along with allspice seeds (*Pimenta dioica*) are pounded, boiled in water, and consumed [Arnason et al., 1980].

To treat **hemorrhaging**, 4 ounces of the decoction described in the second bloody diarrhea treatment above are consumed cool as a tea every 15 minutes for 2 doses. After another 15 minutes, if the bleeding has not stopped, an additional ½ cup is consumed [A484, B3754].

For **high cholesterol**, 9 leaves are boiled in 2 cups of water for 10 minutes and ⅔ cup consumed 3 times daily, before meals, until better. This is said to be very effective [HR].

For **broken bones**, the seeds are pulverized and applied as a poultice daily for 1–2 months until better [B3754].

The sap or resin is used to make **chewing gum**. One tree yields ¼–½ pound of chewing gum, and up to 2 pounds in the rainy season [A484, B3754]. The **latex** from this tree has long been important as a forest product in Belize [MB]. To help the latex to congeal, refer to monkey tail (*Chamaedorea pinnatifrons*) [B3272]. To stamp *chicle* blocks, refer to white China root (*Dioscorea bartlettii*) [LR]. To color chewing gum, refer to *achiote* (*Bixa orellana*) [Thompson, 1927]. To adulterate *chicle*, refer to red fig (*Ficus insipida*) [Lamb, 1946].

Standley and Record (1936), citing Heyder (1930), provided an excellent description for the then-current method for **tapping** these trees:

> The method of tapping Sapodilla differs considerably from methods used in rubber tapping, and is more analogous to the tapping of gutta-percha. There is no continuous flow as in the case of rubber, and the healing of tapping cuts and replacement of latex is extremely slow. After one day's tapping the tree is usually allowed to rest for a period of three years or more, according to the area of bark which has been cut. The method which is used generally in Central America is to make zigzag cuts in the bark, about eighteen inches apart, all the way up the tree, from about two feet above the ground to the first branch. The zigzag pattern of the cuts originates from the fact that it can easily be made with the "machete," which every native carries in the forest of Central America . . . Where the zigzag cuts have been made for more than two-thirds of the way around the stem, or where the cuts have been made too deeply, as frequently happens, the cambium is killed, the bark loosens, and the tree slowly dies. A large percentage of the mature and middle-aged Sapodilla now standing in the forests is in a moribund condition due to these causes.
>
> Tapping is generally done during the early part of the morning between 6 a.m. and 11 a.m. as the air is then still and humid in the forest. The latex coagulates very rapidly on exposure to sun or drying wind, and even without these adverse factors it generally ceases to flow within four to six hours from the time of cutting, so that the chicleros are usually back in their camps soon after midday with the result of their morning's work. Rain does not interfere with tapping as the extra water can easily be evaporated from the latex.

During one morning a chiclero taps perhaps six to eight trees, hunting for these more or less in a big circle around the camp. By the time he has cut his last tree, he is able to return to the first one and remove the bag containing the latex, which will then have ceased to flow. The canvas bags containing the latex are emptied into large tins in the chicleros' camp, and when a sufficient quantity for the purpose has been collected, about 30 gallons or more, the chicle is "cooked," i.e., it is boiled to extract as much of the water content as possible . . . In cooking chicle, a large open cauldron holding about 40 gallons is used, and a small wood fire is placed below it. The chicle bubbles up, giving off a cloud of steam. All through the cooking process, a man stirs the chicle with a paddle, to prevent it from scorching against the sides of the cauldron. When the moisture has been much reduced, and the chicle has become a viscous mass which can hardly be moved with the paddle, it is dumped out of the cauldron on a piece of canvas, previously rubbed with soap to prevent sticking, and there moulded into an oblong or oval block of about 20 pounds' weight. The blocks are set aside to harden for a few days, and then packed into sacks, loaded on mules, and taken to the nearest river bank, whence they are dispatched by boat to the export depot in Belize, the capital town of British Honduras.

By the method of tapping which has been described above, when the cuts have been made on one-half or less than two-thirds of the circumference of the tree, it is generally possible after an interval of about three years to make a second tapping on the remaining area of the stem, provided that the original cuts have healed well and the tree has regained vigor. After a much longer interval, another five years at least, it may be possible to do a re-tapping between the original cuts of the first tapping, but, owing to the occlusion of vessels in the bark around these old wounds, the yield of latex will be much less than from the first two tappings. Under the most favorable conditions the first tapping of a tree at about the middle point of its life may possibly yield 4–5 pounds of latex, a second tapping two pounds, and a re-tapping probably less than two pounds. Such yields are, however, things of the past in British Honduras. Practically speaking, every Sapodilla in the forests, above one foot in diameter, and a great number of smaller trees, have been tapped at least once, most of them twice, and a fair percentage have received re-tappings. This state of things is gradually becoming general in all the more accessible Sapodilla forests of Central America. In some tracts of Guatemala and Mexico there are areas which still yield well, but exhaustion can be visualized at no very distant date. (pp. 41–3)

Thompson (1927) discussed an interesting use of *chicle* gum:

The blowgun used by the Q'eqchi's and Mayas of the Toledo district ranges from four and a half to six and a half feet in length. The wood from which it is constructed is known as *kolmoltše*. A section of wood is chosen and placed in the river under water. It is left there till the soft inner core rots away. A dab of chicle gum serves as a sight. The missile is a small pellet of baked clay, which is placed in the blowgun by means of a piece of hollowed bone. Sometimes hard seeds are employed instead of the clay pellets. The blowgun is useless for shooting anything except birds. (p. 88)

The wood is used in **construction** and to make axe handles, hammer handles, and fence posts. The Maya use the wood in **construction** for beam lintels or cross beams and harvest it during the full moon until 2–3 days after so that it will preserve better [B3754]. Lundell (1938) noted that this wood was well known to the Old Empire Maya, who used it in their **construction**.

Pouteria amygdalina (Standl.) Baehni
silly young [English]
The sap is used for **glue** [A835].

Pouteria belizensis (Standl.) Cronquist
silly young [Creole]
Standley and Record (1936) reported that the **latex** from this species is mixed with *chicle* and the wood used in **construction** for house timbers and all tool handles.

Pouteria campechiana (Kunth) Baehni
mamee ciruela, mamey cerea, mamey cerilla, sapotillo rojo, silillon, zapotillo [Spanish]; *sal tul* [Q'eqchi']; *zac xa nal* [Mayan]; *kanizte* [Additional name recorded]
The ripe fruit is eaten as a **food** by humans and birds [A533]. Lundell (1938) reported that this fruit was flavorless, borne in great abundance, and may have been used in times of food shortages.

To treat **headache**, **sleeplessness**, **exhaustion**, and **skin ailments**, the leaves are mixed with 8 other kinds of leaves and prepared as a bath [B1875].

Pouteria sapota (Jacq.) H.E. Moore & Stearn
mamey, mamey apple, mammy apple [English]; *mamee sapote, sapote, zapote* [Spanish]; *sal tul* [Q'eqchi']; *mame* [Mayan]
The ripe fruit, which is very prized, is eaten as a **food** [B2035, B3753] and tastes like a very

Pouteria campechiana [MB].

Pouteria sapota [MB].

sweet, sweet potato [R84]. The seed kernel is eaten as a **food** and is rich in oil, vitamins, and keratin [B3753].

Steggerda (1943) reported that the seeds were roasted, powdered, mixed with water, and given to treat **vomiting**. To treat **gum infections**, the leaf sap was used by the Yucatan Maya [Flores and Ricalde, 1996]. To treat **dysentery**, the seed is pounded, boiled in water, and consumed [Arnason et al., 1980].

To **stimulate hair growth**, the seed kernel is eaten [B3753]. For **hair loss**, 1 dry seed is well grated, placed in a gauze bag, boiled in 2 quarts of water for 30 minutes, the oil skimmed off the top when cooled, and massaged into the scalp or other area of hair loss 3 times daily [HR].

To make a **soap** or **shampoo**, the flesh of the cohune palm (or the inner "coconut," *Attalea cohune*) is boiled and the oil skimmed off the top, equal parts of this oil and the oil described above for hair loss are mixed and rubbed on the skin or head once daily, in the morning or the evening [B3753].

Standley and Record (1936) reported that the stem was used in **construction** for house frames.

Sideroxylon americanum (Mill.) T.D. Penn.
mol che [Mayan]

The mature, purple fruit is eaten as a **food** [A417].

The wood is used in **construction** to make the bow stem of boats [A417].

(left) *Sideroxylon salicifolium* [RH].

(right) *Capraria biflora* [MB].

Sideroxylon salicifolium (L.) Lam.

silly young [English]; *chachiga, silion* [Spanish]; *faisán, mijico* [Additional names recorded]

The sweet fruit is eaten as a **food** [B1935].

Standley and Record (1936) reported that the tree had abundant **latex** and was commonly tapped by *chicleros*.

Sideroxylon stevensonii (Standl.) T.D. Penn

zapote faisán [Spanish]

Standley and Record (1936) reported that this tree was tapped for its **latex** by *chicleros* and that the product was known as "Chicle Faisán."

SCROPHULARIACEAE

Herbs, subshrubs, sometimes shrubs or vines. Leaves alternate, opposite, or rarely whorled, simple, entire to pinnately lobed. Inflorescences terminal and/or axillary, in thyrses, racemes, spikes, or of solitary flowers. Flowers usually showy. Fruits a capsule, rarely a berry. In Belize, consisting of 16 genera and 30 species. Uses for 3 genera and 3 species are reported here.

Capraria biflora L.

claudiosa, claviosa [Spanish]; *pasmo wa xi uil, tan chi* [Mopan]

This plant is used as a bath and tea for **general ailments** [Arvigo and Balick, 1998].

To treat **fever** and the **chills** from fever, 1 entire plant is heated in the fire and rubbed all over the body [B2304]. For **fever** in an infant, 3 branches with leaves are boiled in 2 quarts of water for 10 minutes and used tepid as a bath once daily for 3 days [B1768].

To treat pasmo (**congested blood; more than the usual amount in an area of the body**), 1 large double handful of branches is boiled in 1 gallon of water or 2 large plants are boiled in 2 gallons of water for 10 minutes, steeped to very warm, poured as a bath over the entire body once daily, at bedtime, and the person is warmly wrapped up for the night [B1768; Arvigo and Balick, 1998].

For **menstrual cramps**, especially in young girls, and for **kidney infections**, 1 handful of fresh leaves is boiled in 3 cups of water for 5 minutes and sipped hot all day long [B1984; Arvigo and Balick, 1998]. To treat **rheumatism, kidney and bladder ailments, cough, tiredness,** or non-insulin-dependent **diabetes** mellitus, the decoction described above for menstrual cramps is sipped all day long [Arvigo and Balick, 1998].

To treat **constipation, urine retention (stoppage of water),** or **kidney stones**, equal parts, about 1 cup each, of fresh or dried leaves from this plant along with little bamboo (*Commelina erecta*), "female" chicken weed (*Euphorbia leucantha*) [B2119], and "male" chicken weed (*Chamaesyce hypericifolia*) are boiled in 1 gallon of water and used as a steam bath. The person sits naked, covered with blankets, on a slotted chair, and the steaming pot is set under the seat, with the steam penetrating the body for 20 minutes once daily, at noon or bedtime [B2116]. For **urinary ailments**, refer to little bamboo (*Commelina erecta*) [B2118]. To treat **kidney stones** or **urine retention (stoppage of water)**, refer to "male" chicken weed (*Chamaesyce hypericifolia*) [B2120]. For **urine retention (stoppage of water)** where there is a blocked urinary flow, refer to bay cedar (*Guazuma ulmifolia*) [B2132].

Russelia sarmentosa Jacq.
la desgracia [Spanish]
The leaves are powdered and used as a **spice** [B2252].

The ground leaves are 1 component of a bath mixture for **general ailments** [B2242].

Scoparia dulcis L.
anise-seed bush [English]; *culantrillo* [Spanish]; *kulantro pim* [Q'eqchi']
To treat **worms** and **stomach infections**, 1 entire plant is boiled in 1 quart of water for 10 minutes and 1 cup consumed once daily,

(left) *Russelia sarmentosa* [MB].

(right) *Scoparia dulcis* [GS].

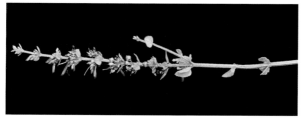

before breakfast, for 3 days [A238]. For **babies who cannot sleep**, 1 teaspoon of the decoction described for worms is consumed and the remainder used warm as a bath once daily at bedtime [A238]. For **fever**, 1 entire plant is mashed in 1 gallon of hot water and poured over the entire body once daily at bedtime [B3637].

SIMAROUBACEAE

Trees or shrubs, rarely herbs. Leaves alternate, compound, rarely simple. Inflorescences axillary or terminal, in panicles, sometimes racemes, cymes, or of solitary flowers. Flowers small. Fruits a drupe. In Belize, consisting of 4 genera and 5 species. Uses for 4 genera and 4 species are reported here.

Alvaradoa amorphoides Liebm. subsp. *amorphoides*

cortacuero [Spanish]

For **curing** and **tanning** cowhide, white lime from a baked and powdered limestone is placed over the hide on a washboard and covered with the mashed or beaten bark [B3680].

Steggerda (1943) reported that the Yucatan Maya used a decoction of this tree bark for **skin diseases**. He also reported the use of leaves for the treatment of **urinary disorders**, leaves in a warm water bath for **rheumatism**, and leaves mixed with honey and corn silk (*Zea mays*) consumed in an unspecified way to treat **hemorrhage**.

Picramnia antidesma Sw. subsp. *fessonia* (DC.) W.W. Thomas

The ripe fruit is eaten as a **food** [A77]. Standley and Record (1936) reported that the leaves and bark are very bitter and have been used as an unspecified **medicine** in Tropical America as well as in Europe.

Quassia amara L.

quassia [English]

Standley and Record (1936) reported that all parts of the plant are bitter and have been used to treat **fever**. "The plant supplies the Quassia or Bitterwood of **commerce**, employed in the manufacture of **insecticides**, as a **substitute for hops** in brewing ale, and in the preparation of proprietary **medicines** and of '**conditioning powders**' for domestic animals."

Alvaradoa amorphoides subsp. *amorphoides* [DA].

(top) *Picramnia antidesma* subsp. *fessonia* showing close-up of flowers [WT].

(right) *Quassia amara* showing flowers and immature fruit and leaves [MB].

Simarouba glauca DC.

dysentery bark [English]; *aceituna, acetuna, negrito* [Spanish]; *pa sak* [Mopan]; *xpazakil* [Mayan]

The black fruit is eaten as a **food** and said to taste like olives [Horwich and Lyon, 1990]. The seeds, which are rich in **oil**, are used for **soap making** or **cooking**, and the "wood burns even when green" [Horwich and Lyon, 1990].

To treat **infected skin sores**, especially if ulcerated, 2 handfuls of ripe fruits are crushed, and the pulp is placed on a clean cloth and then applied as a poultice to the affected area daily until better [A943]. To treat babies with **thrush**, the ripe fruits are crushed and used to wash out the mouth [A943]. For **sores**, the bark or roots are boiled in water and used as a wash over the affected area [Arvigo and Balick, 1998].

Simarouba glauca branch with mature and immature fruit (top) and trunk (bottom) [MB].

As a treatment of **itching** or **fungus**, the juice of 9 blended leaves is mixed with 1 cup of olive oil and applied topically, as a lotion, to the affected area [B2637]. To treat **skin ailments**, one 2-inch × 2-inch piece of bark is boiled in 1 quart of water for 10 minutes and used as a bath [B1812].

For treatment of **dysentery, diarrhea, stomach or bowel ailments, internal infection, internal bleeding, uterine hemorrhage**, and **dysmenorrhea**, one 2-inch × 2-inch piece of bark is boiled in 3 cups of water for 10 minutes, with 1 cup consumed 3 times daily before meals for adults and 1 teaspoon consumed each hour for infants until better [B1812; Arvigo and Balick, 1998]. Alternatively, one 2-inch × 2-inch piece of bark is boiled in 3 cups of water for 10 minutes and used as a bath [Arvigo and Balick, 1998]. In a third reported treatment, 1 small handful of chopped bark and roots is boiled in 3 cups of water for 10 minutes and ½ cup consumed cool every 30 minutes [Arvigo and Balick, 1998].

To treat **diarrhea**, one 2-inch × 2-inch piece of bark is boiled in 1 quart of water for 10 minutes and ⅓ cup consumed 3 times daily [B2637]. Alternatively, the bark may be powdered, soaked in water, and consumed cold [B2637]. For **uterine tumors**, refer to *kibix* (*Bauhinia herrerae*) [B1898, B2220].

The wood is used in **construction** for making house posts, house frames, and broomsticks [B1812; Arvigo and Balick, 1998].

SOLANACEAE

Shrubs, trees, vines, or herbs. Leaves alternate, simple, sometimes pinnately compound. Inflorescences terminal, axillary, or sometimes emergent from stem between nodes, in cymes, panicles, or racemes. Flowers showy. Fruit a berry or capsule. In Belize, consisting of 15 genera and 66 species. Uses for 8 genera and 22 species are reported here.

Capsicum annuum L. var. *glabriusculum* (Dunal) Heiser & Pickersgill
bird pepper, chili pepper, red cayenne pepper, red pepper, wild pepper [English]; *chile colorado* [Spanish]; *ik* [Q'eqchi']
The fruit is eaten as a **spicy addition** to food [B2213; MB]. To treat **anemia** to build up the red blood cell count, refer to pigeon berry (*Phytolacca rivinoides*) [B3670]. To make a **cacao beverage**, refer to cacao (*Theobroma cacao*) [MB, RA].

For **magic**, refer to wild okra (*Parmentiera aculeata*) [RA].

Capsicum chinense Jacq.
habanero pepper [English]
The fruit is eaten as a **spicy addition** to food [B2211; MB].

Capsicum frutescens L.
bird pepper, pepper, wild chili [English]; *chile, chili del monte* [Spanish]; *ik* [Q'eqchi']; *mash àk, mash ík, smash àk* [Mayan]
The fruit is eaten as a **food** and used as a **spice** [B2272, B2391, R62]. For alternative **spice** recipes, refer to amaranth (*Amaranthus dubius*) [Arvigo and Balick, 1998] and mango (*Mangifera indica*) [RA]. Steggerda (1943) observed that the seeds were eaten as a **food** by birds.

It is said that to reduce **parasites** and **cleanse the digestive system**, the fruits are eaten daily with meals [R62]. For babies with an **allergy to cow's milk**, 1 handful of leaves, without fruits, is boiled in 2 quarts of water for 10 minutes, 1 teaspoon consumed, and the remainder used as a sitz bath for 20 minutes once daily for 3 days [HR].

(top) *Capsicum chinense* [MB].

(bottom) *Capsicum frutescens* [IA].

For **magic** and **rituals**, the ripe fruits are roasted, ground into a powder, and sprinkled over objects that have been "enchanted" for an evil purpose. The powder acts as a neutralizer, canceling out the spell placed into the objects [A148]. Thompson (1927) noted that the San Antonio (Toledo District) Maya had a special procedure for putting a spell on an enemy:

> An image of the man that it is desired to harm is made in wax, and a cord smeared in chili and lime is tied tightly round it so as to pass over the heart. Just as the image is constricted by the cord, so the victim will suffer, the chili and lime causing inflammation. The image is often slipped into the house of the victim. (p. 74)

Thompson also noted that clay images were used in the Yucatán to cause ill to an enemy, by rubbing the fruit on the image and burying it in the doorway of the person's dwelling.

According to Thompson (1927), the San Antonio (Toledo District) Maya sowed their crops on or very close to special feast days. He noted that Good Friday was the day on which this plant should be sown.

Steggerda (1943) reported that "it is common knowledge to both doctors and *yerbateros* that the root is highly **poisonous** and it is said to be used sometimes for purposes of deliberate poisoning" among the Yucatan Maya.

Cestrum nocturnum L.

night bloom, nightmare [English]; *dama de noche, sapillo, sauco* [Spanish]; *akul utz, kak yol, ak'ab' kelem* [Q'eqchi'];
julub, ya ax ta [Mayan]; *chacayum* [Additional name recorded]

To treat **skin sores** caused by the dermatitis from black poisonwood (*Metopium brownei*), 1 entire plant is crushed in 2 quarts of water and used cool as a wash over the affected area twice daily [B1970]. To treat **erysipelas**, **sores**, or **swelling**, refer to *coqueta* (*Rivina humilis*) [B2123]. To treat an **abscess** from an infected cut or wound, 1 handful of fresh leaves are mashed, soaked in warm water, and applied as a poultice to the affected area twice daily for 2 days [C22]. For **urinary ailments**, 9 leaves are soaked in 1 quart of water and sipped all day long until better [B1971].

Cestrum nocturnum flowering branch (top) and fruiting branch (bottom) [MB].

To **induce vomiting** to remove poison from the stomach and to **clear the bile**, 9 bitter leaves are crushed in 1 cup of water, strained, and consumed until full. **Caution** is used, as this organ-cleansing decoction should not be taken in excess due to its toxicity [A258].

To treat an infant with a **herniated umbilicus** or **baby colic**, 9 leaves are warmed in oil and placed in a cross formation (one on top of the other, with the top part against the skin) with a piece of nutmeg (*Myristica fragans*) over the navel. This is wrapped up, left on all day, and replaced daily until better [B2097, B2117]. To treat a **baby's cough**, one 16-inch long branch with flowers is soaked in boiling water for 30 minutes, sweetened with sugar, and ¼ cup consumed every 2 hours until better [B1971].

This plant is used in **witchcraft** to make people crazy [A592]. To treat **spiritual ailments**, such as **fright (*susto*)**, **envy (*envidia*)**, **grief (*pesar*)**, and **evil magic (*obeah*)**, refer to bay cedar (*Guazuma ulmifolia*) [RA].

Datura × candida (Pers.) Saff.
chanico [Spanish]
Arnason et al. (1980) reported that the juice obtained from the leaf was put into the ear to treat **earache**.

Datura stramonium L.
chanico [Spanish]
Datura sp. has been the cause of numerous **poisonings** of people experimenting with its **psychoactive** principles, proving fatal in many cases [MB].

To treat **chills**, oil is placed on the leaf and rubbed on the affected area [Arnason et al., 1980].

Lycianthes lenta (Cav.) Bitter
To treat **burns**, the juice from the mature fruits is squeezed onto the affected area [A667].

Datura stramonium showing fruit with inset of flower [MB].

Lycianthes lenta [PA].

Lycopersicon esculentum [MB].

Lycopersicon esculentum Mill. var. *esculentum*

tomato [English]; *tomate* [Spanish]

Thompson (1927) suggested that this plant was known to the ancient Maya and had been cultivated since earliest times as a **food**. Lundell (1938) noted that this species was cultivated as well as grew wild around villages, having escaped from agricultural fields, and had small fruits of "poor quality." To prepare a recipe, refer to prickly pear (*Opuntia cochenillifera*) [BW] and basil (*Ocimum basilicum*) [R70; Arvigo and Balick, 1998].

Nicotiana tabacum L.

tobacco [English]; *tabaco* [Spanish]; *may* [Q'eqchi']; *k'uuȼ* [Yucateca]

Lundell (1938) reported that the plant was very widely cultivated before the Spanish Colonization among many of the peoples of Tropical America.

To treat **swelling**, 1 leaf is warmed in the fire and applied with oil to the affected area [B2103]. For **joint pain**, 1 leaf is warmed, the skin oiled (using any oil), and the leaf wrapped around the joint, tied up with a cloth, and left on until it becomes dry [HR]. To treat **sinus ailments**, 1 leaf is heated, applied to a greased forehead, left on overnight, and the mucus said to flow freely in the morning [HR]. As a **dentifrice** and to treat **gum inflammation**, an exudate from the leaf and stem is used by the Yucatan Maya [Flores and Ricalde, 1996].

To extract **beefworm**, **screwworm**, and **botfly larvae**, 1 leaf is mashed with any type of sticky resin (e.g., dog balls [*Tabernaemontana alba*] or horse balls [*Tabernaemontana arborea*]) and applied to the affected area until the worm stops biting, and after about 4 hours, the worm is squeezed out [A819, B2114, B2173, B2663, B2674, B3132; HR]. Alternatively, refer to horse's balls (*Stemmadenia donnell-smithii*) [B1931].

Steggerda and Korsch (1943) reported that the Yucatan Maya used 2–3 plants cooked in 1 quart of water and consumed to treat **urinary diseases**.

The leaf of tobacco is rolled into a **cigarette** or **cigar** and smoked [MB]. For a tobacco substitute, refer to trumpet tree (*Cecropia peltata*) [A938, B1876, B3527; Arvigo and Balick, 1998] and broomweed (*Sida acuta*) [A895, B2102]. To **give odor and color to tobacco**, refer to *ték ta men* (*Rauvolfia tetraphylla*) [Steggerda, 1943]. To **flavor cigarette tobacco**, refer to contrayerba (*Dorstenia contrajerva*) [Standley and Record, 1936].

Physalis angulata L.

tomatillo [Spanish]

To treat **abscesses** and **boils**, the leaves and roots of 1 entire plant are boiled in 1 quart of water for 15 minutes and applied daily as a plaster to the affected area until better [A543].

Physalis gracilis Miers

cat's balls [English]; *huevo de gato* [Spanish]

The fruit and seeds are eaten as a **food** [A250].

Physalis pubescens L.

miltomate [Spanish]; *top tuč* [Yucateca]

The ripe fruit is eaten as a **food** [A325].

(top) *Nicotiana tabacum* [MB].

(middle) *Physalis angulata* [PA].

(bottom) *Physalis gracilis* [GS].

To treat **swelling**, 1 entire plant is boiled in 1 gallon of water for 15 minutes and used cool as a bath over the affected area [A325].

To treat a **baby's bleeding navel**, the leaf is heated and placed as a poultice over the affected area [Arnason et al., 1980].

Solanum adhaerens Willd. ex Roem. & Schult.

> hark nail, tearing coat [English]; *tomatillo* [Spanish]; *y'ax p'al, k'aham pajl* [Q'eqchi']
>
> To treat **high fever**, 1 quart of the stem and leaves is boiled in 1 gallon of water for 10 minutes and poured tepid as a bath over the entire body [A362].
>
> The fruit is used as a **fish bait** [A575].
>
> **Caution** is used with this plant, as it is not to be taken internally [A362].

Solanum americanum Mill.

> *hierba mora* [Spanish]; *ix cha, maak'uy, ichaj* [Q'eqchi']; *cha yèk, cha yúk* [Mayan]; *bocano* [Additional name recorded]
>
> The mature fruit is eaten as a **food** when prepared in conserves, such as jelly or jam, tarts, or pies [A320, A730, B3671]. The leaves are used to prepare a soup for **invalids** and **weak people** [A320]. The leaves are reported to be rich in iron and may be boiled and eaten as a **food** like a potherb or boiled and eaten fried [A730, B2387].
>
> To treat **lung infection**, 1 entire plant is boiled in 3 cups of water for 10 minutes and 1 cup consumed warm 3 times daily before meals [A320]. To treat adult cases of **herpes zoster**, 1 handful of fresh leaves and fruits is boiled in water for 5 minutes and used warm as a bath over the affected area 3 times daily for 7 days [C49].

Solanum diphyllum L.

> *veneno* [Spanish]
>
> As a treatment for **snakebite**, **spider bite**, or other **venomous bites**, 1 handful of leaves is boiled in 2 quarts of water for 10 minutes and used 3 times daily as a bath for both humans and animals [A255].

(top) *Solanum adhaerens* [GS].

(middle) *Solanum americanum* [MB].

(bottom) *Solanum diphyllum* [GS].

Solanum erianthum D. Don

allay muy, *pito sico* (male), *palo blanco* [Spanish]; *o kutč* [Yucateca]

The leaves are used to prepare a steam bath for **any ailment** [A838, A891].

To relieve **congestion** from a chest cough, 1 cup of the leaves is soaked in 1 quart of water for 1 hour and ½ cup consumed 6 times daily [A838]. To relieve **swelling**, the leaves are warmed and applied warm as a poultice to the affected area [A838]. To relieve **headache**, the leaves are mashed and applied as a poultice to the forehead [A891]. To treat **warts**, the leaves are crushed and applied as a poultice directly to the affected area [Arnason et al., 1980].

The stem is used in schools as a **whip** for children or a **pointing stick** [B1985].

(top) *Solanum jamaicense* [MB].

(bottom) *Solanum mammosum* [MB].

Solanum jamaicense Mill.

susumba [Spanish]

To treat **persistent sores** and **swelling**, the leaves are parched, powdered, and applied to the affected area, causing a burning feeling when used [B2619]. To prepare a less irritating treatment for **sores**, 1 cup of fresh leaves is boiled in 1 quart of water for 5 minutes and used cool as a bath 3 times daily [B2619].

For a **dry cough**, 9 leaves are boiled in 1 quart of water for 10 minutes and 1 cup sipped as hot as possible 3 times daily until better [HR].

Solanum lanceolatum Cav.

The fruits are boiled, cooked a second time with eggs or fried vegetables, and eaten as a **food** [A386].

The leaves are used as a **scouring pad** to wash dishes [A386].

Solanum mammosum L.

ču ču [Yucateca]

To treat **athlete's foot fungus**, the fruit is crushed and applied directly to the feet [Arnason et al., 1980].

Solanum nudum Dunal

diaper wash, Maya washing soap, Maya soap, nightshade [English]; *hoja de puojillo*, *lava paêal*, *lava pañal*, *yerba de barrer* [Spanish]; *saq yol* [Q'eqchi']

For **fever**, 1 quart of leaves is boiled in 1 gallon of water for 10 minutes and poured warm as a bath over the entire body once daily at bedtime [B2720]. To help **foot sores** heal quickly, the leaves and young stems are mashed and applied as a poultice directly to the affected area [B2530].

To rid animals of **lice**, the leaves are placed on the floor of chicken coops [Arnason et al., 1980].

To relieve **joint pain**, 1 cup of leaves is steeped (without boiling) in 1 cup of oil over a low flame for 3 hours and applied as a salve to the affected area as needed. This salve treatment is now obsolete and no longer used by most people; it is a forgotten remedy [A503].

To remove **stains** on baby diapers or other clothing, the leaf juice is squeezed out and applied directly to the stained cloth [A503]. Alternatively, clothes are scrubbed with the leaves, like a soap [A799, W1862].

Caution is used with this plant as it is toxic, like many plants in the Solanaceae family [A46].

Solanum rudepannum Dunal

wild eggplant [English]; *susumba* [Spanish]; *pa al, toom pa'ap* [Mopan]

This plant is 1 of 9 plants combined to make a bath [B1785].

For **skin infections**, **burns**, **boils**, **sores**, and an **abscess**, 1 quart of leaves is boiled in 2 quarts of water for 10 minutes and used tepid as a bath 3 times daily until better [B1785; Arvigo and Balick, 1998]. To promote the healing of **wounds**, the leaves and salt are boiled in water and used as a wash over the affected area [Robinson and Furley, 1983]. Alternatively, the leaves are toasted over a fire, ground into a powder, and sprinkled over the affected area [Robinson and Furley, 1983]. In still another treatment, the leaves are applied as a poultice to the affected area [B1785].

To treat **cough** and **influenza**, 3 small branches with leaves are boiled in 3 cups of water for 5 minutes and 1 cup consumed 3 times daily before meals [Arvigo and Balick, 1998]. To stop the itching of **athlete's foot fungus**, the juice from the crushed fruits is rubbed on the affected area [Arvigo and Balick, 1998]. For **mouth thrush**, 1

Solanum rudepannum [MB].

fruit is boiled in 1 cup of water for 10 minutes and used as a mouthwash [Arvigo and Balick, 1998]. For **arthritis**, refer to fiddle wood (*Vitex gaumeri*) [A936].

To treat **snakebite**, the leaves are boiled in water and used hot as a bath. Additionally, the roots are mashed and applied as a poultice over the affected area [Arvigo and Balick, 1998].

Solanum torvum Sw.

nightshade [English]; *susamba, susumba* [Spanish]; *pach'l, p'al, k'aham pajl* [Q'eqchi']; *tôòm pa'ap* [Yucateca]

The fruit is eaten as a **food**, either green like a vegetable or boiled [A583, B2144].

As an **antiseptic** to treat **external sores** and **wounds**, 1 quart of leaves is boiled in 2 quarts of water for 10 minutes and used tepid as a bath [A802]. Additionally, the dried leaves are powdered and applied to cover the affected area following the bath [A802]. For **wounds**, 1 quart of leaves is mashed in 2 quarts of water and used as a bath over the affected area [B2144]. Additionally, the leaves are dried and powdered, and following the bath, the wound is covered with castor oil (*Ricinus communis*), the powdered leaves, and a clean cloth. This 2-part treatment is repeated later in the day and continued as needed until better [B2144]. To treat **athlete's foot fungus** and **chiggers**, the fruit is crushed and the juice rubbed on the affected area [Arnason et al., 1980]. To treat external **infections** or **wounds**, the leaves, with or without oregano leaves (*Lantana involucrata*), are boiled in water and used as a hot bath to soak the affected area [Mallory, 1991].

To treat **ringworm**, the leaves are roasted, powdered, and applied to the affected area [B3655]. To treat **snakebite**, the leaf is boiled in water and used as a wash over the bite [Arnason et al., 1980].

THEACEAE

Trees or frequently shrubs. Leaves alternate, simple. Inflorescences axillary or of solitary flowers. Flowers white to pinkish. Fruits a capsule. In Belize, consisting of 3 genera and 3 species. A use for 1 genus and 1 species is reported here.

Ternstroemia tepezapote Schltdl. & Cham.

river craboo [English]; *trompito* [Spanish]; *cholol, saq ch'it* [Q'eqchi']

The bark is used for **tanning** cowhide [B2300]. For an average-size cowhide, 4 pounds of bark are soaked and mashed in 10 gallons of water for 15 days with the hide. The hide is then sliced and checked to see if the color has completely penetrated [HR].

ULMACEAE

Shrubs to large trees, rarely lianas. Leaves alternate, usually distichous, rarely opposite, simple. Inflorescences axillary, in cymes or of solitary flowers. Flowers small. Fruits a samara, drupe, or drupelet. In Belize, consisting of 3 genera and 6 species. Uses for 2 genera and 3 species are reported here.

Celtis iguanaea (Jacq.) Sarg.

wild cherry [English]
The fruit is eaten as a **food** by children [B2184].

Steggerda (1943) noted that the Yucatan Maya used the juice of this plant to treat **sore eyes**, but when used in excess, it could aggravate the condition and even cause blindness. He also noted that the leaves were boiled in water and used as a bath to reduce **fever**.

Trema micrantha (L.) Blume var. *floridana* (Britton ex Small) Standl. & Steyerm.

Pine Ridge bay cedar, wild bay cedar [English]; *capulin, mahua blanca* [Spanish]; *ki ik, pixó* [Mayan]
The leaves are used as **fodder** [B1987].

To make sure that the **sugar** comes out well, 1 piece of bark is added to the boiling sugarcane (*Saccharum officinarum*) [B2441].

(top) *Celtis iguanaea* [PA].

(bottom) *Trema micrantha* var. *micrantha* [MB].

To treat **skin rash** caused by having "too much heat in the body, especially in the bladder and kidneys," ½ quart of leaves along with ½ quart of *gumbolimbo* leaves (*Bursera simaruba*) are mashed in 2 quarts of water and used as a bath [A481].

The trunk is used in **construction** to make house posts and provisional huts in the cornfield *milpas* [A769, B1774]. The bark fibers are used as **cordage** for weaving, tying backpacks, and other items [A769, B1987].

Trema micrantha (L.) Blume var. *micrantha*

bastard bay cedar, white capulín, wild bay cedar [English]; *capulin, mahua blanca, pixó* [Spanish]; *chuuk eeg'h* [Mopan]; *klee, q'iib'* [Q'eqchi']; *chapulin, ki ik* [Additional names recorded]
For amoebic **dysentery**, 1 handful of fresh bark and roots is boiled in 1 liter of water for 10 minutes and consumed warm 3 times daily for 5 days [C23].

The fruit is eaten as a **food** by birds [B3202].

The stem is used in **construction** to make house rafters [B3202]. Horwich and Lyon (1990) reported that the fibrous bark was stripped from the tree and used as **cordage** to tie together beams and rafters of houses.

URTICACEAE (INCLUDING CECROPIACEAE)

Herbs, trees, or shrubs, sometimes with stinging hairs, sometimes with dark, sticky sap. Leaves alternate or opposite, simple, sometimes appearing palmately compound. Inflorescences usually axillary, in panicles, glomerules, subumbels, or spikes. Flowers small, green or yellowish. Fruits an achene. In Belize, consisting of 11 genera and 25 species. Uses for 4 genera and 5 species are reported here.

Cecropia obtusifolia Bertol.

trumpet tree [English]; *guarumo* [Spanish]

This plant is a favorite **food** for tapir (*Tapirus bairdii*), monkeys, and deer (*Odocoileus virginianus truei* and *Mazama* spp.) [Marsh, Matola, and Pickard, 1995].

To stop the **bleeding of a baby's navel**, bark shavings are applied directly to the navel for 24–36 hours, until the bleeding stops [B2630]. To treat **burns**, 3 leaves are boiled in 1 gallon of water for 10 minutes, cooled, and poured directly over the affected area twice daily until better [B2630]. To treat non-insulin-dependent **diabetes** mellitus, 1 leaf is boiled in 3 quarts of water, reduced to 1 quart, cooled, and 3 cups consumed cold daily for 3 days. If this does not reduce the blood sugar level, the cold tea is consumed for 3 more days [B2630]. Mallory (1991) reported that 1 entire leaf was boiled in water and consumed hot or cold 3 times daily to treat **high blood pressure**. Alternatively, 3 large pieces of a "healthy leaf" were boiled in water and 1 cup consumed [Mallory, 1991].

Horwich and Lyon (1990) reported that the "'wool' material separated from the leaves and stems is said to be **smoked** by Maya in Mexico."

Cecropia peltata L.

trumpet, trumpet tree [English]; *guarumo, juarumo, warumo* [Spanish]; *cho otz* [Mopan]; *a'kl, pok'jor* [Q'eqchi']

To calm a **cough**, 1 quart of young flowers along with 2 cups of sugar are boiled in 1½ quarts of water for 30 minutes, strained, bottled, and 1 tablespoon consumed every hour as needed and 2 tablespoons consumed at bedtime to sleep better [A938]. To reduce **fever** or **swelling**, the leaves are used to prepare a bath that is washed over the forehead or affected area [Arvigo and Balick, 1998]. For **fever**, the small, young leaves are mashed with oil, wrapped in a cloth, applied as a poultice to the forehead, and removed when better [B1811].

For **high blood pressure**, **swelling**, **sores**, **rash**, **skin infection**, **skin itching**, or **nervousness**, 5 mature leaves are boiled in 1 gallon of water, steeped until cool, and poured over the affected area of skin or over the entire

Cecropia peltata [MB].

body, including the head for blood pressure, once daily until well [B1811; HR]. To treat **high blood pressure**, **edema**, or **kidney disorders**, 1 young leaf is steeped in 3 cups of hot water for 20 minutes and 1 cup consumed warm 3 times daily [B1811, B1876]. For **high blood pressure**, **dropsy**, non-insulin-dependent **diabetes** mellitus, **kidney ailments**, **internal infections**, and use as a **diuretic**, **sedative**, or **mouthwash**, 1 leaf is steeped in 2 cups of hot water for 20 minutes and 1 cup consumed twice daily for 3 days or used twice daily as a mouth gargle [Arvigo and Balick, 1998].

To treat non-insulin-dependent **diabetes** mellitus, 1 young leaf is steeped in 1 quart of hot water for 20 minutes and ½ cup consumed 6 times daily [A938]. Alternatively, three 3-inch young leaves are steeped in 3 cups of hot water for 20 minutes and 1 cup consumed 3 times daily for 2 days [B1811, B2183]. In a third reported treatment, 1 handful of bark from the base of a dead tree is peeled off, steeped in 3 cups of hot water for 20 minutes, and 1 cup consumed 3 times daily until better and then consumed on a regular basis [B2183].

For **liver disorders**, 3 young leaves with stems are boiled in 2 cups of water for 10 minutes and 1 cup consumed cool twice daily, at noon and after supper, for 3 days [B1811]. For **internal infections**, such as **liver ailments**, 1 young leaf is boiled in 1 quart of water for 10 minutes, steeped for 30 minutes, and ½ cup consumed tepid 6 times daily for 3 days [A938]. To tonify the liver when liver ailments cause **bad breath**, the leaves are steeped in boiling water and the tea consumed twice daily for 3–4 days [Williams, 1986]. To treat **diarrhea**, ½ cup of chopped bark is boiled in 3 cups of water for 20 minutes, steeped until cool, and 1 cup consumed 3 times daily, before meals, until better [A938].

To treat **ringworm**, the growing tip of the shoot is mashed with a pinch of salt and applied directly to the affected area [B3282]. Alternatively, the young stems are mashed to extract the juice or "gum," the affected skin scratched to irritate it, the stem juice poured directly over the ringworm, and castor oil (*Ricinus communis*) applied over the top once daily until better [A938]. In a third reported treatment, the leaves are toasted in a dry pan or on a clay *comal* (baking tray) until crispy, powdered through a strainer, and applied directly to affected area once daily until better [A938].

To treat **anxiety**, **nervousness**, or **insomnia**, one 5-inch × 8-inch piece of bark is boiled in 2 quarts of water for 5 minutes, steeped until cool, and 1 cup consumed twice daily, in the morning and in the evening, for 3 days [LR]. For **nervousness**, refer to *jocote* (*Spondias radlkoferi*) [LR]. As a treatment for being **out of breath**, 1 handful of bark is boiled in 1 quart of water for 10 minutes and ½ cup consumed 2–3 times daily [B3683].

To treat **arthritis**, the young leaves are mashed with a pinch of salt, applied as a poultice to completely cover the affected joint, and wrapped well once daily, at bedtime, until better [A938]. Alternatively, 1 young leaf is boiled in 3 cups of water for 10 minutes and 1 cup consumed hot 3 times daily for 3 days [A938]. In a third reported treatment, 3 leaves and a pinch of salt are boiled in 1 gallon of water for 10 minutes and used as a warm bath to soak the sore joints for 15 minutes [A938; RA].

To ease **rheumatism**, 3 leaves are boiled in 1 quart of water for 10 minutes and 1 cup consumed 3 times daily, before meals, for 3 days [B1811]. Alternatively, 2 leaves are boiled in water and used as a wash over the affected area [Arvigo and Balick, 1998]. "To remove **clots** from the vein," 2 leaves are boiled in 1 quart of water for 10 minutes and ½ cup consumed 3 times daily until better [HR].

For **hemorrhages**, ½ cup of chopped bark, 1 leaf, and one 6-inch piece of cinnamon bark (*Cinamomum* spp.) are boiled in 1 quart of water for 20 minutes, steeped until cool, and sipped constantly until the bleeding stops [A938]. For **internal bleeding**, especially **postpartum hemorrhaging**, 1 cup of the chopped inner and outer bark is boiled in 3 cups of water for 15 minutes and sipped cool until the bleeding stops [HR]. To treat **hemorrhaging** in women, 4 ounces of the bark are boiled in 2 cups of water for 10 minutes and ½ cup consumed when cool 3 times daily until the bleeding stops or every 30 minutes, if necessary [HR].

To treat women during difficult **childbirth**, 9 ants running up (not down) this tree are caught, boiled in 1 cup of water for 5 minutes, strained, and the water sipped slowly while still hot [Arvigo and Balick, 1998]. To prevent **infection**, **prolapsed uterus**, or a **swollen belly** after childbirth, the unopened tips of 9 young leaves are boiled in 1 gallon of water for 10 minutes and used as a vaginal steam bath for 3 days. The woman must sit naked over the steam, well covered by blankets, for 20 minutes, and only one 3-day treatment is given unless symptoms call for another [A938]. For **postpartum therapy**, 3 large leaves are boiled in 1 gallon of water for 10 minutes, ½ cup consumed, and the remainder used as a vaginal steam bath for 20 minutes once daily, at bedtime, for 3 days beginning the third day following childbirth [LR]. For **vaginal infections**, 3 leaves, 8–9 inches in diameter, are boiled in 1 gallon of water for 15–20 minutes and used once daily as a douche for 3 days following menstruation [A938].

As a **tobacco substitute**, the leaves are dried, rolled or powdered, and smoked like a cigar. This is commonly practiced by *chicleros* and bushmasters but, when used regularly, may cause persistent coughing [B1876, B3527; Arvigo and Balick, 1998]. Alternatively, the stems and vines are removed from dry leaves that have fallen, the leaves rubbed between the hands to crush well, and then rolled or put into a pipe. **Caution** should be used, as long-term use may be harmful to the lungs [A938].

To make **rafts** that float well and last a long time, the smaller trunks are tied across the larger trunks and used on the river [LR]. Standley and Record (1936) reported that "the name Trumpet sometimes given to the trees alludes to a tradition that the stems were employed for making **trumpets** by the aborigines of tropical America."

Coussapoa oligocephala Donn. Sm.

fig tree [English]; *fruta palo*, *higuera* [Spanish]

To extract **beefworm larvae**, the latex is applied directly to the skin lesions for 1 hour [A833].

(top) *Laportea aestuans* [RH].

(bottom) *Urera baccifera* [MB].

Laportea aestuans (L.) Chew
  ortiga [Spanish]; *làal* [Yucateca]
  To treat **headache**, 1 mature leaf with stinging hairs is placed on the forehead [Arnason et al., 1980].

Urera baccifera (L.) Gaudich. ex Wedd.
  cow-itch, cowitch [English]; *ortiga* [Spanish]; *lah* [Q'eqchi']
  To treat **rheumatism** and **arthritis pain**, the leaves are heated, mashed, and then rubbed over the affected area. It is said that this "will sting at first but then takes out

the pain" [B3215]. For **joint pain**, the leaves are placed in a paper bag, toasted inside an oven, and a melted "soft candle" and 1 teaspoon of kerosene added to 2 ounces of the powdered leaves, which are mixed while still in the bag and then rubbed as a paste on the affected area. A paper bag is used so that no fumes or powder escape [HR]. **Caution** is used when preparing this treatment so as not to get it in the eyes. Toasting and powdering the leaves may cause irritation to orifices, and the eyes, nostrils and mouth *must* be covered during the process. This preparation is best made outdoors so that the wind can blow away fumes and irritating plant parts [MB, RA].

According to Steggerda and Korsch (1943), the juice of this plant was mixed with salt and rubbed over the affected area to treat **pain**. They also reported that the young sour orange leaves (*Citrus aurantium*) were mixed with the hot juice of this plant and consumed to treat **stomach pain**. Additionally, they noted that when the disease was "of cold origin," the "bark" of the roots were mixed with anise liquor (*Pimpinella anisum*) and used to treat **sour stomach**.

VERBENACEAE

Herbs, shrubs, trees, or lianas. Leaves opposite, sometimes whorled or subopposite, simple or sometimes palmately compound. Inflorescences axillary and/or terminal, in racemes or cymes, usually in heads, spikes, umbels, thyrses, or occasionally of solitary flowers. Flowers large and showy to very small, variously colored. Fruits a drupe or schizocarp. In Belize, consisting of 18 genera and 44 species. Uses for 10 genera and 19 species are reported here.

Aegiphila monstrosa Moldenke
stinkin' bush, white julub [English]; *carreta, julum, palo de carreta, tabatillo, hulub* [Spanish]; *julub* [Mayan]
The leaves are boiled and used in baths [A816].

For relief of **pain**, fresh leaves are spread to entirely cover a bed, and a person sleeps naked all night on top of the leaves with a blanket over them [B2126].

For **headache**, "Vicks" (Vicks VapoRub® topical ointment) is rubbed on the forehead, followed by an application of 1–2 heated leaves for 20 minutes [A485, B2126].

Aegiphila monstrosa [GS].

Alternatively, as a type of acupuncture, the spines are used to prick the skin on the head [B2126].

For **rheumatism**, 5 leaves are boiled in 1 quart of water for 10 minutes and 1 cup consumed 3 times daily for 5 days. Treatment is stopped for 7 days and resumed again if needed, with 1 cup consumed 3 times daily [A403]. Additionally, the leaves are heated, mashed, and applied directly over the affected area [A403].

The wood is used in **construction** for building cart wheels [A816].

Callicarpa acuminata Kunth

John Crow bead berries, John Crowfoot [English]; *yok choom, yok osh om* [Mayan]

The leaves are used in baths [B2150, B2454].

For **stomach or abdomen swelling**, 1 handful of leaves are mashed, heated in oil, and applied as a poultice to the affected area [B1936]. Steggerda (1943) reported that the Yucatan Maya crushed the leaves, mixed them with water, and consumed the decoction to treat **dysentery**.

For **menstrual pain**, 1 handful of leaves and vines is boiled in 5 gallons of water for 10 minutes and used as a vaginal steam bath for 20 minutes [B1826]. For **women who have just given birth**, 1 handful of leaves and vines is boiled in 5 gallons of water for 10 minutes and used as a substitute bath if *ix chal che* leaves (*Pluchea odorata*) are unavailable. The bath is by immersion or by pouring over the body; other plants may or may not be included in with the mixture [A660]. To **stimulate lactation** and to help **continue the flow of milk** when feeding young children, the leaves are boiled in water and an unspecified amount consumed [B2443]. This plant should not be taken during pregnancy, as it may have **abortifacient effects**.

The seeds are said to have **psychoactive properties**, but their preparation is not known [RA].

Citharexylum caudatum L.

bird-seed, pigeon berry, pigeon plum, pigeon-feed [English]

The fruit is eaten as a **food** by birds [A663].

To treat **rash** and other **skin ailments**, 1 quart of fresh leaves is boiled in 1 gallon of water for 10 minutes and used cool as a bath twice daily [A349]. For **stubborn**

Callicarpa acuminata [GS].   *Citharexylum caudatum* [GS].

sores on children, especially on the arms, legs, and interior of the mouth, 1 handful of warmed leaves is squeezed and the juice applied directly to the affected area [A663].

Cornutia grandifolia (Schltdl. & Cham.) Schauer
ix chel cher [Q'eqchi']

To treat **cough**, 9 leaves are boiled in 1 quart of water down to ½ quart of water, ½ cup of sugar added, and consumed by the spoonful as a cough syrup as needed [A792].

The leaves are used for **scrubbing** tables to get them clean [A792].

Cornutia pyramidata L.

stinkin' bush [English]; *baston de vieja*, *matasano* [Spanish]; *tzultesnuk* [Additional name recorded]

To relieve **asthma** or **shortness of breath**, the leaves are roasted, squeezed, and 1 tablespoon of leaf juice added to 1 teaspoon of sweet almond oil or olive oil and consumed every 2 hours [A656]. To treat **colds**, 9 leaves are boiled in 1 quart of water for 10 minutes and 1 cup consumed tepid 3 times daily before meals [A656].

To treat **skin rash**, one 12-inch branch of leaves and flowers along with one 12-inch branch of polly red head leaves and flowers (*Hamelia patens* var. *patens*) and one 12-inch branch of wire wis leaves and flowers (*Lygodium venustum*) are boiled in 1 gallon of water for 10 minutes, cooled, and used as a wash over the affected area [B2104]. For **sores** and **rash**, the leaves are rubbed over the affected area [B2104].

Lantana camara L.

sage, wild sage, wood sage [English]; *cinco negritos*, *lantana*, *palabra de cabellero* [Spanish]; *we àch*, *we éch*, *we ích* [Mayan]; *ikilhaxin*, *petekin* [Additional names recorded]

This plant is considered an **ornamental** and has medicinal uses [B3210].

(left) *Cornutia pyramidata* [NYBG].

(right) *Lantana camara* [MB].

Lantana involucrata [MB].

Lantana trifolia [GS].

It is said that this plant is used alone in a bath and not in a mixture [B1887].

For **cough**, the leaves are used to prepare a tea [B2152]. To treat **whooping cough**, three 6-inch branches of stems and leaves are boiled in 3 cups of water for 10 minutes and 1 cup consumed at any temperature 3 times daily before meals [A340]. To treat **itchy skin**, 1 cup of fresh leaves is boiled in 2 quarts of water for 10 minutes, used as a bath or a wash over the affected area, and left on the skin to dry; this treatment is repeated until better. The flowers are not used in this treatment [A945, B1887; Arvigo and Balick, 1998]. Additionally, the leaves are powdered and applied directly to the affected area following the bath [B1887; Arvigo and Balick, 1998].

Steggerda and Korsch (1943) reported that a leaf was roasted and placed over a **wound** to promote healing. To treat **external infections** or **wounds**, refer to nightshade (*Solanum torvum*) [Mallory, 1991].

To treat **venereal diseases** and to **purify the blood**, the plant is soaked in water as part of an unspecified treatment [Robinson and Furley, 1983].

Caution is advised with this plant. "While there are numerous traditional uses that involve ingestion of the leaf, the TRAMIL 4 workshop states that this plant is toxic and should not be taken internally" [Robineau, 1991].

Lantana involucrata L.

oregano, sage, sea sage, spice [English]; *oregano del monte, oregano silvestre, simaron* [Spanish]

The fruits are considered sweet and eaten as a **food**; the leaves are used fresh as a condiment or dried in soups [A407, B1953].

(left) *Lippia alba* [MB].

(right) *Lippia dulcis* [MB].

The stems are used in making the sweeping end for **brooms** and **brushes** [A407, B1953].

To treat **back pain**, said to possibly be **kidney ailments**, the leaves and stems are boiled in water and 1 cup consumed 3 times daily with meals [Mallory, 1991].

Lantana trifolia L.

wild sage [English]; *tu lush* [Q'eqchi']

For **calming nerves**, one 12-inch branch with leaves along with 9 lime leaves (*Citrus aurantiifolia*) and one 12-inch branch of Jack in the bush (*Chromolaena odorata*) are boiled in 1 quart of water for 10 minutes and ½ cup consumed 6 times daily until better [B2096].

Lippia alba (Mill.) N.E. Br. ex Briton & P. Wilson

sea sage [English]

For **stomach pain**, **headache**, or **indigestion**, one 12-inch branch with leaves is boiled in 3 cups of water for 10 minutes and 1 cup consumed tepid twice daily until better [A278].

Lippia dulcis Trevir.

licorice plant [English]; *orasès, orasús* [Spanish]

The leaves are used as a **natural sweetener** in foods and teas [Arvigo and Balick, 1998]. For **toothache**, the flowers are chewed or applied directly to the affected area [Arvigo and Balick, 1998].

For **cough**, 1 entire plant is boiled in 1 quart of water until the mixture thickens and then sipped hot until the mucus comes out. This tastes very sweet, sickeningly so [B1906]. To treat **bronchitis** and **dry, hacking coughs**, 1 handful of fresh plant material along with 1 cup of sugar are boiled in 1 quart of water for 10 minutes, then placed in front of the person's chest with a towel wrapped around to allow the steam to penetrate the chest area. After the steam has stopped, the decoction is strained and sipped hot all day until the mucus comes out [Arvigo and Balick, 1998].

To treat an **asthma attack**, 1 cup of fresh leaves and stems is boiled in 3 cups of water for 10 minutes, steeped for 15 minutes, strained, and slowly sipped hot until better [A807]. To avoid a second episode of **asthma** once an attack has occurred, the same decoction is slowly sipped hot for 3 days [A807].

Lippia graveolens Kunth

oregano, Spanish thyme [English]; *oregano Castillo* [Spanish]

The leaves are eaten as a **food**, **herb**, and **spice** [B2094, B2311; Arvigo and Balick, 1998]. To impart **flavor** and to aid in **digestion**, fresh or dried leaves are added to soups, stews, and sauces [Arvigo and Balick, 1998].

To **maintain healthy skin**, the flowers are eaten [Arvigo and Balick, 1998]. To treat a **baby's cough** or to **put a baby to sleep** when continually coughing, 3 leaves are warmed over a flame, the juice squeezed out, a bit of sugar added, and consumed only once [BW]. To reduce **infection** from a tooth abscess, refer to *golondrina* (*Alternanthera flavogrisea*) [LR]. For **earache**, 1 leaf is heated and 3 drops of leaf juice squeezed into the ear 2–3 times daily [B2389].

For **sprains** and bad **bruises**, 1 cup of fresh or dried leaves is boiled in 1 quart of water for 10 minutes and used cool as a bath over the affected area [Arvigo and Balick, 1998]. To wash **wounds**, **infections**, and **burns**, especially if septic, 1 quart of fresh leaves or 1 handful of dried leaves is boiled in 1 gallon of water for 10 minutes, used cool 3 times daily, and left to air dry [Arvigo and Balick, 1998]. To prevent infection of **wounds**, **cuts**, and **burns**, three 12-inch branches are boiled in 1 gallon of water for 10 minutes and used cool as a wash twice daily until better [A109]. To treat **erysipelas**, refer to *coqueta* (*Rivina humilis*) [B2092, B2123].

For **gas**, **colic**, and **scanty or lacking menstruation**, one 12-inch branch with leaves is boiled in 3½ cups of water for 10 minutes and 1 cup consumed hot 3 times daily, before meals, until better [A109]. To **prevent colic**

Lippia graveolens [MB].

in a newborn following birth, 1 tablespoon of this decoction is consumed [B2382]. To ease the **pain** during childbirth, 1 cup of this same decoction is consumed warm [B2382]. To expel **mucus in the womb** after childbirth, 1 cup of this decoction is consumed and the remainder used as a vaginal steam bath [W1843]. To **cleanse the uterus** following childbirth, 1 cup of this same decoction is consumed warm once daily for 7 days [B2382; Arvigo and Balick,1998]. To expel a **retained placenta** following childbirth, 1 handful of fresh leaves along with 3 teaspoons of soot from above the fire hearth and 1 garlic clove are boiled together in 1 quart of water for 10 minutes and sipped; an abdominal massage may follow to help pass the placenta quickly [Arvigo and Balick, 1998]. To treat a **womb infection** following childbirth, refer to *guanábana* (*Annona muricata*) [LR]. For **uterine ailments**, refer to wild basil (*Ocimum campechianum*) [A941].

To treat **upper respiratory tract infections** and to **induce menstruation** or **increase a scanty menstrual flow**, ½ cup of fresh leaves or 3 tablespoons of dried leaves are steeped in 3 cups of boiling water for 15 minutes, strained, and 1 cup consumed hot 3 times daily, before meals, as needed [Arvigo and Balick, 1998]. To **induce labor**, refer to *achiote* (*Bixa orellana*) [B3573].

Lippia myriocephala Schltdl. & Cham.
baston de vieja (male), *maja rhacco* [Spanish]; *sa ku ch*, *sa ku ché* [Mayan]
For **cough**, 9 leaves are boiled in 1 cup of water for 10 minutes and consumed with sugar 3 times daily [A750].

"In the old days," the leaves were powdered and rolled for **smoking** [B3690].

The wood is used in **construction** for making house posts [B3690].

Lippia strigulosa M. Martens & Galeotti
licorice sweet [English]
For **cough**, 1 entire plant is boiled in 1 quart of water for 10 minutes and sipped warm all day until better [B1898].

Petrea volubilis L.
handkerchief of Sarah, licorice sweet [English]; *girnalda, lava platos, pañuelo de Sarah* [Spanish]
This plant is used in bath mixtures [B3239].

For treatment of a **dry cough**, 9 leaves are boiled in 3 cups of water for 10 minutes and sipped all day long [A336].

Petrea volubilis [IA].

Priva lappulacea (L.) Pers.
cancer herb, sticky sticky burr burr [English]; *hierba del cancer*, *masote*, *mosote*, *pega ropa* [Spanish]; *torâ pim* [Q'eqchi']

For **itchy skin** or **rash**, the leaves are mashed and the juices rubbed over the affected area [Arvigo and Balick, 1998]. To treat **sores**, **ulcers**, **wounds**, and **fungus**, the leaves are dried, powdered, and applied directly to the affected area [A132; Arvigo and Balick, 1998]. For **wounds** or **skin sores**, 1 entire plant is boiled in 1 gallon of water and used very hot as a bath over the affected area [A839]. To treat **bloody dysentery**, 1 double handful of leaves is boiled in 1 gallon of water and used as a steam bath [A132].

To treat **external "cancer" (open sores)**, 1 double handful of leaves is boiled in 1 gallon of

Priva lappulacea [MB].

water and used very hot as a bath once daily [A40]. For **internal "cancer" (growth)**, the branches and stems are boiled in water and consumed very hot as a tea [A40].

For relief from **intestinal parasites** and **stomach ailments**, 1 medium plant with roots is boiled in 1 quart of water for 10 minutes and 1 cup consumed 3 times daily, before meals, for 3 days [A40, A132]. To treat **internal parasites**, 1 handful of leaves is boiled in 3 cups of water for 10 minutes and 1 cup consumed 3 times daily for 3 days, followed by a purge [Arvigo and Balick, 1998].

To relieve **pain**, 3 roots from 3 entire plants are boiled in 1 quart of water for 10 minutes and 1 cup consumed 3 times daily until better [B2384]. To ease **toothache**, especially when the person does not have many teeth to pull, 1 entire plant with roots is boiled in 1 quart of water for 10 minutes and used cool every 30 minutes as a mouth gargle [B2289].

For **cough**, 1 handful of leaves along with 1 *ob'el* leaf (*Piper auritum*) are boiled in 1 quart of water for 10 minutes and 1 cup consumed lukewarm 3 times daily for 2 days [A839].

As a **female contraceptive**, refer to strong back (*Desmodium adscendens*) [B3643, B3644, B3645]. Alternatively, the leaves of 1 entire plant are mashed in 1 quart of water, soaked in the sun for 2 hours, and 3 cups consumed daily, beginning at the onset of menstruation and continuing through its completion [HR].

Stachytarpheta cayennensis (Rich.) Vahl

blue vervain, vervain, vervine, wild verbena [English]; *verbena* (male) [Spanish]; *cot a cam, kaba yax nik* [Mopan]; *xtyay ach bak xu'l* [Q'eqchi']; *camacolal* [Additional name recorded]

To treat **heart ailments**, **nervousness**, **neuralgia**, **overexposure to the sun**, **fever**, **influenza**, **cough**, **colds**, **intestinal parasites**, **stomach ailments**, **liver ailments**, **tumors**, and **postpartum cleansing**, 1 entire plant is boiled in 1 quart of water for 10 minutes and 1 cup consumed 3 times daily, before meals, as needed [B1854, B2627; Arvigo and Balick, 1998].

Stachytarpheta cayennensis [DA].

For **postpartum cleansing**, "to **prevent infection** and to **ensure that the womb returns to its proper position**," 1 entire plant is boiled in 2 quarts of water for 10 minutes and used as a vaginal steam bath for 10 minutes [B1854; Arvigo and Balick, 1998]. For **gonorrhea** and to **cleanse the uterus** after childbirth, a leaf decoction is used as a vaginal douche [Arvigo and Balick, 1998].

As a **laxative** or to **cleanse an upset stomach**, 1 branch is blended in 1 cup of water and 1 tablespoon of the fresh juice consumed as needed. This will either "bring the discomfort up or down," bringing relief [B2627]. For treatment of **intestinal parasites** in adults, ¼ cup of fresh leaf juice is consumed daily for 7 days, following which the stool is retested [Arvigo and Balick, 1998]. To treat **stomach pain** associated with fright (*espanto* or *susto*), one 12-inch piece of the plant is boiled in 3 cups of water for 10 minutes and 1 cup consumed twice daily [B2427]. Standley and Record (1936) reported that a decoction of the plant was used to treat **dysentery**.

For **fever**, 1 double handful of leaves is boiled in a gallon of water for 10 minutes, 1 cup consumed warm 3 times daily, and the remainder used as a bath; both treatments should be continued as needed until better [B3578]. To ease **body pain** that develops from **influenza**, three 12-inch branches with leaves along with three 12-inch branches with leaves of buttonwood (*Piper amalago*) are boiled in 2 gallons of water for 10 minutes and used warm as a bath once daily for 20 minutes [B2290]. Steggerda and Korsch (1943) reported that the Yucatan Maya warmed a handful of the leaves over a fire and placed them warm on the affected area to treat **pain**.

To treat **stubborn sores**, 1 entire plant is boiled in 1 gallon of water for 10 minutes and used twice daily as a warm wash over the affected area [B2627]. For **boils** or **infected sores**, the leaves are mashed and applied as a poultice directly to the affected area [Arvigo and Balick, 1998].

For **headache**, one 12-inch piece of the plant is boiled in 3 cups of water for 10 minutes and used to bathe the head [B2427]. To treat **wind** (viento), this plant is combined with other plants to prepare a bath [B2339].

For two **snakebite** treatments, refer to *contra hierba* (*Blechum pyramidatum*) [A495; LR].

This plant is considered **sacred** by the Maya and used in ritual **ceremonies**. To **repel evil magic**, bunches of the stems are hung in doorways and dried leaves burned as incense [Arvigo and Balick, 1998].

Stachytarpheta jamaicensis (L.) Vahl.

red vervain, vervain [English]; *verbena* [Spanish]

The leaves are used to prepare a bath used for **any ailment** [A896].

For **fever**, **liver ailments**, **colds**, or **uterine ailments**, two 12-inch branches with leaves are boiled in 1 quart of water for 10 minutes, steeped for 1 hour, and 1 cup consumed 3 times daily, before meals, for 3 days [A896]. To treat **gonorrhea** in females, 1 double handful of leaves is boiled in 1 gallon of water for 10 minutes and used as a vaginal steam bath [B1962]. Alternatively, 1 entire plant is boiled in 1 quart of water for 10 minutes and 1 cup consumed 3 times daily [B1962]. To treat **nervousness** and to cool from **overexposure to the sun**, refer to yellow malva (*Malvastrum corchorifolium*) [LR].

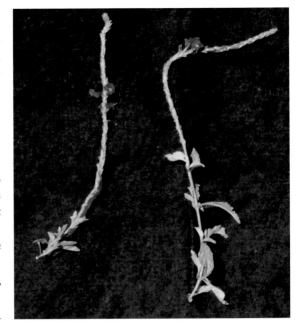

Stachytarpheta miniacea

Moldenke

red vervain [English]

To treat **measles**, **itching**, **sores**, and to **prevent scarring from measles**, 3 roots from 3 different red vervain plants along with 1 large handful of guava (*Psidium guajava*) leaves are boiled in 1 gallon of water for 10 minutes,

Stachytarpheta miniacea [GS].

poured cool over the body, and left to air dry [A378].

Vitex gaumeri Greenm.
blue blossom, blue flowers, dogwood, fiddle wood, second yax nik, walking lady [English]; *flor azul*, *matasano*, *yak nik segunda* [Spanish]; *sak u sol*, *yash nik* [Mopan]
The leaves are used as **fodder**, especially for cows [B3254].

The leaves are used in baths and 9-plant bath mixtures [A110, B2196]. To treat **asthma**, **malaria**, and **chills**, the leaves are boiled in water and used as a bath [Arvigo and Balick, 1998].

For **arthritis**, 9 leaves along with 9 leaves from other plants (e.g., wild eggplant [*Solanum rudepannum*]), are mashed in ¼ cup of oil and applied directly as a poultice to the affected area.

Vitex gaumeri [MB].

Other, unspecified plants can also be used [A936, HR]. For **toothache**, the leaf is bitten, numbing the tongue because of its analgesic properties [A936].

To treat **skin fungus**, **rash**, and **itching** resulting from nervousness, 1 pound of leaves is steeped in 4 quarts of water for 15–20 minutes and used as a wash twice daily [A936]. For **skin fungus**, **infected sores**, and **ringworm**, the bark is toasted, powdered, and applied with a bit of oil to the affected area [Arvigo and Balick, 1998]. To treat **skin sores** and **skin infections**, the fresh bark is pounded into a pulp and applied as a poultice directly to the affected area [A926]. For **skin sores** and **wounds**, the leaves are mashed and applied as a poultice directly to the affected area [Arvigo and Balick, 1998]. To treat **wounds** and **pustules**, ½ pound of bark is boiled in 1½ quarts of water for 5–10 minutes and used as a wash over the affected area [A926; Arvigo and Balick, 1998]. To treat **infected pustules**, 1 pound of bark is boiled in 1½ quarts of water and used as a wash [A926]. Additionally, the bark is grated and applied as a poultice to the affected area following the wash [A926].

To treat **leishmaniasis (baysore)**, 1 piece of bark, equivalent in size to the area needed to cover the sore, is dried, powdered, mashed, mixed with oils, and applied as a poultice directly to the affected area [B1813, B2072].

For **biliousness**, one 1-inch × 6-inch piece of inner bark is boiled in 3 cups of water for 10 minutes and 1½ cups consumed warm once daily for 3 days only [A110;

Arvigo and Balick, 1998]. For **biliousness** and **jaundice**, 1 handful of bark is boiled in 3 cups of water for 10 minutes and 1 cup consumed warm 3 times daily until better [B1877, B1879].

The wood is used for **firewood**, for **carpentry** in making musical instruments (e.g., drums, flutes, rattles, and inexpensive marimbas), and for **construction** in making house posts and oxen yokes [A110, B1813, B2196, B2351, B3254; Arvigo and Balick, 1998]. According to Lamb (1946):

> Recently the wood has been very successfully used for the **heads of polo-sticks** to replace imported bamboo heads. Locally made heads are no heavier than the bamboo head and have been found to have a much longer life. They have little tendency to split and are very difficult to dent. (p. 111)

VIOLACEAE

Trees, shrubs, lianas, or herbs. Leaves alternate or opposite, rarely distichous, rarely simple. Inflorescences axillary or cauliflorous, terminal, in spikes, racemes, panicles, sometimes in cymes, or of solitary flowers. Flowers white, yellow, green, or purple. Fruits a capsule or berry. In Belize, consisting of 5 genera and 12 species. Uses for 1 genus and 2 species are reported here.

Rinorea guatemalensis (S. Watson) Bartlett
  wild coffee [English]; *cafecillo* [Spanish]
  To relieve **headache**, the leaves are mashed in 2 quarts of water at room temperature, soaked for 30 minutes, and used to bathe the forehead [B3735].

Rinorea hummelii Sprague
  coffee bush, green berries, wild coffee, wild plum [English]; *guayabillo* [Spanish]; *kwol kwol ch* [Mayan]
  The fruit is eaten as a **food** by gibnuts (*Agouti paca*), armadillos (*Dasypus novemcinctus*), and other animals [A472, A552].

  To treat **joint pain** or **rheumatism**, 1 large double handful of leaves and twigs is boiled in 1 gallon of water for 10 minutes and used warm to soak the af-

Rinorea hummelii [GS].

fected area for 30 minutes once daily until better [A329].

VITACEAE

Lianas or rarely shrubs, usually with tendrils. Leaves alternate, simple or compound. Inflorescences opposite the leaves or terminal, rarely axillary, in cymes or racemes. Flowers inconspicuous and greenish or showy and red. Fruits a berry. In Belize, consisting of 2 genera and 7 species. Uses for 2 genera and 5 species are reported here.

Cissus cacuminis Standl.
ta kan [Q'eqchi']

To treat **leishmaniasis (baysore)**, **rash**, **ringworm**, and other **fungi**, 1 small handful of leaves and stems is mashed in 1 cup of water to make a strong juice applied as a poultice directly to the affected area and changed twice daily [A597; RA].

(top) *Cissus cacuminis* [GS].

(bottom) *Cissus gossypiifolia* [MB].

Cissus gossypiifolia Standl.
bee rut, rooster's crest [English]; *bejuco parilla*, *cresta de gallo* [Spanish]
To treat babies with **thrush**, one 12-inch branch with leaves is mashed in 2 cups of water, left to ferment into a mucous state, and rubbed in the mouth 4 times daily, including once at bedtime, and repeated until better [A547]. To treat adults with **thrush**, this same decoction is used as a mouthwash [A547].

To aid the healing of **skin sores**, especially **leishmaniasis (baysore)**, the sap from the mature vine is extracted and applied directly to the affected area daily until better [B2031].

Cissus microcarpa Vahl
bejuco de agua [Spanish]; *loc hop*, *roc háp'* [Q'eqchi']
To treat **leishmaniasis (baysore)** and **severe rash**, two 12-inch branches of leaves and stems are mashed in 1 quart of water and freshly applied twice daily to the affected area [A562]. To treat **skin swelling** from the sap of the black poisonwood tree (*Metopium brownei*), 1 handful of leaves are mashed in 3 cups of cold water used to wash the affected area [B3584]. To treat **measles**, 1 handful of the leaves is mashed

(left) *Cissus microcarpa* [MB].

(right) *Cissus verticillata* [GS].

in cold water and used as a bath over the affected area 3 times daily for 7 days [C25]. According to Steggerda (1943), the crushed bark, when mixed with water, was used to wash **sores** and **wounds**, following which the dried, crushed bark was used to cover the affected area.

For **poor blood circulation**, one 12-inch piece of vine is boiled in 1 quart of water or soaked in 1 quart of cold water and 2 cups consumed hot daily for 3 days [B2638]. Alternatively, this decoction is used warm as a bath [B2638].

Cissus verticillata (L.) Nicolson & C.E. Jarvis

coronilla, prenada segunda [Spanish]; *ta kan* [Additional name recorded]

To treat **pimples** and **black heads** that have been scratched, the raw fruits are rubbed into the affected area [Mallory, 1991]. For **wounds**, the vine is soaked in water and used as a wash over the affected area [Mallory, 1991].

To treat **infant bleeding**, **hernia**, or **infections of the umbilicus**, 2–3 leaves are roasted on the fire, made into a powder, mixed with castor oil (*Ricinus communis*), and applied as a plaster twice daily on the navel [A213]. For **joint pain**, 2–3 entire plants are boiled in 1 gallon of water for 10 minutes and used warm as a bath twice daily [A404]. For **stiff knee joints**, especially after a stroke, the leaves are mashed with castor oil (*Ricinus communis*) and applied as a poultice directly to the affected area [A404].

Standley and Record (1936) reported that the tough, flexible stems were used as a **rope substitute** and that the stem sap caused **blisters** on the skin.

Vitis tiliifolia Humb. & Bonpl. ex Roem. & Schult.

water vine, water tie tie, water-wise, wild grape, wild grape vine [English]; *bejuco de uva, bejuco de agua* [Spanish]; *su sú* [Q'eqchi']; *ha ix ak* [Mayan]

(top left) *Vitis tiliifolia* in bud [GS].

(top right) *Vitis tiliifolia* with ripe fruits and showing cut stem with sap consumed as a beverage [FA].

(bottom) *Vochysia hondurensis* [MB].

The fruits are sour but eaten as a **food** [A394, A565, B2438]. The sap in the vine is consumed like **water** [A565, B2438].

As an **antiseptic** to wash the eyes and to bathe an **inflamed umbilicus** of newborns, about ½ cup of the sap from the vine is applied directly to the affected area [A245, A394]. To treat **diarrhea**, one 12-inch piece of vine is boiled in 1 quart of water for 10 minutes and 3 cups consumed daily [A565]. To stop **bleeding** of cuts, 6–10 fresh leaves are mashed and applied as a poultice directly to the affected area [A565].

VOCHYSIACEAE

Medium to large trees. Leaves opposite or whorled, simple. Inflorescences terminal or axillary, usually in large racemes. Flowers frequently showy and bright yellow. Fruits a capsule. In Belize, consisting of 1 genus and 1 species. Uses for 1 genus and 1 species are reported here.

Vochysia hondurensis Sprague

emery, emory, white mahogany [English]; *San Juan* [Spanish]; *emeri, yemeri* [Additional names recorded]

The trunk is used in **construction** for lumber and to make dugout canoes [A619]. Standley and Record (1936) reported that the wood of this tree was useful in **construction** for dugout canoes, furniture, and exterior and interior trim of houses. They also noted that a small amount of it was exported to the United States to produce **veneers**.

BIBLIOGRAPHY

Alcorn, J.B. 1984. *Huastec Mayan ethnobotany*. University of Texas Press, Austin.

Alexiades, M.N. 1996. Collecting ethnobotanical data: an introduction to basic concepts and techniques. *Advances in Economic Botany* 10:53–94.

Alexiades, M.N. 1999. *Ethnobotany of the Ese' eja: plants, health, and change in an Amazonian society*. Ph.D. dissertation. City University of New York.

Altschul, S. 1972. *The genus* Anadenanthera *in Amerindian cultures*. Botanical Museum, Harvard University, Cambridge, Massachusetts.

American Psychiatric Association. 2000. *Diagnostic and statistical manual of mental disorders fourth edition, text revised (DSM-IV-TR)*. American Psychiatric Association, Arlington, Virginia.

Amiguet, V.T., J.T. Arnason, P. Maquin, V. Cal, P. Sanchez Vindas, and L. Poveda. 2005. A consensus ethnobotany of the Q'eqchi' Maya of Southern Belize. *Economic Botany* 59(1):29–42.

Ankli, A., O. Sticher, and M. Heinrich. 1999a. Medical ethnobotany of the Yucatec Maya: healers' consensus as a quantitative criterion. *Economic Botany* 53(2):144–60.

Ankli, A., O. Sticher, and M. Heinrich. 1999b. Yucatec Maya medicinal plants versus non-medicinal plants: indigenous characterization and selection. *Human Ecology* 2(4):557–89.

Arnason, T., F. Uck, J. Lambert, and R. Hebda. 1980. Maya medicinal plants of San Jose, Succotz, Belize. *Journal of Ethnopharmacology* 2:345–64.

Arvigo, R. 1999. *Tree of life: newsletter of the Belize Association of Traditional Healers*. p. 9.

Arvigo, R. and M. Balick. 1993. *Rainforest remedies: one hundred healing herbs of Belize*. Lotus Press, Twin Lakes, Wisconsin.

Arvigo, R. and M. Balick. 1998. *Rainforest remedies: one hundred healing herbs of Belize*. 2nd ed. Lotus Press, Twin Lakes, Wisconsin.

Arvigo, R. with N. Epstein. 1994. *Sastun: my apprenticeship with a Maya healer*. Harper Collins Publishers, San Francisco.

Arvigo, R. with N. Epstein. 1995. *Sastun: my apprenticeship with a Maya healer*. 2nd ed. Harper Collins Publishers, San Francisco.

Baer, R.D. and M. Bustillo. 1993. *Susto* and *mal de ojo* among Florida farmworkers: emic and etic perspectives. *Medical Anthropology Quarterly* 7:90–100.

Bailey, L.H. and E.Z. Bailey. 1976. *Hortus Third, a concise dictionary of plants cultivated in the United States and Canada*. Macmillan Publishing, New York.

Balick, M.J. 1994. Ethnobotany, drug development and biodiversity conservation—exploring the linkages. In D.J. Chadwick and J. Marsh, eds., *Ethnobotany and the search for new drugs*, pp. 4–24. Ciba Foundation Symposium 185. John Wiley & Sons Ltd., Chichester, United Kingdom.

Balick, M.J. 1999. Good botanical practices. In D. Eskinazi, M. Blumenthal, N. Farnsworth, and C. Riggins, eds., *Botanical medicine: efficacy, quality assurance, and regulation*, pp. 121–5. Mary Ann Liebert, Larchmont, New York.

Balick, M.J. and P.A. Cox. 1996. *Plants, people and culture: the science of ethnobotany*. W.H. Freeman, Scientific American Library, New York.

Balick, M.J. and D. Johnson. 1994. The conservation status of *Schippia concolor* in Belize. *Principes* 38:124–8.

Balick, M.J. and R. Lee. 2003. Stealing the soul, *soumwahu en naniak*, and *susto*: understanding culturally-specific illnesses, their origins and treatment. *Alternative Therapies in Health and Medicine* 9(3):106–9.

Balick, M.J. and R. Mendelsohn. 1992. Assessing the economic value of traditional medicines from tropical rain forests. *Conservation Biology* 6(1):128–30.

Balick, M.J., M.H. Nee, and D.E. Atha. 2000. Checklist of the vascular plants of Belize, with common names and uses. *Memoirs of the New York Botanical Garden* 85:1–246.

Balick, M.J., D.E. Atha, S. Canham, and L. Romero. 2002. Capoche: rediscovery of a forgotten febrifuge [*Ocotea veraguensis* (Meisn.) Mez] from Belize, Central America, including a new floristic record. *Economic Botany* 56(1):89–94.

Bennett, B.C. 2007. Doctrine of signatures: an explanation of medicinal plant discovery or dissemination of knowledge? *Economic Botany* 61(3):246–55.

Berlin, B. 1992. *Ethnobiological classification: principles of categorization of plants and animals in traditional societies.* Princeton University Press, Princeton, New Jersey.

Berlin, E.A. and B. Berlin. 1996. *Medical ethnobiology of the Highland Maya of Chiapas, Mexico.* Princeton University Press, Princeton, New Jersey.

Boom, B. 1987. Ethnobotany of the Chacobo Indians, Beni, Bolivia. *Advances in Economic Botany* 4:1–68.

Boster, J. 1973. *K'ekchi' Maya curing practices in British Honduras.* B.A. thesis, Anthropology Department, Harvard University, Cambridge, Massachusetts.

Bye, R.A. 1986. Medicinal plants of the Sierra Madre: comparative study of Tarahumara and Mexican market plants. *Economic Botany* 40:103–24.

Bye, R.A. and E. Linares. 1983. The role of plants found in the Mexican markets and their importance in ethnobotanical studies. *Journal of Ethnobiology* 3(1):1–13.

Carey J.W. 1993. Distribution of culture-bound illnesses in the Southern Peruvian Andes. *Medical Anthropology Quarterly* 7:281–300.

Champy, P., A. Melot, E.V. Guérineau, C. Gleye, D. Fall, G.U. Höglinger, M. Ruberg, A. Lannuzel, O. Laprévote, A. Laurens, and R. Hocquemiller. 2005. Quantification of acetogenins in *Annona muricata* linked to atypical parkinsonism in Guadeloupe. *Movement Disorders: Official Journal of the Movement Disorder Society* 20(12):1629–33.

Clammer, J. 1984. Approaches to ethnographic research. In R. Ellen, ed., *Ethnographic research*, pp. 63–85. Academic Press, London.

Comerford, S.C. 1996. Medicinal plants of two Mayan healers from San Andrés, Petén, Guatemala. *Economic Botany* 50(3):327–36.

Dieterle, J.V.A. 1976. Flora of Guatemala: Cucurbitaceae. Part XI. *Fieldiana: Botany* 24:306–95.

Eliade, M. 1964. *Shamanism, archaic techniques of ecstasy.* Princeton University Press, Princeton, New Jersey.

Farnsworth, N.R., O. Akerele, A.S. Bingel, D.D. Soejarto, and Z.G. Guo. 1985. Medicinal plants in therapy. *Bulletin of the World Health Organization* 63(6):965–81.

Fernandez, B. 1982. *Medicine woman: the herbal traditions of Belize.* Typescript.

Fernandez, B. 1990. *Medicine woman: the herbal traditions of Belize.* National Library Service, Belize City, Belize.

Flores, J.S. and R.V. Ricalde. 1996. The secretions and exudates of plants used in Mayan traditional medicine. *Journal of Herbs, Spices, and Medicinal Plants* 4(1):53–9.

Gann, T. 1918. *The Maya Indians of southern Yucatan and northern British Honduras.* Bureau of American Ethnology, Government Printing Office, Washington, DC.

Gann, T. 1924. *In an unknown land.* Charles Scribner's Sons, New York. Reprinted by the Camelot Press Limited, London and Southampton, United Kingdom.

Gold, L.S. and T.H. Slone. 2003. Aristolochic acid, an herbal carcinogen, sold on the web after FDA alert. *The New England Journal of Medicine* 349:1576–7.

Gomez-Beloz, A., J.C. Rucinski, M.J. Balick, and C. Tipton. 2003. Double incision wound healing bioassay using *Hamelia patens* from El Salvador. *Journal of Ethnopharmacology* 88(2–3):169–73.

Graham, E., D.M. Pendergast, and G.D. Jones. 1989. On the fringes of conquest: Maya-Spanish contact in colonial Belize. *Science* 246:1254–9.

Hartshorn, G.S., L. Nicolait, L. Hartshorn, G. Bevier, R. Brightman, J. Cal, A. Cawich, W. Davidson, R. Dubois, C. Dyer, J. Gibson, W. Hawley, J. Leonard, R. Nicolait, D. Weyer, H. White, and C. Wright. 1984. *Belize: country environmental profile: a field study.* Robert Nicolait Associates Ltd., Belize City, Belize.

Heyder, H.M. 1930. *Sapodilla* tapping in British Honduras. *Empire Forestry Journal* 9(1):107–13.

Holdridge, L.R. 1967. *Life zone ecology.* Tropical Science Center, San Jose, Costa Rica.

Horwich, R.H. and J. Lyon. 1990. *A Belizean rainforest: the Community Baboon Sanctuary.* Orangutan Press, Gays Mills, Wisconsin.

Klein J. 1978. *Susto*: the anthropological study of diseases of adaptation. *Social Science & Medicine* 12:23–8.

Lamb, A.F.A. 1946. *Notes on forty-two secondary hardwood timbers of British Honduras.* Forest Department Bulletin 1. Forest Department, British Honduras.

Larme A.C. 1998. Environment, vulnerability, and gender in Andean ethnomedicine. *Social Science & Medicine* 47:1005–15.

Lewis, C.J. and S. Alpert. 2000. Aristolochic acid: FDA concerned about botanical products, including dietary supplements, containing aristolochic acid. Available online at http://www.fda.gov/Food/RecallsOutbreaks Emergencies/SafetyAlertsAdvisories/ucm095302.htm (accessed February 7, 2014).

Lewis, M.P., ed. 2009. *Ethnologue: languages of the world.* 16th ed. SIL International, Dallas, Texas.

Lundell, C.L. 1938. Plants probably utilized by the old empire Maya of Petén and adjacent lowlands. *Michigan Academy of Science Arts and Letters* 24:37–56.

Mallory, J.M. 1991. *Common medicinal ethnobotany of the major ethnic groups in the Toledo District of Belize, Central America.* M.A. thesis. California State University, Fullerton.

Martin, G.J. 1992. Searching for plants in peasant marketplaces. In M. Plotkin and L. Famolare, eds., *Sustainable harvest and marketing of rainforest products*, pp. 212–23. Island Press, Washington, DC.

Marsh, L., S. Matola, and V. Pickard. 1995. *The ABC's to the vegetation of Belize: a handbook.* The Belize Zoo and Tropical Education Center, Belize.

McSweeney, K. 1993. *The palm landscape of Belize: human interaction with the cohune palm.* M.S. thesis. University of Tennessee, Knoxville, Tennessee.

McSweeney, K. 1995. The cohune palm (*Orbignya cohune*, Arecaceae) in Belize: a survey of uses. *Economic Botany* 49:162–71.

Morris, D. 1883. *The colony of British Honduras: its resources and prospects; with particular reference to its indigenous plants and economic productions.* Edward Stanford, London.

Munro, D.M. 1989. *Ecology and environment in Belize: an account of the University of Edinburgh expedition to Belize, C.A., 1986.* Occasional Paper No. 12. Department of Geography, University of Edinburgh, United Kingdom.

Mutchnick, P. and B.C. McCarthy. 1997. An ethnobotanical analysis of the tree species common to the subtropical moist forests of the Petén, Guatemala. *Economic Botany* 51(2):158–83.

Nash, D. and L.O. Williams. 1976. Flora of Guatemala. *Fieldiana: Botany*, 24(XII):1–603.

Nelson, L., R. Shih, and M.J. Balick. 2007. *Handbook of poisonous and injurious plants.* 2nd ed. Springer and the New York Botanical Garden Press, New York.

Nyazema, N.M., J. Ndamba, C. Anderson, N. Makaza, and K.C. Kaondera. 1994. The doctrine of signatures or similitudes: a comparison of the efficacy of praziquantel and traditional herbal remedies used for the treatment of urinary schistosomiasis in Zimbabwe. *International Journal of Pharmacognosy* 32(2):142–8.

Pendergast, D.M. 1972. The practice of *primicias* in San Jose Succotz, British Honduras. *Ethnos* 1(4):88–96.

Peters, C.M. 1983. Observations on Maya subsistence and the ecology of a tropical tree. *American Antiquity* 48(3):610–5.

Phillips, O.L. 1996. Some quantitative methods for analyzing ethnobotanical knowledge. *Advances in Economic Botany* 10:171–97.

Pieroni, A. 2000. Medicinal plants and food medicines in the folk traditions of the upper Lucca Province, Italy. *Journal of Ethnopharmacology* 70:235–73.

Pinheiro, C.U.B. 1997. Jaborandi (*Pilocarpus* sp., Rutaceae): a wild species and its rapid transformation into a crop. *Economic Botany* 51(1):49–58.

Rebhun L.A. 1994. Swallowing frogs: anger and illness in Northeast Brazil. *Medical Anthropology Quarterly* 8:360–82.

Record, S.J. and C.D. Mell. 1924. *Timbers of tropical America.* Yale University Press, New Haven, Connecticut.

Risser, A.L. and L.J. Mazur. 1995. Use of folk remedies in a Hispanic population. *Archives of Pediatrics & Adolescent Medicine* 149:978–81.

Robineau L. 1991. *Towards a Caribbean pharmacopoeia. TRAMIL 4.* Enda-Caribe, Santo Domingo, Dominican Republic.

Robinson, G.M. and P.A. Furley. 1983. *Resources and development in Belize: an account of the University of Edinburgh expedition to Central America, 1981.* Department of Geography, University of Edinburgh, United Kingdom.

Rodriguez, E. and R. Wrangham. 1993. Zoopharmacognosy: the use of medicinal plants by animals. *Recent Advances in Phytochemistry* 10:89–105.

Rosengarten, F.A. 1978. A neglected Mayan galactagogue, Ixbut (*Euphorbia lancifolia*). *Botanical Museum Leaflets, Harvard University* 26:277–309.

Roys, R.L. 1931. *The ethno-botany of the Maya.* Department of Middle American Research, Tulane University of Louisiana, New Orleans.

Schultes, R.E. 1941. *Aechmea magdalenae* and its utilization as a fiber plant. *Botanical Museum Leaflets, Harvard University* 9(7):117–22.

Smith, N., S.A. Mori, A. Henderson, D.W.M. Stevenson, and S.V. Heald. 2004. *Flowering plants of the Neotropics*. Princeton University Press, Princeton, New Jersey.

Staiano, K. 1981. Alternative therapeutic systems on Belize: a semiotic framework. *Social Science and Medicine* 15B:317–32.

Standley, P.C. and S.J. Record. 1936. *The forests and flora of British Honduras*. Field Museum of Natural History Press, Chicago.

Standley, P.C. and J.A. Steyermark. 1946–68. The flora of Guatemala. *Fieldiana: Botany* 24(1–8).

Standley, P.C. and L.O. Williams. 1961–69. The flora of Guatemala. *Fieldiana: Botany* 24(7–8).

Steggerda, M. 1943. Some ethnological data concerning one hundred Yucatán plants. *Bulletin* 136(29):189–226.

Steggerda, M. and B. Korsch. 1943. Remedies for diseases as prescribed by Maya herb doctors. *Bulletin of the History of Medicine* 13:54–82.

Stepp, J.R. and D.E. Moerman. 2001. The importance of weeds in ethnopharmacology. *Journal of Ethnopharmacology* 75:19–23.

Stevenson, N.S. 1932. The cohune palm in British Honduras. *Tropical Woods* 30:3–5.

Surgeon General. 1999. *Mental health: a report of the Surgeon General*. Available online at http://profiles.nlm.nih.gov/ps/retrieve/ResourceMetadata/NNBBHS (accessed September 9, 2013).

Thomas, E., I. Vandebroek, P. Van Damme, L. Semo, and Z. Nosa. 2009. *Susto* etiology and treatment according to Bolivian Trinitario people: a "masters of the animal species" phenomenon. *Medical Anthropology Quarterly* 23:298–319.

Thompson, J.E. 1927. *A correlation of the Mayan and European calendars*. Field Museum of Natural History Press, Chicago.

Thompson, J.E. 1930. *Ethnology of the Mayas of southern and central British Honduras*. Field Museum of Natural History Press, Chicago.

Vandebroek I., E. Thomas, S. Sanca, P. Van Damme, L. Van Puyvelde, and N. De Kimpe. 2008. Comparison of health conditions treated with traditional and biomedical healthcare in a Quechua community in rural Bolivia. *Journal of Ethnobiology and Ethnomedicine* 4:1.

Voeks, R.A. 2004. Disturbance pharmacopoeias: medicine and myth from the humid tropics. *Annals of the Association of American Geographers* 94:868–88.

Walker, S.H. 1973. *Summary of climatic records for Belize*. Land Research Division, Surbiton, Surrey, England.

Wang, J.-H. 1998. Personal communication.

Wilk, R.R. 1986. Mayan ethnicity in Belize. *Cultural Survival Quarterly* 10(2):73–7.

Wilk, R.R. 2010. Personal communication, February 25.

Williams, L.O. 1981. The useful plants of Central America. *Ceiba* 24:1–342.

Williams, R.J. 1986. *A collection of Belizean herbs: their uses and applications*. The Belize Herb Center, Belize City, Belize.

Young, J.C. 1981. Non-use of physicians: methodological approaches, policy implications, and the utility of decision models. *Social Science and Medicine* 15(4):499–507.

INDEX

Photos, illustrations, and figures are indicated with **bold** page numbers.